Illustrated Dictionary of

Architecture

3rd Edition

Illustrated Dictionary of
Architecture

Ernest Burden

New York Chicago San Francisco
Lisbon London Madrid Mexico City
Milan New Delhi San Juan Seoul
Singapore Sydney Toronto

ISBN 978-0-07-177293-8
MHID 0-07-177293-6

The sponsoring editor for this book was Joy Evangeline Bramble, the editing supervisor was Stephen M. Smith, and the production supervisor was Pamela A. Pelton. The art director for the cover was Jeff Weeks.

This book is printed on acid-free paper.

McGraw-Hill books are available at special quantity discounts to use as premiums and sales promotions, or for use in corporate training programs. To contact a representative, please e-mail us at bulksales@mcgraw-hill.com.

Preface

Architecture throughout the ages has provided shelter from the elements, refuge and safety from intruders, palaces for royalty, shops for merchants, and shrines for religions. Throughout history all of these structures have been constructed with different materials, components, forms, and architectural styles. These items were all part of an integrated system of building, which represented the current customs of each culture.

The Egyptians had a relatively integrated system of building using simple parts. This produced a monolithic style, which featured extensively carved ornamentation on the otherwise simple, massive forms. The Greeks developed an architecture derived from wooden prototypes, which consisted of a kit of parts. It is this system which became the standard adopted by western civilizations, and modified by many succeeding generations into infinite variations of these basic forms. Other civilizations, such as those in China, Japan, Thailand, and India, developed similar stylistic features indigenous to their own cultures and religions. This dictionary describes these styles, and illustrates many of them with photographs of their typical structures.

The number of individual building components has not increased significantly over the past centuries of building. In fact, the number of building components has decreased as building designs became simplified. On the other hand, buildings have become more complex on the technical and functional aspects of the interior and mechanical systems. Many of these new technical terms have been included here, including the new language of green, sustainable, and ecological architecture.

Architecture is a tangible product, and the numerous photographs contained in this edition add a dimension that is not possible using word definitions alone. However, there are many intangible aspects involved in contemporary practice as well, and these have also been listed. They include not only aspects of the design and building process, but many new terms relating to building renovation and restoration that are so prevalent in today's practice.

The typical function of a dictionary is to isolate and define individual elements, and to provide specialized information. This dictionary carries it to another level, by illustrating many of the definitions with photographs of these elements in their position within the structures. In addition, this dictionary illustrates several variations of the same element, including both historical and contemporary examples.

The photographs in this book were selected from the author's library of building sites from around the world. Some examples are well known, while other images provided the clearest illustration of the definition. No attempt was made to identify any of the illustrated components by building type, location, date, or architect, except for the listings of the architects themselves and some of their most well-known buildings.

Another distinctive feature of this dictionary is the use of color photographs to illustrate the definitions. The first two editions of this work were in black and white. The addition of full-color illustrations not only makes this a departure from most other works of this kind, but may become the universal lexicon of the future.

Acknowledgments

A special thanks to my editor, Joy Evangeline Bramble, who suggested a new and updated look for this edition, and the inclusion of green, sustainable, and ecological terminology. Steve Chapman, publisher, McGraw-Hill Professional, encouraged the revision with the addition of color images. Pamela Pelton expertly handled the production of this edition, as she has done with previous editions, and coordinated the effort with the printer in China. Stephen Smith managed the editing and proofing of the text and was helpful in coordinating a complex, staged production process. A very special thanks to my wife, Joy Arnold Burden, for her untiring assembling of the files for the editing, enhancement of the color images, graphic design, and layout of the entire book. She provided a high level of quality control throughout the final production. All the photographs in this work are from the extensive image library of the author, plus the numerous photographs taken by my wife, Joy, on our many photographic excursions around the world. Thanks also to my son Ernest III for his photographs of Santiago Calatrava's Ciutat de la Arts in Valencia, Spain.

About the Author

Ernest Burden pursued an architectural career at the Rhode Island School of Design, transferring one year later to study with pioneering architect Bruce Goff at the University of Oklahoma. While at Oklahoma he was a finalist for the Prix de Rome in Architecture, and upon graduation recieved the Senior Faculty Design Award. Mr. Burden began his professional career as an architect and also provided other architects with architectural renderings and video presentations as they sought approval for their projects. The compilation of a decade of this presentation work was featured in his first book, *Architectural Delineation: A Photographic Approach to Presentation,* which was first published by McGraw-Hill in 1970, followed by two additional editions. Another popular book, *Entourage: A Tracing File,* was originally published in 1980, and is now in its 5th edition, with all color images. Other book ventures with McGraw-Hill include *Building FACADES: Ornamental Details,* and *Visionary Architecture: Unbuilt Works of the Imagination.* Mr. Burden also has written and illustrated two other dictionaries, the *Illustrated Dictionary of Architectural Preservation,* and the *Illustrated Dictionary of Building Design and Construction,* both published by McGraw-Hill.

About the Book

The first edition of this dictionary was produced using a 1 MB Mac Plus for the text, and black and white photographs produced on a photocopier scanner, and the mechanicals for the printer were produced by hand. This third edition was produced using scanned images of the original photographs or direct digital images, and electronically generated text, and assembled in a desktop publishing program. Each single page of this edition tops 1 MB and demonstrates the leap in technology over a decade.

Introduction

This book was designed to be an educational experience through the study of the color photographs, as well as providing technical information through the written definitions, and by clustering and cross-referencing of similar elements. There are over 40 grouped categories of definitions, where one can find many related definitions in the same general location in the book. This includes specific types of arches, doors, joints, moldings, roofs, walls, or windows. The format throughout the book is structured so that it can be used as an easy reference guide, consisting of two main categories: definitions that are illustrated, and those that are not illustrated.

ILLUSTRATED DEFINITIONS
The definitions that are illustrated are shown in a narrow column, with the first illustration adjacent or following, as shown in the following example.

Abacus
The flat area at the top of a capital, dividing a column from its entablature. It usually consists of a square block, or is enriched with moldings. In some orders the sides are hollowed and the angles at the corner are truncated.

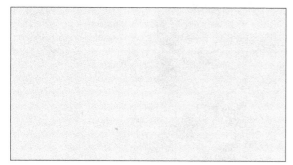

DEFINITIONS ONLY
Definitions that are not illustrated are shown in a wide column, as shown in the following example.

Abstraction
The omission or severe simplification of details in a drawing, leaving only massing, form and solids, so that the basis of the design can be explained.

HISTORIC STYLES
All historic styles are set in a wide column, whether they are illustrated or not, due to the lengthy descriptions of those entries, as shown in the following example.

Absolute architecture
The opposite of Functionalism, as its forms were to be created by imagination rather than by consideration of need. It was proposed as a purposeless architecture by Walter Pickler and Hans Hollien in the 1960s.

ARCHITECTS
Abbreviated biographies of well-known architects and their major works and dates are shown in the wide-column format, as in the following example.

Aalto, Alvar (1898-1976)
Finnish architect who designed the Viipuri Public Library with an undulating timber roof built in 1929. His Paimio Sanitorium, Paimio, Finland, was one of the first hospitals to be built in the International style. Later work includes the Baker House, Massachusetts Institute of Technology, Cambridge, MA (1948), and the Public Library, Rovaniemi, Finland (1963), and Finlandia Hall (illus.), Helsinki, Finland (1971).

CROSS-REFERENCES
All cross-referenced listings of illustrated entries that are subcategories are listed alphabetically throughout the text where they would normally appear, as shown in the following example.

Acute arch See Arch

SUBCATEGORIES
The cross-referenced definitions included in the grouped subcategories of illustrated entries are listed using all lower case letters. They follow the main listing of that category in alphabetical order, as shown in the following examples.

Arch
A basic architectural structure built over an opening, made up of wedge-shaped blocks, keeping one another in position, and transferring the vertical pressure of the superimposed load into components transmitted laterally to the adjoining abutments.

acute arch
A sharply pointed two-centered arch, whose centers of curvature are farther apart than the opening.

Illustrated Dictionary of
Architecture

A a

Aalto, Alvar (1898–1976)

Finnish architect who designed the Viipuri Public Library with an undulating timber roof, built in 1929. His Paimio Sanitorium, Paimio, Finland, was one of the first hospitals to be built in the International style. Later work includes the Baker House, Massachusetts Institute of Technology, Cambridge, MA (1948), and the Public Library, Rovaniemi, Finland (1963), and Finlandia Hall (illus.) Helsinki, Finland (1971).

Abacus

The flat area at the top of a capital, dividing a column from its entablature. It usually consists of a square block, or is enriched with moldings. In some orders the sides are hollowed and the angles at the corner are truncated.

Abadie, Paul (1812–1889)

Designed Sacré-Coeur Church in Paris and is best known for French Neo-Romanesque churches.

Abandonment

The relinquishment of ownership or control of a piece of property, which often causes or accelerates the deterioration of the property. Also, when a contractor stops work on a job and removes all personnel and equipment before it is completed.

Abatement

The removal or reduction of a health risk to an acceptable level from an identified hazardous material, such as those containing lead or asbestos. It includes demolition and removal and protection of the surrounding environment.

Abat-jour

A skylight in a roof that admits light from above; any beveled aperture.

Abat-vent

A louver placed in an exterior wall opening to admit light and air, but offering a barrier to the wind.

Abbey

The monastic buildings of religious bodies governed by an abbot or abbess.

Above-ground archeology

The study of a building or other visible artifact by careful scrutiny, as distinct from conventional archeology's focus beneath the surface of the ground.

Abraham, Raimund (1933–2010)

Austrian architect and educator renowned for his drawings. Abraham was responsible for several significant buildings, including the ultra-modern Austrian Cultural Forum (2002) in New York City.

Abrahamson, Bruce A. (1925–2008)

A partner at the Minneapolis, MN firm of Hammel, Green, and Abrahamson (HGA). Their many projects included the Minneapolis Clinic of Psychiatry and Neurology in Golden Valley, the Siebens Education Building at the Mayo Clinic in Rochester, and the downtown St. Paul Skyway system, all in Minnesota.

Abramovitz, Max (1908–1963)

American architect educated at the École des Beaux-Arts in Paris. He was a partner of Wallace K. Harrison. He absorbed the Modern movement working on Rockefeller Center, New York City (illus.), and designed the interior of Radio City Music Hall in Rockefeller Center. His designs for the U.S. embassy in New Delhi, India, built in 1954, and the Kennedy Center for the Performing Arts in Washington, D.C., built in 1961, were axial and symmetrical, and they paraphrased the Classical movement.

Absolute architecture

The opposite of Functionalism; as its forms were to be created by imagination rather than by consideration of need. It was proposed as a purposeless architecture by Walter Pickler and Hans Hollien in the 1960s.

Absorber

A component of a solar heater that soaks up heat from the sun and helps transmit it to the water or heating system.

Absorption

The process by which incident light energy is converted to another form of energy, usually heat.

Absorption chiller

A water-chilling process in which cooling is accomplished by the evaporation of a fluid (usually water), which is then absorbed by a different solution (usually lithium bromide), then evaporated under heat and pressure. The fluid is then condensed with the heat of condensation ejected through a cooling tower.

Abstraction

The omission or severe simplification of details in a drawing, leaving only massing, form and solids, so that the basis of the design can be explained.

Abut

To meet or touch with an end, as on a construction member meeting an adjoining one.

Abutment

A masonry mass, pier, or solid part of a wall that takes the lateral thrust of an arch.

Abutment arch See Arch

Academy
A place of study to advance the arts or sciences; named after the Akademia in Athens where Plato taught.

Acanthus
A common plant of the Mediterranean, whose leaves, stylized, form the characteristic decoration of capitals of the Corinthian and Composite orders. In scroll form it appears on friezes and panels.

Accelerated weathering
Determination of the weather-resisting properties of materials by testing cycles that imitate natural weathering conditions as closely as possible.

Accent fixtures
Lighting fixtures that focus on specific targets such as art, retail display, lecterns, and features on building facades.

Accent lighting
Any directional lighting that emphasizes a particular object or draws attention to a particular area.

Acceptable indoor air quality
Air in an occupied space toward which a substantial majority of occupants express no dissatisfaction and in which there are not likely to be known contaminants at concentrations leading to exposures that pose a significant health risk.

Access door
Any door which allows access to concealed equipment or parts of a building not often used, such as a door to concealed plumbing parts.

Access to public transportation
Locations where public transportation is a viable alternative to private automobiles for access to work, school, recreation, and commerce. It gives the affordable housing developer room to argue for a reduced parking requirement, thereby making more of their land purchase available for housing units and reducing the land cost per unit.

Accessibility
Refers to items and conditions specified in the Americans with Disabilities Act (ADA) that provide access for physically handicapped persons.

Accessibility (handicapped)
The provision of access to and through a building or site for physically impaired individuals.

Accessibility audit
Evaluation of the compliance of an existing building to the building code as it relates to handicapped accessibility.

Accessory building
A secondary structure on a site, such as a detached garage, guest house, or shed.

Accolade
A rich ornamental treatment made up of two ogee curves meeting in the middle, as found over a door, window, or arch.

Accordion door See Door

Accouplement
Pairs of closely spaced columns or pilasters; typically the abacuses of the capitals touch each other without being joined.

Acetone
A solvent which evaporates quickly; used in paint removers, thinners, and lacquers.

Achromatic
Having no color; a neutral such as black and white.

Acid deposition
The deposit of acid constituents to a surface. This occurs not only through precipitation but also by the deposition of atmospheric particulate matter and the incorporation of soluble gases.

Acid rain
The precipitation of dilute solutions of strong mineral acids, formed by the mixing in the atmosphere of various industrial pollutants, primarily sulfur dioxide and nitrogen oxides, with naturally occurring oxygen and water vapor.

Acorn ornament See Ornament

Acoustic construction
Any building method that reduces sound either entering, leaving, or transferring through a structure by means of dense construction, use of absorbent materials, or discontinuous construction.

Acoustic plaster
Plaster with a high degree of sound absorption properties.

Acoustical correction
Planning, shaping, and equipping a space to establish the best possible hearing conditions for faithful reproduction of sound within a space.

Acoustical door See Door

Acoustical materials
Sound-absorbing materials for covering walls and ceilings; a term applied to special plasters, tile, or any material composed of mineral wool, wood, or cork, used to deaden sound.

Acoustical tile See Tile

Acropodium
An elevated pedestal bearing a statue that is raised above the substructure.

Acropolis
Elevated stronghold or group of buildings serving as a civic symbol: those of ancient Greek cities usually featured the temple of a deity, such as at Athens.

Acroteria
A pedestal for statues and other ornaments placed on the apex and the lower angles of a pediment; or often refers to the ornament itself.

Acrylic
A plastic which in solid form is rigid, clear, and transparent often used as glazing in areas for additional safety as it does not shatter.

Acrylic fiber See **Plastic**

Active closed-loop solar water heater
Solar water heater in which an electric pump circulates a freeze-protected heat-transfer fluid through the collector and heat exchanger within a storage tank.

Active diffuser
An air supply outlet with a local fan to deliver air from the plenum through the diffuser into the conditioned space.

Active drain-back solar water heater
Solar water heater in which water or another heat-transfer fluid is pumped through the collector and drains back to a tank in the house when the pump turns off.

Active noise control
Electronic masking of sound to cover unwanted or intrusive sound, such as speech or equipment noise.

Active solar
A solar application, which uses electrical or mechanical equipment to assist in the collection and storage of solar energy for the purpose of heating, cooling, or making electricity.

Active solar energy
The energy from sunlight that is absorbed by photovoltaic collectors and transferred directly to the building interior; stored using a battery, a hydrogen fuel cell, or some other system for later use; or sold to a utility.

Active solar heating
Systems that collect and absorb solar radiation, then transfer the solar heat directly to the interior space or to a storage system, from which the heat is distributed. There are two types of systems: liquid-based and air-based systems. Both air and liquid systems can supplement forced-air systems.

Active solar power
A solar electric photovoltaic or PV system that converts the sun's energy into electricity for the home. It is usually done with PV panels installed on the roof.

Active solar water heater
Heat from the sun is absorbed by collectors and transferred by pumps to a storage unit. The heated fluid in the storage unit conveys its heat to the domestic hot water of the house through a heat exchanger. Controls regulating the operation are needed.

Active system
A traditional HVAC system that uses mechanical means to artificially cool, heat, or ventilate the air supply in a building and that draws power for these processes from electricity or gas.

Acute angle
An angle of less than 90 degrees.

Acute arch See **Arch**

ADA Accessibility Guidelines
A document illustrating and describing barrier-free handicapped access as required by the Americans with Disabilities Act.

Adam style (1728–1792)
An architectural style based on the work of Robert Adam and his brothers, predominantly in England and strongly influential in the United States and elsewhere. It is characterized by a clarity of form, use of color, subtle detailing, and unified schemes of interior design. Basically Neoclassical, it adopted Neo-Gothic, Egyptian, and Etruscan motifs.

Adam, Robert (1728–1792)
British architect whose designs returned to the Classical forms of antiquity, not their Renaissance-derived imitations. Adam excelled in using the early Classical forms in domestic settings.

Adaptability
Design strategy that allows for multiple future uses in a space as needs evolve and change. Adaptable design is considered a sustainable building strategy because it reduces the need for major renovations or tearing down a structure to meet future needs.

Adaptation
In lighting design, the process by which the human visual system becomes accustomed to more or less light, resulting from a change in the sensitivity of the eye to light.

Adaptation compensation controls
The opposite of daylighting. It increases interior light, as exterior light decreases.

Adapted plants
Plants that reliably grow well in a given habitat with minimal attention from humans in the form of winter protection, pest protection, water irrigation, or fertilizer once root systems are established in the soil. Adapted plants are considered to be low maintenance and not invasive.

Adaptive abuse
Inappropriate conversion of a building to a new use that is detrimental to the structure's architectural character, or undesirable in relation to its neighbors.

Adaptive use
Changing an existing building to accommodate a new function; the entire process may involve the removal of some existing building elements.

Addendum
Written or graphic document issued before the receipt of bids that modifies or interprets the bidding documents, by additions, deletions, clarifications, or corrections. They usually become part of the final contract documents.

Addition
Construction that increases the size of the original structure by building outside the existing walls or roof.

Addorsed
Animals or figures that are placed back to back and featured as decorative sculpture over doors, and in pediments, medallions, and other ornamental devices.

Adhesion
The property of a material that allows it to bond to the surface to which it is applied.

Adhesive
A substance, such as glue, paste, mastic, or cement, that is capable of bonding materials together; force is then required to separate the members.

Adjacency
The location of elements in a planning diagram, placed and ranked according to their relative importance, including their interconnections.

Adjustable luminaires
Fixtures with a flexible, adjustable mechanism to permit the principal focus to be altered. Examples are track lights, flood lights, and accent lights.

Adler, Dankmar (1844–1900)
German-born engineer; moved to Chicago in 1854 and became a partner of Louis H. Sullivan. The Auditorium Building in Chicago, built in 1886, was their first joint commission. In 1889, the firm employed the young Frank Lloyd Wright.

Adler, Dankmar

Admixture
A material other than water, aggregate, and cement, used as an ingredient in concrete or mortar; may add coloring or control strength or setting time.

Adobe
Large, roughly molded, sun-dried clay units of varying sizes.

Adsorbent
Material that is capable of binding and collecting substances or particles on its surface without chemically altering them.

Adsorption factor
Adsorption refers to the process whereby soft furnishings, such as carpet, foam-covered fabric chairs and partitions, harbor resting VOCs (volatile organic compounds) as the indoor temperature cools.

Advanced framing
A construction method that uses less material in the framing of a home and can reduce material costs and improve energy performance of the building envelope.

Advanced new home construction program
A proposed U.S. Environmental Protection Agency (EPA) program to promote the construction of "better than Energy Star" homes. The proposal calls for builders to comply with a "builder option package" of measures including super-insulated walls, triple-glazed windows, ducts within the conditioned space, compact duct layout, a furnace with sealed plenums and a variable-speed energy-efficient blower motor, an efficient hot-water distribution system, and a solar water heater or heat pump.

Advertisement for bids
A published public notice soliciting bids for a construction project. Often conforms to legal requirements for procuring public work.

Advisory Council on Historic Preservation
An independent agency that reviews proposed federal projects affecting properties listed in the U.S. National Register of Historic Places, or determined eligible for listing; provides suggestions for ensuring their preservation; and assists other federal agencies by reviewing Environmental Impact Statements. Also advises the president and congress on preservation matters.

Aedicule
A canopied niche flanked by colonnettes, intended as a shelter for a statue or a shrine; a door or window framed by columns or pilasters and crowned with a pediment.

Aegricranes See Ornament: animal forms

Aeolic capital see Capital

Aerated autoclaved concrete
Precast concrete that is cured by steam pressure inside a kiln called an autoclave. The material is lighter weight than conventional concrete and has good insulation properties.

Aeration
Exposing water to the air; often results in the release into the atmosphere of gaseous impurities found in polluted water.

Aerial photo-mosaic
A composite of aerial photographs depicting a portion of the earth's surface; basic mapping information such as the name of towns and cities is usually added.

Aerosol adhesive
An adhesive packaged as an aerosol product in which the spray mechanism is permanently housed in a nonrefillable can designed for handheld application without the need for ancillary hoses or spray equipment.

Aesthetic Historic Site
A display in period room settings of examples of furniture and furnishings.

Affleck, Raymond (1922–1989)
A Canadian architect whose best multipurpose building is the Place Bonaventure in Montreal (1968), a vast complex with many internal spaces, but overall representing a forbidding example of Brutalism.

Affronted

Figures or animals that are placed facing each other, as decorative features over doors and in pediments.

A-frame

A house constructed of wood, with a steep roof that extends down from the ridge to near the foundations; the roof is supported by a framework in the shape of the capital letter A.

Agate See **Stone**

Agenda 21

A program run by the United Nations related to sustainable development. It represents a comprehensive blueprint for global action drafted by the 172 governments present at the 1992 Earth Summit organized by the UN in Rio de Janeiro. It is a comprehensive archetype of action to be taken globally, nationally, and locally by organizations of the UN, governments, and major groups in every area in which humans impact the environment. The number 21 refers to the 21st century.

Aggregate

Any of a variety of materials, such as sand and gravel, added to a cement mixture to make concrete.

Agora

An open public meeting place for assembly surrounded by public buildings, or a marketplace in an ancient Greek city; the Roman forum is a typical example.

Agraffe

The keystone section of an arch, especially when carved with a cartouche or human face.

Agricultural by-products

Products developed in agriculture that were not a primary goal of the agricultural activity. The most commonly used as a building product is straw, which is used in wall panels or as bales in a technique called strawbale construction, with the bales used as building blocks.

Agricultural fibers

Fibrous materials resulting from agricultural operations, such as cotton fibers used in insulation applications.

Agricultural waste

Materials left over from agricultural processes, such as wheat stalks and shell hulls. Some of these materials are finding new applications as building materials and finishes. Examples include structural sheathing and particleboard alternatives made from wheat, rye, and other grain stalks, and panels made from sunflower seed hulls.

Agrifiber board

A composite panel product derived from recovered agricultural waste fiber from sources including, but not limited to, cereal straw, sugarcane bagasse, sunflower husk, walnut shells, coconut husks, and agricultural prunings. The raw fibers are processed and mixed with resins to produce panel products with characteristics similar to those derived from wood fiber.

Air barrier

Building assembly components that work as a system to restrict air flow through the building envelope. Air barriers may or may not act as a vapor barrier. The air barrier can be on the exterior of the assembly, the interior, or both.

Air change

The replacement of air contained within a room with an equivalent volume of fresh air.

Air changes per hour (ACH)

A metric of the air-tightness of a structure, often expressed as ACH50, which is the air changes per hour when the house is depressurized to–50 pascals during a blower door test. The term ACHn or NACH refers to "natural" air changes per hour, meaning the rate of air leakage without blower door pressurization or depressurization. ACHn or NACH is used by many in the residential HVAC industry for system sizing calculations.

Air cleaner

A filtering device that actively removes impurities from the air.

Air cleaning

Indoor-air quality-control strategy to remove various airborne particulates and/or gases from the air. Most common methods are particulate filtration and electrostatic precipitation.

Air conditioning

A system that extracts heat from an area using a refrigeration cycle and treats the air to meet the requirements of a conditioned space by controlling its temperature, humidity, cleanliness, and distribution. A complete system of heating, ventilation, and air conditioning is referred to as HVAC.

Air Conditioning and Refrigeration Institute (ARI)

A trade association representing manufacturers of more than 90 percent of the air-conditioning and commercial refrigeration equipment installed in North America that develops standards for and certifies the performance of these products.

Air drying

Drying a material such as timber in the air instead of seasoning it in a kiln.

Air duct

A duct usually fabricated of metal, fiberglass, or cement, used to transfer air from a heating or cooling source to locations throughout a facility.

Air flow

The movement of air within a room, duct, or plenum.

Air handler

A fan that a furnace, central air conditioner, or heat pump uses to distribute heated or cooled air throughout the house.

Air inlet

Apertures such as grilles, diffusers, or louvered openings through which air is intentionally drawn from a conditioned space.

Air lock

A lobby or small room with self-closing doors to allow access between two other spaces while restricting the amount of air exchanged between them.

Air outlet

Apertures such as grilles, diffusers, or louvered openings through which air is intentionally delivered into a conditioned space.

Air plenum

Any space used to convey air in a building, furnace, or structure. The space above a suspended ceiling is often used as an air plenum.

Air pollution

The presence of contaminants or pollutant substances in the air that interfere with human health or welfare, or produce other harmful environmental effects.

Air quality

The level of particulates, gases, vapors, pollens, and micro-organisms in the air; achieving and maintaining indoor air quality are sustainable design/building management objectives required to mitigate sick building syndrome, enhance amenity, and promote work environments.

Air quality construction management plan

A systematic plan for addressing construction practices that can impact air quality during construction and continuing on to occupation.

Air quality standards

The level of pollutants prescribed by regulations that are not to be exceeded during a given time in a defined area.

Air rights

A privilege or right, protected by law, to build in, occupy, or otherwise use a portion of air space above real property at a stated elevation, in conjunction with specifically located spaces on the ground surface for the foundation and supporting columns.

Air rotation units
The units that use outside air to condition a building. However, the units only bring in outside air during occupied times. When the building is not occupied, the ARU rotates the higher warm air with the lower cool air in the space, keeping the building satisfied without taking into account the necessary air quality requirements based on occupancy.

Air sealing
The sealing of cracks and holes in a home's envelope to prevent uncontrolled movement of air.

Air source heat pump
A heat pump that relies on outside air as the heat source and heat sink; not as effective in cold climates as ground source heat pumps.

Air space
A hollow space between the inner surfaces within a building, or building component, such as between the panes of insulating glass, inner surfaces of masonry walls, between ceiling and floor in steel truss construction, and between an inner and outer dome.

Air supply volume
The volume of supply air flowing through a cross-sectional plane of a duct per unit time, found by multiplying air velocity by the cross-sectional area of the duct.

Air temperature
A measure of the heat energy contained in ambient air.

Air toxics
Any air pollutant for which a standard in the National Ambient Air Quality Standards (NAAQS) does not exist that may reasonably be anticipated to cause serious or irreversible chronic or acute health effects in humans.

Air trap
A U-shaped pipe filled with water and located beneath plumbing fixtures to form a seal against the passage of gas and odors.

Air-change effectiveness
The ability of an air-distribution system to provide outside air where occupants breathe. It is defined as the age of air that would occur throughout the space if the air was perfectly mixed, divided by the average age of air where occupants breathe.

Air-dried lumber
Lumber that has been piled in yards or sheds for a specified period of time.

Air-exchange rate
The rate at which outside air replaces indoor air in a given space.

Air-flow sensor
A device that measures air velocity by way of differential pressure inside a duct.

Air-handling unit
Heating and/or cooling distribution that channels warm or cool air to different parts of a building. Equipment includes a blower or fan; heating and/or cooling coils; and related equipment such as controls, condensate drain pans, and air filters. It does not include ductwork, registers, grilles, boilers, or chillers. Conditioning may include particle filtering, adding or removing heat and moisture. A varying portion of the return air from the conditioned space may be recirculated and mixed with incoming air.

Air-inflated structures See **Pneumatic structures.**

Air-side economizer
An air economizer cycle reduces the load on the chilled water system by increasing the flow of outside air above the minimum required for ventilation when the outside air temperature is favorable in comparison to return air temperature; also known as free cooling. For example, if there is cool, dry air outside, it will be drawn into the building to cool it, instead of using chillers to provide cooling.

Airtight drywall
Use of drywall with carefully sealed edges and joints that serves as an interior air barrier in building assemblies.

Aisle
The circulatory space flanking and parallel to the nave in a church, separated from it by a row of columns; a walkway between seats in a theater, auditorium, or other place of public assembly.

Alabaster See **Stone**

Albedo
Percentage of light reflected off a surface; a material with a high albedo is very reflective.

Alberti, Leon Battista (1404–1472)
Italian Renaissance architect and author who designed the marble facade of San Maria Novella (illus.), Florence, Italy, from 1456 to 1470, which contains Classical details in an otherwise Gothic church. From Vitruvius, via Alberti, came the concept that buildings should be in proportion to the human body and all their dimensions should be related. In 1452, Alberti wrote *De re Aedificatoria*, the first architectural treatise of the Renaissance.

Alcove
A small recessed space, connected to or opening directly into a larger room.

Alhambra
One of the most exquisite, elaborate, and richly ornamented of all Moorish palaces in Spain; consisting of a series of joined pavilions with two great courts set at right angles; channels of water, linking pools with fountains, add to the overall effect.

Allen, Rex Whitaker (1915–2008)
A leader in healthcare architecture, and one of the first to advocate for and design residential-style nursing homes and specialized facilities, such as those for Alzheimer's patients. He was the author of one of the first books on healthcare facility design, *Hospital Planning Handbook,* and many journal and magazine articles.

Alignment
An arrangement or adjustment of forms or spaces according to a specific line.

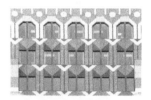

Alkaline resistance
The degree to which a paint resists reaction with alkaline materials such as lime cement or plaster; a necessary property for paints used in bathrooms, kitchens, and laundries.

Allegory
A figurative representation or sculpture in which the meaning is conveyed by the use of symbols.

Allegory

Alley
A passageway providing access to a rear yard or central courtyard.

Alligatoring
A defect in a painted surface, appearing like alligator hide, from the application of a hard finishing coat over a soft primer coat, when the new coat cracks and slips over the old coat, exposing it to view.

Alloy steel See **Metal**

Altarpiece
A panel, sculpted or painted, situated above and behind an altar.

Alteration
A term in the building code referring to any change in a structure which does not increase any of its exterior dimensions, or any modification in construction or grade of occupancy.

Alternative energy
The energy from a source other than conventional fossil-fuel sources, e.g., oil, natural gas, and coal, such as wind, running water, the sun, geothermal, or biomass. Also referred to as alternative fuel.

Alternative fuel vehicles
A vehicles that use low-polluting, nongasoline fuels such as electricity, hydrogen, propane or compressed natural gas, liquid natural gas, methanol, and ethanol. Efficient gas-electric hybrid vehicles are included in this group for LEED (Leadership in Energy and Environmental Design) purposes.

Alternative fuels
Substitutes for traditional liquid, oil-derived motor vehicle fuels like gasoline and diesel. These include mixtures of alcohol-based fuels with gasoline, methanol, ethanol, compressed natural gas, and others.

Alternative refueling station
Location that provides the service of refueling an alternative fuel vehicle, such as electricity or compressed natural gas. This is considered a sustainable building strategy, encouraging the use of alternative fuels.

Alternative-fueled vehicle
A vehicle that is powered by hybrid-electric, electric, natural gas, bio-diesel, or fuel cells.

Alto-relievo See **Relief**

Aluminum door See **Door**

Aluminum foil
A thin sheet of aluminum; commonly used for reflective insulation.

Aluminum See **Metal**

Amalaka
A type of capital found in Hindu architecture.

Ambient air
The surrounding air.

Ambient lighting
A form of lighting that illuminates a room in a uniform, unfocused, indirect manner. Task-ambient lighting provides both focused lighting, and general illumination at a lower level, thus conserving energy.

Ambient noise
The background noise level in a space, which is not identifiable as being from a specific source, such as a nearby piece of equipment.

Ambient temperature
The temperature of the surrounding air or other medium.

Ambulatory corridor
A passageway around the apse of a church, or for walking around a shrine; the covered walk of an atrium in a cloister.

Amenities
The activities provided by a facility regarding comfort and convenience.

Amenity
A building, landscape feature, or the like that makes an aesthetic contribution to the environment, rather than one that is purely utilitarian.

American Association for State and Local History
Serves state and local historical societies in the U.S. and Canada, house and history museums, professional and amateur historians and educators.

American Institute of Architects (AIA)
Founded in 1857 as a national professional society whose members are registered architects. The AIA sponsors the Committee on Historic Resources with State Preservation Coordinators, and is a signatory to the Historic American Buildings Survey (HABS). It promotes public awareness of architectural and environmental issues; supports programs for continuing education; maintains a library at its headquarters in Washington, D.C.; and coordinates state and local chapters in every state.

American National Standards Institute (ANSI)
Oversees the creation, promulgation, and use of thousands of norms and guidelines that directly impact businesses in nearly every sector: from acoustical devices to construction equipment, energy distribution, and many more. ANSI is also actively engaged in accrediting programs that assess conformance to standards—including globally recognized cross-sector programs such as the ISO 9000 and ISO 14001 quality environmental management systems.

American Order
A capital resembling that of the Corinthian order, with the acanthus leaves replaced with corncobs, corn ears, and tobacco leaves; invented by the architect Benjamin Latrobe for the U.S. Capitol in Washington, D.C.

American Plastics Council (APC)
A major trade association for the U.S. plastics industry working to ensure that plastics are recognized as a preferred material by actively demonstrating they are a responsible choice in a more environmentally conscious world.

American Plywood Association (APA)
Now called APA—The Engineered Wood Association, it is a nonprofit trade association for manufacturers of engineered wood products, including glue-laminated timber, composite panels, wood I-joists, and laminated veneer lumber. APA and APA EWS (Engineered Wood Systems) trademarks identify products that meet the organization's manufacturing and performance guidelines.

American School style (1940–1959)
This style , characterized by the later work of Frank Lloyd Wright and the early work of Bruce Goff, represents the association of organic principles, such as relationship of the part to the whole, self-sufficiency, rejection of tradition, freedom of expression, and passion for the land.

American School style

American Society of Heating, Refrigerating and Air-Conditioning Engineers (ASHRAE)

A worldwide organization that promotes the arts and sciences of and publishes standards for heating, ventilation, air conditioning, and refrigeration. Particularly important to green building construction is ASHRAE 90.1, Energy Standard for Buildings Except Low-Rise Residential Buildings, a code setting requirements for energy efficiency and methods of determining compliance.

American Society of Interior Designers (ASID)

A trade association representing the interior design professional community. ASID partnered with USGC to develop the REGREEN program, a green residential remodeling program

American Society of Landscape Architects (ASLA)

A professional association for members who practice landscape architecture and are concerned with the natural environment rather than the built environment.

Americans with Disabilities Act (ADA)

A federal law that defines requirements for handicapped access to public facilities, as described in the ADA Guidelines. It requires removal of existing barriers, except any that would be harmful to the historic significance of a structure.

Amorini

In Renaissance architecture and derivatives, a decorative sculpture or painting, representing chubby, usually naked infants; also called putti.

Amorini

Amorphous

Those forms that do not have a definite or specific shape; or a distinctive crystalline, geometric, angular or curvilinear structure.

Amortizement

The sloping top portion of a buttress or projecting pier that is designed to shed water.

Amp

Abbreviation for ampere, the basic unit for electrical current.

Amphiprostyle

A temple featuring porticos at both ends.

Amphitheater

A circular, semicircular, or elliptical auditorium in which a central arena is surrounded by rising tiers of seats; originally for the exhibition of combat or other public events.

Analglyph See **Relief**

Anamorphic image

A distorted image that must be viewed in a special mirror in order to become recognizable.

Anchor

A metal device fastened on the outside of a wall and tied to the end of a rod or metal strap connecting it with an opposite wall, to prevent bulging; often consisting of a fanciful decorative design.

Anchor bolt

A bolt with its head embedded in the structure; used to attach a structural member, such as securing a sill to a foundation wall.

Anchor plate

A metal plate on a wall that holds the end of a tie rod; used in masonry construction.

Anchorage

A device used for permanently securing the ends of a post-tensioned member, or for temporarily securing the ends of a pretensioned member during hardening of the concrete.

Ancone

A scrolled bracket or console in classical architecture, which supports a cornice or entablature over a door or window.

Ando, Todao (1941–)

Internationally recognized Japanese architect; largely self-educated, who founded his office in Osaka. He uses traditional materials, vernacular style, and modern techniques of construction. Works include Church of Light, Osaka (1989); Japanese Pavilion, Seville Expo (1992), and Naoshima Contemporary Art Museum, Kagawa Prefecture, Japan (1995).

Andrews, John (1933–)

Australian-born architect who made his name with Scarborough College (illus.), Toronto, Canada (1964), a megastructure using the raw materials and the chunky forms of New Brutalism. He designed the George Gund Hall, Harvard University, Cambridge, MA (1972). He also designed the CN Tower in Toronto, Canada (1975).

Angle brace See **Brace**

Angle bracket

A bracket at an inside corner of a cornice; usually presenting two perpendicular decorative sides.

Angle buttress See **Buttress**

Angle capital See **Capital**

Angle cleat

A small bracket formed of angle iron, which is used to locate or support a member of a structural framework.

Angle column See **Column**

Angle iron

A steel section, either hot-rolled or cold-formed, consisting of two legs, almost always at a right angle.

Angle joint See **Joint**

Angle joist

A joist running diagonally from an internal girder to the corner intersection of two wall plates; used to support the feet of hip rafters.

Angle niche See **Niche**

Angle of incidence

In terms of solar energy, the angle that the sun's rays make with an imaginary line perpendicular to a surface. The angle of incidence determines the intensity of the energy that any surface experiences.

Angle of maximum candela

The direction in which the luminaire emits the greatest luminous intensity.

Angle of reflection

The angle that a reflected ray of light makes with the surface that is reflecting it. The angle of reflection is the same as the angle of incidence.

Angle post See **Post**

Angle volute

One of the four corner volutes of a Corinthian capital; with an axis at 45 degrees to the face of the abacus.

Angled bay window See Window

Anglo Saxon architecture (800–1066)

The pre-Romanesque architecture of England before the Norman conquest; it is characterized by its massive walls and round arches and by timber prototypes later translated into stone.

Angular

Areas formed by two lines diverging from a common point, two planes diverging from a common line, and the space between such lines or surfaces, whether on the exterior or interior of a structure.

Annually renewable resource

A resource that is capable of being restored or replenished annually.

Annulet

A small molding, usually circular in plan and angular in section, encircling the lower part of a Doric capital above the necking; a shaft or cluster of shafts fitted at intervals with rings.

Anshen, Robert (1910–1964)

Partner in the firm of Anshen and Allen, San Francisco, CA. Designed the Chapel of the Holy Cross (illus.), Sedona, AZ (1956), the International Building (illus.), San Francisco, CA, (1961), and the Bank of California Building (illus.), San Francisco, CA, (1967).

Anta

A pier or pilaster formed by a thickening at the end of a wall, most often used on the side of a doorway or beyond the face of an end wall.

Antechamber

A room that serves both as a waiting area as well as an entrance to a larger room.

Antefix

A decorated upright slab used in classical architecture and other derivatives to close or conceal the open end of a row of tiles that covers the joints of roof tiles.

Antepagment

The stone or stucco decorative dressings enriching the jambs and head of a doorway or window; a door Jamb.

Anthemion

A common Greek ornament based on the honeysuckle or palmette, used in a radiating cluster, either singly on a stele or antefix, or as a running ornament on friezes.

Anthemion band

A Classical Greek style decorative molding with bas-relief or painted anthemion leaf clusters; often, two alternating designs are used along the band.

Antic

A grotesque sculpture consisting of animals, human and foliage forms incongruously run together and used to decorate molding terminations and other parts of medieval architecture.

Anticorrosive paints
Coatings formulated and recommended for use in preventing the corrosion of ferrous metal substrates.

Anti-neglect ordinance
A local regulation that provides penalties for owners who allow historic properties to deteriorate as a means of undermining preservation efforts.

Apadana
The hypostyle hall of Persian kings, such as that found at Persepolis, in Iran.

Apartment
A room or group of rooms designed to be used as a dwelling; usually one of many similar groups in the same building.

Apartment hotel
A building with multiple dwelling units that do not have private kitchens and are usually rented on a short-term basis.

Apartment house
A building containing a number of individual residential dwelling units.

Aperture
An opening for the purpose of admitting light.

Apex
The highest point, peak, or tip of any structure.

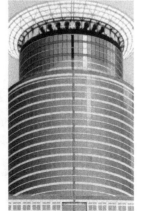

Apex stone
The uppermost stone in a gable, pediment, vault, or dome, usually triangular, often highly decorated.

Apophyge
The concave curve formed where the base or top of a classical style column curves inward to meet the shaft; tangent to the shaft and perpendicular to the fillet above.

Applied trim
Supplementary and separate decorative strips of wood or moldings applied to the face or sides of a frame.

Applique
An accessory decorative feature applied to a structure. In ornamental work, one material affixed to another.

Apprentice
A young person who is legally bound to a craftsman for a specified period of time in order to learn the skills of a particular trade.

Appropriate vegetation
Noninvasive plants suited to the site's climate and geology that do not require excessive care or resources to maintain plant vigor.

Approved document
Various documents associated with building regulations which give detailed guidance on structure, fire, site preparation, and conservation of power.

Approved equal
Material, equipment, or method approved by the architect or engineer for use in the work as being acceptable as an equivalent to those originally specified in the contract documents.

Apron
A flat piece of trim below the interior sill of a window, limited to the width of the window.

Apse
A semicircular or polygonal space, usually in a church, terminating an axis and intended to house an altar.

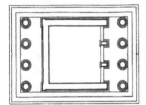

Apteral
Classical building with columns across the width of the portico at one or both ends, but without columns on the sides.

Aquatic systems
Ecologically designed treatment systems that utilize a diverse community of biological organisms, such as bacteria, plants, and fish, to treat wastewater at advanced levels.

Aqueduct
An elevated masonry structure with a channel for carrying water that is supported on arches; it was invented by the ancient Romans.

Aquifer
Any underground water-bearing rock formation or group of formations that supplies groundwater, wells, or springs.

Arabesque
Generic term for an intricate and subtle ornate surface decoration based on a mixture of intermixed geometrical patterns and natural botanical forms used in Muslim countries.

Arabesque style
Imitation of the Moorish architectural style of Spain, characterized by masonry walls with multicolored Mooresque decoration in plaster, precious stones, or glazed tiles on most surfaces; horseshoe arches, interior courtyards surrounded by colonnades; pointed dome roofs, wide horizontal stripes, and stalactite ornaments.

Arakawa, Shusaku (1937–2010)

With his wife, artist Madeline Gins, they called their philosophy "reversible destiny," and it manifested itself architecturally in buildings that challenged its occupants. In the couple's 2008 Bioscleave 1 house (Lifespan Extending Villa) built on Long Island, floors sloped and undulated like dunes, windows were placed slightly higher or lower than expected, switches were oddly placed, more than 36 paint colors were used, and no doors separated interior spaces. In addition to visual art, Arakawa and Gins wrote several books, including *The Mechanism of Meaning* (1979), *Reversible Destiny: We Have Decided Not to Die* (1997), and *Making Dying Illegal* (2006).

Arbitration

A method of resolving disputes between parties to a contract that are deadlocked. Independent arbitrators are selected for their specialized knowledge in the field, to hear the evidence and render a binding decision.

Arbor

A light open latticework frame, often used as a shady garden shelter or bower.

Arboretum

An informally arranged garden, usually on a large scale, where trees are grown for display, educational, or scientific purposes.

Arcade

A line of arches along one or both sides, supported by pillars or columns, either freestanding or attached to a building. Applies to a line of arches fronting shops, and covered with a steel and glass skylight usually running the length of the arcade.

blind arcade

A row of continuous arches applied to a wall and used as a decorative element.

interlaced arcade

An arcade formed by a series of columns supporting arches with overlapping archivolts.

intersecting arcade

Arches resting on alternate supports in one row, meeting on one plane at the crossings.

Arcaded arch See Arch

Arcading

A series of arches, raised on columns, that is represented in relief as decoration on a solid wall.

Arcature

An ornamental arcade on a miniature scale.

Arch

A basic architectural structure built over an opening, made up of wedge-shaped blocks, keeping one another in position, and transferring the vertical pressure of the superimposed load into components transmitted laterally to the adjoining abutments.

abutment arch

The first or last of a series of arches located next to an abutment.

acute arch

A sharply pointed two-centered arch, whose centers of curvature are farther apart than the opening.

arcaded arch

An arch that occurs where a vault intersects a wall.

back arch

An arch that supports an inner wall where the outer wall is supported in a different manner, such as a brick arch behind a stone lintel.

barrel arch

An arch that is formed by a curved solid plate or slab, as contrasted with one formed with individual members or curved ribs.

basket handle arch

A flattened arch designed by joining a quarter circle to each end of a false ellipse; a three-centered arch with a crown whose radius is greater than the outer pair of curves.

bell arch

A round arch resting on two large corbels with curved profiles.

blind arch

An arch within a wall that contains a recessed flat wall rather than framing an opening. Used to enrich an otherwise unrelieved expanse of masonry.

blind arch

blunt arch

An arch rising only to a slight point struck from two centers within the arch.

broken arch

A form of segmental arch in which the center of the arch is omitted and is replaced by a decorative feature usually applied to a wall above the entablature over a door or window.

camber arch

A flat segmental arch with a slightly upward curve in the intrados and sometimes also in the extrados.

catenary arch
An arch that takes the form of an inverted catenary, i.e., the curve formed by a flexible cord hung between the two points of support.

cinquefoil arch
A five-lobed pattern divided by cusps; a cusped arch with five foliations worked into the intrados; a cinquefoil tracery at the apex of a window.

circular arch
An arch whose intrados takes the form of a segment of a circle.

composite arch An arch whose curves are struck from four centers, as in the English Perpendicular Gothic style.

compound arch
An arch formed by concentric arches set within one another.

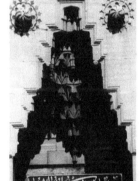

corbel arch
A false arch constructed by corbeling courses from each side of an opening until they meet at a midpoint; a capstone is laid on top to complete it.

counterarch
Arch that resists the thrust of another arch.

crescent arch
A type of horseshoe arch.

23

cusped arch
An arch which has cusps or foliations worked into the intrados.

depressed arch
A flat-headed opening with the angles rounded off to segments of circles; it was frequently used in the perpendicular style.

diminished arch
An arch having less rise or height than a semicircle.

discharging arch
An arch, usually segmental and often a blind arch, built above the lintel of a door or window to discharge the weight of the wall above the lintel to each side.

drop arch
A pointed arch which is struck from two centers that are nearer together than the width of the arch so that the radii are less than the span; a depressed arch.

elliptical arch
A circular arch in the form of a semi-ellipse.

equilateral arch
A pointed arch with two centers and radii equal to the span.

extradosed arch
An arch in which the extrados is clearly marked, as a curve exactly or roughly nearly parallel to the intrados; it has a well-marked archivolt.

false arch

A form having the appearance of an arch, though not of arch construction, such as a corbeled arch.

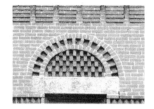

flat arch

An arch with a horizontal, or nearly horizontal intrados, with little or no convexity; an arch with a horizontal intrados with voussoirs radiating from a center below.

flat keystone arch

Flat arch with a distinctive keystone at its center.

Florentine arch

An arch whose entrados is not concentric with the intrados, and whose voussoirs are therefore longer at the crown than at the springing line, common in Florence in the early Renaissance.

foiled arch

An arch incorporating foils in the intrados, such as a two-cusped or trefoil arch.

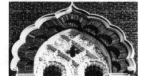

French arch

A flat arch with voussoirs inclined to the same angle on each side of a keystone.

gauged arch

Arch of wedge-shaped bricks which have been shaped so that the jambs radiate from a common center.

Gothic arch

A pointed arch, especially one with two centers and equal radii.

groin arch
An arch formed by the intersection of two simple vaults; an arched extrusion of a cross vault.

half arch
One half of a full arch; used with a quarter-round window or in the lower part of a flying buttress.

haunch arch
An arch having a crown of different curvature than the haunches, which are thus strongly marked; usually a basket-handle or three-centered arch.

horseshoe arch
A rounded arch whose curve is wider than a semi-circle, so that the opening at the bottom is narrower than its greatest span.

inverted arch
An arch with its intrados below the springline, especially used to distribute concentrated loads in foundations.

lancet arch
Same as an acute arch.

Mayan arch
A corbeled arch of triangular shape common in the buildings of the Maya Indians of Yucatan.

miter arch
A structural support over an opening in a masonry wall formed by two stones that meet at a 45-degree angle and form a right-angle miter in the center; technically not a true arch.

Moorish arch
The Islamic arch of North Africa and of the region of Spain under Islamic domination.

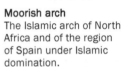

multi-centered arch
An arch having a shape composed of a series of circular arcs with different radii, making an approximate ellipse.

obtuse-angle arch
A type of pointed arch, formed by arcs of circles that intersect at the apex; the center of the circles are nearer together than the width of the arch.

ogee arch
A pointed arch composed of reversed curves, the lower concave, the upper convex; a pointed arch, each haunch of which is a double curve with the concave side uppermost.

parabolic arch
An arch similar to a three-centered arch, but whose intrados is parabolic with a vertical axis.

pier arch
An arch supported by piers rather than columns.

pointed arch
Any arch with a point at its apex, characteristic of but not limited to Gothic architecture.

proscenium arch
Large opening in a proscenium wall that surrounds the stage; may be rectangular or the shape of an arch.

rampant arch
An arch in which the impost on one side is higher than that on the other.

rear arch
An inner arch of an opening which is smaller in size than the exterior arch of the opening and which may be a different shape.

recessed arch
An arch with a shorter radius set within another of the same shape.

relieving arch See **discharging arch**

ribbed arch
An arch composed of individual curved members or ribs.

rigid arch
An arch without joints that is continuous and rigidly fixed at the abutments.

Roman arch
A semicircular arch in which all units are wedge-shaped.

rough arch
An arch constructed with rectangular bricks and tapered mortar joints; usually found on relieving arches.

round arch
An arch having a semi-circular intrados.

rowlock arch
A segmental arch composed of full rowlock bricks; especially when formed with concentric rows.

rustic arch
An arch laid up with regular or irregular stones; the spaces between them are filled with mortar.

segmented arch
An circular arch in which the intrados is less than a semicircle; an arch struck from one or more centers below the springing line.

semi-arch
An arch in which only one half of its sweep is developed, such as in a flying buttress.

semicircular arch
A round arch whose intrados is a full semicircle.

shouldered arch
A lintel carried on corbels at either end: a square headed trefoil arch.

splayed arch
An arch opening which has a larger radius in front than at the back.

squinch arch
A small arch across the corner of a square room, that supports a superimposed mass above it.

stepped arch
An arch in which the outer ends of some or all of the voussoirs are cut square to fit into the horizontal courses of the wall at the sides of the arch.

stilted arch
An arch whose curve begins above the impost line; one resting on imposts treated as a downward continuation of the archivolt.

surbased arch
An arch having a rise of less than half the span.

Syrian arch
A round arch with the springline almost at ground level; commonly found in Richardsonian Romanesque style doorways.

three-centered arch
An arch struck from three centers: the two on the sides have short radii, the center has a longer radius, and the resultant curve of the intrados approximates an ellipse.

three-hinged arch
An arch with hinges at the two supports and at the crown.

three-pinned arch
An arch with two pin joints at the supports and a third pin at the crown.

transverse arch
An arched construction built across a hall or the nave of a church, either as part of the vaulting or to support or stiffen the roof.

trefoil arch
An arch having a cusped intrados with three round or pointed foils.

triangular arch
A primitive form of arch consisting of two stones laid diagonally to support each other over an opening.

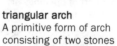
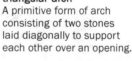

trussed arch
An arch-shaped truss with radial posts between the parallel arched top and bottom chords

Tudor arch
A four-centered pointed arch, common in the architecture of the Tudor style in England; a four-centered arch whose inner pair of curves is greater than that of the outer pair.

two-centered arch
An arch struck from two centers, resulting in a pointed arch.

two-hinged arch
An arch with two hinges at the supports at both ends.

Venetian arch
A pointed arch in which the intrados and extrados are further apart at the peak than at the springing line.

Arch order
A Roman architectural style characterized by a series of arched openings with an entablature above the head of the arch and engaged columns that appear to support the entablature between adjacent arches

Archaic
Antiquated or old fashioned, but when used in connection with Greek architecture refers to a specific period, c. 600-500 B.C.

Arched
Shapes formed by the curved, pointed, or rounded upper part of openings or supporting members.

Arched beam See **Beam**

Arched dormer See **Dormer**

Arched impost
An impost block with moldings that continue around the archivolt.

Arched truss See **Truss**

Archeological artifact
Objects made by humans including utensils, tools, or human remains that are recovered from an archeological site, such as those found in the tombs of Egypt, pyramidal tombs of ancient Mexico, ruins at Ephesus and Thira, Greece and Pompeii, Italy (illus.), or those of Native Americans.

Archeologist
A professional with experience in research, excavation, and analysis of archeological sites and artifacts.

Archeology

The systematic recovery by scientific methods of material evidence remaining from human life and culture in past ages, through evidence found in the ground, and the detailed study of this evidence. Archeology evidence case study.

Archigram (1960–1975)

Group formed by Peter Cook, Ron Herron, Dennis Crompton, and others, who publicized their ideas through seductive graphics, exhibitions, and the magazine *Archigram*.

Archiphobia

A term denoting the fear of architecture; a malady of those who fear to look at or understand the built environment.

Architect

An individual who is engaged in the design of buildings and who often supervises the construction.

Architect's scale

A ruler that uses a series of small measuring units, each representing one foot. This provides an accurate scaled-down version of the actual measurement on the job.

Architectonic

Related or conforming to technical architectural principles.

Architectural

Pertaining to architecture, its features, characteristics, or details; also to materials used to build or ornament a structure such as mosaic, bronze, wood and the like.

Architectural and historical evaluation

Part of a feasibility study which includes space utilization and circulation; listing elements to be preserved, those needing repair; historical designation benefits, and design reviews or special permits.

Architectural conservation

The process of maintaining and/or repairing the materials of a building or structure to reduce or reverse physical deterioration; it includes cleaning, repointing of masonry joints, and reattaching any loose elements.

Architectural design

A process which includes analysis of a program that results in the creation or alteration of a building or similar structure.

Architectural drawings

Plans, elevations, and details of the building to be constructed, consisting of foundation plans, floor plans, roof framing plan, electrical, plumbing and HVAC diagrams; exterior and interior elevations, details of door and window installations and structural connection and details.

Architectural element

Portion of a building or its ornamentation.

Architectural engineering

The art and science of engineering functions that relate to buildings or structures.

Architectural historian

Specialist in the history of the built environment, with special expertise in architecture.

Architectural history

The field of study of architectural style which includes the theoretical basis of design and the evolution of design vocabularies and construction techniques.

Architectural Review Board

An appointed local body that reviews proposed new construction and alterations to existing buildings in a historic district for conformance to established design guidelines and/or good design practice.

Architectural significance

The importance of a particular structure based on its design, materials, form, style, or workmanship; particularly if used for a nomination for the U.S. National Register.

Architectural style

The overall appearance of the architecture of a building, including its construction, form, and ornamentation; may be a unique individual expression or part of a broad cultural pattern.

Architectural woodwork
Finish carpentry for casework, cabinets and ornamental carvings.

Architecture
The art and science of designing and building structures, or groups of structures, in keeping with aesthetic and functional criteria.

Architrave
The lowest of the three divisions of a classical entablature, the main beam spanning from column to column, resting directly on the capitals.

Architrave cornice
An entablature in which the frieze is eliminated and the cornice rests directly on the architrave

Archival paper
A paper that is made with a slightly alkaline or neutral pH so it will not deteriorate, yellow, or turn brittle over time. Archival papers must meet national standards for permanence: They must be acid-free and alkaline with a pH of 7.5 to 8.5; include 2 percent calcium carbonate as an alkaline reserve; and not contain any ground wood or unbleached wood fiber.

Archivolt
The ornamental molding running around the exterior curve of an arch, around the openings of windows, doors, and other openings.

Archway
A passageway through or under an arch, especially a long barrel vault.

Arcology
A conception of architecture (1969) involving the fusion of architecture and ecology, proposed by Paolo Soleri, an Italian architect living in America. Arcology is Soleri's solution to urban problems. He proposed vast vertical structures capable of housing millions of inhabitants. One of Soleri's' visionary projects, Arcosanti (illus.), is being constructed north of Phoenix in Meyer, AZ.

Arcology

Arcuated

Based on, or characterized by, arches or arch-like curves or vaults; as distinguished from trabeated (beamed) structures

Area drain

A receptacle designed to collect surface or rainwater from an open area and connect it to a drainage system.

Areaway

An open subsurface space adjacent to a means of access to a basement window, door, or crawl space under the structure.

Areaway wall

Any wall built to hold back earth around an areaway; usually built of concrete, concrete blocks, brick, or stone.

Arena

A space of any shape surrounded by seats rising in tiers surrounding a stage; a type of theater without a proscenium.

Armory

A building used for military training or for the storage of military equipment.

Arquitectonica

High-style designers in the Postmodern style, such as the Atlantis apartments in Miami, FL (1982), their first major project (illus.). Their latest is the E Walk hotel and commercial development in Times Square, New York City (2000).

Arris

An external angular inter-section between two sharp planar or curved faces, as in moldings; or between two flutes in a Doric column.

Arroyo, Nicolás (1918–2008)

A Washington, D.C., architect, Arroyo had been a leading modernist architect in pre-Castro Cuba. With Gabriela Menendez y Garcia Beltran, his wife and professional part-ner of more than 60 years, he designed several prominent modernist buildings in Cuba, including the Havana Hilton Hotel, known as the Habana Libre following the 1959 revo-lution; the Teatro Nacional; and the Ciudad Deportiva sports complex.

Art Deco style (1925–1940)

Stimulated by an exhibition in Paris, this style drew its inspiration from Art Nouveau, Native American art, Cubism, the Bauhaus, and Russian ballet. The stylistic elements were eclectic, including the use of austere forms. It was characterized by linear, hard edge, or angular composition with stylized decoration. It was the style of cinemas, ocean liners, and hotel interiors. It was called "modernistic," and reconciled mass production with sophisticated design. Facades were arranged in a series of setbacks emphasiz-ing the geometric form. Strips of windows with decorative spandrels add to the composition. Hard-edged, low-relief ornamentation was common around door and window openings and along roof edges or parapets. Ornamental detailing was either executed in the same material as the building, in contrasting metals, or in glazed bricks or mosaic tiles. The style was also used for skyscraper de-signs such as the Chrysler building (illus.) in New York City.

Art glass See Glass

Art metal

Decorative metal elements, such as sheet-metal cornices or pressed-tin ceilings.

Art Moderne

A design style characterized by horizontal elements, rounded corners, flat roofs, glass blocks, smooth walls, windows that wrap around corners without posts, and asymmetrical massing, all intended to look streamlined. Also called Style Moderne and Streamline Moderne.

Art Moderne style (1930–1945)

A modern style characterized by rounded corners, flat roofs, and smooth wall finishes devoid of surface orna-mentation. A distinctive streamlined look was created by wrapping curved window glass around corners. Ornamen-tation consisted of mirrored panels and cement panels with low relief. Aluminum and stainless steel were used for door and window trim, railings and balusters. Metal or wooden doors often had circular windows.

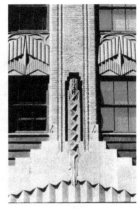

Art Moderne style

Art Nouveau architecture

Art Nouveau architecture (1880–1910)

A movement in European architecture and applied arts, developed principally in France and Belgium, characterized by flowing and sinuous organic and dynamic forms, naturalistic ornament and a strict avoidance of any historical traits. Other names for the style include Le Modern Style (France); Jugendstil (Germany), after the German term for youth style; Stile Liberty (Italy), named after the Liberty and Company store in London; Modernismo (Spain); and Sezession (Austria), named after its proponents seceded from the Academy of Art in Vienna. The style drew primarily on Baroque, Gothic and Moorish traditions, but was mainly unbounded by rules. Art Nouveau exploited the machine and reveled in the possibilities of decorative tiles and wrought iron. This was a deliberate attempt to put an end to imitations of past styles. In its place was a free type of architecture that integrated arts and crafts and architectural forms.

Articulation

Shapes and surfaces that have joints or segments which subdivide the area or elements; the joints or members add scale and rhythm to an otherwise plain surface.

Artifact

Individual product of human manufacture, such as cutlery, glassware, pottery, textiles, tools, and weapons.

Artificial light

Illumination produced by an electrical process, or the burning of fuel.

Artificial lighting

Any light source other than natural light. Artificial light sources include those with a continuous spectrum such as candles, normal electric light bulbs (tungsten lighting), special photographic light bulbs, photoflood bulbs, as well as discontinuous light sources such as fluores-

Artificial sky

A hemisphere or ceiling painted and illuminated to imitate the natural sky. Lighting effects can be used to simulate any time of day or night.

Artificial stone

A mixture of stone chips or fragments, usually embedded in a matrix of mortar, cement, or plaster; the surfaces may be ground, polished, molded, or otherwise treated to simulate natural stone.

Artificial wood See wood

Arts and Crafts movement (1880–1891)

A movement which restored creativity to the decorative arts and indirectly to architecture. Architects such as Henry Van de Velde, Joseph Hoffman, and Charles Rennie Mackintosh had a very strong influence on this movement. It abandoned the stylistic imitation of the nineteenth century and laid the groundwork for the creative works of the Art Nouveau styles that followed.

Asam, Egid Quirin (1692–1750)

One of two brothers who created a distinctive Baroque style in Bavaria in southern Germany. They designed church interiors in Munich and Regensberg, using fresco and stucco to create fantastically rich effects.

Asbestos

A noncombustible, flexible mineral fiber that is able to withstand high temperatures; it is fabricated into many forms, either alone or mixed with other ingredients. Once commonly used in many building materials, including insulation, fireproof siding, and resilient flooring. The use of asbestos in some products has been banned by the EPA (Environmental Protection Agency) and the U.S. Consumer Products Safety Commission; manufacturers also have adopted voluntary limitations on its use. When found in older buildings, most commonly in floor tiles, pipe and furnace insulation, or asbestos shingles, the product's friability is a major determinant in how it must be handled during renovations.

Asbestos abatement

Removal of material that contains asbestos, which is considered a potential health hazard, in a way that minimizes risk to the abatement workers and the public through encapsulation, repair, enclosure, encasement, and operations and maintenance programs.

Asbestos shingle

A roofing shingle that is composed of cement reinforced with asbestos fibers, manufactured in various shapes and sizes.

Asbestos slate

An artificial roofing slate that is manufactured with asbestos-reinforced cement.

Asbestos-cement board

A dense, rigid board containing a high percentage of asbestos fiber bonded with Portland cement; noncombustible; used in sheet or corrugated sheathing.

As-built drawings

Plans that incorporate the changes made during construction to record accurately the actual construction of the structure, as opposed to the initial construction documents.

As-built schedule
The final project schedule that depicts the start and completion date, duration, costs, and consumed resources for each activity.

Ashlar masonry See Masonry

ASHRAE 62.2
A standard for residential mechanical ventilation systems established by the American Society of Heating, Refrigerating and Air-Conditioning Engineers. Among other requirements, the standard requires a home to have a mechanical ventilation system capable of ventilating at a rate of 1 cubic foot per minute for every 100 square feet of occupied space plus 7.5 cubic feet per minute per occupant.

ASHRAE 90.1
A standard that establishes minimum requirements for energy-efficient design of buildings except low-rise residential buildings. This standard addresses construction, operation, and maintenance issues to minimize the use of energy while not compromising building functions or the comfort and productivity of building occupants.

Aspect
The point from which one looks, a point of view: a position facing a given direction, an exposure.

Aspect ratio
In any rectangular configuration, the ratio between the longer dimension and the shorter dimension.

Asphalt
A mixture of bitumens obtained from native deposits or as a petroleum by-product used for paving, waterproofing, and roofing applications.

Asphalt roofing
A roofing material manufactured by saturating a dry felt sheet with asphalt, and then coating the saturated felt with a harder asphalt coating, usually in roll form.

Asphalt shingles
Shingles manufactured from saturated roofing felt that is coated with asphalt, with mineral granules on the side that is exposed to the weather.

Asphalt tile
Resilient floor tile that is composed of asbestos fibers with asphalt binders; set in mastic and installed over wood or concrete floors.

Asplund, Eric Gunnar (1885–1940)
One of the most eminent of Swedish architects. He adopted Neoclassicism and designed the City Library, Stockholm, in 1920, which had architectonic shapes reduced to basic rectangles and cylinders.

Assembly room
A room in a hotel or town hall, where social gatherings can be held.

Associate
Closely connected as in function or office, but having secondary or subordinate status; an architect who has an arrangement with another architect to collaborate in the performance of service for a project, or a series of projects.

Association for Preservation Technology International (APTI)
A joint Canadian-American association founded (1968), by professional preservationists, architects, museum curators, architectural educators and archeologists devoted to promoting preservation research and disseminating technical information.

Association of Energy Engineers (AEE)
A trade organization for certification and information on energy efficiency, utility deregulation, facility management, plant engineering, and environmental compliance.

Assyrian architecture (900–700 B.C.)
Large palaces and temple complexes with ziggurats characterize this style: the exterior walls were often ornamented in carved relief or polychrome bricks. Doorways had semicircular arches with glazed brick around the circumference; windows were square-headed and high up in the wall. Interior courts were filled with slender columns with high molded bases, fluted shafts and capitals of recurring vertical scrolls. The bracket form of the topmost part was fashioned with the heads of twin bulls. They were widely spaced to support timber and clay roofs.

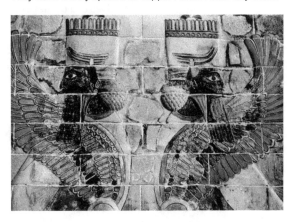

Assyrian architecture

ASTM International
A nonprofit organization originally established in 1898 that sets standard tests and specifications for various con-struction materials and methods; formerly known as the American Society of Testing and Materials (ASTM).

Astragal
A member or combination of members, fixed to one of a pair of doors or casement windows to cover the joint between the meeting stiles and to close the gap in order to prevent drafts, passage of light, air or noise.

Astylar
Buildings with Classical style features, but without any of the traditional orders or pilasters.

Asymmetry
Not symmetrical, with the parts not arranged correspondingly identical on both sides of a central axis.

Atelier
A place where artwork or handicrafts are produced by skilled workers; a studio where the fine arts, including architecture, are taught.

Atlas
A figure of a man used in place of a column to support an entablature; also called Atlantes and Telemon.

Atrium
The forecourt of an early Christian basilica, with colonnades on all four sides, and usually a fountain in the center. It was derived from the entrance court or hall of a Roman dwelling, roofed to leave a large opening to admit light. Rain was received in a cistern below. The modern version is a common vertical space with skylights in an office or hotel complex.

Atrium

Attached greenhouse

A structure located on a home's south side can provide passive solar heat to the home. Heat collected by the greenhouse at the lower level rises into the interior of the home by way of convection.

Attached house

A house which shares one or more common walls with another house, including row houses and semidetached houses.

Attainment area

An area considered to have air quality as good as or better than the NAAQS as defined in the Clean Air Act. An area may be an attainment area for one pollutant and a nonattainment area for others.

Attic

The top story or stories of a building; the structure's termination against the sky.

Attic base

A circular Ionic column base with an upper and lower torus joined by a concave scotia molding bordered with a pair of quadra fillets.

Attic fan

A fan typically mounted on the roof to create positive air flow through an attic that does not rely on wind or require excessive passive venting. It is connected to a thermostat and operates automatically.

Attic order

Small pillars or pilasters decorating the exterior of an attic story; in design and size they are subordinate, but related to the main order.

Attic ventilator

A mechanical fan in the attic of a house, which removes hot air from the roof space and discharges it to the outside.

Auditorium

That part of a theater, school, or public building that is set aside for the listening and viewing audience.

Aulenti, Gai (1927-)

Italian architect. lighting and interior designer, and industrial designer, She is well known for several large-scale museum projects, including the Contemporary Art Gallery at the Centre Pompidou in Paris, and the d'Orsay Museum (illus.) also in Paris. The d'Orsay was originally a train station, Gare d'Orsay, and transformed by Aulenti, The museum contains replicas of architectural ornamentation and sculptures on buildings throughout Paris.

Aulenti, Gai

Autoclaved aerated concrete

Masonry building material made of Portland cement, sand, and water in an autoclaving process heated under pressure; results in the production of air pockets in the material, making it less dense and better insulating.

Automatic door See **door**

Automatic fire alarm

A detector that automatically sounds a signal notifying the presence of a fire.

Automatic fixture sensors

Motion sensors that automatically turn on and off lavatories, sinks, water closets, and urinals. Sensors may be hardwired or battery operated.

Awning

A roof-like cover of canvas or other lightweight material, extending in front of a doorway or window, or over a deck, providing protection from the sun or rain.

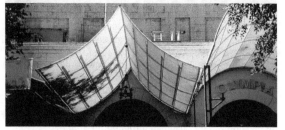

Awning

Awning window See **Window**

Axial composition

A design with a central axis that features bilateral symmetry.

Axial plan

The placing of several buildings or rooms longitudinally, such as along a single line.

Axis

An imaginary straight line, about which parts of a building, or group of buildings, can be arranged or measured.

Axonometric perspective

Form of orthographic projection in which a rectangular object, projected on a plane, shows three faces.

Axonometric projection See **Projection drawing**

Aztec architecture (1100–1520)

An architecture that emerged from the austere forms of the Toltecs, characterized by strong grid plans, monumental scale, and brightly colored exteriors, often with highly stylized surface carvings of human figures, floral patterns and images of gods. Pyramids often supported two temples with parallel stairways. Destruction by the Spanish left very few remains, as the Aztec capital of Tenochtitlan is entirely buried under modern Mexico City.

B b

Babcock, Charles (1839–1913)
American architect who worked with architect Richard Upjohn; one of the founders of the American Institute of Architects. He later designed buildings for Cornell University, Ithaca, New York.

Babylonian architecture (2000–1600 B.C.)
An architecture characterized by mud-brick walls articulated by pilasters and faced with glazed brick. The city of Babylon contained the famous Tower of Babel and the Ishtar Gate, decorated with enameled brick friezes of bulls and lions, and the Hanging Gardens of Semiramis. The ruins of the Assyrian Palace of Khorsabad show evidence of monumental sculptural decoration. The Palace of Darius at Persepolis featured magnificent relief carvings.

Back arch See Arch

Back building
A detached or contiguous subsidiary structure behind the main building.

Back hearth
The fireproof floor under the grate area upon which a fire is built.

Back vent
A pipe for ventilating purposes, attached to a waste pipe on the sewer side of its trap to prevent the siphoning of waste.

Backband molding See Molding

Backdrafting
Potentially hazardous condition in which the exhaust from combustion appliances does not properly exit the building. This can be due to a number of factors including a blocked flue or a pressure difference within the building.

Backfill
Crushed stone or coarse soil placed around the foundation walls to provide drainage for water collecting in the soil behind the wall.

Backing brick See Brick

Backlighting
Lighting of an object from the rear, used to provide drama and emphasis.

Backup material
A material placed at the back of a curtain wall for fire or insulation purposes, or behind a finished face of masonry.

Bacon, Henry (1866–1924)
American architect, associated with McKim Mead and White at the World's Columbian Exposition in Chicago. He specialized in commemorative buildings and public monuments, such as the Lincoln Memorial, Washington, D.C., built in 1922.

Baffle
An opaque or translucent element used to diffuse or shield a surface from direct or unwanted light.

Bailey

Stockade or walled enclosure surrounding a castle; also an open court in a medieval fortification.

Baked enamel

A hard, glossy metal finish composed of synthetic resins baked in an oven; used at one time for the spandrel panel of curtain wall construction.

Bakeout

A technique for reducing the exposure of occupants to emissions of new construction. The building temperature is raised to a high level, generally around 100 degrees to enhance emissions of volatile compounds from furniture and new materials for several days before occupancy, using 100 percent outside air to exhaust the emissions.

Balance

A harmonizing or satisfying arrangement, or proportion of various parts, as in a design or composition; the state of equipoise between different architectural elements.

formal balance

Designs that are almost always characterized by symmetrical elements.

informal balance

Designs where the forms are mostly asymmetrical.

Balance point

The outdoor temperature at which a building's heat loss to the environment is equal to internal heat gains from people, lights, and equipment. Typical balance point temperatures are in the range of 27–35 degrees Fahrenheit.

Balanced ventilation

Mechanical ventilation system in which separate, balanced fans exhaust stale indoor air and bring in fresh outdoor air in equal amounts; often includes heat recovery or heat and moisture recovery.

Balancing

Process of measuring and adjusting equipment to obtain desired flows within the system.

Balconet

A pseudo-balcony; a low ornamental railing to a window, projecting but slightly beyond the plane of the window, threshold or sill, having the appearance of a balcony when the window is fully open.

Balcony

A projecting platform usually on the exterior of a building, sometimes supported from below by brackets or corbels, or cantilevered by projecting members of wood, metal or masonry. They are most often enclosed with a railing, balustrade, or other parapet.

Balcony beam

Horizontal beam which supports a balcony.

Bald cypress See **Wood**

Baldachino

A permanent canopy over a throne or altar, supported by four columns.

Balistraria

A loophole or similar aperture in a medieval wall, most often cruciform in shape, through which bowmen fired arrows.

Ballast

Power-regulating device that modifies input voltage and controls current to provide the electrical conditions necessary to start and operate gaseous discharge lamps, especially fluorescents and HID (high-intensity discharge) lamps.

Ballflower

A spherical ornament composed of three conventionalized petals enclosing a ball, usually in a hollow molding; popular in the English Decorated style.

Balloon framing

A system of framing a wooden building wherein all vertical studs in the exterior bearing walls and partitions extend the full height of the frame from sill to roof plate; the floor joists are supported by sills.

Ballroom

A large room in a dwelling, hotel, or public building such as a town hall designed for dancing, concerts, and similar entertainment.

Balsam fir See **Wood**

Balteus

The vertical bead between the scrolls at the side of the Ionic capital.

Baluster

One of a number of short vertical members used to support a stair railing.

Baluster column See **Column**

Baluster side

The rounded side of an Ionic capital.

Balustrade

An entire railing system, as along the edge of a balcony, including a top rail, bottom rail and balusters.

Balustrade order
One of the Neoclassical orders of balusters and rails.

Ban, Shigeru (1957–)
Japanese and international architect, most famous for his innovative work with paper, particularly recycled cardboard paper tubes used to quickly and efficiently house disaster victims. Ban's work is known for his humanitarianism and attraction to ecological architecture. Ban's use of paper and other materials is heavily based on its sustainability and that it produces very little waste. As a result of this, Ban's DIY (do-it-yourself) refugee shelters, used in Japan after the Kobe earthquake, and in Turkey and Rwanda, are very popular and effective for low-cost disaster relief housing. Ban created the Japanese pavilion building at Expo 2000 in Hanover, Germany, in collaboration with the architect Frei Otto and structural engineering firm Buro Happold. The 72-meter-long grid-shell structure was made with paper tubes. After the exhibition the structure was recycled and returned to paper pulp. Ban fits well into the category of "ecological architects," but he also can make solid claims for being a modernist, a Japanese experimentalist, as well as a rationalist.

Band
A flat horizontal fascia, or a continuous member or series of moldings projecting slightly from the wall plane, encircling a building or along a wall, that makes a division in the wall.

Band molding See Molding

Banded architrave
A door or window architrave that is broken at regular intervals by a series of projecting blocks.

Banded rustication
Alternating smooth ashlar and roughly textured stone.

Banderole ornament See Ornament

Banding
Horizontal subdivisions of a column or wall using profile or material change.

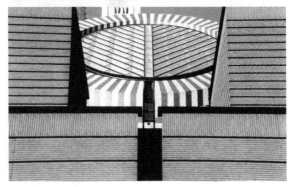

Banister
A slender pillar turned on a lathe, used to support the handrail of a stair.

Bank barn
A two-level barn built against a bank or hillside, with an upper level that can be reached directly from the hill slope.

Baptistry
A building for Christian baptismal rites containing the font; often located in a structure separate from the church.

Bar
A long, thin strip of material, especially iron, used for a variety of building purposes.

Bar joist
An open web joist with diagonal struts made of round or square steel bars with the top and bottom chords made up from pairs of steel angles.

Bar molding See **Molding**

Bar tracery See **Tracery**

Barbican
An outer defense of a castle or city, in particular a tower above a gate or drawbridge.

Barge cornice
A raked cornice at the top of a gable end.

Bargeboard
A trim board used on the edge of gables where the roof extends over the wall; it either covers the rafter or occupies the place of a rafter. Originally it was ornately carved.

Bargecourse
The coping of a wall, formed by a course of bricks set on edge.

Bargestone
One of the projecting units of masonry, which forms the sloping top of a gable wall.

Barn
A building for housing animals and storing farm equipment, hay, and other agricultural produce.

Barn

Barn doors
Adjustable hinged panels mounted on the rim of a lighting fixture to direct the beam of light; usually consisting of four panels.

Barnes, Edward Larrabee (1915–2004)
His most noted early works is the Crown Center, Kansas City, Missouri (1968). Later works include the IBM Building, New York City (1983), and the Walker Art center, Minneapolis (1971).

Baroque architecture (1600–1760)
A style named for the French word meaning bizarre, fantastic or irregular. It was the most lavish of all styles, both in its use of materials and the effects that it achieved. Mannerist styles were often adopted and carried to the extreme as bold, opulent and intentionally distorted. Pediments are broken and facades designed with undulating forms, while interiors are more theatrical, exhibiting a dramatic combination of architecture, sculpture, painting and the decorative arts.

Baroque architecture

Barracks
Temporary or permanent housing erected for soldiers or groups of workers.

Barrel arch See **Arch**

Barrel roof See **Roof**

Barrel vault See **Vault**

Barrigan, Louis (1902–1988)
Self-taught Mexican architect and landscape architect, educated originally as an engineer. Won the Pritzker Prize in 1980. He had been a pioneer in the use of color in modern architecture and used raw materials such as stone and wood.

Barroco-mudejar style
An architectural style of the Mexican colonial period that employs Italian Baroque features and Moorish decorative motifs: thick adobe walls, scrolled parapet copings, elaborately carved window and door surrounds and concave shell-shaped ornamentation over the door head.

Barry, Sir Charles (1795–1860)
English architect of the early Victorian period, inspired by Renaissance models. The Houses of Parliament, London, is his most important work.

Basalt See **Stone**

Bascule
A structure that rotates about an axis, as a seesaw, with a counterbalance equal to the weight of the structure at one end, used for movable bridges.

Bascule bridge See **Bridge**

Base

The lowest and most visible part of a building, often treated with distinctive materials, such as rustication; also pertains to the lowest part of a column or pier that rests on a pedestal, plinth or stylobate.

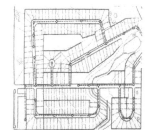

Base map

Graphic representation of a defined area, such as a particular site, neighborhood, town, region, or state, showing legal boundaries and physical features.

Base molding See **Molding**

Baseboard

A flat projection from an interior wall or partition at the floor, covering the joint between the floor and wall, and protecting the wall: it may be plain or molded.

Base bid

Amount of money stated in the bid as the sum for which the bidder offers to perform the work required by the project documents. Does not include the work for which alternate bids are submitted.

Base bid specifications

The specifications describing only those materials, equipment, and methods of construction upon which the base bid must be predicted, exclusive of any alternate bids.

Baseline

A standard measure or perceived common level of performance at a given point in time, against which design objectives and/or performance targets may be set against to achieve improved design outcomes or performance standards.

Baseline building performance

The annual energy cost for a building design intended for use as a baseline for rating above standard design, as defined in ASHRAE 90.1–2004.

Basement

Usually the lowest story of a building, either partly or entirely below grade.

Basement window

Frames and sash of either wood or metal for use in basement openings. Usually such windows do not have more than two or three lights.

Base block

A block of any material, generally with little or no ornament, forming the lowest member of a base at the foot of a door or window.

Baseplate

A steel plate for transmitting and distributing a column load to the supporting foundation material.

Base cabinet

Kitchen cabinet that rests on the floor and supports the counter.

Base cap

Molding that rests on or overlaps the top of a baseboard.

Base coat

The first coat of plaster finish, typically composed of lime, cement, and sand; the first layer of any coating applied in a liquid or plastic state, such as paint.

Base course

The lowest course on a masonry wall or pier, may also be part of the footing.

Basic services

The services performed by an architect during the following five phases of a project: schematic design, design development, construction documentation, bidding or negotiation, and contract administration.

Batt insulation

Insulation, usually of fiberglass or mineral wool and often faced with paper, typically installed between studs in walls and between joists in ceiling cavities. Correct installation is crucial to performance.

Basilica
A Roman hall of justice with a high central space lit by a clerestory with a timbered gable roof. It became the form of the early Christian church, with a semicircular apse at the end preceded by a vestibule and atrium.

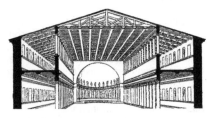

Basis of design
Includes design information necessary to accomplish the owner's project requirements, including system descriptions; indoor environmental quality criteria; other pertinent design assumptions such as weather data; and references to applicable codes, standards, regulations, and guidelines.

Basket capital See **Capital**

Basket handle arch See **Arch**

Basketweave bond See **Bond**

Bas-relief See **Relief**

Bastion
A projecting portion of a fortification designed to defend the adjacent curtain wall; it typically approximates a semi-hexagon, with the two outer faces meeting at an acute angle, and two flanks abutting a curtain wall on each side.

Batch solar water heater
Solar water heater in which the potable water is heated in the same place it is stored.

Bath
An open tub used as a fixture for bathing; the room containing the bathtub. The Roman public bathing structure consisted of hot, warm and cool pools; sweat rooms, athletics and other related facilities.

Bathhouse
A building equipped with bathing facilities: a small structure containing rooms or lockers for bathers, as at the seaside.

Bathroom
A room with bathing facilities; typically containing a bathtub, water closet, and lavatory.

Batt
A length of insulation that is precut to fit certain wall cavity dimensions. Batt insulation is typically sold in a prepackaged roll.

Batt insulation
Loosely packed fibrous insulation more than 1 inch thick; types of fiber include fiberglass and wood.

Batten
A narrow strip of wood that is applied over a joint between parallel boards in the same plane. In roofing, the standing seam of a metal roof gives the same appearance of a batten,

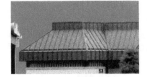

Batten door See **door**

Batter
A wall that gently slopes inward toward the top.

Battered
Forms that slope from a true vertical plane from bottom to top, as in the outside surface of a wall.

Battlement
A parapet having a regular alternation of solid parts and openings, originally for defense, but later used as a decorative motif.

Bau-biologie
Literally "building biology," a term coined in Germany to describe the use of holistic, healthy-building principles—particularly those focused on indoor air quality and electromagnetic fields—to safeguard the well-being of a building's occupants.

Bauhaus style
An architectural style developed at the school of design established by Walter Gropius in 1919 in Weimar, Germany. It moved to Dessau in 1926 and closed in 1933. The term became virtually synonymous with modern teaching methods in architecture and the applied arts and with the functional aesthetics for the industrial age. It epitomized the marriage of modern design, mass production and industrial design.

Bay
A principal compartment or division in the architectural arrangement of a building, marked either by buttresses or pilasters in the wall, by the disposition of main arches and pillars, or by any repeated spatial units that separate it into corresponding portions.

cant bay
A bay erected on a canted outline.

Bay leaf garland See **Garland**

Bay window See **Window**

Bead and reel molding See **Molding**

Bead molding See **Molding**

Beaded joint See **Mortar joint**

Beak molding See **Molding**

Beam
A rigid structural member whose prime function is to carry and transfer transverse loads across a span to the supports; as a joist, girder, rafter, or purlin.

arched beam
A beam whose upper surface is slightly curved, similar to a camber beam.

bolster beam
A timber or steel beam that supports the end of a bridge truss on an abutment or pier; it is set perpendicular to the trusses.

box beam
One or more vertical plywood webs laminated to seasoned wood flanges. Vertical spacers separate the flanges at intervals along the length of the beam to distribute the loads and to provide stiffness.

built-up beam
A beam composed of multiple parts, such as a box beam, compound beam, lattice beam, and angle girder.

camber beam
A beam curved slightly upward toward the center.

channel beam
A channel used as a beam, typically with the web in the vertical position.

collar beam
A horizontal member that ties together two opposite common rafters, usually at a point halfway up the length of the rafters below the ridge.

continuous beam
A beam that is continuous over intermediate supports and thus statically indeterminate; as opposed to a simply supported beam.

crossbeam
A beam that runs transversely through the centerline of a structure; any transverse beam in a structure such as a joist.

edge beam
A beam at the edge of a shell plate structure, providing stiffness that provides an increase in the load-bearing capacity.

encased beam
Iron and steel beams that are encased in a variety of materials for protection against fire.

grade beam
Reinforced concrete beam or slab that is normally placed directly on the ground.

hammer beam
One of a pair of short horizontal members, attached to the foot of a principal rafter, located within a roof structure in place of a tie beam.

I-beam
An iron or steel beam with a symmetrical I-shaped cross section.

laminated beam
A wood beam composed of a series of overlapping wood members that are glued together.

open web beam
A truss with parallel top and bottom chords formed by a pair of angles, employing a web of diagonal struts and used as a beam; the struts connecting the top and bottom chords are also composed of steel bars.

plate girder
A steel beam built up from vertical web plates and horizontal flange plates.

simple beam
A beam without restraint or continuity at the supports, as opposed to a fixed-end beam.

summerbeam
A large horizontal beam in the ceiling of an American colonial timber-framed house.

tie beam
In roof framing, a horizontal timber connecting two opposite rafters at their lower ends to prevent them from spreading.

trussed beam
A beam or purlin stiffened with a tie rod.

welded beam
A large steel I-section, fabricated by welding from plates instead of hot rolling; same as a plate girder.

Beam ceiling
A ceiling formed by the underside of the floor, exposing the beams that support it; also applies to a false ceiling imitating exposed beams.

Bearing pile
A pile that carries a vertical load, as compared with a sheet pile, which resists earth pressure.

Bearing plate
A metal plate at the end of a beam or bottom of a column to distribute the load over a larger surface.

Bearing wall See Wall

Beaux-Arts Classicism (1890–1920)
Grandiose compositions with exuberant ornamental detail and a variety of stone finishes characterize this style Classical colossal columns were grouped in pairs on projecting facades with enriched molding and free-standing statuary; pronounced cornices and enriched entablatures are topped with a tall parapet, balustrade, or attic story. It fostered an era of academic revivals, principally public buildings featuring monumental flights of steps.

Beaux-Arts style (1860–1883)
Historical and eclectic design on a monumental scale, as taught at the École des Beaux-Arts in Paris, typified this style. It was one of the most influential schools in the nineteenth century, and its teaching system was based on lectures combined with practical work in studios and in architectural offices. Its conception of architecture lies in the composition of well-proportioned elements in a symmetrical and often monumental scheme.

Becket, Welton (1902–1964)

Head of his successful Los Angeles architectural firm. Designed the Capitol Records Building, Los Angeles (1956); the Kaiser Center, Oakland, CA (1960); the Performing Arts Center of Los Angeles County (1964); Disney's Contemporary Resort, Lake Buena Vista, FL (1971); the Hyatt Regency Hotel and Reunion Tower (illus.), Dallas (1978); and the Dorothy Chandler Music Pavilion (illus.), Los Angeles, CA, (1964),

Bed

In masonry and bricklaying, the side of a masonry unit on which it lies in the course of the wall, which is the underside placed horizontally; also, the layer of mortar on which a masonry unit is set.

Bed joint

The horizontal joint between two masonry courses; one of the radial joints in an arch.

Bed molding See Molding

Bed mortar joint See Joint

Bedplate

A plate, frame, or platform which supports a wall above.

Bedroom

A room used for sleeping; a bedchamber.

Beeby, Thomas (1941–)

American architect who was a partner in the firm Hammond, Beeby and Babka, whose major project was winning the competition for the Harold Washington Library Center (illus.), Chicago, IL (1991).

Beehive tomb

A conical-shaped subterranean tomb constructed as a corbeled vault and found on pre-Archaic Greek sites.

Behnisch, Günter (1922–2010)

A prominent German architect credited with helping re-shape the face of postwar Germany. His airy modern buildings include the Munich Olympic Stadium, with engineer Frei Otto, for the 1972 Olympic Games; the Plenary Complex of the German Parliament in Bonn (2002); Genzyme Center in Cambridge, MA (2003); and the Center for Cellular and Bimolecular Research in Toronto, Canada (2005).

Behrens, Peter (1869–1938)

Designed the AEG Turbine Factory (Illus.), Berlin, in 1908. It was a strictly functional factory, constructed of concrete, steel, and glass, without ornamentation. He also designed the German Embassy, in St. Petersburg, Russia, in the Classical idiom, devoid of all ornamentation.

Belcast eaves

A curve in the slope of a roof at the eaves; used not only because of its aesthetic appearance but also because it protects the exterior walls from rainwater running off the roof.

Belfry

A room at or near the top of a tower that contains bells and their supporting timbers.

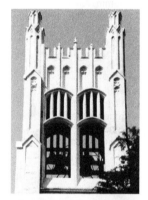

Belgian block See Stone

Bell

The body of a Corinthian capital or a Composite capital without the foliage.

Bell arch See Arch

Bell cage

Timber framework which supports the bells in a belfry or steeple.

Bell chamber

A room containing one or more large bells hung on their bell cage.

Bell gable

A small turret placed on the ridge of a church roof to hold one or more bells.

Bell roof See Roof

Bell tower

A tall structure either independent or part of a building used to contain one or more bells; it is also called a campanile.

Bell turret

A small tower, usually topped with a spire or pinnacle, containing one or more bells.

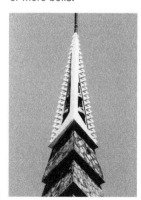

Bell-cast roof See **Roof**

Bell-shaped dome See **Dome**

Belluschi, Pietro (1899–1994)
Italian-born architect and engineer; he showed an inclination to the International Style. His Equitable Life Assurance building in Portland, Oregon (1944), was one of the first examples of an aluminum and glass curtain wall enclosing a concrete frame tower. Later works include the cathedral of Saint Mary of the Assumption (Illus.), with Pier Luigi Nervi, San Francisco, CA, 1969.

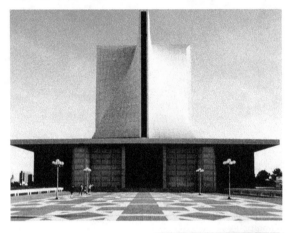

Beltcourse
A projecting horizontal course of masonry, of the same or dissimilar material used to throw off water from the wall; usually coincides with the edge of an interior floor.

Belvedere
A building, architectural feature, or rooftop pavilion from which a vista can be enjoyed.

Bench
A seat with or without a back; most often it is constructed of wood.

Benchmark
A permanent reference mark, fixed to a building or to the ground, whose height above a standard datum level has been accurately determined by survey.

Benefit/cost analysis
An economic method for assessing the benefits and costs of achieving alternative health-based standards at given levels of health protection.

Benjamin, Aaron (1932–2010)
An urban planner and housing specialist who worked for the U.S. Agency for International Development (USAID) as a Housing and Urban Development officer. He was involved worldwide, often following a disaster, in reconstruction, export development, and disaster preparedness response programming.

Benjamin, Asher (1773–1845)
American architect in Boston, MA. He wrote several pattern books that helped disseminate Georgian, Federal, and Greek Revival styles. He designed the Charles Street Meeting House, Boston, MA (1807).

Bent
A framework, which is transverse to the length of a framed structure; usually designed to carry both a lateral and a vertical load.

Bent sash
A window sash that is curved in plan.

Bent window See **Window**

Berlage, Hendrik P. (1856–1934)
Dutch architect who designed the Amsterdam Stock Exchange (Illus.), Amsterdam, The Netherlands, using brick with a stone trim, which was a fresh interpretation of the Romanesque style.

Berlage, Hendrik P.

Berm
The horizontal surface between a moat and the exterior slope of a fortified rampart; a continuous bank of earth piled against one or more exterior walls of a building as a protection against the elements.

Bernini, Gian Lorenzo (1598–1680)
Italian architect and sculptor whose works include the Trevi Fountain (Illus.) Rome, in 1632. He designed St. Peter's Piazza (Illus.), the Vatican, and the Colonnade (Illus.), a huge elliptical space surrounded by a colonnade with columns four deep in (1656).

Best practice
Best practice in construction refers to construction that goes beyond compliance in terms of sustainability and actively attempts to have a positive environmental impact during the construction process, rather than simply a minimized negative impact.

Bestiary See **Ornament: animal forms**

Bethune, Louise (1856–1913)
American architect, and the first woman Fellow of the American Institute of Architects; opened her office at age 25 and designed apartment buildings, schools, stores, and industrial structures.

Beton brut
Concrete textured by leaving the impression of the form in which it was molded, as when wood is used to create a grained effect on the surface.

Bevel
A sloped or canted surface resembling a splay or other chamfer, where the sides are sloped for the purpose of enlarging or reducing them.

Beveled joint See **Joint**

Beveled molding See **Molding**

Beveled siding See **Wood**

Beyer Blinder Belle
American historic preservation firm in New York. Projects include preservation of the Ford Center for the Performing Arts, Ellis Island Visitors' Center (illus.), and the interior of Grand Central Terminal, (illus.), all in New York City.

Bezant An ornament shaped like a coin or disc, sometimes used in a series in decorative molding designs.

Bibiena, Guiseppe (1657–1748)
Italian artist from an important family of painters, theatrical designers, and architects; produced fantasy scenes of classical antiquities.

Bicycle storage
Covered and/or secured storage for building occupants commuting by bicycle. This amenity is considered a sustainable building technique in that it encourages human-powered transportation options. Some local governments offer subsidies or incentives to include bicycle storage in an existing or proposed building project.

Bid
A complete and properly signed proposal to do all or part of the work for the sum stipulated in the bid; supported by data called for by the bidding requirements.

Bid bond
A form of bid security executed by the bidder as principal.

Bid opening
The time, place, and date set for receiving and opening the bids for a specific project, may be either public or private.

Bidding
In construction, the process of extending invitations and receiving price proposals from contractors or subcontractors for a defined scope of work, typically based on the construction documents and specifications.

Bidding documents
Contains the advertisement or invitation to bid; instruction to bidders; bid form; and the proposed contract including any addendum issued before receipt of bids.

Bidding requirements
Documents providing information and establishing procedures and conditions for the submission of bids. They consist of the notice to bidders, instruction to bidders, and sample bid forms.

Bifold door See **Door**

Bifron capital See **Capital**

Big basin
The first approach to stormwater management, still widely used in the United States, which involves large retention or detention basins or ponds. However, these systems do not manage the stormwater where it falls, often preventing infiltration and taking away available space for recreation or other site design needs.

Big five energy sources
Coal, oil, natural gas, large-scale hydroelectric, and nuclear power.

Bilateral symmetry See **Symmetry**

Billet molding See **Molding**

Binder
Glue used in manufactured wood products, such as medium-density fiberboard (MDF), particleboard, and engineered lumber. Some binders are made with formaldehyde.

Bio-based material
A material made from living matter, such as agricultural crops. Bio-based materials are usually biodegradable.

Biocide
Chemicals toxic to micro-organisms. Biocides, which include pesticides and antimicrobial agents, are used in paint, building materials, and floor coverings to kill bacteria, mold spores, and insects.

Biodegradable
Waste material composed primarily of constituent parts that occur naturally, are able to be decomposed by bacteria or fungi, and are absorbed into the ecosystem. Wood, for example, is biodegradable, while plastics are not.

Biodeterioration
Agents that contribute to the deterioration of stone. Modes of disintegration are both physical and chemical, and are affected by bacteria, algae, lichen, fungi, mosses, and guano from birds and other mammals.

Biodiversity

The tendency in ecosystems, when undisturbed, to have a great variety of species forming a complex web of interactions. Human population pressure and resource consumption tend to reduce biodiversity dangerously; diverse communities are less subject to catastrophic disruption.

Bioengineering

The use of living plants, or a combination of living and nonliving materials, to stabilize slopes and drainage ways.

Bio-fuel

Fuel such as methane produced from renewable resources, especially plant biomass and treated municipal and industrial wastes.

Biological contamination

The contamination of a building environment caused by bacteria, molds and their spores, pollen, viruses, and other biological materials (e.g., animal and human dander insect and arachnid excreta). It is often linked to poorly designed and maintained HVAC systems. People exposed to biologically contaminated environments may display allergic-type responses or other physical symptoms.

Biological productivity

Nature's capability to reproduce and regenerate, thereby accumulating biomass.

Biomass

Any organic matter that is available on a renewable or recurring basis, excluding old-growth timber, including dedicated energy crops and trees, agricultural food and feed crop residues, aquatic plants, wood and wood residues, animal wastes, and other waste materials.

Biomaterials

Include organic materials such as bamboo, cordwood and straw bales, rammed earth, mud and clay, and other natural composites.

autoclaved aerated concrete blocks

Concrete blocks that are stronger, lighter, and a green alternative to conventional masonry blocks. They can be produced from fly ash, which make this composite eco-friendly and nontoxic, fire and insect resistant, with excellent sound absorbing properties and insulation values.

bamboo

An organic, natural material that is lightweight, durable, flexible, biodegradable, and recyclable. It is fast growing and rapidly renewable. When treated, it becomes as strong as steel and may be used as the main structural material in small dwellings. It is a more renewable material than wood because it is a fast-growing grass/reed that can be harvested after only 4–6 years of growth, much shorter than the 30–60 years required for comparable wood species. Replanting is not necessary, as it regenerates on its own. It is stronger than oak, which is widely considered the most durable hardwood. When laminated, bamboo is nearly as strong as soft steel. Bamboo doesn't swell or shrink as hardwoods do, making it a perfect material for furniture and flooring.

cordwood

A practical, resource-efficient, renewable material that is usually obtained from timber and trees unsuitable for building purposes. Deadfall and standing dead timber are preferred, because they are partially seasoned and contain minimal rot.

cork composites

A durable, low-maintenance, decay-resistant material with excellent thermal and acoustical insulation properties made of cork and recycled rubber. Its cellular structure traps air inside, which reduces noise and vibration, and gives floors natural shock absorption. A naturally occurring substance in cork repels insects and molds and protects cork from rotting when moist.

green concrete

An environmentally friendly version of concrete produced with either non-CO_2-emitting materials and/or non-CO_2-producing methods. One of the most commonly used methods is to use indus-trial waste by-products such as fly ash from coal combustion and blast furnace slag from iron manufacturing to constitute the cement mixture used in producing concrete. One of the main advantages of this method is that it prevents these waste materials from entering landfills.

green form-release agents

These form-release biocomposites are non-petroleum alternatives to conventional oil products. They are typically water-based, nontoxic, biodegradable composites that contain very low VOCs (volatile organic compounds). These composites can be used to coat various types of concrete forms and liners, such as steel, aluminum, fiberglass, and plywood, and they are usable at below-freezing points.

green PVC alternatives

Green alternatives to the widespread use of PVC materials can be found in many building products. Clay, cast iron, and high-density polyethylene replace PVC piping. Wood, acrylic, and fiber-cement boards replace vinyl siding. Electrical insulation can be replaced with low-density polyethylene. Sheathing can be replaced with halogen-free, low-density polyethylene. Roofing can be replaced with soil and grass, or light metal. Natural biomaterials such as wood and bamboo, as well as ceramic composite tiles, can replace PVC flooring. Biofiber and polyethylene can replace PVC wall coverings. PVC window and door frames can be replaced with wood, fiberglass, and aluminum. PVC carpet fibers can be replaced with recycled biomaterials and natural fiber backing.

hybrid composite panels

Composite panels that are made with an aluminum core, coupled with wood, polymers, woven fiberglass, and fibers. They are very rigid, lightweight, high-strength materials with excellent impact resistance.

hybrid frame systems

A new generation of wood–polymer composites that are extruded into a series of shapes for window frames and sash members. These composites can be treated like regular wood. They are very stable and have the same

hybrid frame systems

(or better) structural and thermal properties as conventional wood, but with better moisture and decay resistance.

linoleum

A green surface composite made from organic, biodegradable materials consisting of linseed oil as the binder, lime as the filler, and organic pigments as the color for the material. Natural fibers such as jute are used for stabilizing the material. Organic linoleum has significantly low-embodied energy, creates little waste during manufacturing, does not require constant maintenance, and can be fully recycled and reused. It has very low-VOC emissions when installed with low-VOC adhesives, and it does not contain formaldehyde, asbestos, or plasticizers.

mud and clay

Included in the most basic building materials are mud and clay. Mud is typically used as a type of concrete and for insulation. Mud and clay have excellent thermal-mass qualities and keep the indoor temperatures at a constant level. Overall, mud and clay are highly available, low-cost, energy-efficient, high-performance, ecologically green materials, but they may require constant maintenance and structural support.

rammed earth

A damp mixture of soil, sand, gravel, clay, and other stabilizers pressed against an external frame that produces a solid earth wall. It is an economical and versatile material, which is as effective for nonlinear surfaces, corners, curves, and arches as it is for straight walls. Rammed earth is a clean, recyclable, biodegradable, resource-efficient material with a high thermal-mass value. Its inherent insulation and radiation capability reduces the energy required for heating. Earth can be dug locally, thereby reducing transportation and manufacturing requirements.

recycled glass/ceramic

A composite composed of recycled glass products, such as light bulbs, glass waste, and automobile windshields. Crushed glass is mixed with ceramic materials to produce the composite, creating a material that is cost-effective, durable, scratch and wear resistant, and easy to clean.

recycled rubber composites

A green composite for indoor and outdoor uses, such as sidewalks and vehicular roads. Indoor rubber composites are quite sustainable, slip resistant, and require very low maintenance. Recycled rubber composites are also used for wall panels, insulation, carpet underlayment, and roofing. They are water- and rust-resistant, and withstand extreme temperatures that cause curling, splitting, cracking, and rotting. In addition, rubber composite materials are lightweight, impact resistant, and energy efficient with high insulation values.

recycled wood/plastic composites

Composed mainly of wood fibers and waste plastics that include high-density polyethylene, the material is formed into both solid and hollow profiles and used to produce

recycled wood/plastic composites

building products such as decking, door and window frames, and exterior moldings.

rice hulls

The hard, protective shells formed over rice grains resist moisture penetration and are excellent insulation materials, with a thermal resistance of about R 3.0 per inch. They are fire resistant and resistant to rot, molds, mildew, insects, rodents, and fungus. Rice hull interlocking panels exhibit excellent load bearing capability, sound absorption and fire resistance.

sod

A thin block of grass held by its roots, usually used for turf and lawns, but can be used as a temporary building material. Like brick, sod is cut and laid in regular block shapes. The walls of a sod building are usually protected with a layer of stucco or wood panels.

straw bale

An agricultural by-product made from the stems of cereal crops, sugar cane, wheat, oats, rye, rice, barley, and others. Low cost and general availability make straw bales a highly desirable, natural green material. A post-and-beam framework is the most common non-load-bearing construction method, where the framework supports the structure and straw bales are used as infill. Straw bale is a very economical material, but it must be protected from getting wet both during and after construction.

Biome

An entire community of living organisms in a single major ecological area.

Biomimicry

A new science that studies nature's models and then imitates or takes inspiration from these designs and processes to solve human problems, e.g., a solar cell inspired by a leaf. The application of methods and systems found in nature to the study and design of engineering systems and modern technology. It uses an ecological standard to judge the rightness of our innovations. After 3.8 billion years of evolution, nature has learned what works, what is appropriate, and what lasts.

Biophilia

Theory developed by biologist Edward O. Wilson suggesting that humans have an innate affinity for nature.

Bioplastic

Plastics made from corn, potato, or other renewable resources that are biodegradable.

Bioremediation

A process that uses biological agents, such as bacteria, fungi, or green plants, to remove or neutralize contaminants, as in polluted soil or water. Plants can be used to aerate polluted soil and stimulate microbial action. The cleanup of a contaminated site using biological methods, i.e., bacteria, fungi, and plants, is a form of bioremediation. Organisms are used to either break down contaminants in

Bioremediation

soil or water, or accumulate the contaminants in their tissue for disposal. Many bioremediation techniques are substantially less costly than traditional remediation methods using heat, chemical, or mechanical means.

Biosphere

The ecosystem composed of the earth and its atmosphere in which living organisms exist or that is capable of supporting life.

Bioswale

A landscape element, often a planted strip along a street or parking lot, for the purpose of capturing surface water run-off and filtering out silt and pollution before the stormwater enters the drainage system or groundwater and retains and cleanses runoff from a site, roadway, or other source.

Biota

Collectively, the plants, micro-organisms, and animals of a region.

Biotecture (1966–1970)

A term combining "biology" and "architecture" coined by Rudolph Doernach. It denotes architecture as an artificial "super system," live, dynamic, and mobile.

Birch See Wood

Bird See Ornament: animal forms

Birdsmouth

A V-shaped cut at the end of a structural member; typically found where a rafter meets the top sill or plate of a stud-framed wall.

Birkerts, Gunnar (1925–)

Latvian-born American architect; much influenced by Eero Saarinen. He designed the Federal Reserve Bank (Illus.), Minneapolis, MN (1973), and the Corning Museum of Glass, Corning, NY (1976).

Bissell, George (1928–2010)

A California-based architect, his projects included the San Francisco Solano Catholic Church in Rancho Santa Margarita, Our Lady of the Rosary Cathedral in San Bernardino, and a redesign of the Bowers Museum in Santa Ana, all in California.

Bizarre architecture

Strikingly unconventional in both style and appearance. Eccentric and bizarre architecture should be preserved for its historical significance and value.

Black mortar

Mortar with the addition of ash, either because a black color is desired for pointing, or to reduce cost.

Blackall, Clarence H. (1857–1942)

American architect; designer of the Copley Plaza Hotel, Boston, and an estimated 300 theaters.

Black-figure technique

The silhouetting of dark figures against a light background of natural reddish clay, found on early Greek pottery.

Blackwater

Nonindustrial wastewater containing significant food residues, high concentrations of toxic chemicals from household cleaners, and/or toilet flush water. Wastewater from toilets and urinals is always considered blackwater; wastewater from kitchen sinks, showers, or bathtubs may be considered blackwater by state or local codes. After neutralization, blackwater is typically used for nonpotable purposes, such as flushing or irrigation.

Blank door See Door

Blank window See Window

Blanket insulation

Thermal insulation in a rolled sheet form, with a flexible lightweight blanket of mineral wool or a similar material, often backed with a vapor barrier, such as felt-treated paper or vinyl sheeting.

Bleaching

In wood finish, cleansing or whitening by the use of acid.

Blemish

A minor defect in appearance that does not affect the durability or strength of wood, marble, or other material.

Blending

A gradual merging of one element into another.

Blight
A term applied to a deteriorating influence or condition which affects the value of a property or real estate.

Blind
A device to obstruct vision or keep out light, consisting of a shade, screen, or an assemblage of panels or slats.

Blind arcade See **Arcade**

Blind arch See **Arch**

Blind door See **Door**

Blind hinge
A hinge for a cabinet or door designed so that it is not visible when the door is closed.

Blind joint See **Joint**

Blind pocket
A pocket in the ceiling at a window head to accommodate a venetian blind when raised.

Blind stop
The molding used to stop an outside door or window shutter in the closed position.

Blind story
A floor level without exterior windows.

Blind tracery See **Tracery**

Blind window See **Window**

Block
A large piece of stone, taken from the quarry to the mill for sawing and further working.

Block modillion See **Modillion**

Blocking
Pieces of wood used to secure, join, or reinforce framing members or to fill spaces between them.

Blower door
A large fan placed in an exterior doorway to pressurize or depressurize a building to determine its air leakage rate expressed in air changes per hour or cubic feet per minute.

Blower door test
A powerful fan mounts into the frame of an exterior door and pulls air out of the house, lowering the air pressure inside. Outside air then flows in through all unsealed cracks and openings. By measuring the force needed to maintain a certain pressure difference, a measure of the home's air-tightness can be determined.

Blow-in
Method of introducing loose fiberglass, cellulose, or mineral wool to framing cavities or attic space, typically using a machine with an attached hose.

Blowing agent
Compound used in producing foam insulation. Mixed as a liquid with the foam ingredients under pressure, the blowing agent evaporates, creating gas bubbles that provide the insulation.

Blown-in batt
A method of installing loose insulation in wall cavities, using a powerful blower and a fabric containment screen.

Blueprint
A reproduction of a drawing by a contact printing process on light-sensitive paper, producing a negative image of white lines on a blue background; refers to architectural working drawings for construction.

Bluestone See **Stone**

Blunt arch See **Arch**

BNIM
National leading firm in sustainable design, practice, and technology. BNIM (Berkibile, Nelson, Immenshuh, McDowell) is known by their work on buildings such as the Lewis and Clark State Office Building in Jefferson City, MO (LEED Platinum), the School of Nursing and Student Community Center at the University of Texas Health Science Center at Houston, TX (LEED Gold), and the Omega Center for Sustainable Living in Rhinebeck, NY (LEED Platinum).

Board
A long thin piece of lumber cut from a log; typically with a rectangular cross section; can be hand-hewn, hand-sawn, or mill-sawn.

Board and batten
A form of sheathing for wood frame buildings consisting of wide boards, usually placed vertically, whose joints are covered by narrow strips of wood over the joints or cracks.

Boardwalk
A walkway, usually above a beach, whose surface is constructed of parallel planks; often set on a diagonal to the framing timbers below.

Boathouse

A structure for storing boats when not in use; generally built at the water's edge or partly over water and sometimes including provisions for social activities.

Bodhika

The capital of a column, found in Indian architecture.

Boehm, Gottfried (1920–)

German architect and son of Dominikus Boehm (1880–1955), one of the most prominent Catholic church builders of his time in Germany. He is known for his typological and constructive innovations as well as his expressive architecture. He won the Pritzker Prize in 1986.

Bofill Levi, Ricardo (1939–)

Barcelona-born architect who designed a series of large Postmodern housing blocks, such as Les Espaces d'Abraxes, Marne-La-Valiee (1979), near Paris. These are typical examples of his monumental stripped Neoclassical style.

Bogardus, James (1800–1874)

American architect in New York. Awarded a patent for the first complete iron building, Laing Stores in 1848. He also designed Harper and Brothers printing plant in 1854, both in New York City.

Boiler

A system used to heat water for hydronic heating. Most boilers are gas-fired or oil-fired, although some are electric or wood-fired; a boiler can also heat water for domestic uses through a tankless coil or an indirect water heater. In most boilers, the fluid is water in the form of liquid or steam.

Boiler feed water

Input water used by a boiler.

Boiserie

Wood paneling decorated with carvings in shallow relief.

Bolection molding See Molding

Bollard

A low single post, or one of a series, usually made of stone or concrete, set upright in the pavement, closely spaced to prevent motor vehicles from entering an area.

Bollard

Bollman truss See Truss

Bolster

A horizontal piece of timber that caps a column, pillar, or post to provide a greater bearing area to support a load from above; often has a carved profile or ornamental detail.

Bolster beam See beam

Bolt

A rod or pin, with a permanent head on one end, that holds parts of a building or structure together.

Bolted connection

A connection between structural members made with plates and bolts, as opposed to a riveted or welded construction.

Bond

An arrangement of masonry units to provide strength, stability and in some cases beauty by setting a pattern of overlapping units on one another to increase the strength and to enhance the appearance, or by connecting them with metal ties.

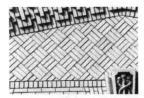

basketweave bond
A checkerboard pattern of bricks, laid either horizontally and vertically, or on the diagonal.

bull header bond
A brick header unit which is laid on edge so that the end of the masonry unit is exposed vertically.

bull stretcher bond
Any stretcher which is laid on edge exposing its broad face horizontally.

chain bond
A masonry wall bond formed by horizontal metal bars or pieces of lumber built into a wall.

common bond
A bond in which every fifth or sixth course consists of headers, the other courses by stretchers

course
A horizontal row containing brick headers in a masonry structure.

cross bond
A brick bond with courses of Flemish bond with alternating courses of stretchers and with the joints of every other row of stretchers centered on a vertical line of headers.

diagonal bond
A type of raking bond in masonry walls, consisting of a header course with the bricks laid at a diagonal in the face of the wall.

diagonal bond

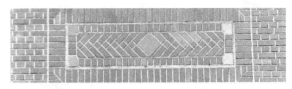

dogtooth course
A stringcourse of bricks laid diagonally so that one corner projects beyond the face of the wall.

Dutch bond
Same as English cross bond or Flemish bond.

English bond
Brickwork that has alternate courses of headers and stretchers, forming a strong bond which is easy to lay.

Flemish bond
In brickwork, a bond in which each course consists of headers and stretchers laid alternately, each header is centered with respect to the stretcher above and the stretcher below it.

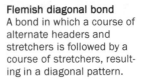

Flemish diagonal bond
A bond in which a course of alternate headers and stretchers is followed by a course of stretchers, resulting in a diagonal pattern.

header bond
A pattern of brickwork consisting entirely of headers; usually displaced by one-half the width of one header in the course above and below.

herringbone bond
A brick wall bond with concealed diagonal headers laid at right angles to each other in a herringbone pattern.

raking bond
A method of bricklaying in which the bricks are laid at an angle in the face of the wall; either diagonal bond or herringbone bond pattern.

running bond
Same as stretcher bond.

skintled bond
Brickwork laid so as to form a wall with an irregular face, produced by the rough appearance of the skintled joints.

soldier bond
A brick course laid vertically with the longer narrow face exposed.

stack bond
In brickwork, a patterned bond where the facing brick is laid with all vertical joints aligned; in stone veneer masonry, a pattern in which single units are set with continuous vertical and horizontal joints.

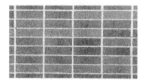

stack bond

timber bond
Horizontal timbers once used as a bond for a brick wall.

Bond, J. Max Jr. (1936–2008)
A principal of Davis Brody Bond Aedas, he was the partner in charge of the museum portion of the National September 11 Memorial and Museum at the World Trade Center. This firm designed the Martin Luther King Jr. Center for Nonviolent Social Change in Atlanta, the Schomburg Center for Research in Black Culture in Harlem, and the Birmingham Civil Rights Institute in Alabama.

Bondone, Giotto di (c. 1266–1337)
Italian architect; designed the Florence Cathedral campanile in 1334, which combines Romanesque, Classical, and Gothic elements.

Bonnet
A small, self-supporting protective hood or roof over an exterior doorway; may be constructed of any exterior material.

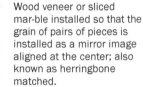

Book matched
Wood veneer or sliced mar-ble installed so that the grain of pairs of pieces is installed as a mirror image aligned at the center; also known as herringbone matched.

Boom
A cantilevered or projecting structural member, such as a beam or spar, which is used to support, hoist, or move a load.

Bootleg conversion
An illegal change in a structure contrary to zoning regulations, as in the partitioning of a single-family dwelling into two or more units.

Border
A margin, rim, or edge around or along an element; a design or a decorative strip on the edge of an element.

Bored pile
A pile formed by pouring concrete into a hole in the ground, usually containing some steel reinforcing, as opposed to a pre-cast pile driven into the ground with a pile driver.

Borehole sample
A core sample obtained by boring or drilling for the purpose of determining the nature of the foundation material.

Borescope
A device for examining hidden areas in an existing structure, consisting of a flexible rod of fiber optics; one bundle of fibers carries light to the ends of the fibers, the other bundle is used for viewing or recording the image with a video camera.

Borromini, Francesco (1599–1677)
Italian Baroque architect who designed San Carlo Alle Quattro Fontane (Illus), in Rome, in 1638 to 1671, a church in which convex and concave wall surfaces are juxtaposed both on the facade and on the interior.

Borromini, Francesco

Boss
A projecting ornament, usually richly carved and placed at the intersection of ribs, groins, beams, or at the termination of a molding.

Bossage
In masonry work, a projecting, rough-finished stone left during construction for carving later into final decorative form.

Botanical garden
Greenhouse where a variety of plants are grown for recreational viewing

Botanical garden

Botta, Mario (1943–)
Swiss architect; worked with Le Corbusier and Louis Kahn. He designed a series of private houses in Lugano, Switzerland, which were set alone in the landscape, and had strong geometric forms. His late work includes the San Francisco Museum of Modern Art (Illus.), in San Francisco, CA (1994).

Bottom chord
A chord along the lower perimeter of a truss; one of a pair with a top chord.

Boullée, Etienne-Louis (1728-1799)
French Neoclassical architect. His designs were extreme reactions against the Baroque style; they stressed plain shapes of enormous size without softening elements.

Boundary
The outer limits of an area, such as a piece of property; which may be defined by a series of markers, fence, stone wall, or other natural feature.

Boundary marker
May consist of a wooden stake, surveyor's marker, or monument located at the points where the perimeter changes directions, as indicated on a plot

Bouquet ornament See **Ornament**

Bovine See **Ornament: animal forms.**

Bow girder
A girder curved horizontally in plan; an arch turned through a right angle, to serve as a spandrel on a curved facade or to support curved balconies.

Bow knot ornament See **Ornament**

Bow window See **Window**

Bower
A shelter or covered place in a pleasure ground or garden, usually made with boughs of trees bent and twined together for shade; also a crude dwelling made from sticks, bark, and natural materials.

Bowstring truss See **Truss**

Box beam See **Beam**

Box column See **Column**

Box frame

A rigid frame formed by load-bearing walls and floor slabs, most suitable for structures that are permanently divided into small repetitive units.

Box girder

A hollow beam with either a square, rectangular or circular cross section; sometimes vertical instead of horizontal, and attached firmly to the ground like a cantilever.

Box stair See **Stair**

Boxed cornice See **Cornice**

Boxed out

Rectangular or square framing around an opening or penetration, such as around a vertical pipe.

Box-head window See **Window**

Brace

A metal or wood member used to stiffen or support a structure; a strut that supports or fixes another member in position, or a tie used for the same purpose.

angle brace

Supporting member across the corner of a rectangular frame or structure.

counterbrace

A subordinate diagonal brace, crossing the main brace of a truss, which resists variable live loads and helps to dampen any vibration.

cross brace

A pair of braces crossing each other to stabilize a structural frame against lateral force

diagonal bracing

A system of inclined members for bracing the angles between the members of a structural frame against horizontal forces, such as wind.

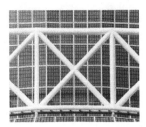

knee brace

A diagonal corner member for bracing the angle between two joined members; being joined to each other partway along its path serves to stiffen and strengthen the joint.

lateral bracing

Stabilizing a wall beam or structural system against lateral forces by means of diagonal or cross bracing either horizontally by roof or floor construction or vertically by pilasters, columns or cross walls.

sway brace

A diagonal member designed to resist wind loads or other horizontal forces acting on a light structural frame.

x-brace

A truss panel, or similar structure, with a pair of diagonal braces from corner to corner that form a crossed shape; may be either struts in compression or tie rods in tension.

Braced frame

The frame of a building in which resistance to lateral forces is provided by diagonal bracing, knee-bracing or cross-bracing; sometimes using girts that are mortised into solid posts which are full frame height.

Bracket

A projection from a vertical surface providing structural or visual support under cornices, balconies, windows, or any other overhanging member.

Bracket

Bracket capital See **Capital**

Bracketed cornice See **Cornice**

Bracketed hood
A projecting surface that is supported on brackets above a door or window; provides shelter or serves as ornamentation.

Bracketed hood

Bragdon, Claude Fayette (1866–1946)
An American architect, writer, and stage designer, Bragdon enjoyed a national reputation as an architect working in the progressive tradition associated with Louis Sullivan and Frank Lloyd Wright. He soon became a leading participant in the Arts and Crafts movement, working with Harvey Ellis, Gustav Stickley, and other Arts and Crafts artists. Accordingly, he promoted regular geometry and musical proportion as ways for architects to harmonize buildings with one another and with their urban context. Bragdon was well regarded for his ink-rendering talent. In 1915 he created a new ornamental vocabulary he called "projective ornament," a system for generating geometric patterns. In his books on architectural theory, *The Beautiful Necessity* (1910), *Architecture and Democracy* (1918), and *The Frozen Fountain* (1932), he advocated a theosophical approach to building design, urging an "organic" Gothic style over the "arranged" Beaux-Arts architecture of the classical revival.

Bramante, Donato (1444–1514)
High Renaissance architect, based in Milan, Italy. He designed the Tempietro of St. Pietro, which became a source work for designs for St. Peter's in Rome built in 1506.

Branch tracery See **Tracery**

Branch, Melville C. (1914–2008)
Urban planner who taught for many years at the University of Southern California, was the author of more than 20 books, and served on the Los Angeles Planning Commission through the 1960s.

Brass See **Metal**

Brattice
A tower or bay of timber construction, erected at the top of a wall on medieval fortifications during a siege.

Breast

That portion of a wall between the floor and a window above; a defensive wall built about breast high.

Breathing zone

The region within an occupied space between planes, 3 and 6 feet above the floor and more than 2 feet from the walls or fixed air-conditioning equipment.

Breezeway

A covered passageway, open to the outdoors; connecting either two parts of a building or two buildings.

Breuer, Marcel Lajos (1902–1981)

An Modernist, born in Hungary, who taught at the Bauhaus at Weimar, Germany, and invented a series of furniture designs using bent steel tubes finished in chrome. Typical of his style is the Armstrong Rubber Company office building (illus.), New Haven, CT, the HUD building (illus.), Washington, D.C. His last work was the Whitney Museum of Art (illus.), in New York City.

Brick

A solid or hollow masonry unit of clay mixed with sand, that is molded into a small rectangular shape while in a plastic state, then baked in a kiln or dried in the sun.

adobe brick

Roughly molded, sun-dried clay brick of varying sizes.

backing brick

A relatively low-quality brick used behind the face brick or behind other masonry.

brick bat

A broken or cut brick that has one complete end remaining and is less than half a full brick in length.

bull header

A brick made with one long corner rounded or angled, used for sills and corners.

clinker brick

A very hard burnt brick whose shape is distorted by nearly complete vitrification; used mainly for paving or as ornamental accents.

closer

The last stretcher brick that encompasses a course of brickwork; types include a king closer and queen closer.

common brick

Brick for building purposes not treated for texture or color.

compass brick

A wedge-shaped brick used in constructing arches or building curved walls; has the two large faces inclined toward each other.

face brick

Brick made or selected to give an attractive appearance when used without rendering of plaster or other surface treatment of the wall: made of selected clays, or treated to produce the desired color.

firebrick

Brick made of a ceramic material which will resist high temperatures; used to line furnaces, fireplaces and chimneys.

gauged brick

Brick that has been cast, ground, or otherwise manufactured to exact and accurate dimensions.

glazed brick

Brick or tile having a ceramic glazed finish.

king closer

A three-quarter brick used as a closer; a diagonal piece is cut off one corner to keep the bond straight at corners of brick walls.

modular brick

A brick with nominal dimensions based on a 4-inch module.

molded brick
Any specially shaped brick, usually for decorative work.

pressed brick
A masonry unit without holes made by pressing a relatively dry clay mix into a mold, as opposed to one that is extruded.

queen closer
A brick cut in half along its length to keep the bond correct at the corner of a brick wall.

rowlock
A brick laid on its edge so that its end is exposed; used on a sloping window sill, or to cap a low brick wall.

rustic brick
A fire-clay brick having a rough-textured surface, used for facing work: often multicolored.

sailor
A brick laid vertically with the broad face exposed.

shiner
A brick laid horizontally on the longer edge with the broad face exposed.

soldier
A brick laid vertically with the longer, narrow face exposed.

wirecut brick
Bricks shaped by extrusion and then cut to length by a set of wires.

tapestry brick
Face brick that is laid in a decorative pattern with a combination of vertical, horizontal, and diagonal elements, such as a basket-weave bond.

veneer brick
A layer of bricks built outside a masonry backing or timber frame; the frame supports the load.

Bridge
A structure that spans a depression or provides a passage between two points at a height above the ground: affording passage for pedestrians and vehicles.

bascule bridge
A drawbridge with one or two balanced leaves that pivot vertically on a trunion located at one end of the span.

Cable Stay Bridge See **Cable Stay**

drawbridge
At the entrance to fortifications, a bridge over the moat or ditch, hinged and provided with a raising and lowering mechanism so as to hinder or permit passage.

footbridge
A narrow bridge structure that is designed to carry pedestrians only.

sidewalk bridge
A lightweight structural covering over a sidewalk to protect pedestrians from construction or cleaning of the structures overhead.

skywalk
A walkway that is located over the ground level and the street; and often connects buildings across a street.

suspension bridge
A bridge hung from cables that are strung between two towers or a tower and abutment.

swing bridge
A bridge that opens by turning horizontally on a turntable supported on a pier.

Bridging
A continuous row of stiffeners between floor joists or other parallel structural members to prevent rotation about their vertical axes; it includes cross bracing.

Bridle joint See **Joint**

Brightness

The amount of light energy reflected from a surface. The degree of brightness depends on the color value and texture on its surface. The surface brightness of a task should be the same as its background or brighter. The maximum brightness ratio should be 3:1, and the brightness between the task and the darkest background area should be 5:1 to avoid objectionable glare.

Brise-soleil

A fixed or movable device, such as a fin or louver, designed to block the direct entrance of sun into a building.

British thermal unit (BTU)

The amount of heat required to raise one pound of water 1 degree Fahrenheit in temperature—about the heat content of one wooden kitchen match.

Broach

A half pyramid above the corners of a square tower to provide a transition to an octagonal spire.

Broken arch See **Arch**

Broken gable See **Gable**

Broken joint See **Joint**

Broken pediment See **Pediment**

Broken rangework masonry See **Masonry**

Bronze See **Metal**

Bronze Age

A period of human culture between the Stone Age and the Iron Age (4000–3000 B.C.), characterized by the use of bronze implements and weapons.

Brown, Terry (1955–2008)

Organic architect, best known for his own home and studio in Cincinnati Ohio, the Mushroom House, built between 1992 and 2006 and also called the Tree House. The eccentric residence is a multilayered handcrafted exercise in distorted perspective. Like much of Brown's later work, the home is made of a variety of materials—clear and colored glass, metal, shells, ceramics, and various woods and siding, and its features include round windows, bark-like exterior, and low roof line.

Brownfield

The designation of the U.S. Environmental Protection Agency (EPA) for existing facilities or sites that are abandoned, idled, or otherwise underused real property where expansion or redevelopment is complicated by real or perceived environmental contamination that needs to be cleaned up before the site can be used again. Examples are former dry-cleaning establishments and gas stations. The use of brownfields typically reduces land cost by using land that is less desirable. However, lower land costs must be balanced against the cost of any required remediation and possible health risks to residents. The EPA sponsors an initiative to help mitigate these health risks and return the facility or land to renewed use. Many green guidelines and standards provide points for building in brownfield areas.

Brownstone See **Stone**

Bruder, Will (1946–)
American architect whose major buildings includes the Central Library, Phoenix (1995).

Brunelleschi, Filippo (1337–1446)
Florence-born architect who designed the dome of the Florence Cathedral from 1420 to 1434; the Foundling Hospital, Florence (1421); and St. Lorenzo, Florence (1425).

Bruno, Victor (1921–2011)
Modernist architect who designed homes and businesses throughout the New Orleans area. One of Bruno's highly visible buildings is the Gallery Apartments complex in the Garden District of New Orleans, LA. It features a glass-encased lobby and a concrete screen that runs up the five-story facade between jutting balconies. An admirer of Frank Lloyd Wright.

Brutalism (1945–1960)
An uncompromisingly modern style which was expressed in large scale using raw and exposed materials emphasizing stark forms. It was distinguished by its weighty, textured surfaces and massiveness; created mainly by large areas of patterned concrete. Windows consist of tiny openings, and the combination of voids and solids gave walls an egg-crate appearance. Mechanical systems are left exposed on the interior of the bare structure.

Bryant, Gridley James Fox (1816–1899)
American architect who made his fame as the winner of the Boston City Hall Competition in 1865.

Bucklin, James C. (1801–1890)
American architect in Rhode Island who designed monumental buildings in the Greek Revival style, including the Providence Arcade in 1828 and Manning Hall, Brown University, in 1833 both in Providence, RI.

Bucranium See **Ornament: animal forms**

Bud
An Egyptian style column capital in the form of a stylized lotus bud; also, a small, foliated portion of a Corinthian capital located between the base of two acanthus leaves above the caulicoles.

Builder
A person who directs or performs the construction or renovation of a house or other structure.

Building
An enclosed and permanent structure for residential, commercial, industrial, institutional or office use, as distinguished from mobile structures or those not intended for occupancy.

Building alteration
Construction in a building that changes the structure, equipment or location of openings, without increasing the overall area or dimensions; as distinct from additions to an existing structure.

Building area
The total area of a site covered by buildings, as measured on a horizontal plane at ground level; terraces and uncovered porches are usually not included in this total.

Building artifact
An element on a building demonstrating human crafting, such as a stained glass window or an ornament of archaeological or historic interest.

Building automation
Optimizes the startup and performance of HVAC equipment and alarm systems. Building automation greatly increases the interaction of mechanical subsystems within a building, improves occupant comfort, lowers energy use, and allows offsite building control.

Building automation system (BAS)
An integration of digital, electronic, and/or pneumatic controls and devices to provide unattended and automatic operation of buildings systems. Systems may include HVAC, elevators, fire suppression, smoke control, security, alarm systems, lighting, and other subsystems .

Building code
A collection of regulations by local authorities having the jurisdiction to control the design and construction of buildings, alterations, repairs, quality of materials, use and occupancy; it contains minimum architectural, structural, and mechanical standards for sanitation, public health, safety, and welfare, and the provision of light and air.

Building commissioning
The startup phase of a new or remodeled building. This phase includes testing and fine-tuning of the HVAC and other systems to ensure proper functioning and adherence to design criteria. Commissioning also includes preparation of the system operation manuals and instruction of the building maintenance personnel. It is a systematic process of ensuring, through documented verification, that all building systems perform interactively according to the documented design intent and the owner's operational needs, beginning in the design phase, lasting at least one year after construction, and including the preparation of operating staff.

Building component
An element manufactured as an independent unit, which can be joined with other elements, including electrical, fire protection, mechanical, plumbing, structural and all other systems affecting health and safety.

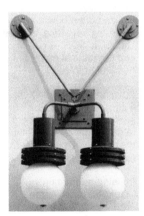

Building conservation
The management of a building to prevent its decay, destruction, misuse, or neglect; may include a record of the history of the building and conservation measures applied.

Building construction
The fabrication and erection of a building by the process of assembly, or by combining building components or systems.

Building control system (BCS)
A system that controls the comfort and safety of a building's assets and internal environment.

Building cooling load
The hourly amount of heat that must be removed from a building to maintain indoor comfort, measured in British thermal units (BTUs).

Building density
The floor area of the building divided by the total area of the site (square feet per acre).

Building ecology
The physical environment and systems found inside the building.

Building envelope
The entire outer shell of a building enclosed by its roof, walls, doors, windows, and foundation. The envelope can minimize temperature gain or loss and moisture infiltration, and protection from colder and warmer outdoor temperatures and precipitation, and it includes both an insulation layer and an air infiltration layer. It is a key factor in the "sustainability" of a building. Common measures of the effectiveness of building envelopes include protection from the external environment, indoor air quality, durability, influx of natural light, and energy efficiency.

Building environment
The combination of conditions that affect a person, piece of equipment or system in a building, such as lighting, noise, temperature and relative humidity.

Building footprint
The area on a project site that is used by the building structure and is defined by the perimeter of the building plan. Parking lots, landscapes, and other nonbuilding facilities are not included in the building footprint.

Building for Environmental and Economic Sustainability (BEES)
Software program developed by the NIST (National Institute of Standards and Technology). It is aimed at designers, builders, and product manufacturers. It provides a way to balance the environmental and economic performance of building products. BEES measures the environmental performance of building products by using an environmental life-cycle assessment approach specified in the latest versions of ISO 14000 draft standards. All stages in the life of a product line are analyzed: raw material acquisition, manufacture, transportation, installation, use, and

Building for Environmental and Economic Sustainability

recycling, and waste management. Economic performance is measured using the ASTM standard life-cycle cost method, which covers the costs of initial investment, replacement, operation, maintenance and repair, and disposal.

Building grade
The ground elevation; established by a regulating authority, that determines the height of a building in a specific area.

Building height
The vertical distance measured from the grade level to a flat roof or to the average height of a pitched, gable, hip, or gambrel roof, not including bulkheads or penthouses.

Building industry
The general term used to include building, civil, mechanical, and electrical engineering.

Building information modeling (BIM)
The physical and functional characteristics are set forth in a relational database format rather than in traditional drawings. BIM data reside inside the objects of building components and products, which are assembled and configured to produce a virtual model of the building. These objects contain highly accurate geometric data such as spatial coordination and a clash detection system, where interferences are identified virtually and corrected before causing expensive field changes. A BIM object also contains a variety of relevant nongraphical data, such as material strength, insulation value, finish, surface reflectivity, light transmission, or fire rating. They also carry links to other associated documentation such as technical data, warranty, and maintenance information.

generic objects
Carries geometric and nongraphical data representing the functional need but does not carry specific information that a manufacturer's product model does.

objects
Contain different types of BIM content, related to the characteristics and installed use of the products and materials they represent.

parametric objects
BIM data available from building products manufacturers in a variety of configurations, heights, widths, thicknesses, materials finishes, hardware options, fire ratings, and energy and sound transmission.

proprietary objects
Represents a specific manufacturers' product with its particular specifications and functional characteristics.

Building inspector
A member of the building department who inspects construction to determine conformity to the building code and the approved plans, or one who inspects occupied buildings for violations of the building code.

Building life cycle
The amortized annual cost of a building, including capital costs, installation costs, operating costs, maintenance costs, and disposal costs discounted over the lifetime of the building.

Building line
A line or lines established by law or agreement, usually parallel to the property lines, beyond which a structure may not extend; it usually does not apply to uncovered entrance platforms or terraces.

Building maintenance
Actions ensuring that a building remains in working conditions by preserving it from deterioration, decline, or failure.

Building management system
A system for centralizing and optimizing the monitoring, operating, and managing of a building. Services may include heating, cooling, ventilation, lighting, security, and energy management.

Building material
Any material used in the construction of buildings, such as steel, concrete, brick, masonry, glass, and wood.

Building monitoring systems
Computerized monitoring of engineering services, security, and other building systems for the purposes of recording, reporting, and implementing operational control. It provides a method to maximize safety, security, and operational performance for overall cost minimization and efficiency.

Building occupancy
A general classification of the type of use of a structure, such as residential or commercial; used in building codes to determine the level of fire hazard, and consequently, the size and type of construction permitted, including egress requirements, and other restrictions.

Building orientation
Proper building orientation can have an impact on heating, lighting, and cooling costs. By maximizing southern exposure, for example, you can take optimal advantage of the sun for daylight and passive solar heating. Cooling costs can be lowered by minimizing western exposures, where it is most difficult to provide shade from the sun.

Building paper
A heavy, waterproof paper usually impregnated with tar and used over sheathing and subfloors to prevent the passage of moisture.

Building paper

Building permit
A written document that is granted by the municipal agency having jurisdiction, authorizing an applicant to proceed with construction of a specific project after the plans have been filed and reviewed.

Building preservation
The process of applying measures to maintain and sustain all of the existing materials, integrity, and form of a building, including its structure and building artifacts.

Building pressurization
The air pressure within a building relative to the air pressure outside. Positive building pressurization is usually desirable to avoid infiltration of unconditioned and unfiltered air. Positive pressurization is maintained by providing adequate outdoor makeup air to the HVAC system to compensate for exhaust and leakage.

Building program
A written, detailed account of overall project goals, explaining the context, conditions, requirements, and objectives for a design project, inclusive of design concept and budget.

Building reconstruction
The reproduction by new construction of the exact form and details of a building that no longer exists or artifacts as they once appeared.

Building rehabilitation
To restore a building to a useful life, by repair, alteration or modification.

Building remodeling
To replace all or a portion; to reconstruct or renovate.

Building renovation
To restore to an earlier condition, or improve by repairing or remodeling.

Building Research Establishment Environmental Assessment Method (BREEAM)
A comprehensive tool for analyzing and improving the environmental performance of buildings through design and operations. The UK-based Building Research Establishment has developed this methodology.

Building restoration
The process of returning a building as nearly as possible to its original form and condition, usually as it appeared at a particular time; may require removal of later work or reconstruction of work that had been previously removed.

Building retrofit
The addition of new building materials or new building elements, and components not provided in the original construction to upgrade or change its functioning.

Building services
The utilities and services supplied and distributed within a building, including heating, air conditioning, lighting, water supply, drainage, gas and electric supply, fire protection and security protection.

Building shell
Outer framework of a building.

Building subsystem
An assembly of components that performs a specific function in a building; for example, an air-conditioning system consisting of its components, such as ductwork, a fan, air diffusers, and controls.

Building survey
Detailed record of the present condition of a structure.

Building systems
Includes architectural, mechanical, electrical, and control systems along with their respective subsystems, equipment, and components, all of which must be commissioned.

Building tuning
The ongoing process of adjustment of building equipment and systems to optimize performance, maintain environmental ratings, and continuously improve occupant satisfaction with the building.

Building user guide
A building user guide provides details regarding the everyday operation of a building and ways in which occupants and facility managers encounter and interact with the facility. The ESD (ecologically sustainable development) section of the guide provides guidance on the green initiatives designed into the building. It is provided to ensure that the ESD initiatives continue to be utilized correctly by building occupants, so that the benefits will be realized long after the design and delivery team have completed their work.

Building-integrated photovoltaic system
The system that incorporates PV cells as part of the building fabric.

Building-integrated photovoltaics
Solar cells that convert sunlight directly into electricity, which are integral in materials that replace conventional building materials in parts of the building envelope such as the roof, skylights, spandrels, glazing, or shading devices.

Building-related illness
Diagnosable illness whose cause and symptoms can be directly attributed to a specific pollutant source within a building, such as Legionnaire's disease, hypersensitivity, pneumonitis, attributed directly to airborne building contaminants.

Buildings reborn
The adaptation of old buildings for uses different from their original purpose.

Built environment
That portion of the physical surroundings created by humans such as roads, bridges, and building structures, as opposed to the natural environment.

Built-up
A structural member made up of two or more parts fastened together so they act as a single unit.

Built-up beam See Beam

Built-up column
A column which is composed of more than one piece.

Built-up roofing See Roof

Bulkhead
A horizontal or inclined door over a stairway giving access to a cellar; a structure on the roof of a building covering a water tank, shaft or service equipment.

Bull header bond See Bond

Bull stretcher bond See Bond

Bullfinch, Charles (1763–1844)
American architect of Boston churches and the Massachusetts State House, Boston, MA, built in 1795, a combination of Palladian and pure Classicism styles.

Bull-nosed step See Step

Bungalow
A one-story frame house, or a summer cottage, often surrounded by a large covered veranda, with a widely bracketed gable roof; often built of rustic materials.

Bungalow door See Door

Bungalow style (1890–1940)
This residential style typified by a one-story house with gently pitched gables, had its roots in the Arts and Crafts movement. A lower gable usually covers a screen porch, which features battered piers at the corners. Rafters extend beyond the roof and are often exposed. Wood shingles are the favored exterior covering and are left natural. Windows are sash or casement with numerous lights.

Bungalow window See Window

Bunshaft, Gordon (1909–1990)
Partner of the firm Skidmore, Owings & Merrill. Designed Lever House in New York City in 1952, a 21-story curtain-walled skyscraper slab set on a lower podium.

Burgee, John (1933–)
American-born architect who became a partner with Philip Johnson (1968–1991), and participated in a series of office towers that included Pennzoil Place (Illus.), Houston (1975), and the Crystal Cathedral (Illus), Garden Crove, CA, with Philip Johnston, in 1980.

Burl See Wood

Burnham, Daniel H. (1846–1912)

An American architect who partnered with John Wellborn Root. The firm was responsible for starting the Chicago School of Skyscraper Designs. He first designed the Monadnock Building (Illus.) in 1891, then the Reliance Building in 1890, both in Chicago. The latter is based on a metal steel skeleton and terra-cotta cladding. He also designed the Flatiron Building (Illus.) in New York City in 1902.

Bush-hammered concrete

Concrete having an exposed aggregate finish, usually obtained with a power-operated hammer which removes the sand-cement matrix around the aggregate particles to a depth of one-quarter inch.

Bush-hammered finish

A stone or concrete surface dressed with a bush-hammer; used decoratively, or to provide a roughened surface for treads, floors, and pavement requiring greater traction.

Business and Institutional Furniture Manufacturers Association (BIFMA)

Organization that creates and manages standards for evaluating the safety, durability, ergonomics, toxicity, and structural adequacy of contract and commercial furniture.

Butt joint See **Joint**

Butt splice See **Splice**

Buttress

An exterior mass of masonry projecting from the wall to absorb the lateral thrusts from roof vaults; either unbroken in their height or broken into stages, with a successive reduction in their projection and width. The offsets dividing these stages are generally sloped at a very acute angle. They terminate at the top with a plain slope ending at the wall or with a triangular pediment.

diagonal buttress

A buttress that bisects the 270-degree angle at the outside corner of a building.

angle buttress

One of the two buttresses at right angles to each other; forming the corner of a structure.

flying buttress

A characteristic feature of Gothic construction in which the lateral thrusts of a roof or vault are carried by a segmental masonry arch, usually sloping, to a solid pier or support that is sufficiently massive to receive the thrust.

flying buttress

Buy locally

In addition to the physical attributes of green products, buying locally considers the origin of products as well. Buying regional products reduces transportation costs and energy and keeps dollars in the local economy. A major tenet of sustainability is making use of resources from the immediate region. This results in a greater understanding of the region, its characteristics and resources, and the impact of using or producing that resource on that region.

BX cable

Trade name for an electrical cable consisting of a spiral of interlocking steel carrying insulated wires inside. The addition of a separate ground wire as required by codes has replaced the original BX cable. Now used only for temporary work.

Byard, Paul Spencer (1940–2008)

A land-use lawyer turned architect, credited with the preservation and restoration of some of New York's most prominent landmarks, such as Carnegie Hall, the Cooper Union Foundation Building, the New York State Supreme Court's Appellate Division Courthouse, and the old Customs House. Byard's designs for new buildings included the New 42nd Street Studios and the Chanel 57 building in Manhattan. He was director of Columbia University's historic preservation program from 1998 to 2008. Byard wrote several books, including *The Architecture of Additions: Design and Regulation* (1998) and *Architecture and Social Policy: Learning from the Twentieth Century* (2001).

By-product

Material, other than the principal product, generated as a consequence of an industrial process or as a break-down product in a living system.

Byrne, Barry (1883-1968)

American architect; once employed by Frank Lloyd Wright. He took over Walter Burley Griffin's work after Griffin left for Australia. He later designed modernistic churches in the Midwest.

Byzantine architecture (300–1450)

When the seat of the Roman Empire moved to Byzantium, a new style became the official architecture of the church.

Byzantine architecture

Plans were based on a Greek cross, with a large cupola rising from the center and smaller ones crowning the four small arms. The style was characterized by large domes supported on pendentives, circular or horseshoe arches, elaborate columns and richness in decorative elements. Doorways were square-headed with a semicircular arch over the flat lintel. The round arch, segmented dome, extensive use of marble veneer and rich frescoes with colored glass mosaics are also characteristic of this style. The most well-known examples are the Hagia Sophia in Istanbul, and St. Marks (illus.), in Venice, Italy.

Byzantine Revival style

A late nineteenth-century style using Byzantine forms and decoration, including domes supported on high drums, arcades on columns with plain shafts, and bas-relief decorative work.

C c

Cabin
A small, crudely constructed dwelling, which may have a living room with a fireplace, plus one or more small rooms; also a room aboard a vessel.

Cabinet
A built-in or freestanding piece of furniture fitted with drawers and/or shelves, typically behind one or a pair of doors.

Cabinet window See **Window**

Cabinetmaker
A craftsperson specializing in fine joinery, with the skills, materials, and tools necessary to make furniture and other pieces of woodwork.

Cable molding See **Molding**

Cable stays
Straight cables connected directly to a roof or floor structure, and anchored to a mast, allowing greater clear spans within the structure. They transfer loads from the structure to the top of the mast, which transmits them to the foundations.

Cable tray
In electrical installations, a continuous support, usually overhead, on which electrical cables or conduits are placed. Usually constructed of sheet metal with low sides, resembling a tray.

Cabled fluted column See **Column**

Cabled fluting molding See **Molding**

CADD
Abbreviation for computer-aided design and drafting.

Caisson
A large-diameter steel pipe used to buttress foundation sheet pilings until a slurry wall can be built.

Caisson pile
A cast-in-place concrete pile, made by driving a steel tube into the earth, excavating it, and filling it with concrete.

Calatrava, Santiago (1951–)
Spanish-born architect and engineer who combined architecture and engineering in structures which resemble elegant skeletal sculptures. Works include Alamillo Bridge, Seville, Torre Telifonica, Barcelona (1991), Ciutat de la Arts, Valencia (illus.), all in Spain. Later work includes the Milwaukee Art Museum (illus.), Milwaukee, WI, 2001.

Calatrava, Santiago

Caldarium
The hot room in a Roman bath.

Calendar
A sculptured or painted emblematic series depicting the months of the year, often including the signs and symbols of the zodiac.

Calf's tongue molding See **Molding**

Calibration
Process of adjusting equipment to ensure that operation is within design parameters.

California Corner
A value engineering technique that uses two studs (instead of the usual three or four) to make an exterior corner. The result is better insulation and use of fewer resources, in addition to cost savings, variations are possible.

California Green Building Initiative
California is a progressive state with its energy efficiency, conservation, sustainability, and green-building and green-purchasing practices. In addition, California's Title 23 specifies a number of energy-saving requirements for building. The Green Building Initiative calls for state buildings to be 20 percent more energy efficient by 2015 and encourages the private sector to do the same.

Callicrates (c. 500 B.C.)
Leading Athenian architect; famous for the design of the Parthenon along with Ictinus. He also designed the Temple of Nike Apteros and part of the defensive Long Walls connecting Athens to the port at Piraeus.

Calthorpe, Peter
In 1989 Calthorpe proposed the concept of a "pedestrian pocket": up to 110 acres of pedestrian-friendly, transit-linked, mixed-use urban area with a park at its center, mixing low-rise, high-density housing, commercial, and retail uses. The concept had a number of similarities with Ebenezer Howard's Garden City. Calthorpe has been a visionary leader in the field of urban regional planning for his work in walkable communities and land preservation.

Camber
A slightly convex curvature intentionally built into a beam, girder, or truss to compensate for an anticipated deflection so that it will not sag under load; any curved surface designed to facilitate runoff water.

Camber arch See **Arch**

Camber beam See **Beam**

Camber window See **Window**

Came
A slender rod of cast lead, with or without grooves, used in casements and stained-glass windows to hold the panes or pieces of glass together.

Campaniform capital See **Capital**

Campanile
A bell tower detached from the main body of a church.

Campbell, Wendell J. (1927–2008)
A founder of the National Organization of Minority Architects, Campbell served as NOMA's first president and was a mentor to many African American architects. The firm's hundreds of projects included an expansion of McCormick Place, the DuSable Museum of African American History, the New Bronzeville Military Academy, and the Metcalf Federal Building, all in the Chicago vicinity. The firm also prepared redevelopment planning schemes for New Orleans, Las Vegas, Detroit, Chicago, Milwaukee, and Gary, Indiana.

Campen, Jacob van (1595–1657)
Dutch Classical architect who introduced a version of the Palladian style into Holland, which became very popular. Works include the Town Hall, Amsterdam.

Canal
A channel or groove as in the recessed portions of the face of a triglyph.

Canalis
The concave area of the Ionic volute, located in between the two lines of the spiral and the cushion above the cymatium.

Candela
The unit of luminous intensity.

Candela, Felix (1910–1981)
Designed the Church of the Miraculous Virgin, Narvarte Maxico, an expressionist building made of concrete in the form of a hyperbolic paraboloid.

Canephora
An ornament representing a young maiden bearing a basket of ceremonial offerings on her head, used either as a column support or as a freestanding garden ornament.

Canopy
A decorative hood above a niche, pulpit, or stall; a covered area that extends from the wall of a building, protecting an enclosure.

Canopy roof See **Roof**

Cant
A salient corner; a line or surface angled in relation to another, as in a wall or surface sloped away from the perpendicular.

Cant bay See **Bay**

Cant molding See **Molding**

Cant strip
A filler piece cut at an angle, most often at 45 degrees, to make a transition from a horizontal roof to a vertical parapet.

Cant wall See **Wall**

Cant window See **Window**

Cantilever
A structural member or any other element projecting beyond its supporting wall or column and weighted at one end to carry a proportionate weight on the projecting end.

Cantilever retaining wall See **Wall**

Cantilevered
Refers to forms that have rigid structural members or surfaces that project significantly beyond their vertical support.

Cantilevered

Cantilevered step See **Step**

Cap
The top member of any vertical architectural element; often projecting, with a drip for protection from the weather; the coping of a wall, the top of a pedestal or buttress, or the lintel of a door.

Cap molding See **Molding**

Capillary forces
The forces that lift water or pull it through porous materials, such as concrete.

Capital
The upper member of a column, pillar, pier, or pilaster; usually decorated. It may carry an architrave, arcade, or impost block. The classical orders each have their own distinctive representative capitals.

Capital

aeolic capital
A primitive type of Ionic capital featuring a palm-like design, evolved by the Greeks in Asia Minor.

angle capital
A capital occurring at a corner column, especially an Ionic capital where the four volutes project equally on the diagonals, instead of along two parallel planes.

basket capital
A capital with interlaced bands resembling the weave of a basket, found in Byzantine architecture.

bifron capital
A capital with two fronts or faces looking in two directions, similar to a double herm.

bracket capital
A capital extended by brackets, lessening the clear span between posts, often seen in Near Eastern, Muslim, Indian, and some Spanish architecture

campaniform capital
A bell-shaped Egyptian capital representing an open papyrus profile.

composite capital
One of the five classical orders which combines acanthus leaves of the Corinthian order with the volutes of the Ionic order.

cushion capital
A capital resembling a cushion that is weighted down; in medieval architecture, a cubic capital with its lower angles rounded off.

Doric capital
The uppermost member of a column or pilaster of the Doric order, consisting of the necking, fillets, and echinus; located under the abacus.

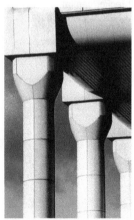

lotus capital
In ancient Egyptian architecture, a capital having the shape of a lotus bud.

Moorish capital
The capital in the style developed by the Moors in the late Middle Ages.

papyriform capital
A capital of an Egyptian column with the form of a cluster of papyrus flowers.

geminated capital
Coupled or dual capitals.

palm capital
A type of Egyptian capital resembling the spreading crown of a palm tree; a column capital resembling the leaves of a palm tree.

quadrafron capital
Having four fronts or faces looking in four directions.

scalloped capital
A medieval block or cushion capital, when each lunette is developed into several truncated cones.

Capital cost
The prime cost of construction, including acquisition of the land or existing structure, design, materials, equipment and erection or renovation.

Capital improvement
An improvement made to real estate that has an extended lifetime and increases the property's value, such as new construction, rehabilitation, or replacement of mechanical equipment, as opposed to maintenance work.

Capital value
The market value of a parcel of land or building, less the mortgage or other debts against it.

Capitol
The building in which a state legislature assembles. An important building type, seat of all state governments, almost all centered on a high dome with flanking lower wings, built of masonry in a classical style, and need continual restoration to maintain the proper civic image.

Capstone
The top row of stones on a retaining wall; the top stone on a pier or on top of a corbeled vault.

Captured rainwater
Through the use of appropriate roofing materials and gutter systems, rainwater can be harvested or collected, then stored for future nonpotable use such as showers and hand washing.

Carbon credit
A term that refers to three types of units of greenhouse gas reductions defined under the Kyoto Protocol.

Carbon dioxide (CO$_2$)
A colorless, odorless, nonpoisonous gas that exists in trace quantities (less than 400 parts per million) within ambient air. Carbon dioxide is a product of fossil-fuel combustion. Although carbon dioxide does not directly impair human health, it is a greenhouse gas that traps terrestrial (i.e., infrared) radiation and contributes to the potential for global warming.

Carbon dioxide equivalent (CO$_2$e)
Emissions of greenhouse gases are typically expressed in a common metric so that their impacts can be directly compared, as some gases are more potent, i.e., have a higher global warming potential than others. The international standard practice is to express greenhouse gases in carbon dioxide equivalents.

Carbon dioxide monitoring
A method for determining indoor air quality by using the concentration of carbon dioxide as an indicator. Although the level of CO$_2$ is a good general indicator of air quality, it relies on the presence of certain conditions and must be applied accordingly.

Carbon dioxide sensor
Device for monitoring the amount of carbon dioxide in an air volume.

Carbon footprint

A measure of the amount of carbon dioxide emitted through the combustion of fossil fuels that a community, industry, or other entity contributes to the atmosphere through energy use, transportation, and other means. A carbon footprint is often expressed as tons of carbon dioxide or tons of carbon emitted, usually on an annual basis. It is the total set of greenhouse gas emissions caused directly and indirectly by an individual, organization, event, or product.

Carbon footprint/neutral

Measured in units of carbon dioxide, a measurement of impact on the environment. Carbon neutral is emitting no carbon dioxide into the atmosphere, or alternately adopting practices that absorb or offset the carbon dioxide that is produced. Planting trees is one way to help offset your carbon footprint.

Carbon monoxide (CO)

A colorless, odorless, poisonous gas produced by incomplete fossil-fuel combustion, usually associated with incomplete combustion of gas stoves, fireplaces, kerosene appliances, tobacco smoke, and automobile exhaust. Proper ventilation is important to prevent negative health effects such as fatigue, dizziness, nausea, and even death.

Carbon neutral

A scenario in which the net discharge of carbon dioxide into the atmosphere is zero. Carbon neutrality can be achieved by planting trees so that CO_2 emissions as a result of combustion would be offset by CO_2 absorption by the plants. In the presence of water and light, trees convert CO_2 into sugar and oxygen through the process of photosynthesis. The average tree absorbs 22 pounds of CO_2 per year. Carbon neutral is also referred to as "net zero carbon." It is also the result of balancing the amount of carbon dioxide produced from activities like driving a car or producing electricity with an equal amount offset by planting trees or supporting renewable energy from solar and wind.

Carbon offset

A system intended to equalize carbon production around the globe by trading greenhouse gas emissions—typically produced through fossil-fuel consumption—for environmentally friendly actions, such as planting trees and using clean energy sources. For example, a factory or production facility may not be able to reduce its own carbon footprint any further through its own actions, so it may voluntarily purchase credits for another party to offset their actions. The goal of carbon offsets is to attain a carbon-neutral overall effect.

Carbon rationing

Limiting the amount of carbon you introduce into the environment each year. Carbon rationing action groups (CRAGS) help you reduce your carbon footprint.

Carbon sequestration

The uptake and storage of carbon. Trees and plants, for example, absorb carbon dioxide, release the oxygen, and store the carbon.

Carbon sink

The carbon reservoirs and conditions that take in and store more carbon (i.e., carbon sequestration) than they release.

Carbon-neutral design

Refers to a design process that produces no carbon release through any of its stages, and therefore will not contribute to greenhouse gases or global warming.

Carbon-neutral house

A house that, on an annual basis, does not result in a net release of carbon dioxide, a greenhouse gas that contributes to global warming, into the atmosphere.

Cardboard models

Used in the design process to show formal and spatial relationships without regard to the materials or functions of the final built structure.

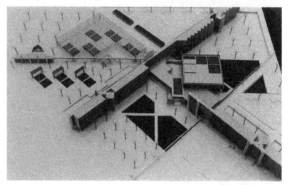

Cardinal, Douglas (1934–)

A Canadian-born architect of Native American descent, whose best-known work is the Candian Museum of Civilization (illus.), Gatinu, Quebec (1989). It features powerful curved forms in the style of Organic architecture. Later work includes the National Musem of the Indian (illus.), Washington, D.C. (2004).

Cardinal, Douglas

Cardinal points
North, South, East, and West; elevations facing these directions are called cardinal fronts.

Carolingian architecture (750–980)
The early Romanesque architecture located in France and Germany, based on an attempt by the Emperor Charlemagne to re-create Imperial Roman styles and forms.

Carpenter
A craftsperson who is skilled in the transformation of lumber into framing and enclosing a structure.

Carpenter Gothic style (1800–1880)
A style characterized by the application of Victorian Gothic motifs, often elaborate, by skilled artisan woodworkers using modern machinery to produce ornamentation for building facades.

Carpenter Gothic style (1800–1880)

Carpet America Recovery Effort (CARE)
A voluntary initiative of the carpet industry and government to prevent carpet from burdening landfills, focusing on developing carpet reclamation and recycling methods.

Carpet and Rug Institute (CRI) Green Label
A nonprofit trade organization that sets standards for the carpet industry. The CRI Green Label tests carpeting, cushions, and adhesives to help those who specify carpet to identify products with low-VOC emissions.

Carpool
An arrangement in which two or more people share a vehicle for transportation.

Carport
A roofed automobile shelter adjoining a house and open on two or more sides.

Carrere, John Merven (1858–1911)
American architect with Thomas Hastings (1860–1929); specialized in Beaux-Arts designs, such as the Ponce de Leon Hotel, St. Augustine, FL, built in 1888; the New York Public Library, built in 1911; Manhattan Bridge and Approaches, built in 1911; Henry Clay Frick House, built in 1914; and the Standard Oil Building, built in 1926, all in New York City.

Carriage
The framing members that support the treads of a staircase. Also called a stringer.

Carriage house
A building, or part thereof, for housing carriages or automobiles when they are not in use.

Carriage porch
A roofed structure over a driveway at the door to a building, protecting those entering or leaving, or getting in or out of vehicles, from the weather.

Carrying capacity
In terms of the built and urban environment, refers to the upper limits of development beyond which the quality of

Carrying capacity

human life, health, welfare, safety or community character and identity might be unsustainably altered; a measure of the ability of a region to accommodate growth and development within the limits defined by existing infrastructure and natural resource capabilities. It is a finite quantity that equates to the ecosystem resources of a defined area such as a locality, habitat, region, country, or place.

Cartoon See Design drawing

Cartouche
A decorative ornamental tablet resembling a scroll of paper with the center either inscribed or left plain; but framed with an elaborate scroll-like carving.

Carved work
In stonework, any hand-cut ornamental features that cannot be applied from patterns.

Carved work

Carver
A craftsperson who is skilled in the ornamental engraving and cutting of wood or stone, in various forms of relief.

Caryatic order
A repeated series of caryatids and the entablature that they support.

Caryatid
A supporting member serving the function of a pier, column, or pilaster, and carved in the form of a draped, human figure; in Greek architecture.

Caryatid

Cascade

An artificial waterfall that breaks the water as it flows over stone steps, usually found in a garden setting.

Cased frame
The wood window frame of a double-hung window, including the sash pockets.

Cased opening
An interior window or doorway trimmed with casing but without a sash or door.

Casein paint
A water-based paint, used in fresco painting on wet plaster.

Casement window See **Window**

Casework
High-quality shelving and display cases such as for a store; also used to describe cabinetwork.

Casing
A trim member, molding, framing or lining around door and window openings which give a finished appearance. They may be flat or molded.

Casino
A building used for public recreation and gambling activities.

Cast iron See **Metals**

Cast stone
A mixture of fine stone chips and Portland cement in a matrix. Once cast, it may be ground, polished or otherwise treated to simulate natural stone.

Castellated
Bearing the external fortification elements of a castle, in particular, battlements, turrets, and crenellated patterns.

Castellation
A notched or indented parapet, originally used for fortifications, but afterwards used on church facades; was intended as ornament.

Cast-in-place concrete See **Concrete**

Cast-iron architecture

Buildings originally designed to emulate stone buildings in Renaissance style designs. They eventually developed a recognizable style of their own. They housed manufacturing companies who have since vanished, leaving large uninterrupted spaces with heavy timber and iron construction in its wake. Many have now been converted to living lofts above commercial stores.

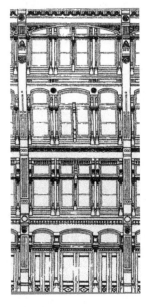

Cast-iron facade

A load-bearing facade composed of prefabricated parts, commonly used on buildings around 1850–1870.

Cast-iron lacework

A panel of ironwork employing an intricate ornamental design that is formed by a mass-produced casting process.

Cast-iron pipe

Consists of various shapes of pipes and fittings made of cast-iron, and used for drains and sewer pipes.

Castle

A stronghold, building, or group of buildings intended primarily to serve as a fortified post; a fortified residence of a nobleman.

Castle style

A type of eighteenth-century architecture that employed battlements, loopholes, and turrets to create the impression of a fortified dwelling; some of these same elements were also used to create follies, picturesque cottages, and fake ruins.

Castle style

Catacombs
Subterranean burial places consisting of galleries with niches for sarcophagi and small chapels for funeral feasts and commemorative services.

Catalano, Eduardo (1918–2010)
Argentine architect and professor, born and educated in Buenos Aires, he was awarded scholarships to study at the University of Pennsylvania and later the Harvard Graduate School of Design under Walter Gropius. He is best known for the modernist home he designed for himself in Raleigh, North Carolina (razed in 2001). The home, with its innovative undulating parabolic roof over a glass-walled structure, was named House of the Decade in the 1950s by House and Home magazine.

Catch basin
A masonry or concrete-walled pit that collects surface stormwater and directs it into a storm sewer.

Catchment area
Surface, typically on a roof, where rainwater is caught and directed into a rainwater harvesting system.

Category
A component of the LEED Green Building Rating System. Each LEED (Leadership in Energy and Environmental Design) prerequisite and credit falls within one of six categories (five Sustainable categories and one Innovation and Design Process category).

Catenary
The curve assumed by a flexible uniform cable suspended freely between two points. When uniformly loaded, the curve takes the form of a parabola.

Catenary arch See Arch

Cathedral
The principal church of a diocese, which contains the home throne of a bishop, called the cathedra.

Cathedralized attic
An unvented attic with insulation installed between the rafters or above the roof sheathing. Moving the insulation from the attic floor to the roof plane turns the attic into conditioned or semiconditioned space; this is especially beneficial in homes with attic ductwork. Usually refers to an attic that does not include finished space.

Cathri
A pierced screen or metal railing formed by superimposing perpendicular and diagonal crosses.

Catwalk
A narrow walkway located at the side of a bridge or near the ceiling of a building, as in a theater.

Caudill, William (1914–1983)
Partner in the firm Caudill, Rowlett Scott (CRS), most well known for its method of involving the "user groups" to participate in the design "Charrette" process from 1958 to 1973.

Caulicoli
The eight stalks that support the volutes in a Corinthian capital.

Caulk

To render a joint tight against the elements by filling the seam with a malleable substance such as tar, lead, or putty.

Caulking

The process of making a joint watertight; includes mastics, rubber, silicone and other flexible sealants.

Caulking compound

A soft material intended for sealing joints in buildings, preventing leakage or providing a seal at expansion joints.

Cavetto molding See Molding

Cavity wall

A wall assembly constructed of two wythes of masonry bonded with wall ties and separated by an air cavity, which is sometimes filled with insulation.

Cavity wall masonry See Masonry

Cavity-fill insulation

The insulation installed in the space created by wall, ceiling, roof, or floor framing, most commonly fiberglass-batt, spray-applied or dense-pack cellulose, or spray polyurethane.

Cavo-relievo See Relief

Cedar See Wood

Ceiling

The undercovering of a roof or floor; generally concealing the structural members from the room or roof above, or the underside surface. It may have a flat or curved surface, and be self-supporting, suspended from the floor above, or supported from hidden or exposed beams.

exposed ceiling

A ceiling in which all the structural and mechanical systems are left exposed, either in their natural state or painted.

false ceiling

A ceiling suspended or hung from the floor above, which hides the underneath structure and provides a space for the mechanical systems, wires and ducts.

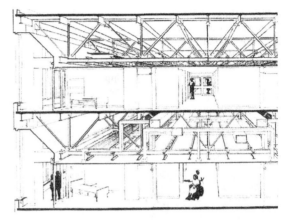

luminous ceiling

A system in which the whole ceiling is translucent with lamps that are installed above and suspended from a structural ceiling.

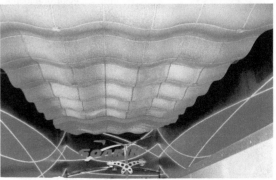

suspended ceiling

A nonstructural ceiling suspended below the overhead structural slab or from the structural elements of a building and not bearing on the walls.

Ceiling cavity

The cavity formed by the ceiling, the plane of the luminaires, and the wall surfaces between them.

Ceiling diffuser

An air outlet from an air-conditioning duct, which diffuses the air over a large area to produce an even distribution and avoid drafts.

Ceiling fan

A slowly rotating overhead fan with a wide sweep, moving large volumes of air at low speed; used in hot, humid climates to improve thermal circulation.

Ceiling fixture

Requires an electrical source in the ceiling such as an outlet box covered with a metal cap. The box may be connected to a wall switch for operation, and a dimmer can be installed at the switch.

Ceiling grid

A rectangular grid of metal supports for a suspended ceiling, especially for acoustical tile; typical the supports are inverted

Ceiling joist See Joist

Ceiling medallion

A bas-relief ornament on a ceiling, especially a relatively large, circular one located in the center of a room.

Ceiling rosette

A circular element that covers the connection of a pendant fixture to the electrical supply.

Cell

A small compartment, such as a room in a dormitory, but especially a confined study-bedroom allotted to a monk or nun in a monastery.

Cella

The sanctuary of a classical temple containing the cult statue of the god.

Cellar

That part of a building, the ceiling of which is entirely below grade; or having half or more of its clear height below grade.

Cellular construction

Construction with concrete elements in which part of the interior concrete is replaced by voids.

Cellulose

A fibrous part of plants used to manufacture paper and textiles.

Cellulose insulation

Plant fiber that is used in wall and roof cavities to separate the inside and outside of the building thermally and acoustically. Typical materials used to manufacture the product include old newspapers and telephone directories; borates and ammonium sulfate are included to retard fire and pests. Four major types of loose-fill cellulose products have been developed under a variety of brand names, generally characterized as dry cellulose, spray-applied cellulose, stabilized cellulose, and low-dust cellulose.

Cellulose insulation with borates

Cellulose insulation is made from recycled newspaper. The borates provide fire and vermin protection. Most cellulose insulation now uses chemical fire retardants as opposed to the natural borates. Fiberglass-batt insulation often contains formaldehyde, which can adversely affect indoor air quality and human health. Cellulose insulation with high recycled content provides maximum environmental benefit.

Celtic Revival
A revival in the nineteenth century of Celtic art and ornament, found mostly in Britain and Ireland, which influenced the Arts and Crafts movement, as well as the development of Art Nouveau.

Cement
A material, or a mixture of materials without aggregate, which when in a plastic state, possesses adhesive and cohesive properties and hardens in place.

Cement paint
A paint that is a mixture of cement and pigment, or one based on an alkali-resistant vehicle, such as casein, and used over cement surfaces.

Cement plaster
A plaster with Portland cement as the binder; sand and lime are added on the job before installation.

Cement siding
Same as stucco made with Portland cement.

Cement stucco dash brush
A stiff brush used to give a stucco effect to a plaster wall; the brush is dipped into the material and thrown onto the surface using varying techniques.

Cementitious foam insulation
A magnesium-oxide-based material blown with air to create an inert, effective insulation. It is especially good for people with chemical sensitivities.

Cenotaph
An empty tomb; a commemorative monument not intended for burial.

Centaur See **Ornament: animal forms**

Center of vision See **Perspective projection**

Center to center
A linear measurement taken from the middle of one member to the middle of the next member.

Center-hung door See **Door**

Centering
A temporary wooden framework placed under vaults and arches to sustain them while under construction; the form was covered with mortar so that the undersurface of the vault showed an impression of the boards used.

Centerpiece
An ornament placed in the middle of an area, such as a decoration in the center of a ceiling.

Central air conditioning
An air-conditioning system in which the air is treated by equipment at one or more central locations and conveyed to and from these spaces by means of fans and pumps through ducts and pipes.

Central business district (CBD)
The main commercial area with concentrated office and retail stores located in an urban area.

Central heating
A system where heat is supplied to all areas of a building from a central plant through a network of ducts or pipes.

Central plant
An area or building in which the chillers and boilers for a building or group of buildings are located.

Central visual axis See **Perspective projection**

Centralized organization See **Organization**

Centralized structure
Building in which all the principal axes are of the same length.

Centrally planned
A structure that radiates from a central point, as opposed to one based on an axial plan.

Ceramic
Burnt clayware, consisting of a mixture of sand and clay, shaped, dried and finally fired in a kiln. Main types include terra-cotta, used mainly for unglazed air bricks, chimney pots and floor tiles; fire clay, used for flue linings since it has a fire resistance; vitreous china, used for plumbing fixtures and sanitary appliances.

Ceramic mosaic tile See **Tile**

Ceramic veneer
Architectural terra-cotta with either ceramic vitreous or glazed surfaces, characterized by large face dimensions and their sections; the back is either scored or ribbed.

Ceramics
A brittle, noncorrosive and nonconductive product made of clay or similar material, fired during its manufacture to produce porcelain or a hard terra-cotta.

Certificate of appropriateness
A document awarded by a preservation commission or architectural review board allowing an applicant to proceed with a proposed alteration or new construction following a determination of the proposal's suitability according to applicable criteria; may also permit demolition.

Certificate of insurance
A memorandum issued by an authorized representative of an insurance company stating the types, amount, and effective dates of insurance in force for a designated insured person.

Certificate of Occupancy (C of O)
A permit from a local government agency granting permission to use a building or site for a particular purpose; issued after the satisfactory completion of construction.

Certification
The process of evaluating the accuracy of the documentation on the eligibility of a property for listing on the National Register of Historic Places.

Certified Historic Structure
For the purposes of the federal preservation tax incentive, any structure subject to depreciation as defined by the U.S. Internal Revenue Code that is listed individually in the National register of Historic Places, or located in a registered historic district and certified by the Secretary of the Interior as being of historical significance to the district. A structure that the U.S. National Park Service has determined is eligible to obtain a Certificate of Significance.

Certified lumber
General shorthand term for lumber that has been certified sustainable harvest by an independent certification authority. The Forest Stewardship Council (FSC) is the independent authority recognized in the U.S. Green Building Council's Leadership in Energy and Environmental Design (LEED) standards. The underlying guidelines are for preservation of a diverse sustainable forest that exhibits the same ecological characteristics as a healthy natural forest.

Certified rehabilitation
Any rehabilitation of a certified historic structure that the U.S. Secretary of the Interior has determined is consistent with the historic character of the property or the district in which the property is located.

Certified sustainably managed
Wood determined to have been harvested from a sustainable forest that exhibits the same ecological characteristics as a healthy natural forest. A number of certifying organizations have been established to oversee the harvesting of wood for lumber and provide guidelines for preservation of forests.

Certified wood
Under the guidance of the Forest Stewardship Council (FSC), wood-based materials used in building construction that are supplied from sources that comply with sustainable forestry practices, thereby protecting trees, wildlife habitat, streams, and soil.

CFC-free
A product does not contain or use chlorofluorocarbons.

Ch'in architecture (221–206 B.C.)
A dynasty in China marked by the construction of the Great Wall of China.

Chain bold See **Bond**

Chain molding See **Molding**

Chain-link fence
A fence composed of steel wire woven in a diamond pattern, with steel-pipe posts and top rails; typically without any decorative elements.

Chain-of-custody
A document that tracks the movement of a wood product from the forest to a vendor or end user, and is used to verify compliance with FSC (Forest Stewardship Council) guidelines. A "vendor" is defined as the company that supplies wood products to project contractors or subcontractors for onsite installation.

Chain-of-custody certification
A product that has met certain requirements throughout its life, beginning from its extraction and production all the way to its distribution and sale.

Chain-of-information
Wood-based materials that are supplied from sources that comply with sustainable forestry practices, protecting trees, wildlife habitat, streams, and soil as determined by the Forest Stewardship Council (FSC).

Chains
Vertical strips of rusticated masonry rising between the horizontal moldings and the cornice, dividing the facade into bays or panels.

Chair rail
A horizontal wood strip affixed to a plaster wall at a height that prevents the backs of chairs from damaging the wall surface.

Chaitya arch
A recurring motif in Hindu architecture depicting miniature arches derived from a cross section of a vaulted chaitya hall.

Chaitya hall
A structure, or artificial cave, with an apse encircling a stupa.

Chalet

A timber house in the style of those traditionally built in the Swiss Alps, distinguished by exposed and decorative use of structural members, balconies, and stairs; the upper floors usually project beyond the story below.

Chalking
Disintegration of paint and other coatings, which produces loose powder at, or just beneath, the surface.

Chamber
A room used for private living, conversation, consultation or deliberation, in contrast to more public and formal activities.

Chambranle

A structural feature, enclosing the sides and top of a doorway, window, fireplace or similar opening, often highly ornamental.

Chamfer

The groove or oblique surface made when an edge or corner is beveled or cut away, usually at a 45-degree angle.

Chamfer stop

Any ornamentation which terminates a chamfer.

Chamfered rustication

Rustication in which the smooth face of the stone parallel to the wall is deeply beveled at the joints so that, when the two meet, the chamfering forms an internal right angle.

Chancel

The part of a large church that is located beyond the transept, containing the altar and choir.

Chancery

A building or suite of rooms designed to house any of the following: a low court with special functions, archives, a secretarial, a chancellery.

Chandelier

A fixture with multiple arms hung from the ceiling to support lights; originally for candles, but later manufactured for gas, then electric lights.

Change order

A written order to the contractor signed by the owner and architect or engineer, issued after the execution of a contract, authorizing a change in the work or an adjustment in the contract sum or contract time.

Channel

A rolled iron or steel or extruded aluminum shape with a vertical flange and horizontal top and bottom webs that project on the same side as the flange.

Channel beam see Beam

Channel column See Column

Channeling

A decorative groove in carpentry or masonry: a series of grooves in an architectural member, such as the flutes in a column.

Chapel

A small area within a larger church, containing an altar and intended for private prayer; a small secondary church in a parish; a room designated for religious use within the complex of a school, college, or hospital.

Chapter house

A place for business meetings of a religious or fraternal organization; usually a building that is attached to a hall for gatherings; occasionally contains living quarters for members of such groups.

Charrette See Design

Charrette process

A French word that means "cart" and is often used to describe the final, intense work effort expended by art and architecture students to meet a project deadline. This use of the term is said to originate from the École des Beaux-Arts in Paris during the 19th century, where proctors circulated a cart, or "charrette," to collect final drawings while students frantically put finishing touches on their work. Early involvement of the entire project team is fundamental to the successful use of a systems

Charrette process

approach to green building; thus "charrette" is also used to denote a meeting held early in the design phase of a project, in which all the participants, including the design team, engineers, contractors, clients, end users, community stakeholders, and technical experts, are brought together to develop goals, strategies, and ideas for maximizing the environmental performance of the project. Research and experience have indicated that early involvement of all interested parties increases the likelihood that sustainable building will be incorporated as a serious objective of the project and reduces the soft costs sometimes associated with a green design project.

Chase

A covered recess in a wall that forms a vertical shaft, in which plumbing pipes or electrical wires are inserted.

Chashitsu

A small Japanese structure, or room, for the tea ceremony.

Chateau

A castle or imposing country residence of nobility in old France; any large country estate.

Chateau style (1860–1898)

A style characterized by massive and irregular forms, steeply pitched hipped or gable roofs with dormers, towers, and tall elaborately decorated chimneys featuring corbeled caps. Windows are paired and divided by a mullion and transom bar. Renaissance elements such as semicircular arches or pilasters are mixed with Tudor arches, stone window tracery, and Gothic finials.

Chateauesque style

A house style based on sixteenth-century French chateau mixed with Gothic and Renaissance details, popular for large houses and mansions of the late nineteenth and the early twentieth centuries; the most characteristic details include steeply pitched mansard roofs, dormers with pinnacles, and parapeted gables and turrets.

Chatri

An Indian pavilion which consists of a horizontal slab carried on four colonnettes, recurring in Hindu architecture.

Chattra

A stone umbrella on top of a stupa, symbolizing dignity, composed of a stone disk on a vertical pole.

Chaumukh

In Indian architecture, four images, each facing a cardinal point, which are placed back-to-back.

Chavin style (900–200 B.C.)

A Peruvian style based on the worship of the jaguar god and characterized by grandiose terraced platforms constructed of stone, which were grouped around large sunken plazas, excellent stone sculpture, elaborate gold work, and remarkable ceremonies. The style is named after a town in central Peru, where a complex of massive stone buildings with subterranean galleries was surrounded by courtyards.

Checkdam
Low dam of stone, wood, or other material used for holding and spreading runoff and sediment in a swale.

Checker
One of the squares in a checkered pattern, which is contrasted to its neighbor by color or texture; often only two of the effects are alternated, similar to a chessboard.

Checkered
Those forms that are marked off with a pattern of checks or squares that is divided into different colors, or variegated by a checked or square pattern of different materials.

Checkerwork
In a wall or pavement, a pattern formed by laying masonry units so as to produce a checkerboard effect.

Checking
A defect in a coated surface characterized by the appearance of fine cracks in all directions; also cracking of wood grain caused by improper drying.

Checklist
A detailed list of all the elements of a particular job; used to ensure that details are not overlooked in preparing an estimate, or other times during the course of a job.

Cheek
A narrow upright face, forming the end or side of an architectural or structural member or one side of an opening.

Cheneau
A gutter at the eaves of a building, especially one that is ornamental; an ornament, crest, or cornice.

Chermayeff, Peter (1936–)
The first project of the founding principal of Cambridge Seven Associates was the New England Aquarium, Boston, MA. It opened in 1969. He also worked on the Osaka Aquarium, the Tennessee Aquarium, the Aquarium of Genoa, the Lisbon Oceanarium, and the Alaska SeaLife Center in Seward, Alaska.

Chermayeff, Serge (1900–1996)
Born in Russia, he immigrated to England in 1910 and worked as a designer before joining Eric Mendelsohn. He immigrated to America in 1940 and designed many Modern movement buildings.

Chernikov, Lakov (1898–1951)
Russian-born architect and teacher; his work focused on architectural compositions in perspective, illustrating forms and structures of the imagination, "following construcivist" principles.

Cherry See **Wood**

Cherubs
A decorative sculpture or painting representing chubby, usually naked infants: They are also called amorini or putti.

Chestnut See Wood

Chevet
The rounded east end of a Gothic cathedral, including the apse and ambulatory.

Chevron
A symmetrical "V" shape that represents a triangle with its third side removed. It can be bordered and interlaced and is often repeated in various patterns that point up or down, with an angle between 60 and 75 degrees.

Chiaroscuro
The effect of light and shadow within an area or composition, brought about by the use of deep variations to enhance the forms.

Chiattone, Mario (1891–1957)
Studied in Milan, became a member of the Futurist movement, and with fellow student Sant'Elia exhibited a collection of drawings in Milan called *Structures for a Modern Metropolis* (1914).

Chicago School
A group of active architects working in Chicago (1880–1910) known for major innovations in high-rise construction and for the development of modern commercial building design. The group included Daniel H. Burnham, John W. Root, William Lee Baron Jenney, W. B. Mundie, William Holabird, and Louis Sullivan. The emphasis was on verticality and having the structural frame expressed on the exterior in the form of a skeleton structure, with the fenestration treated as an infill within the framework.

Chicago window See **Window**

Chigi
A pair of crossed timbers that are placed at the end of the ridge of a roof of a Shinto shrine; also called forked finials.

Chiller
A device that generates a cold liquid that is circulated through an air-handling unit's cooling coil to cool the air supplied to the building; consists of a compressor, condenser, and evaporator.

Chimera See **Ornament: animal forms**

Chimney
A vertical noncombustible structure, containing one or more flues to carry smoke from the fireplaces to the outside, usually rising above the roof.

Chimney breast
A projection into a room of the fireplace walls that form the front portion of the chimney stack.

Chimney cap
A cornice forming a crowning termination of a chimney.

Chimney cricket
A small false roof built over the main roof behind a chimney, used to provide protection against water leakage where the chimney penetrates the roof.

Chimney hood
A covering which protects a chimney opening.

Chimney lining
Rectangular or round tiles placed within a chimney for protective purposes. The glazed surface of the tile provides resistance to the deteriorating effects of smoke and gas fumes.

Chimney piece
An ornamental embellishment above or around the fireplace opening.

Chimney pot
Cylindrical shape placed on top of a chimney to increase its height; often of terra-cotta and treated as an ornamental device.

Chimney pot

Chimney stack
That part of a chimney that is carried above the roof of a building: a group of chimneys carried up together.

Chimney throat
That part of a chimney directly above the fireplace where the walls of the flue are brought close together as a means of increasing the draft.

Chimu architecture (1150–1400)
A style dominant in northern Peru that featured houses built in rows along symmetrically laid out streets inside high city walls. Buildings were constructed of adobe brick with wooden lintels. Walls were decorated with wide moldings featuring geometrical designs.

Chinese architecture (400–1600)
A homogeneous traditional architecture that was repeated over the centuries in structures consisting of a wooden framework of columns and beams; stone and brick were used for permanent structures such as fortifications. The most prominent feature was tile-covered gabled roofs, with widely overhanging and upward curving eaves resting on complex multiple brackets. In the design of pagodas, each floor was articulated in a distinctive rhythmical, horizontal effect.

Chinese architecture (400–1600)

Chinking
The material used to fill a space between the logs of a log house; composed of clay mixed with sand and hair, then covered with daubing.

Chinoiserie
A Western style of architecture and decoration, utilizing Chinese design elements.

Chip carving
Hand decoration of a wooden surface by slicing away chips, resulting in incised geometric patterns.

Chipperfield, Sir David Alan (1953–)
Uncompromisingly modernist in outlook, Chipperfield designed the award-winning River and Rowing Museum in Henley-on-Thames, England, using green oak cladding, concrete, and glass. He designed the Figge Art Museum in Davenport, Iowa; the Central Public Library in Des Moines, Iowa, commissioned in 2001; and the Anchorage Museum Expansion in Anchorage, Alaska. Chipperfield was architect for the reconstruction of the destroyed Neues Museum in Berlin.

Chlorofluorocarbons
Stable, artificially created chemical compounds containing carbon, chlorine, fluorine, and sometimes hydrogen. Chlorofluorocarbons, used primarily to facilitate cooling in refrigerators and air conditioners, have been found to deplete the stratospheric ozone layer that protects the earth and its inhabitants from excessive ultraviolet radiation.

Choir
That part of a church where the religious service isaccompanied by singing, usually part of the chancel and often separated by an ornamental screen.

Chord
A principal member or pair of members of a truss extending from one end to the other, to resist bending.

Christ-Janer, Victor (1916–2008)
Christ-Janer worked alongside the group of architects associated with the Harvard Graduate School of Design known as the "Harvard Five"—Phillip Johnson, Marcel Breuer, Landis Gores, John Johansen, and Eliot Noyes—to change the landscape of colonial New Canaan, CT. Christ-Janer's work included his own home, a market, stables, and a post office in the town, as well as many regional religious buildings. His work is described as sleek, preferring concrete and glass as materials and strong horizontal forms. He also worked on college campuses.

Christo (1935–)
Born Christo Vladimirov Javacheff in Bulgaria and married to Jeanne-Claude (1935–2009), who together created environmental works of art. Their work included wrapping of the Reichstag in Berlin, and the Pont-Neuf Bridge in Paris, the 24-mile Running Fences in Marin County, CA, and The Gates (illus.), in New York's Central Park.

Chroma
The attribute of a color that allows the observer to judge how much color it contains.

Chromated copper arsenate
Type of wood preservative that has now been largely eliminated from residential wood products because of concerns about leaching and toxicity. Huge quantities of chromated copper arsenate treated wood remain in use, especially in residential decks.

Church
An edifice or place of assemblage specifically set apart for Christian worship.

Churn
The frequency with which a building's occupants are moved, either internally or externally, including those who move but stay within an organization and those who leave a company and are replaced.

Churrigueresque style (1700–1750)
A lavishly ornamented Spanish Baroque style named after architect Jose Churriguera; the style was also adapted in South America.

CIAM (Congrès Internationaux d'Architecture Moderne) (1928–)
A declaration signed by 24 architects, representing France (6), Switzerland (6), Germany (3), Holland (3), Italy (2), Spain (2), Austria (1), and Belgium (1), emphasizing "building" rather than architecture. Advocated the introduction of efficient production methods in the building industry.

Cincture
A ring of moldings around the top or bottom of the shaft of a column, separating the shaft from the capital or base; a fillet around a post.

Cinquefoil arch See **Arch**

Cinquefoil See **Tracery**

Circle
The simplest and most fundamental of geometric shapes; a continuous curved line, every point of which is equidistant from a central point.

Circuit
The path taken by an electric current in flowing through a conductor through one complete run of a set of wires from a power source, such as a panelboard, to various electrical devices and back to the same power source. The wires used for various circuits are prescribed by codes, such as the National Electrical Code.

Circuit breaker
A device such as the electromagnetic opening of a spring-loaded latch, or the heating of a metallic strip, which stops the flow of current by opening the circuit automatically when more electricity flows through the circuit than it is capable of carrying; resetting can be automatic or manual.

Circular arch See **Arch**

Circular barn
A barn that has a circular plan; similar to that built by Shakers in Hancock, Massachusetts, in 1826; also called a round barn.

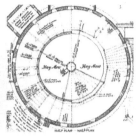

Circular stair See **Stair**

Circular window See **Window**

Circulation space
A space within a facility that provides access between functional areas for people, goods and services.

Circus
In ancient Rome, a large oval arena surrounded by rising tiers of seats, for the performance of public spectacles.

Cistern
A tank used to store rainfall that has been collected from a roof or some other catchment area, usually located underground. The water is generally used for watering lawns and gardens, washing cars, and similar uses, but it can also be used for flushing toilets and, with treatment, for all other uses. Cisterns help conserve water and prevent pollution of nearby streams from runoff.

Citadel
A fortress or castle in or near a city; a refuge in case of a siege, or a place to keep prisoners.

City Hall
Building that houses the administrative offices of a city.

City plan
Large-scale map of a city that depicts streets, buildings, and other urban features.

Cityscape
Represented by the silhouette of groups of urban structures that make up a skyline, including distinguished landmarks as well as natural elements, such as rolling hills, mountains or large bodies of water.

Cityscape

Civic center
An area of a city containing a grouping of municipal buildings; often includes the city hall, courthouse, public library, and other public buildings.

Civil engineering
A branch of engineering that encompasses the conception, design, construction, and management of residential and commercial buildings and structures, water supply facilities, and transportation systems for goods and people, as well as control of the environment for the maintenance and improvement of the quality of life. Civil engineers include planning and design professionals in both the public and private sectors, contractors, builders, educators, and researchers.

Cladding
The process or the resulting product produced by the bonding of one metal to another, to protect the inner metal from weathering.

Clapboard siding See **Wood products**

Classical architecture

The architecture of Hellenic Greece and Imperial Rome on which the Italian Renaissance and subsequent styles were based. The five orders, the Doric, Ionic, Corinthian, Tuscan, and Composite, are a characteristic feature.

Classical Revival (1790–1830)

A rebirth of art and architecture in the style of the ancient Greeks, Romans, and Italian Renaissance, popular styles in the United States.

Classical orders

A set of styles developed over the centuries based on Greek and Roman adaptations. The elements of each order consist of a base, column and capital, and an entablature with architrave, frieze and cornice. The most common types are the Tuscan, Doric, Ionic, Corinthian and Composite order.

Clathri

A lattice of bars as gratings for windows.

Clay tile flooring

Burnt clay tiles made in various sizes and thickness, usually laid on a concrete base; most often used for kitchens and bathrooms, porches, vestibules and fireplace hearths.

Clay tile See Tile

Clean Air Act

A federal statute enacted in 1963 that was the first of a series of acts and amendments that exerted increasing federal pressure on air polluters to clean up their emissions.

Clean Air Act 1972

Groundbreaking legislation administered by the EPA (Environmental Protection Agency) that mandates specific measures to protect the air quality and respiratory health of U.S. inhabitants.

Clean design

The systematic incorporation of life-cycle environmental considerations into product design.

Clean energy

The energy created from renewable sources with low environmental impact.

Clean Water Act

A federal statute enacted in 1972 that has been successful in improving the water quality of lakes and rivers.

Cleaner production

Implies improvements to a production process so that the process uses less energy, water, or other inputs, or generates less waste or less environmentally harmful waste.

Cleanout

A unit with removable plate or plug allowing access into plumbing or other drainage pipes for cleaning out extraneous material.

Clear span

Distance between the inside faces of the supports on both sides.

Cleat

A small piece of timber or metal fixed to one member and used to reinforce, and positively locate or support another member, a strip of wood or other material applied to the wall, for the purposes of supporting another member fastened to it.

Clerestory

An upper story or row of windows rising above the adjoining parts of the building, designed as a means of admitting increased light into the inner space of the building.

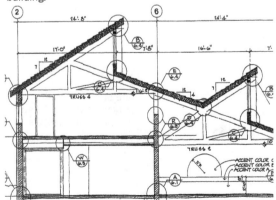

Client
The person or organization who has a need for a project; responsible for the overall financing of the work and directly or indirectly employs the entire design and building team.

Cliff dwelling
Dwellings built by natives in the southwestern United States under cliff overhangs and on cliff tops, typically constructed of natural stone slabs with mud for mortar; roofs were constructed of earth-covered poles, or the underside of the natural ledge above was used as a roof.

Climate change
Attributed directly or indirectly to human activity that alters the composition of the global atmosphere and is in addition to natural climate variability over comparable time periods; often used to describe global warming with environmental implications including temperature and sea-level rises; changes in rainfall and the intensity and frequency of extreme weather events; groundwater, atmospheric, and ocean circulation patterns and locations; and displacement of ecosystems and commercial resources.

Climate control
An HVAC system.

Climate neutral
No net production of greenhouse gases.

Climate, Community and Biodiversity Alliance (CCBA)
The CCBA has developed and is promoting rigorous standards for evaluating land-based carbon projects. The CCBA Standards identify land-based climate change mitigation projects that simultaneously generate climate, biodiversity, *and* sustainable development benefits.

Climate-dominated building
A building in which energy consumption is driven by the heat loss or gain that moves across the building's envelope. The internally generated energy requirements from machines, appliances, or people are smaller than the energy requirements created by heat or cold moving through the building's envelope.

Climax
A number of design ideas so arranged that each succeeding one makes a stronger statement than its predecessor. The final culmination or highest point is the summation of the process.

Clinic
A medical facility; independent or part of a hospital in which ambulatory patients receive diagnostic and therapeutic medical and surgical care.

Clinker brick See **Brick**

Clip angle
A short piece of angle iron attached to one piece of a column or beam with bolts, in preparation for attaching to another piece upon erection.

Clock
Any instrument for measuring or indicating time, especially a mechanical device with a numbered dial and moving hands or pointers.

Clocktower
Any instrument for indicating time, such as a mechanical device with a numbered dial and moving hands or pointers, positioned in a single tower or a tower-like portion of a structure.

Cloister
A square court surrounded by an open arcade, a covered walk around a courtyard, or the whole courtyard.

Close grain
Lumber which has narrow and inconspicuous annual rings, producing a fine-grained pattern.

Closed loop
A processing cycle in which wastes are completely recycled or reused and never enter the environment.

Closed string stair See Stair

Closed-loop control
A control system that uses measurement of a controlled variable for feedback. Based on the measured feedback, a closed-loop control system alters its output in an attempt to force the controlled variable to reach a given set point.

Closed-loop process
Part of an industrial production process, but not part of a waste management process. Materials reclaimed and returned in a closed-loop process are not classified or defined as, nor do they operate as, a waste. Materials in a closed-loop process are treated as commodities in a manner designed to avoid loss or release to the environment.

Closed-loop recycling
When a used product is recycled into a similar product; a recycling system in which a particular mass of material is remanufactured into the same product, for example, glass bottles into glass bottles.

Closer See Brick

Cluster
Any configuration of elements that are grouped or gathered closely together.

Cluster development
A development design that concentrates buildings and infrastructure in specific areas on a site to allow remaining land to be used for recreation, common open space, or the preservation of historical or environmentally sensitive features.

Cluster housing
Suburban, medium-density apartment or row-house complexes.

Clustered column See Column

Clustered housing
Homes that are grouped, collected, or gathered closely together and related by proximity to each other.

Clustered organization See Organization

CO_2-based high-limit ventilation control
A ventilation strategy that monitors the CO_2 concentration in a zone or in the return air duct from the zone. If the CO_2 concentration approaches a predetermined high limit, the outdoor air-flow controller is reset to provide additional ventilation. This process supplements standard ventilation-control strategies by providing additional ventilation for unexpected occupancy.

Coach house
A building or part for housing carriages or automobiles when they are not in use. See also carriage house.

Coalition for Environmentally Responsible Economics (CERES)
The leading U.S. coalition of environmental, investor, and advocacy groups working together for a sustainable future. It has committed to continuous environmental improvement by endorsing the CERES Principles, a 10-point code of environmental conduct.

Coarse grain
Wood or lumber having wide and conspicuous annual rings in which there is considerable difference between the springwood and summerwood.

Coat of arms
A tablet containing a representation of a heraldic symbol.

Coaxial cable
A round, flexible, two-conductor cable that consists of a copper wire at the center, a layer of protective insulation, a braided mesh sleeve, and an outer shield or jacket.

Cobb, Henry Ives (1859–1931)
American architect in Chicago, known for personalized adaptations of Romanesque and Gothic styles. Designed the Fisheries Building in 1893, World's Columbian Exposition, and the plan for the University of Chicago in 1895, all located in Chicago, IL.

Cobble See **Stone**

Cobblestone See **Stone**

Code compliance
Review of plans and specifications by a building official to ensure that all building code requirements are met.

Code enforcement
A principal tool used to ensure neighborhood upkeep, by enforcing local regulation of building practices, and safety and housing code provisions; includes onsite inspection by building officials and legal sanctions against owners of property that are in violation of the codes.

Codes regulations
Ordinances or statuary requirements of a governmental unit relating to building construction and occupancy, adopted and administered for the protection of the public health, safety, and welfare.

Coefficient of performance
Compares the heating capacity of a heat pump to the amount of electricity required to operate the heat pump in the heating mode. Performance varies with the outside temperature; as the temperature falls, the performance also falls, given that the heat pump is less efficient at lower temperatures.

Coefficient of utilization
A measure of the efficiency of a light fixture: the ratio of the luminous flux reaching a surface to the total lumen output of the fixture. This is found in the photometric test report as the percentage of total lamp lumens generated

Coefficient of utilization
within the fixture that actually reach the work plane. It is based on the efficiency of the fixture.

Coffer
A recessed box-like panel in a ceiling or vault, usually square, but often octagonal or lozenge-shaped, sometimes dressed with simple moldings or elaborately ornamented.

Coffering
Ceiling with deeply recessed panels, often highly ornamental, executed in marble, brick, concrete, plaster, or stucco; a sunken panel in a vaulted ceiling.

Cogeneration
The simultaneous production of electrical or mechanical energy (power) and useful thermal energy from the same fuel/energy source such as oil, coal, gas, biomass, or solar panels. Conventionally, heat is emitted through cooling towers or flue gas is lost, but cogeneration makes use of that heat. A cogeneration plant is often referred to as a combined heat and power plant.

Cogged joint See **Joint**

Coheat test
A test that measures the distribution of heating and cooling systems throughout a building. Measures the overall heat loss factor. The home is alternately heated with the furnace and an array of small heaters (coheaters) to calculate heat-delivery efficiency.

Cohousing

Development pattern in which multiple (typically 8–30) privately owned houses or housing units are clustered together with some commonly owned spaces, such as a common workshop or greenhouse. Automobiles are typically kept to the perimeter of the community, creating a protected area within the community where children can play. Usually, residents are closely involved in all aspects of the development, from site selection to financing and design.

Coliseum

Any large amphitheater;
any large sports arena,
open or roofed.

Collaborative for High-Performance Schools (CHPS)

A national nonprofit organization that offers resources, assessment tools, and recognition programs for building and operating high-performance, healthful, green schools. It offers state-specific rating systems designed by school facilities experts to create schools that are both good for the environment and healthful for students and faculty. Schools can receive third-party verification, and they also can self-verify.

Collage See Design drawing

Collar beam See Beam

Collar beam roof

A roof in which the rafters are tied together and stiffened by collars.

Collar brace

A structural member which reinforces a collar beam in medieval roof framing.

Collar joint

The joint between a roof structure and a collar beam.

Collar tie

In wood construction, a timber that prevents the roof framing from changing shape.

Collard, Max Ernest (1909–2008)

A modernist, responsible for many prominent public buildings in Australia. In 1963 the Australian government relocated all the country's public servants to Canberra from Melbourne, and the need for housing and public buildings there resulted in a multiyear building boom. In 1971 he

Collard, Max Ernest

completed Russell Building 14, part of a defense complex for the navy. Featuring a dramatic lobby and facade that became a local landmark.

Collegiate Gothic

A term for the version of Gothic architecture that was characteristic of the colleges at Oxford and Cambridge, England, and adapted as the style for a number of American colleges in the late nineteenth and early twentieth centuries.

Colonial architecture

A classification pertaining to any architectural style that is transplanted from a country to its overseas colonies. Examples are the Portuguese Colonial in Brazil, Dutch Colonial in New York, French Colonial in New Orleans, and English Colonial in all the former North American colonies.

Colonial panel door

A door characterized by recessed panels that are framed by stiles, rails, and muntins.

Colonial Revival style

The reuse of Georgian and Colonial design in the United States toward the end of the nineteenth and twentieth centuries, typically in bank buildings, churches, and homes.

Colonial styles

The styles most popular in America (1600–1780) were English Colonial (1607–1620); Dutch Colonial (1625–1830); and French Colonial (1665–1803).

Colonnade

A combination or grouping of columns paced at regular intervals, and arranged with regard to their structural or ornamental relationship to the building. They can be aligned either straight or arced in a circular pattern.

Colonnette

A small column, usually decorative, found at the edge of windows; a thin round shaft to give a vertical line in elevation, or as an element in a compound pier.

Color

The appearance of an object or surface, distinct from its form, shape, size, or position; depends on the spectral composition of incident light, spectral reflectance of the object, and spectral response of the observer.

Color chronology

Determining the approximate date of historic structures through the examination of colors, pigments, paint types with scientific means.

Color of light

The spectral distribution of light varies according to the nature of the source. The apparent color of a surface is the result of its reflection of its hue and its absorption of the other colors of the light shining on it.

Color rendering

The effect of a light source on the color of a surface, compared to the same color lit by daylight.

Color temperature

The temperature to which one would have to heat a "black body" to produce light of similar spectral characteristics. Low color temperature implies warmer (yellow-red) light, and high color temperature implies a colder (blue) light. Daylight has a rather low color temperature near dawn, and a higher one during the day. Therefore an electrical lighting system should supply cooler light to supplement daylight when needed and fill in with warmer light at night. The standard unit for color temperature is Kelvin (K).

Colored cement

Portland cement blended with a pigment which does not react chemically with the component of the concrete.

Colored glass See Glass

Color-rendering index

A method for describing the effect of a light source on the color appearance of objects, compared to a reference source of the same color temperature. The highest CRI attainable is 100. Typical cool white fluorescent lamps have a CRI of 62. Lamps having rare-earth phosphors are available with a CRI of 80 and above. In a daylighting context, the CRI defines the spectral transmissive quality of glasses or other transparent materials. In this case, values of 95 or better are considered acceptable.

Colossal order

A Classical order consisting of giant columns rising from the ground through more than one story.

Colossus

A statue which is larger than life-size.

Column

A vertical structural compression member or shaft supporting a load, which acts in the direction of its vertical axis and has both a base and a capital, designed to support both an entablature or balcony.

angle column
A freestanding or engaged column placed outside the corner of a building or portico.

baluster column
A short, thick-set column in a subordinate position, as in the windows of early Italian Renaissance facades.

banded column
A column or pilaster with drums alternately larger and smaller, alternately plainer and richer in decoration, or alternately protruding.

Bethlehem column
Rolled-steel H-section used as a column; it was manufactured by the Bethlehem Steel Company as a replacement for columns built up from steel plate and angle irons.

box column
A hollow, built-up column constructed of wood, usually rectangular or square in section.

cabled column
A column in the shape of twisted strands of a large-fiber rope.

cabled flute column

A column or pilaster with semicircular convex moldings in the center of the flutes, usually starting at the base of the column and stopping at one-third or one-half the column height; the moldings may be plain or ornamental.

channel column

A hollow steel column composed of a pair of channels that are attached with plates so that the channels are facing outward.

clustered column

A column or pillar composed of a cluster of attached or semiattached additional shafts, grouped together to act as a single structural or design element.

composite column

A column in which a metal structural member is completely encased in concrete containing special spiral reinforcement.

coupled column

Columns set as close pairs with a much wider space between the pairs.

coupled column

diminished column

A column with a greater diameter at its base than at its capital.

elephantine column

A large round or square column that has a broad base and tapers toward the top, found in many bungalow-style homes.

embedded column
A column that is partly, but not wholly, built into a wall.

engaged column
A column that is attached and appears to emerge from the wall, as decoration or as a structural buttress.

grouped columns
Three or more closely spaced columns or pilasters forming a group, often on one pedestal.

half-column
An engaged column projecting approximately one-half its diameter, usually slightly more.

insulated column
A column which is entirely detached from a main building or structure.

knotted column
A column with a shaft carved to resemble a knot or two intertwined ropes.

lally column
A proprietary name for a cylindrical column that is filled with concrete; used as a structural column to provide support to beams or girders.

long column
A column whose slenderness ratio is high, making it liable to failure by buckling.

mid-wall column A column that carries a portion of a wall above it that is much thicker than its own diameter.

monolithic column
A column whose shaft is of one piece of stone, wood, or marble as opposed to one made up from components.

mushroom column
A column with an enlarged head supporting a flat slab.

pipe column
A column made from steel tubing, frequently filled with concrete to increase its strength and stiffness.

solomonic column
A twisted column with a helical shaft and Classical capital and base; often used in Baroque architecture.

tension column
A column subjected to tensile stresses only.

transfer column
A column in a multistory building that does not go down to the foundation but is supported by a transfer girder, which transfers its load to the adjacent columns.

wall column
A column that is embedded or partially embedded in a wall.

wreathed column
A column shaft that is entwined with a raised spiraling molding.

Column baseplate
A horizontal plate beneath the bottom of a column which transmits and distributes the column load to the supporting materials below the plate.

Columniation
A system in Classical architecture of grouping columns in a colonnade based on the diameters of the columns.

Combination system
Heating system that uses the domestic water heater for both water and space heating. Hot water is typically piped to a heat exchanger coil, where a fan blows air over the coil to produce heated air.

Combination windows
Windows having an inside removable section so the same frame serves both summer and winter. In warm weather a screen may be inserted, and in winter a storm window can be installed in place of the screen.

Combined heat and power
The generation of electricity and the capture and use of otherwise wasted heat energy by-products. A combined heat and power system is also referred to as cogeneration.

Combined sewer overflow
A combined sewer system conveys both sanitary sewage and stormwater in one piping system. During normal dry weather conditions, sanitary wastewater collected in the combined sewer system is diverted to the wastewater treatment plant before it enters natural waterways. During periods of significant rainfall, the capacity of a combined sewer may be exceeded. When this occurs, excess flow, a mixture of stormwater and sanitary wastewater, is discharged at CSO points, typically to rivers and streams. Release of this excess flow is necessary to prevent flooding in homes, basements, businesses, and streets.

Combined sewer system
A sewer system that carries both sewage and stormwater runoff. Normally, its entire flow goes to a waste treatment plant, but during a heavy storm, the volume of water may be so great as to cause overflows of untreated mixtures of stormwater and sewage into receiving waters.

Combustion efficiency
The efficiency at which a fuel is burned in a combustion appliance when operating at its rated output; the combustion efficiency is always higher than the annual fuel utilization efficiency (AFUE).

Comfort
An important design objective in sustainable building. Designing for comfort aims to create a space that people enjoy; such qualitative, performance-based objectives are a hallmark of sustainable building.

Comfort criteria
Specific original design conditions that shall at a minimum include temperature (air, radiant, and surface), humidity, and air speed as well as outdoor temperature design conditions, outdoor humidity design conditions, clothing (seasonal), and activity expected.

Comfort envelope
The range of conditions in mechanically ventilated buildings in which the majority of occupants are likely to feel comfortable. ASHRAE (American Society of Heating, Refrigerating and Air-Conditioning Engineers) Standard 55 defines a comfort zone based on the six variables of air temperature, air velocity, relative humidity, radiant temperature, and occupant's clothing insulation and activity level.

Commercial building
Typically refers to any nonresidential building such as a shopping center, office tower, business park, industrial property, or tourism and leisure asset.

Commercial style (1890–1915)
A skeletal, rectangular style of the first five-to fifteen-story skyscrapers brought to full form in Chicago, New York, and Philadelphia. It was characterized by flat roofs, and minimal ornament, except for slight variations in the spacing of

Commercial style
windows. Extensive use of glass was made possible by its steel-frame construction, which could bear the structural loads that masonry could not.

Commission
A formal written assignment for a work of art or architecture, for which the designer is paid.

Commissioning
The process of ensuring that installed systems function as specified, performed by a third-party commissioning authority. Elements to be commissioned are identified, installation is observed, sampling is conducted, test procedures are devised and executed, staff training is verified, and operations and maintenance manuals are reviewed and systems capable of being operated and maintained to perform in conformity with the owner's project requirements. Typically commenced at completion of construction, often including initial user occupancy, intended to allow designers and managers to check functional subsystems, to determine that the facility is functioning properly, and undertake any remedial action.

Commissioning agent
This is the person or company engaged for the specific task of commissioning a building. Preferably, this agent should be an independent party retained on behalf of the party that owns the building at the point of handover and should be appointed during the design stages of the project.

Commissioning authority
The commissioning authority is the qualified person, company, or agency that plans, coordinates, and oversees the entire commissioning process. The commissioning authority may also be known as the commissioning agent.

Commissioning final report
The commissioning final report is the document prepared during the acceptance phase of the commissioning process, after all functional performance tests have been completed. It includes the executive summary, a building description, and the completed commissioning plan, as well as all the documentation generated during the process, including the completed commissioning test plans.

Commissioning issues log
The commissioning issues log lists all equipment or systems malfunctions that require team resolution.

Commissioning plan
The commissioning plan is a document prepared for each project that describes all aspects of the commissioning process, including schedules, responsibilities, documentation requirements, and functional performance test requirements. The level of detail depends on the scope of commissioning specified.

Commissioning report
The document that records the results of the commissioning process, including the as-built performance of the HVAC system and unresolved issues.

Commissioning specification

The contract document that details the objective, scope, and implementation of the construction and acceptance phases of the commissioning process as developed in the design-phase commissioning plan.

Commissioning team

Includes those people responsible for working together to carry out the commissioning process.

Commissioning test plan

The commissioning test plan is a document that details the prefunctional performance test, the functional performance test, and the necessary information for carrying out the testing process for each system, piece of equipment, or energy efficiency measure. The test plans are included as an appendix to the final report.

Committee on the Environment (COTE)

The American Institute of Architects (AIA) energy committee was formed in 1973 to study energy building issues and to develop energy guidelines for passive solar design and building energy performance. The AIA formed COTE in 1990 to advance, disseminate, and advocate to the profession, the building industry, academic institutions, and the public, design practices that integrate built and natural systems and that enhance the environmental performance of buildings.

Common bond See Bond

Common brick See Brick

Common rafter

A sloped roof member that is smaller than the principal rafter, which spans from the top plate of the exterior wall to the roof ridge rafter.

Common services

A combination of systems for services designed to provide lighting, heating, plumbing and air conditioning for several families occupying the same building or several connected buildings.

Community

An interacting population of individuals living in a specific area with increased emphasis on sustainable building and sustainable development. Design and building-related practices enhancing and supporting community ideals and functions are considered more sustainable than those that do not, all else being equal.

Community Development Block Grant

A revenue-sharing program administered by the U.S. Department of Housing and Urban Development (HUD) to provide grants for approved community development projects.

Community development corporation

Community-owned venture capital funds set up by local groups in low-income areas for economic revitalization and rehabilitation.

Comold

Compression molded wood. It is composed of post-consumer wood waste that is combined with a binding agent with no added formaldehyde and molded into structural components used in many chairs.

Compact development

Increased number of housing units on a given piece of land reduces the cost of affordable housing. Compact development is not new; higher density housing is a long-standing tradition, whether it comprises midrise, multi-family, or single-family buildings on small lots.

Compact fluorescent lamp

Fluorescent light bulb in which the tube is folded or twisted into a spiral to concentrate the light output. Compact fluorescent lamps (CFLs) are typically three to four times as efficient as incandescent light bulbs and last 8–10 times as long. CFLs combine the efficiency of fluorescent light with the convenience of an Edison or screw-in base, and new types have been developed that better mimic the light quality of incandescents. Not all CFLs can be dimmed, and frequent on-off cycling can shorten their life. Concerns have been raised over the mercury content of CFLs, and though they have been deemed safe, proper recycling and disposal are encouraged.

Company town

Towns that were established at or near the raw materials for the manufacturing processes that the company owned, such as the coal mining towns in the south around the eighteenth century. Some provided homes, recreational halls, churches, hospitals, and stores for the workers.

Comparative risk analysis

An environmental decision-making tool used to systematically measure, compare, and rank environmental problems. The process typically focuses on the risks to human health, the natural environment, and quality of life, and results in a list of issues ranked in terms of relative risk.

Compartment

A small space within a larger enclosed area, often separated by partitions.

Compass brick See Brick

Compass roof See Roof

Compass window See Window

Compass-headed roof See Roof

Complementary colors

Those pairs of colors, such as red and green, that together embrace the entire spectrum. The complement of one of the three primary colors is a mixture of the other two.

Complete street

A multimodal street that is designed and operated to enable safe access for all users. Pedestrians, bicyclists, motorists, and bus riders of all ages and abilities are able to safely move along and across a complete street.

Completed design area
The total area of finished ceilings, finished floors, full height walls and demountable partitions, interior doors, and built-in case goods in the space when the project is completed; exterior windows and exterior doors are not considered.

Completion The act of bringing a structure to the point of construction when it is physically ready for use and occupancy, legally considered a completed condition.

Complexity
Consisting of various parts united or connected together, formed by a combination of different elements; intricate, interconnecting parts that are not easily disentangled.

Compliance
Meeting legislated performance standards; "beyond compliance" means voluntarily exceeding those standards.

Compluvian
The opening in the center of the roof of the atrium in an ancient Roman house; it slopes inward to discharge rainwater into a cistern or tank in the center of the atrium.

Component
The various parts or materials that go together to form the elements of a building.

Composite arch See **Arch**

Composite capital See **Capital**

Composite column See **Column**

Composite construction
A type of construction made up of different materials; specifically structural steel and reinforced concrete designed as a single structural system.

Composite lumber
Lumber, typically decking, made from plastic, often high-density polyethylene, and wood fiber or other agricultural by-products. Composite lumber often contains recycled content. Also called composite decking.

Composite materials
A complex material made up of two or more complementary substances. They can be difficult to recycle. Plastic laminates are an example. Composite materials are best applied in situations where they can be removed for reuse, not requiring remanufacture.

Composite order
One of the five classical orders; a Roman elaboration of the Corinthian order; the acanthus leaves of its capitals are combined with the large volutes of the Ionic order and set on the diagonal in plan view.

Composite wood
Man-made wood is produced using adhesives to bind together the strands, particles, fibers, or veneers of wood to form a composite product. This wood can be "green" when

Composite wood
the wood fiber is provided by scrap or waste wood, when the adhesives are low or no VOC (volatile organic compound), and when the resulting product—often used for outdoor decks—eliminates the need for initial and periodic painting, staining, or waterproofing. The substitution of vegetable fibers from potentially sustainable sources, such as rye, wheat, and rice straw, for wood fibers can also result in a similar engineered green cellulosic product. Also known as engineered wood.

Composition board
A building board fabricated of wood fibers in a binder, compressed under pressure at an elevated temperature.

Composition See **Design**

Composition shingle
Any shingle made with a mixture of binder materials and fibers; types include asbestos shingle and asphalt shingle.

Compost
Process whereby organic wastes, including food wastes, paper, and yard wastes, decompose naturally, resulting in a product rich in minerals and ideal for gardening and farming as a soil conditioner, mulch, resurfacing material, or landfill cover

Composting
Controlled biological decomposition of organic material in the presence of air to form a humus-like material. Controlled methods of composting include mechanical mixing and aerating, ventilating the materials by dropping them through a vertical series of aerated chambers, or placing the compost in piles out in the open air and mixing it or turning it periodically.

Composting system
Outdoor bins for converting vegetable scraps, garden trimmings, and other plant matter into a rich, high-organic-content soil amendment.

Composting toilet systems
Dry plumbing fixtures that contain and treat human waste through microbiological processes.

Compound arch See **Arch**

Compound beam
A timber beam or rafter built up from a number of pieces either nailed, glued or bolted with connectors.

Compound pier
A pier that has several engaged shafts against its surface, used often in Romanesque and Gothic structures.

Compound vault See **Vault**

Compound wall
A wall constructed in two or more skins of different materials; e.g., a timber frame wall with brick veneer.

Comprehensive Environmental Response, Compensation, and Liability Act (CERCLA)
Commonly known as Superfund, CERCLA addresses abandoned or historical waste sites and contamination. It was enacted in 1980 to create a tax on the chemical and petroleum industries and provide federal authority to respond to releases of hazardous substances.

Compression
Direct pushing force, in line with the axis of the member: the opposite of tension.

Computer-aided design
The analysis and/or design, modeling, simulation, or layout of building design with the aid of a computer.

Concave
Forms that are curved like the inner surface of a hollow sphere or circular arc.

Concave mortar joint See **Mortar joint**

Concealed gutter
A gutter which is constructed in such a manner that it cannot be seen; also called a hidden gutter.

Concealed sprinkler
A sprinkler head that does not hang below the surface and is installed flush with the ceiling.

Concentrated load
A load acting on a very small area of the structure's surface; the exact opposite of a distributed load.

Concentric
Having a common center.

Concentric walls
Fortification consisting of one complete system of defense inside another.

Conception See **Design**

Conceptual architecture (1960–1993)
A form of architecture representing plans and drawings for buildings and cities never constructed. It can also be regarded as architecture arrested at the conceptual stage of development. The term also applied to work that could be realized but lacked the funds to construct it, or work done as an end in itself. It was also defined as a limitless activity devoid of dogma; that is, pure research or speculation.

Conch
Semidome vaulting of an apse or eastern end of a church.

Concourse
An open space where several roads or paths meet; an open space for accommodating large crowds in a building, such as in a railway terminal or airport.

Concrete
A composite artificial building material consisting of an aggregate of broken stone mixed with sand, water and cement to bind the entire mass; fluid and plastic when wet and hard and strong when dry.

cast-in-place concrete
The concrete that is deposited in the place where it will harden as an integral part of the structure, as opposed to precast concrete.

precast concrete
Material that reduces the need for onsite formwork with a process known as "tilt-up" construction, in which precast panels are lifted into a vertical position and then attached to the structural frame.

prestressed concrete
A process of anchoring steel rods into the ends of forms, then stretching them before the concrete is poured, putting them under tension. When the concrete hardens, they spring back to their original shape, providing additional strength.

Concrete block
A hollow concrete masonry unit, rectangular in shape, made from Portland cement and other aggregates.

decorative concrete block
A concrete masonry unit having special treatment of its exposed face for architectural effect; may consist of exposed aggregates or beveled recesses for a patterned.

Concrete frame
A structure consisting of concrete beams, girders, and columns, which are rigidly joined.

Concrete grille
An openwork barrier used to conceal, decorate, or protect an opening.

Concrete masonry See **Masonry**

Concrete masonry unit
Precast concrete block used to build walls. It has hollow cores that can be filled with concrete onsite for additional reinforcement. The use of stronger, more lightweight types of concrete such as autoclaved aerated concrete is popular in the manufacture of these units.

Concrete paint
A specially prepared thin paint, consisting of a mixture of cement and water, applied to the surface of a concrete wall to give it a uniform finish and to protect the joints against weathering.

Concrete panel
A panel that is precast and prefabricated elsewhere and placed in the structure, rather than cast in place.

Concrete shell
A curved thin membrane which is usually poured or sprayed over forms with a network of steel rods and wire mesh; most often a lightweight aggregate is used to decrease the weight-to-strength ratio.

Condemnation
A pronouncement by a legally constituted authority provided with police power, declaring a structure unfit for use or occupancy because of its threatened danger to persons or other property. Also, the judicial exercise of the right of eminent domain; taking over private property for public use, with just compensation to the owner.

Condensation
The formation of water on a surface, caused by the air temperature falling below its dew point. Most likely to occur in cool weather, when the temperature drops.

Condenser
HVAC component used to convert a vapor or gas to a liquid.

Condition assessment
The visual inspection of an structure to identify existing conditions prior to the design of a restoration, renovation, or adaptive reuse; may involve testing or monitoring of structural elements to determine the safety of existing members.

Conditioned air
The air that serves a space and has had its temperature and/or humidity altered to meet design specifications.

Conditioned space
The part of a building that is heated or cooled, or both, for the comfort of occupants.

Conditions of contract
A contract document that describes the rights and obligations of the client and contractor; may be original or adapted from a standard form of contract.

Condominium
An apartment house, office building or other multiple-unit complex; the units are individually owned, and there is joint ownership of common elements such as hallways, elevators, and all mechanical systems.

Conduction
Movement of heat through a material. R-value is a measure of resistance to conductive heat flow.

Conductive floor
A concrete floor deck with metal particles embedded in the surface to prevent sparking due to static electricity. Used to prevent explosions of gas in hospital operating rooms.

Conductor
Material capable of transmitting electricity, heat, or sound.

Conduit
Any round, tubelike enclosure to protect insulated electrical wires from external mechanical damage. Conductors are pulled through by fishtape and connected to outlet boxes and fixtures. Flexible conduit consists of a spiral of flexible interlocking aluminum through which wires can be run. Rigid conduit consists of straight sections of aluminum, steel, or plastic pipe joined by fittings, which can be bent into curved sections. Thin wall conduit consists of steel tubing with a thinner wall than rigid conduit, which can be used where damage potential is minimal.

Cone of vision See **Perspective projection**

Configuration
The form of a figure as determined by the arrangement of its parts, outline, or contour.

Conic sections
Circle, ellipse, parabola, and hyperbola; all produced by cutting a plane through a cone at different angles.

Conical
Pertaining to a cone shape, generated by rotating a right triangle around one of its legs.

Conical roof See **Roof**

Conical vault See **Vault**

Conifer See **Wood**

Conservation
Preserving and renewing, when possible, human and natural resources. The use, protection, and improvement of natural resources according to principles that will ensure their highest economic or social benefits.

Conservation
The management of a building to prevent its decay, destruction, misuse, or neglect; may include a record of the history of the building and the conservation measures applied.

Conservation archeology
The field of archeology concerned with limiting excavations to an absolute minimum consistent with research objectives, and with the preservation of architectural sites for future scientific investigation.

Conservation easement
The easement restricting a landowner to land uses that are compatible with long-term conservation and environmental values. Also, a preservation tool that may be used by a land trust or conservation group to limit development.

Conservation society
A local private organization gathered for a common preservation cause.

Conservation subdivision
A form of land regulation that permits flexibility of design in order to promote environmentally sensitive and efficient uses of the land. With land subdivided through a conservation subdivision regulation, local government can preserve unique or sensitive natural resources such as groundwater, floodplains, wetlands, streams, steep slopes, woodlands, and wildlife habitat. Conservation subdivisions enable clustering of houses and structures on less environmentally sensitive soils, which reduces the amount of infrastructure including paved surfaces and utility easements necessary for residential development.

Conservatory
A school for teaching music, drama, or other fine arts; a glass-enclosed room of a house for the cultivation and display of plants.

Console
A vertical decorative bracket in the form of a scroll, projecting from a wall to support a cornice, window, or a piece of sculpture.

Console table
A table attached to a wall and supported on consoles.

Constant air volume
A type of air-handling system that supplies air to a conditioned space at a constant air flow and modulates heating and cooling by varying the air temperature.

Constructed wetland
An engineered system that has been designed and constructed to utilize the natural processes involving wetland hydrology, vegetation, soils, and their associated microbial assemblages to assist in treating wastewater. It is designed to take advantage of many of the same processes that occur in natural wetlands, but within a more controlled environment.

Construction
The onsite work done in building or altering structures, from land clearance through completion, including excavation, erection and the assembly and installation of components and equipment.

Construction administration

The special management services performed by the architect or others during the construction phase of the project: under either a separate or special agreement with the owner.

Construction and demolition debris

Nonhazardous materials such as asphalt, concrete, brick, lumber, wallboard, roofing materials, ceramics, and plastics resulting from construction, deconstruction, remodeling, repair, cleanup, or demolition operations.

Construction and demolition waste

The waste building materials, dredging materials, tree stumps, and rubble resulting from construction, remodeling, repair, and demolition of homes, commercial buildings, and other structures and pavements. May contain lead, asbestos, or other hazardous substances.

Construction checklist

A construction checklist ensures that the specified equipment has been provided, is properly installed, and has initially started and been checked out adequately in preparation for full operation and functional testing.

Construction cost

The cost of all of the construction portions of a project, generally based upon the sum of the construction contracts and other direct construction costs. Not included is compensation for professional services, land, rights of way, or other costs, specified as the responsibility of the owner outlined in the contract.

Construction documents

The third phase of architectural basic services wherein the approved design development documents are used to prepare the working drawings, specifications and the bidding information for approval by the owner.

Construction IAQ management plan

A document specific to a building project that outlines measures to minimize indoor air quality contamination in the building during construction and to flush the building of contaminants prior to occupancy.

Construction joint See Joint

Construction loads

The loads imposed on the building during construction due to the erection, assembly, and installation of building components and equipment.

Construction loan

An interim loan of money secured by a negotiable bond or mortgage, or trust deed. The money obtained is intended to defray the cost of the building to be erected. Usually advanced in specified sums during the progress of construction. After completion, the loan is often converted into a long-term loan.

Construction manager

A construction manager in an organization has the role of managing the construction team and various contractors to

Construction manager

build and test the building systems for the project, represents the owner in taking bids of subcontractors and coordinating their activities, and administering all of the construction contracts for a fee or guaranteed maximum price. The construction manager also works with the commissioning authority to identify and correct deficiencies.

Construction observation

An onsite visit by the design architect or the engineer to determine if the construction materials and installations are in accordance with the construction documents and specifications, which includes review of testing reports.

Construction phase

The final phase of the architect's basic services, which includes the architect's general administration of the construction contract.

Construction Specifications Institute (CSI)

A national U.S. organization devoted to assisting specifications writers; publishes standard model specifications and a standardized format and numbering system.

Construction waste management

General term for strategies employed during construction and demolition to reduce the amount of waste and maximize reuse and recycling. Construction waste management is a sustainable building strategy in that it reduces the disposal of valuable resources, provides materials for reuse and recycling, and can promote community industries.

Construction waste management plan

A plan that diverts construction debris from landfills through the processes of recycling, salvaging, and reusing.

Constructivism (1920–1935)

A movement originating in Moscow based on order, logic, structure, abstraction and geometry, primarily in sculpture but with broad applications to architecture. An expression of construction was the base for all building design with emphasis on functional machine parts. Vladimir Tatlin's monument is the most notable example of this style. The industrial fantasies of Jacob Tchernikhov, published in 1933, show buildings perched on cantilevered structures, suggesting construction for construction's sake. The movement can be regarded as part of the broader movement of functionalism, with an accent on constructional expression. All traditional accessories, such as ornamental details, were discarded in favor of mass and space in relation to the new sculptural forms.

Consulate
A building where a consul conducts official business.

Consultant
A specialist who is consulted on building design and engineering; restoration and preservation specialists are types of consultants.

Consulting engineer
A person retained to give expert advice in regard to all engineering problems.

Contaminant
An unwanted airborne constituent that may reduce acceptability of the air.

Contamination
Introduction into water, air, and soil of micro-organisms, chemicals, toxic substances, wastes, or wastewater in a concentration that makes the medium unfit for its next intended use. Also applies to surfaces of objects, buildings, and various household and agricultural use products.

Contemporary style
A term loosely applied to any of a number of styles from the 1940s to the 1980s, often referring to "modern architecture."

Contextual
Any doctrine emphasizing the importance of the context in establishing the meaning of terms, such as the setting into which a building is placed, its site, its natural environment, or its neighborhood.

Contextualism
An approach to urban planning that considers the city in its totality, the view that the experience of a city is greater than the sum of its parts. According to proponents, all architecture must fit into, respond to, and mediate its immediate surroundings.

Contextualism

Contingency
A sum of money identified contractually to provide for unforeseen activities during the course of the building project.

Contingency planning
The process of anticipating and developing alternative actions, should a preferred course of action be disrupted or terminated.

Contingent valuation method
A method that attempts to objectively measure the dollar value of changes in environmental quality; often uses questionnaires and other surveys that ask people what they would pay for various environmental improvements.

Continuing professional development
A specific form of training that usually involves short seminar or classroom-style programs focused on updating the knowledge and skills of practicing industry professionals, as opposed to tertiary programs, which are of longer duration and lead to formal qualifications.

Continuous beam See **Beam**

Continuous commissioning
An ongoing program of structured commissioning throughout the lifetime of a building.

Continuous footing See **Footings**

Contour map
A topographic map that portrays relief by the use of contour lines which connect points of equal elevation; the closer the spacing of lines, the greater relative slope in that area.

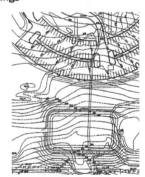

Contract administration
Architectural services that are commonly provided during construction of a project; these typically include construction observation, review of shop drawings and materials samples, processing change orders, and approving certificates of payment to the contractor.

Contract documents
The documents that a contract comprises, including plans and/or drawings, specifications, all addenda, modifications and changes, together with any other items stipulated as being specifically included.

Contractor
One who undertakes the performance of construction, provides labor and materials in accordance with plans and specifications, and contracts for a specific cost and schedule for completion of the work.

Contrapposto
The disposition of the human figure in which one part is turned in opposition to another, creating a counterpositioning of the body about its central axis.

Contrast
A juxtaposition of dissimilar elements to show the differences of form or color, or to set in opposition in order to emphasize the differences.

Control date
The date within the control period that marks the end of restoration efforts on a historic property.

Control joint See **Joint**

Control period
The duration of restoration work on a historic structure.

Controlled photography
Photography that yields distortion-free scaled photographs with perspective parallax control, most easily done with large format cameras, although there are perspective-control lenses for 35-mm cameras as well.

Controls
An instrument or set of instructions for operating or regulating building systems.

Convalescent home
A medical care institution for patients recovering from acute or postoperative conditions who do not require the skilled services provided by an extended-care facility, such as those services provided by a nursing home.

Convection
Movement of heat from one place to another by physically transferring heated fluid molecules, usually air or water. Natural convection is the natural movement of that heat; forced convection relies on fans or pumps.

Conventional irrigation
Refers to the most common irrigation system used in the region where a building is located. A common conventional irrigation system uses pressure to deliver water and distributes it through sprinkler heads above the ground.

Conventional power
The power produced from nonrenewable fuels such as coal, oil, nuclear, and gas, also known as traditional power.

Conversion
In adaptive reuse, change of use of a property, such as from a railroad station to a commercial facility.

Convex
Forms that have a surface or boundary that curves outward; as in the exterior or outer surface of a sphere.

Cook, Sir Peter (1936–)
Founder of Archigram, and former Director of the Institute of Contemporary Arts, and of the Bartlett School of Architecture at University College, all in London. Cook's achievements with radical experimentalist group Archigram have been the subject of numerous publications and public exhibitions and were recognized by the Royal Institute of British Architects in 2004, when members of the group were awarded the RIBA's highest award, the Royal Gold Medal. Cook was knighted in 2007 by the Queen for his services to architecture and teaching. He is also a Royal Academician and a Commandeur de l'Ordre des Arts et Lettres of the French Republic. He is a Senior Fellow of the Royal College of Art, London.

Cool beam bulb
A type of bulb that reflects visible light, but transmits infrared radiation so that heat is greatly reduced.

Cool color
Blue, green, and violet; they generally appear to recede.

Cool roof
A white roof reflects and emits the sun's heat back to the sky instead of transferring it to the building below.

Coolidge, Charles (1858–1936)
Partner in the firm Coolidge, Shepley, Bullfinch, and Abbott. He opened a Chicago office in 1892, which employed many of the colleagues who later worked for Frank Lloyd Wright.

Cooling load
The amount of heat generated within a building space from occupants, electrical equipment, artificial lighting, and solar radiation that the HVAC system must remove.

Cooling tower
A structure; usually located on the roof of a building, over which water is circulated so as to cool it by evaporation.

Cooling tower
Device that dissipates the heat from water-cooled systems by spraying the water through streams of rapidly moving air. Cooling towers can be substantial water users and, as such, present an opportunity for water conservation.

Cooling tower blow-down water
The water released from a cooling tower to maintain proper water mineral concentration.

Cooling tower makeup water
The water added to a cooling tower to replace water lost to evaporation or blow-down.

Coop Himmelb(l)au (1968–)
A member in a cooperative architectural design firm primarily located in Vienna and now also in Los Angeles and Guadalajara, founded by Wolf Prix, Helmut Swiczinsky, and Michael Holzer. The firm gained international acclaim alongside Peter Eisenman, Zaha Hadid, and Frank Gehry with the 1988 exhibition *Deconstructivist Architecture* at the Museum of Modern Art in New York City. Their work ranges from commercial buildings to residential projects.

Cooperative apartment
A unit in a building owned and managed by a nonprofit corporation that sells shares in the building, entitling the shareholders to occupy apartments in the building.

Cope, Walter (1860–1902)
American architect in Philadelphia, PA, partnered with John Stewardson (1858–1986), known for their campus designs in the Collegiate Gothic style. Designed buildings at Bryn Mawr College; University of Pennsylvania, Philadelphia, PA; Princeton University, Princeton, NJ; and Washington University, St. Louis, MO.

Coped
Cut to conform to the irregular outline of an abutting piece, such as where two moldings meet at an inside corner.

Coping
A protective covering over the top course of a wall or parapet, either flat or sloping on the upper surface to throw off water. If it extends beyond the wall, it may be cut with a drip to protect the wall surface below.

raking coping
A coping set on an inclined surface, as at a gable end.

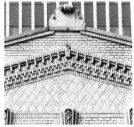

Coping course
The row of roofing tiles at the edge of a roof that forms the coping of the wall below.

Coping stone
The stone block that forms a coping to a wall.

Copper pipe
Small hollow pipe, connected using soldered joints; typically used for water supply due to its resistance to corrosion.

Copper See Metal

Copper wire
The primary metal used for electrical conductors, designated in standard wire gauges.

Coquillage
A representation of the forms of seashells used as a decorative carving over doors and windows, and in friezes and architraves.

Corbel
In masonry construction, a series of projections; each one stepped progressively outward from the vertical face of the wall as it rises up to support a cornice or overhanging member above.

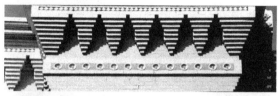

127

Corbel arch See Arch

Corbel table

A raised band composed of small arches resting on corbels; a projecting course of masonry supported on corbels near the top of a wall, such as a parapet or cornice.

Corbel vault See Vault

Corbeled chimney cap

The termination of a chimney using a series of successive courses of bricks that step out with the increasing height.

Corbeling

Masonry courses wherein each is extended out farther from the one below to form a rough arch-shaped lintel, vault, or dome.

Corbett, Harvey H. (1873–1954)

American architect who set up a studio for teaching young architects the methods of the École des Beaux-Arts. He was involved in the design of Rockefeller Center, New York City, and was named director of planning for the 1933 Chicago World's Fair, with Raymond Hood, Paul Cret, and Arthur Page Brown, Jr.

Corbiestep

Stepped edge of an incline that terminates a masonry gable end-wall, masking the surface of a pitched roof beyond; found in northern European masonry construction.

128

Corbiestep

Corinthian order

The most ornamental of the three orders of architecture used by the Greeks, characterized by a high base, pedestal, slender fluted shaft with fillets, ornate capitals using stylized acanthus leaves, and an elaborate cornice.

Cork

The lightweight elastic bark of the cork oak tree; used primarily to produce floor tile and sound insulation board.

Cork tile

Tile made from compressed cork; the tiles form a resilient floor covering with good insulating properties.

Corner

The position at which two lines or surfaces meet; the immediate exterior of the angle formed by the two lines or surfaces, as in the corner of a building or structure. The corner is one of the most important zones expressing the junction of two facades. Corners can take many forms such as recessed, rounded, retracted, framed, or stepped in shape. They can be angular, curved, or articulated in many different ways.

Corner

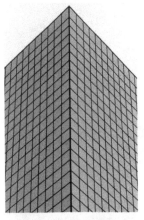

Corner bead
A vertical molding used to protect the external angle of two intersecting wall surfaces; a perforated metal strip used to strengthen and protect an external angle in plasterwork or gypsum wallboard construction.

Corner bead molding See **Molding**

Corner blocks
Wood blocks positioned at the top corners of either window or door casings; often enhanced with design elements, such as concentric oval disks.

Corner board
A board which is used as trim on the external corner of a wood frame structure and against which the ends of the siding are fitted.

Corner drop ornament See **Ornament**

Corner lot
A lot which abuts two or more streets at their point of intersection.

Cornerstone
A stone that is situated at a corner of a building uniting two intersecting walls, usually located near the base and often carrying information about the structure.

Cornice

The uppermost division of an entablature; a projecting shelf along the top of a wall supported by a series of brackets; the exterior trim at the meeting of a roof and wall, consisting of soffit, fascia and crown molding.

boxed cornice

A hollow cornice, built up of boards and moldings, resulting in a soffit under the eaves.

bracketed cornice

A deep cornice having large, widely spaced ornamental brackets supporting an overhanging eave; it is common in the Italianate style.

horizontal cornice

The level cornice of the pediment under the two inclined cornices.

modillion cornice

A cornice supported by a series of modillions, often found in Composite and Corinthian orders.

open cornice

Overhanging eaves where the rafters are exposed at the eaves and can be seen from below.

raking cornice

A cornice following the slope of a gable, pediment, or roof.

raking cornice

Cornice lighting
A method of concealing the light source behind a dropped cornice from the edge of the ceiling.

Cornice return
The extension of a cornice in a new direction, particularly where the raked cornice of a gable end returns a short distance in a horizontal direction.

Cornucopia ornament See **Ornament**

Corona
The overhanging vertical member of a classical cornice supported by the bed moldings and crowned by the cymatium, usually incorporating a drip.

Corporate economic development
A strategy by which local development organizations mobilize local resources for a multifaceted development campaign. Pursuing opportunities, risk taking in investment, innovation, and creativity are collective community strategies that promote development by a community for itself. Government programs may be used and corporate support attracted, but organizations representing the interests of the local community launch and direct the initiatives.

Corporate social responsibility
A business outlook that acknowledges responsibilities to stakeholders not traditionally considered, including suppliers, customers, and employees, as well as the local and international communities and their natural environments

Corporate social responsibility
in which the business operates. As a major part of society, businesses must become more accountable for every aspect of life, work, and the environment upon which their policies and practices have an impact.

Corrected design ventilation rate
Design ventilation rate divided by the air-change effectiveness.

Corridor
A narrow passageway or gallery connecting several rooms or apartments within a residence, school, hospital, office building or other structure.

Corrosion
The deterioration of marble or concrete by a chemical reaction resulting from exposure to weathering, moisture, chemicals, or other agents in the environment in which it is placed.

Corrugated
Forms that are shaped into folds of parallel and alternating ridges and valleys; either to provide additional strength or to vary the surface pattern.

Corrugated glass See **Glass**

Corrugated metal
Sheet metal that has been drawn or rolled into parallel ridges and furrows to provide additional mechanical strength; aluminum and galvanized sheet metal are the most widely used.

Cortile
An interior courtyard enclosed by the walls of a palace or other large building.

Costa, Lucio (1902–1998)

Brazilian architect and urban planner who reconciled traditional Brazilian forms and construction techniques with international modernism, particularly the work of Le Corbusier. His works include the Brazilian pavilion at the 1939 World's Fair in New York, designed with Oscar Niemeyer; the Parque Guinle residential complex in Rio de Janeiro; and the Hotel do Park São Clemente in Nova Friburgo (1948). Among his major works are also the Ministry of Education and Health, in Rio (1943), designed with Le Corbusier, Oscar Niemeyer, and Roberto Burle Marx, and the Pilot Plan of Brasília.

Cost-benefit analysis

A method of evaluating projects or investments by comparing the present value or annual value of expected benefits to costs; the practical embodiment of discounted cash flow analysis; a useful technique for making transparent the benefits of upfront investments in sustainable design features or technologies.

Cottage

A small rustic country house of the late eighteenth century.

Cottage Residence style (1840–1890)

A small Gothic style country house popularized in the 12 editions of *Cottage Residences: A Builder's Handbook* published by horticulturist Andrew Jackson Davis. The style was characterized as picturesque with pointed gables and arched windows including bargeboards; mostly built of wood, brick, and stucco.

Cotton

A renewable material that can be used in place of wood to manufacture paper. Cotton fiber papers tend to be stronger and more durable than wood-based papers and are known to last several hundred years without fading, discoloring, or deteriorating, making it an excellent eco-friendly alternative to wood fiber papers.

Cotton insulation

An insulation made from recycled cotton-textile trimmings that is treated with a nontoxic fire-retardant and sold in the form of batting sized to fit between framing studs.

Council on Environmental Quality

An independent U.S. federal agency that reviews environmental impact statements and oversees federal efforts to achieve the goals of the National Environmental Policy Act of 1969.

Counter

A horizontal work surface, display, or serving surface, such as in a store, in a restaurant, or on top of a kitchen cabinet.

Counter arch See Arch

Counter brace See Brace

Counterflashing

Waterproof material, such as sheet metal, installed on the face of a wall that overlaps and seals the top of the vertical flashing below.

Counterpoint

A contrasting but parallel element or theme.

interweaving counterpoint

The forms or elements are integrated, with each one being a part of the other.

overlapping counterpoint

The forms are in contact but are not connected to each other.

parallel counterpoint

The forms run together, but do not cross or interweave, as in bands running in the same direction.

Counterpoise

The disposition of the parts of the body so that the weight-bearing leg, or engaged leg, is distinguished from the raised leg, or free leg, resulting in a shift in the axis between the hips and shoulders.

Countersunk

Hardware installed in a recess that allows it to be flush with or below the surface on which it is installed.

Countertop

The top surface of a counter, installed above the cabinets; may be of wood, plastic laminate, or marble.

Countertop materials

The options for countertops made from renewable and recycled materials with low or no VOC (volatile organic compound) emissions are numerous and growing, including recycled glass, aluminum, and paper products. There are also simple and attractive concrete countertops.

Counterweight

A heavy component used to counterbalance the weight of a movable element; connected either with a cable over a pulley as in elevators or at one end of a lever as in a bascule bridge.

Coupled column See Column

Coupled window See Window

Course

A layer of masonry units running horizontally in a wall or over an arch that is bonded with mortar. The horizontal joints run the entire length; the vertical joints are broken so that no two form a continuous line.

Coursed masonry See Masonry

Coursed rubble See Masonry

Court

An open space about which a building or several buildings are grouped, completely or partially enclosing the space. They may be roofed over with glass or open to the air.

Courthouse

A building designed to contain one or more courtrooms and related facilities, such as judge's chambers, administrative offices, and jury rooms.

Courtyard

An open area within the confines of other structures, sometimes as a semipublic space.

Coussinet

The stone placed on the impost of a pier to receive the first stone of an arch.

Cove

Concave surface that connects a wall and ceiling.

Cove lighting

Fixtures installed in a wall or ceiling behind a concave molding that conceals them, but allows light to shine upward onto the ceiling. As an architectural detail built into the structure, the fixtures cannot be changed easily.

Cove molding See Molding

Coved base
A trim piece at the base of a wall forming a concave rounded intersection with the floor.

Coved ceiling
A ceiling having a cove at the wall line or elsewhere.

Coved eave
Eaves of a building that are enclosed with a concave curved surface, so that the rafters are not exposed.

Cover
In reinforced concrete, the thickness of concrete over-lying the steel bars nearest the surface. An adequate layer is needed to protect the reinforcement from rusting and from fire.

Cover molding See Molding

Covered way
A passageway connecting two buildings, or parts of a building that is roofed over, but may have open sides.

Cowl
A metal cover fitted with louvers, often capable of rotating, fixed on a roof ventilator or chimney to improve the natural ventilation or draft.

Cradle-to-cradle
Term used to describe the recycling of waste materials and manufactured products into new products rather than permanently disposing of them. The concept and its societal implications was the focus of the 2002 book *Cradle to Cradle: Remaking the Way We Make Things* by chemist Michael Braungart and architect William McDonough.

Cradle-to-cradle certification
Provides a company with a means to tangibly, credibly measure achievement in environmentally intelligent design and helps customers purchase and specify products that use environmentally safe and healthy materials; design for material reutilization that uses renewable energy and is energy efficient (e.g., recycling or composting); have efficient use of water; maximize water quality associated with production; and institute strategies for social responsibility.

Cradle-to-cradle design
At a fundamental level, the new paradigm proposes that human design can learn from nature to be effective, safe, enriching, and delightful. Cradle-to-cradle design models human industry on nature's processes, in which materials are viewed as nutrients circulating in healthy, safe metabolisms. Industry must protect and enrich ecosystems nature's biological metabolism while also maintaining a safe, productive technical metabolism for the high-quality use and circulation of minerals, synthetics, and other materials.

Cradle-to-cradle philosophy
A philosophy established by architect William McDonough based on the idea that products and the built environment should be designed in a closed system so that when they are no longer useful, they provide fuel for new products or natural cycles, eliminating waste. This framework seeks to create production techniques that are not just efficient, but essentially waste-free. In cradle-to-cradle production, all material inputs and outputs are seen either as technical or biological nutrients. Technical nutrients can be recycled or reused with no loss of quality, and biological nutrients can be composted or consumed.

Cradle-to-gate
Analysis of a partial product life cycle from manufacturer to the factory gate (before the product is transported to the consumer). The use and disposal phase of the product is omitted. Cradle-to-gate assessments are usually the basis for environmental product declarations in the products most people purchase for their homes.

Cradle-to-grave
With no consideration for sustainability, these types of products are used for a period of time and then discarded, often long before their useful life is actually complete

Cradle-to-grave analysis
An analysis of the impact of a product from the beginning of its source-gathering processes, through the end of its useful life, to disposal of all waste products. *Cradle-to-cradle* is a related term signifying the recycling or reuse of materials at the end of their first useful life.

Cradle-to-grave or manifest system
A term used in life-cycle analysis to describe the entire life of a material or product up to the point of disposal. Also refers to a system that handles a product from creation through disposal.

Craftsman style
A style of house that was popular in the early 1900s, influenced by the Arts and Crafts movement; Gustav Stickley popularized the style in his magazine, *The Craftsman*, from 1901 to 1916.

Cram, Ralph Adams (1863–1942)
A leading Gothic Revivalist in the United States; influenced by William Morris and John Ruskin.

Crane

A piece of construction machinery containing a mechanical device for lifting or lowering a load and moving it horizontally; the hoisting mechanism is an integral part of the boom.

Crane, Charles Howard (1885–1952)
Crane specialized in the design of movie palaces in North America, including 250 theaters in total, more than 50 of them in the Detroit area. His 5,174-seat Detroit Fox Theatre was the largest of the Fox Theatres. Crane designed many cinemas across Britain, but with much tamer designs than his American movie palaces.

Crawl space
The space under a suspended floor needed for access to services.

Crawling
In painting, a defect that appears during the painting process, where the film breaks, separates, or raises, as a result of applying the paint over a slick or glossy surface.

Crazing
In painting, a minute random cracking of a finish coat of paint due to uneven shrinking of the paint. In masonry, the appearance of very fine cracks while the surface is drying due to uneven contraction.

Creativity
Ability or power to create: to bring into existence, to invest with a new form, to produce through imaginative skill, or to make or bring into existence something new.

Credit
LEED (Leadership in Energy and Environmental Design) Green Building Rating System component. Compliance is optional, and meeting credit criteria results in earning points toward certification.

Credit interpretation ruling
Used by design team members experiencing difficulties in the application of a LEED (Leadership in Energy and Environmental Design) prerequisite or credit to a project. Typically, difficulties arise when specific issues are not directly addressed by LEED information or guides, or a special conflict exists that requires resolution.

Crenel
The open space between the solid members of a battlement producing a pattern of repeated and identical indentations.

Crenelet
A small crenel, used as a decorative design.

Crenellated molding See **Molding**

Crenellation
A pattern of repeated depressed openings in a fortification parapet wall.

Creosote
A distillate of coal tar, used as a wood preservative.

Creosoting
The process of injecting creosote into timber as a means of increasing its durability when it is to be exposed to wetting and drying.

Crescent
Shape similar to the visible part of the moon in its first or last quarter; also the shape of a row of townhouse fronts constructed in the approximate shape of a circular or elliptical arc.

Crescent arch See **Arch**

Crespidoma
The solid mass of masonry at the base; forming the stepped platform upon which a classical temple is constructed.

Crest
Ornament on a roof, a roof screen or wall, which is frequently perforated, and consists of rhythmic and identical decorative patterns.

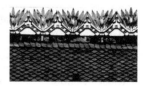

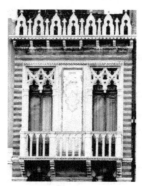

Crest tile See **Tile**

Cret, Paul Phillippe (1876–1945)
An American Modern Classicist architect and educator in Philadelphia. Designed the Indianapolis Public Library in 1917, and the Detroit Institute of Arts in 1927, with Zantzinger, Borie and Medary, and the Folger Shakespeare Library, Washington, DC, in 1932.

Cricket
A small element with two slopes, in the form of a miniature gable roof, placed behind a chimney that penetrates a sloping roof, to shed water.

Cripple
Any member shorter than most of the others in a structure, such as a stud beneath a window.

Cripple studs
The studs in a wall system that support headers above (and below) windows or doors; indiscriminately placed, these additional studs can result in extra heat loss because they do not insulate as well as the insulation in the wall cavity.

Cripple window See **Window**

Criteria pollutants
A list of air pollutants identified in the 1970 Clean Air Act Amendments deemed to be critical in controlling air pollution and for which National Ambient Air Quality Standards (NAAQS) were established. Criteria pollutants include sulfur dioxide (SO_2), nitrogen dioxide (NO_2), volatile organic compounds (VOCs), particulate matter, carbon monoxide (CO), and lead (Pb).

Critical path method (CPM)
A construction scheduling device that diagrams the interrelationships between activities and identifies the critical path that optimizes the sequence of operations to minimize the construction period.

Critical root zone
The area of undisturbed natural soil around a tree defined by a concentric circle. A code-based standard for site design and/or construction practices, used in site plan review by planning departments of municipal and county governments. Also called tree protection zone.

Critical zone
Any location in a building with contaminant sources sufficiently strong enough that proper control of ventilation, with no margin for error, is crucial for maintaining the immediate comfort of occupants. Critical zones may include conference rooms, smoking rooms, cafeterias, washrooms, auditoriums, or anywhere occupancy can rapidly change.

Crocket capital See **Capital**

Crocket ornament See **Ornament**

Cross
Two lines intersecting each other at right angles so that the four arms are of equal length.

Cross bond See **Bond**

Cross bracing See **Brace**

Cross vault See **Vault**

Cross ventilation
The technique of using natural air movement from the out-

Cross ventilation
side and drawing it inside without the aid of ventilation systems to cool buildings. Positioning windows in line with each other on opposite walls will create the maximum air flow and cooling effect.

Crossbeam See **Beam**

Crossette
A lateral projection of the architrave moldings of Classical doors and windows at the extremities of the lintel or head; a small projecting part of an arch stone, which hangs upon an adjacent stone.

Crossing
The square space of a cruciform church, created by the intersection of the nave and chancel with the transept. Intersection of two elements in the form of a cross, such as the ridges of a cross gable.

Crossing square
The area in a church formed by the intersection of a nave and a transept of equal width.

Cross-linked polyethylene
Specialized type of polyethylene plastic that is strengthened by cross-linking, that is, chemical bonds formed in addition to the usual bonds in the polymerization process. Primarily used as tubing for hot and cold water distribution and radiant-floor heating.

Cross-sectional area
The area of a section cut transversely to the longitudinal axis of a member.

Crown
Any uppermost or terminal features in architecture; the top of an arch including the keystone; the corona of a cornice, often including the elements above it.

Crown

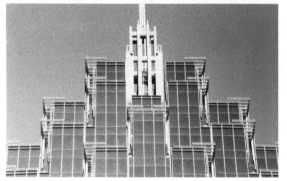

Crown glass
An early form of window glass, cut from blown disks.

Crown molding See Molding

Crown post
Vertical member in a roof truss, especially a king post.

Crowstep gable See Gable

Crowstone
The top stone of the stepped edge of a gable.

Cruciform
The characteristic cross-shaped plan for Gothic and other large churches that is formed by the intersection of nave, chancel, and apse with the transepts.

Cruck
One of a pair of large curved structural timbers, forming the wall posts and roof rafters of timber-framed houses; they are joined at the top of the frame where they support a ridge beam.

Crypt
A story in a church, below or partly below ground level, and under the main floor, often containing chapels and sometimes tombs; a hidden subterranean chamber or complex of chambers and passages.

Crypta
In ancient Roman architecture, a long, narrow vault sometimes below ground level, for the storage of grain.

Crystalline
A three-dimensional structure consisting of periodically repeated, identically constituted, congruent unit cells; found abundantly in natural objects.

138

Cube
A solid figure, bounded by six squares, and hence also called a hexahedron.

Cubic feet per minute
A measure of the volume of a substance flowing through air within a fixed period of time. With regard to indoor air, refers to the amount of air, in cubic feet, that is exchanged with outdoor air in a minute's time; i.e., the air exchange rate.

Cubit
The principal measure of length used in Ancient Egypt, Babylon, Israel and Greece. It was based on the length of the forearm, and varied from 525 to 445 mm. This forearm measure was called "braccia" in medieval and Renaissance Italy, and varied from city to city.

Cul-de-sac
A street, lane, or alley that is closed at one end; usually having an enlarged, somewhat circular area for turning around.

Cull
Any building material rejected as being below standard quality.

Cullet
Crushed waste glass that is returned for recycling. Also, large irregular chunks of glass, produced by the slag from a glass furnace. When the glass cools, it is chipped off the sides of the furnace and later used as a catalyst, in the mixture of the new batch.

Cummings, Charles A. (1833–1905)
American architect who, with partner William T. Sears, secured the commission for the New Old South Church, Boston, MA, in 1871.

Cuneiform
Designs having a wedge-shaped form; especially applied to characters, or to the inscriptions in such characters, of the ancient Mesopotamians and Persians.

Cupola
A tower-like device rising from the roof, usually terminating in a miniature dome or turret with a lantern or windows to let light in.

Cupola

Curb roof See Roof

Curing of concrete
The maintenance of an appropriate humidity and temperature in freshly placed concrete to ensure the satisfactory hydration of the cement, and proper hardening of the concrete. In cold weather it may be necessary to provide a source of heat, such as a salamander or electric heater.

Current loop
In electrical wiring, a situation in which separation of hot and neutral leads results in higher than normal electromagnetic fields.

Curtail
A spiral scroll-like termination of any architectural member, as at the end of a stair rail.

Curtail step See Step

Curtain truss
Nonstructural truss that extends from a structural wall system solely for the purpose of holding cavity-fill insulation

139

Curtain truss

known also as a Larson truss. Often used on timber-frame houses and in super-insulation retrofits, curtain trusses may be as much as 12 inches deep, providing an insulating value greater than R-40. Since they aren't structural, curtain trusses are often constructed from 2×2s with plywood reinforcement flanges to minimize wood use.

Curtain wall See Wall.

Curtain wall addition
Protection of a building from air and water infiltration. A glass curtain wall allows daylight in while reducing infiltration. A glass curtain wall can be fitted with operable windows or vents.

Curvilinear
Forms that are bounded by or characterized by curved lines, whether geometric or free-flowing.

Curvilinear

Curvilinear tracery See Tracery

Curvilles, Francois (1695–1768)
Belgian-born architect who became the principal exponent of the Rococo style in Southern Germany; the style combined fantastic exuberance with delicacy and elegance.

Cushion capital See Capital

Cusp
The intersection of two arcs or foliations in a tracery; the figure formed by the intersection of tracery arcs or foliations.

Cusped arch See Arch

Cuspidation
Any system of ornamentation that consists of, or contains cusps.

Cuspidation

Cut and fill
Excavating soil from one area of a site and depositing it on a different area, to change the grades in accordance with a landscaping or foundation plan.

Cut glass
A glass which has been decorated by grinding figures or patterns on its surface by abrasive means, followed by polishing it.

Cut stone
Any stone cut or machined to a specified size and shape to conform to drawings, for installation in a designated place; it can also be carved by the intaglio method.

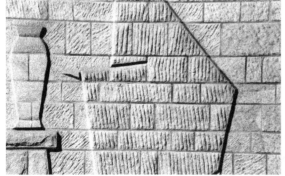

Cutaway corner
A corner formed by the meeting of three wall surfaces; produced by one short surface cutting across on the diagonal; often embellished with corner brackets.

Cutaway See Projection drawing

Cutoff angle
The critical viewing angle beyond which a source can no longer be seen because of an obstruction such as a baffle or overhang.

Cyclopean masonry See Masonry

Cylinder lock
A lock with a central cylinder, which rotates when the key lifts the internal tumblers.

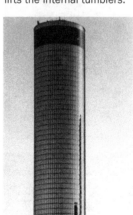

Cylindrical
Having the shape of a cylinder, generated by rotating a rectangle around one of its sides.

Cylindrical

Cylindrical shell

A roof structure that forms part of a cylinder; its cross section is generally a circular arc, though elliptical and catenary cylindrical shells have been built.

Cyma molding See **Molding**

Cyma recta molding See **Molding**

Cyma reversa molding See **Molding**

Cymatium molding See **Molding**

Cypress See **Wood**

D d

Dado
A rectangular groove cut across the full width of a piece of wood to receive the end of another piece.

Dagoba
In Buddhist architecture, a monumental structure containing relics of Buddha or a Buddhist saint.

Dakin, James H. (1806–1852)
American architect, associated with Town and Davis. Specialized in in Greek, Gothic, and Egyptian Revival works; New York University, New York City, in 1837; Bank of Louisville, Louisville, KY, in 1836; and Louisiana State Capitol, Baton Rouge, LA, in 1852.

Damper
An adjustable metal plate inside a flue or air duct, which regulates the draft of a stove, fireplace, or furnace, or controls the flow of air inside a duct; operates automatically in the event of a fire to close off the duct.

Dampproofing
The special preparation of a wall to prevent moisture from oozing through it; material used for this purpose must be impervious to moisture.

Dart
The vertical element shaped like an arrow in between the egg-shaped motif in the egg-and-dart molding.

Date of substantial completion
The date certified by the architect when the work or a designated portion of it is sufficiently complete, in accordance with the contract documents, so that the owner may occupy it.

Date stone
A stone that is carved with the date of completion of the structure and embedded in the walls; found in many colonial buildings.

Datum
A point with a given coordinate in space from which other heights and depths can be measured.

Datum level
A basic level or line used as a reference for determining heights and depths of points or surfaces in building construction. Within any given city, the established datum level is recorded in the building codes, which control the building standards for that city.

Daubing See Plaster

Davis, Alexander (1803–1892)
American architect who produced mostly Greek Revival style buildings.

Davis, Eldon Carlyle (1917–2011)
American architect, considered largely responsible for the creation of Googie architecture, a form of modern architecture originating in Southern California, largely influenced by that region's car culture and the Space Age of the mid-20th century. One of the best preserved examples of Davis's work is the Pann's coffee shop and its neon sign in the Westchester neighborhood of Los Angeles, CA. Davis also designed the early prototypes for local Big Boy and Denny's restaurants in Los Angeles.

Day, Frank Miles (1861–1918)
American architect; won the competition for the Philadelphia Arts Club, Philadelphia, 1887.

Daylight factor
The ratio of interior illuminance at a given point on a given plane, usually the workplane, to the exterior illuminance under known overcast sky conditions. This is one of the key values when analyzing the quantitative aspects of daylighting. Since the outside illuminance varies a lot with weather conditions (8,000–25,000 lux), the interior illuminance alone does not provide much useful information. The acceptability of the amount of daylight reaching an interior space for a specific task can only be determined in relation to the situation outside. There is little use in computing the relation of outside and inside illuminances under sunny sky conditions.

Daylight harvesting
Uses available daylight to reduce the amount of electrical light used in a room. Using a combination of windows, skylights, and an automatic control, electric lights are turned off when there is enough daylight to sufficiently light the room and turned on when the daylight levels are too low.

Daylighting
Using natural light in an interior space to substitute for artificial light. Daylighting is considered a sustainable building strategy in that it can reduce reliance on artificial light and reduce energy use in the process, and when well designed, it contributes to occupant comfort and performance. Common daylighting strategies include the proper orientation and placement of windows, use of light wells, light shafts or tubes, skylights, clerestory windows, light shelves, reflective surfaces, and shading, and the use of interior glazing to allow light into adjacent spaces. South-facing windows are most advantageous for daylighting and for moderating seasonal temperatures. They allow the most winter sunlight into the home but little direct sun during the summer.

De Stijl

Term meaning "The Style," derived from the name of a group of Dutch artists and the journal founded by the painter Theo van Doesberg in 1917; other members of the group included Piet Mondrian, Reitveld, and Oud. It was influenced by Cubism, and proposed an abstracted expression divorced from nature; instead, advocating straight lines, pure planes, right angles, and primary colors. It had a profound influence on the Bauhaus movement.

Dead bolt

A bolt with a square head controlled directly by the key when moved in either direction.

Dead load

The weight of all permanent and stationary construction materials or equipment in a building. See also live load.

Deadening

Installing materials in a building that will inhibit the transfer of sound waves through the use of insulation and other construction techniques, such as discontinuous construction of walls and floors.

Deck

Substrate over which roofing is applied. Usually plywood, wood boards, or planks.

Decking

Thick floor boards or planks used as a structural flooring, usually for long spans between joists. Also, light-gauge sheets of metal, which are ribbed, fluted or stiffened for use in construction of concrete floors and roofs.

Deconstruction

A process to carefully dismantle or remove usable materials from structures, as an alternative to demolition. It maximizes the recovery of valuable building materials for reuse and recycling and minimizes the amount of waste going to a landfill. Deconstruction options may include reusing the entire building by remodeling; moving the structure to a new location; or taking the building apart to reuse lumber, windows, doors, and other materials.

Deconstructivism (1984–)

An architectural style known as "Neomodernism," or "Poststructuralism," It takes many of its forms from the work of the Constructivists of the 1920s, such as Tchernikhov and Leonidou. It takes modernist abstraction to an extreme and exaggerates already known motifs. It is an antisocial architecture, based on intellectual abstraction. Some examples are Bernard Tschumi's designs for the Parc de la Villette, Paris; Peter Eisenman's Wexner Center for the Visual Arts, Ohio; and work by Frank Gehry, Architectonica, SITE, and Morphosis.

Deconstructivism

Decor

The combination of materials, furnishings, and objects used in interior decoration to create an atmosphere or style.

Decorated style (1280–1350)

The second of three phases of English Gothic was characterized by rich decoration and geometric tracery and by the use of multiple ribs in the vaulting. The earliest development was geometric, while the later forms were curvilinear, with complicated rib vaulting and naturalistic carved foliage that displayed a refinement of stonecutting techniques.

Decoration

A treatment applied to the surface of a structure with the intent of enhancing its beauty; includes gilding, stenciling,

Decorative concrete block

A concrete masonry unit having special treatment of its exposed face for architectural effect, which may consist of exposed aggregates or beveled recesses for a patterned appearance, especially when Illuminated obliquely.

Decorative glass See **Glass**

Decorator
Person who performs the interior decoration of residences or commercial offices.

Dedication
Ceremony that officially begins the occupation and use of a building; also applies to the donation of land to the public for development of a tract of land, such as a road or other form of easement.

Deed restriction
A statement included in the deed to a piece of land, placing limitations upon the use of the property.

Defect
In lumber, an irregularity occurring in or on wood that will tend to impair its strength, durability, or utility value.

Deflection
The deformation or displacement of a structural member as a result of loads acting on it.

Deforestation
To cut down and clear away the trees or forests from an area.

Deformation
An act of deforming or changing the shape or an alteration in form that a structure undergoes when subjected to the action of a weight or load.

Deformed bar
A reinforcing bar made with a pattern of protruding ridges to produce a better bond between the bar and the concrete.

Degradation
A loss of the original characteristics, or weakening of an element by erosion; a disintegration of paint by heat, moisture, sunlight or natural weathering. Also harmful action caused by human activity, such as vandalism.

Degree-day
Measure of how cold or warm a location is over a period of time relative to a base temperature, typically 65 degrees F (although other base temperatures, such as 75 degrees F, can be used for cooling). To calculate the number of heating degree-days (HDDs) of a given day, average the maximum and minimum outdoor temperatures and subtract that from 65 degrees F. The annual number of HDDs is a measure of the severity of the climate and is used to determine expected fuel use for heating. Cooling degree-days (CDDs), which measure air-conditioning requirements, are calculated by subtracting the average outdoor temperature from an indoor base temperature.

Dehumidifier
A mechanical device that removes water vapor in the air inside a building, by lowering it below the dew point.

De-inking
The process of removing ink, dyes, and other contaminants from paper so the pulp can be recycled.

Delamination
Coming apart layer by layer; as in a separation of plies in a plywood panel, either through failure of the adhesive or through failure at the interface of the adhesive and the lamination.

Delano, William Adams (1903–1941)
American architect with Chester Holmes Aldrich (1871–1940); designed estates for wealthy owners. Also designed the Walters Art Gallery in Baltimore, in 1910; John D. Rockefeller House, Pocantico Hills, NY, in 1908; and La Guardia Air Terminal in Queens, NY, 1940.

Delivered product
Green power fed into the same electric transmission and distribution system that serves the end user.

Delta-T
Difference in temperature across a divider; often used to refer to the difference between indoor and outdoor temperatures.

Demand control ventilation
The ventilation provided in response to the actual number of occupants in a building and their activity.

Demand controlled ventilation
Outdoor air-flow rate is determined by CO_2 monitors within occupied spaces.

Demand hot-water system
Hot-water heaters designed to provide instantaneous hot water, rather than storing preheated hot water in a tank. Such devices can serve an entire home, or be point-of-use, serving an individual water use. Benefits include elimination of standby losses, or energy wasted keeping stored water warm. With point-of-use devices, there is reduction or elimination of water wasted while waiting for water to get warm, as well as conductive losses as water travels through pipes. Electric-demand systems tend to use a large amount of energy; gas-fired units with standing pilot lights lose much of their efficiency due to the ongoing pilot light.

Demand limit controller
The way a demand controller controls works is called load control strategy. It is the definition of each load's importance in relation to all other loads being controlled by the system. Generally, there are three load control strategies: fixed, rotating, and combination.

Demand management
Lighting can be gradually dimmed up to 20 percent with little effect on productivity, but a profound impact on the overall building load. By sensing incoming electric service for peak loads, lighting can be dimmed when other building systems are peaking in load. The result is a flattening of the energy use curve, which lowers energy cost usually computed using peak demand volumes.

Demand water heater
Water heater that heats water only as needed; there is no storage tank and thus no standby heat loss. Also called a tankless water heater.

DeMars, Vernon (1908-2005)
American architect who was apprenticed to Frank Lloyd Wright; set up his own practice in the San Francisco Bay area, and designed several buildings on the UC Berkeley campus.

Demolition
The intentional destruction of all or a part of a structure; may include removal of structural elements, partitions, mechanical equipment, and electrical equipment and wiring.

Demolition by neglect
The exact opposite of preservation by maintenance; any building or site that is not taken care of on a regular basis is a potential candidate for the eventual disuse, disrepair, and ultimate need for demolition.

Demolition delay
A temporary halt or stay in the planned razing of a property, sometimes resulting from a court injunction obtained by preservationists to allow a period of negotiation.

Demolition permit
Written legal authorization by the appropriate building authority to proceed with demolishing any or all of a structure.

Demonstration Cities and Metropolitan Development Act of 1966
Also called the Model Cities Act; sets up urban renewal demonstration grants to make it possible for communities to identify, acquire, and restore historic structures and sites and allocates funds to assist those relocated due to urban renewal activity.

Denat de Gutllebon, Jeanne-Claude (1935-2009)
The wife of conceptual artist Christo Jayacheff. The pair is responsible for some of the largest, most dramatic public art installations in the world, including covering Germany's Reichstag building in fabric, "mummifying" the Pont Neuf in Paris, surrounding a string of islands off the coast of Honda in pink nylon, and erecting 7,500 fabric gates in New York's Central Park (illus.). Their temporary installations often relied on millions of yards of fabric and a permitting process that could take years, if it happened at all. Together they completed 19 projects; 37 had to be abandoned because permission was refused. The larger the project, the more publicity it generated. Six of their projects were the subject of documentary films by Albert Maysles, including "Christo's Valley Curtain" in 1974, which was nominated for an Oscar. (See also Christo.)

Densitometer
A photometer for measuring the optical density (the opposite of transmittance) of materials.

Density
A planning or zoning unit of measurement of the ratio between buildings per acre, or occupants per gross square foot of floor area, according to the type of zoning for that particular area under consideration, such as commercial residential, rural, and the like.

Dentil
A series of closely spaced ornamental rectangular blocks resembling teeth, used as moldings; most often found in continuous bands just below the cornice.

146

Dentil

Dentil band
A plain and uncarved band; occupying the position in a cornice where dentils would normally occur.

Department of Energy (DOE)
A department of the U.S. government responsible for energy policy and nuclear safety, including setting industry efficiency standards and monitoring the consumption of energy sources.

Depreciated property
Real estate used in business typically depreciates over the years, unless improved along the way, due to wear and tear, and changing development patterns.

Depreciation
The reduction in the value or worth of an asset, such as a building, through physical deterioration over time, and general obsolescence.

Depressed arch See Arch

Depression Modern style (1935–1945)
Designs from the decade of the Great Depression are represented in this style, which marked a reaction against the Art Deco style. It was characterized by simplicity, smoothness of forms, clarity of line, horizontality, stream-lining and functional expressiveness.

Depressurization
A condition that occurs when the air pressure inside a structure is lower that the air pressure outdoors. Depres-surization can occur when household appliances that consume or exhaust house air, such as fireplaces or fur-naces, are not supplied with enough makeup air. Radon may be drawn into a house more rapidly under depressur-ized conditions. Backdrafting of furnaces and vented appliances can also occur with depressurization, introduc-ing exhaust gases into the house.

Depth
The extent, measurement or distance from top to bottom (downward) or from front to back (inward), or an element consisting of several layers.

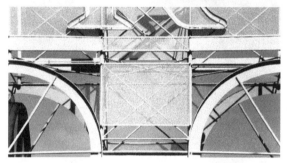

Derelict
Land or buildings that have been deserted or abandoned, or damaged by serious neglect or other processes, which in their existing state are unsightly and incapable of use without restoration or renovation or other treatment.

Desertification
The spread of desert-like conditions due to human exploi-tation and misuse of the land.

Design
To compose a plan for a building; the architectural con-cept of a building as represented by plans, elevations, renderings, and other drawings; any visual concept of a constructed object, as of a work of art.

charrette
The intense effort to com-plete an academic architec-tural problem within a specified time; from the French word meaning the "cart" that was used to carry the student work at the École des Beaux-Arts to be judged. See also charrette process.

composition

The combining of various elements into proper position; to form an entity in terms of structure or organization.

conception

A drawing of something that does not yet exist.

image

Any representation of form or features, especially one of the entire figure of a person; a statue, effigy, bust, relief, or intaglio.

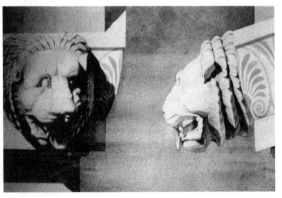

isometric drawing

A three-dimensional projection in which all planes are drawn parallel to the corresponding axes at true dimensions; all the horizontals are at 30 degrees from the normal horizontal axis; verticals are still parallel to the vertical axis.

model

A representation or reproduction, usually at a small scale: for studying or to illustrate construction.

perspective drawing

A graphic representation of a project or portion thereof, as it would appear in three dimensions.

mock-up

A model of an object in the course of design, as in a cross section of a window or its parts; built to scale or full size; for studying construction details, judging its appearance, and/or testing performance.

parti

Any scheme or concept for the design of a building that is represented by a diagram.

preliminary drawing
Drawings prepared during the early stages of the design phase of a project.

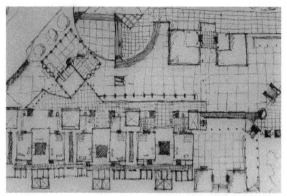

presentation drawing
Any of a set of design drawings made to articulate and communicate a design concept or proposal; such as for an exhibition, review, or publication.

Design build
A method of contracting whereby the building contractor provides both design and construction services for the client.

Design controls
Regulations by local governments regarding alterations to existing structures in historic towns or districts; usually restricted to the exterior use of materials and overall design style.

Design development
The second phase of the architect's basic services; drawings that describe the character of the project as to structural, mechanical, and electrical systems; materials and all other essentials, and probable construction costs.

Design document
The design document records all the details of the design intent.

Design drawing
Any of the drawings made to aid in the visualization, exploration, and evaluation of a concept in the design process.

cartoon
A drawing or painting made as a detailed model of an architectural embellishment, often full-scale, to be transferred in preparation for a fresco, mosaic or tapestry.

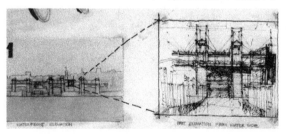

collage
An artistic composition of often diverse materials and objects in unlikely or unexpected juxtaposition, which are pasted over a surface; often with unifying lines and color.

detail
A small or secondary part of a painting, statue, building, or other work of art, especially when considered or represented in isolation.

diagram
A plan, sketch, drawing, chart or graph, not necessarily representational, that explains, demonstrates or clarifies the arrangement and relationship of the parts to a whole.

diagram

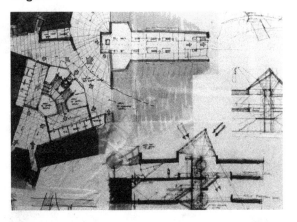

draft
A preliminary sketch of a design or plan, especially one executed with the idea of potential revision or refinement.

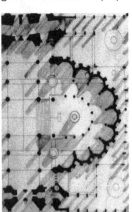

esquisse
A first sketch or very rough design drawing showing the general features of a proposed project.

rendering
A drawing, especially a perspective of a building or interior space, artistically delineating materials, shades and shadows, done for the purpose of presentation and persuasion.

scheme
The basic arrangement of an architectural composition; a preliminary sketch for a design.

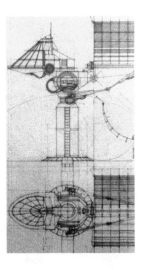

sketch
A rough drawing that represents the main features of a plan or building; used as a preliminary study.

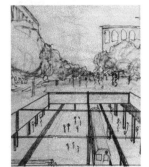

study
A drawing executed as an educational exercise, produced as a preliminary to a final work or made record observations.

vignette
A drawing that is shaded off gradually into the surrounding background so as to leave no definite line at the border.

Design for disassembly
A design that facilitates future change and the eventual dismantlement, in part or whole, for recovery of systems, components, and materials. This design process includes developing the assemblies, components, materials, construction techniques, and information and management systems to accomplish this goal.

Design for environment
An engineering perspective in which the environmentally related characteristics of a product, process, or facility design are optimized.

Design for the environment
A design concept that focuses on reducing environmental and human health impacts through thoughtful design strategies and careful selection of materials.

Design guideline recommendations
Recommendations that are typically adopted and published by local regulating agencies, which control new construction, additions, and alterations in historic towns or districts.

Design guidelines
Criteria developed by preservation commissions to identify design concerns in an area and help property owners ensure that rehabilitation and new construction respect the character of designated buildings or districts.

Design intent
A detailed technical description of the ideas, concepts, and criteria defined by the building owner to be important. It should include the functional and environmental needs of the facility. The design intent is developed by the design team from descriptions provided by the building owner.

Design professionals
Design professionals are the architects, engineers, or other parties responsible for the design and preparation of documents for the various building systems.

Design review
The process of determining whether modifications to historic structures or settings meet standards of appropriateness established by a review board.

Design temperature
Expected minimum or maximum temperature for a particular area; used to size heating and cooling equipment.

Design-build firm
Company that handles house design and construction. Since both services are provided by the same firm, integrated design can often be more easily achieved.

Designer
The person who draws, lays out, or prepares a design.

Destruction
The partial or complete loss of a structure, generally connoting a sudden or unplanned occurrence such as a fire, earthquake, or accident, as opposed to overt demolition.

Destructive testing
Testing materials by methods that destroy them in the process, such as compression tests of concrete and loading structures to failure.

Desuperheater
Device that takes waste heat extracted by heat pumps or air conditioners and uses it to heat domestic hot water.

Detail See **Design drawing**

Detention
In stormwater management, ponding of runoff in pools and basins for water-quality improvement and flood prevention.

Deterioration
A worsening of a structure's condition, generally attributable to exposure, weather, normal wear and tear, aging, or lack of maintenance.

Developer
A person or organization that controls and manages the process of the construction of buildings or other facilities by the arrangement of financing for the project, hiring the architect or contractor, obtaining zoning, regulatory approvals, and building permits, and that leases, sells, or manages the completed property.

Development
The process of improving property through the construction of roads, sewers, electrical service, and construction of residential, commercial or industrial buildings.

Development footprint
The area on the project site that has been impacted by any development activity. Hardscape, access roads, parking lots, nonbuilding facilities, and building structure are all included in the development footprint.

Development potential
The projected future use of a property compared to the existing use.

Devrouax, Paul (1943–2010)
With business partner Marshall Purnell, founded the firm of Devrouax+Purnell, which was one of the largest black-owned firms in Washington. Their projects included the Walter E. Washington Convention Center, Washington, DC, the Pepco Building, Arlington, VA, the Frank D. Reeves Municipal Center, Washington, DC, and Nationals Park with HOK Sport + Venue + Event. He had also been a past president of the National Organization of Minority Architects.

Diaglyph See **Relief**

Diagonal
Joins two nonadjacent sides of a polygon with a slanted or oblique direction from one corner to the other; their use in a square or rectangle produces two triangular shapes.

Diagonal bond See **Bond**

Diagonal brace See **Brace**

Diagonal buttress See **Buttress**

Diagonal chimney stack
A group of chimney stacks that are square in cross section and set on a diagonal alignment; they are often corbeled and joined at the top.

Diagonal compressive stress
One of the stresses that results from the combination of horizontal and vertical shear stresses in a beam.

Diagonal rib See **Rib**

Diagonal sheathing
A covering of wood boards placed over exterior studding at a diagonal with respect to the vertical; provides a base for the application of wall cladding.

Diagonal tensile stress
One of the stresses that results from the combination of horizontal and vertical shear stresses in a beam or slab.

Diagram See **Design drawing**

Diamond fret
A molding that is usually continuous, consisting of fillets that intersect to form a diamond shape, or rhombus

Diamondwork masonry See **Masonry**

Diaper patterns
Flat patterns based on grids, containing either straight or curved lines; the grid may overlap or produce figures by connecting the diagonals and by combining them with circles, arcs and segments.

Diaphragm
A relatively thin element in a structural member, which is capable of withstanding shear in its plane; it stiffens the structural member.

Dichromatic brickwork
Patterns formed by using two different colors of brick.

Diffuse wash light
Reduces the likelihood that surface textures will be noticed, strengthens the impression of surface smoothness, and removes shadows. Suitable for a gypsum board wall or acoustical tile ceiling.

Diffused transmission
A luminaire that dispenses light in all directions and eliminates any directional quality of the beam. Full diffused transmission is achieved by using opal glass and plastics that incorporate microscopic particles. Semidiffused transmission consists of translucent materials that emit light at wider angles due to the configuration on at least one side of the material. A greater degree of diffusion is achieved by etches, sandblasting, and matting aerosol sprays. Provides concealment of the lamp and glare control.

Diffuseness
A measure of a light's dispersion as it travels from the source. Diffused light is useful for general lighting purposes.

Diffuser
In a forced-air heating/cooling system, the diffuser is a register or grille attached to ducting through which heated or air-conditioned air is delivered to the living space. In a tubular skylight or an electric light fixture, the diffuser is a cover plate through which scattered light is delivered.

Diglyph
A rectangular block on the frieze of a Doric entablature, with two V-shaped vertical grooves instead of the three more commonly found in the triglyphs of other classical styles.

Dimension
The measured distance between two points, which when shown on a drawing is to become the precise distance between two points in a building.

Dimension work
Masonry built with stone cut to exact size and shape, as in ashlar masonry walls.

Dimensional coordination
The design of building components to conform to a dimensional standard, such as a module.

Dimensional stability
Applies to a material that has little moisture movement and creep, since thermal and elastic deformation are unavoidable.

Dimensional timber See Wood

Dimensioning
Measurement and placement of dimensional information during the process of drawing the plans, elevations and details.

Dimetric projection See Projection drawing

Diminished arch See Arch

Diminished column See Column

Diminution
The decrease in size of a column toward the top; typically employed as a device to overcome or correct an optical appearance of the top being larger than the bottom.

Dimmer
A solid-state device used to vary the voltage to a light fixture and thereby lower or raise the intensity of the light. Used mainly for incandescent lamps and fluorescent lamps with special ballasts. Also called a rheostat, it can be installed on a wall switch or directly into a lamp and can easily be added to existing installations.

Diner

A restaurant with a long counter and booths, originally sloped like a railroad car. Diners were designed as stationary evolutions of the railroad dining car minus the wheels. They usually had a counter with stools and a row of booths opposite. They featured large windows around the exterior above the level of the booths.

Dining hall
A large room in a school or other institution for group dining.

Dining room
A room in a residence for eating while seated at a table.

Dinwiddie, James (1902–1959)
American architect who was briefly in partnership with Eric Mendelsohn.

Diorama
A large painting, or series of paintings, intended for exhibition in a darkened room in a manner that produces an appearance of reality created by optical illusions; a building in which such paintings are exhibited.

Dipteros
Temple surrounded by a double row of columns.

Direct component
That portion of light energy, from sources such as the sky or sun, that reaches a specified location without any significant diffusion.

Direct digital controls
The application of microprocessor technology to environmental building controls. DDC systems make it possible to control heating and cooling functions with software that takes into account a wide range of variables, thereby achieving greater efficiency.

Direct lighting
Lighting provided from a source without reflection from other surfaces. In daylighting, this means that the light has traveled on a straight path from the sky (or the sun) to the point of interest. In electrical lighting, it usually describes an installation of ceiling-mounted or suspended luminaires with mostly downward light-distribution characteristics.

Direct luminaire

Emits over 90 percent of light downward. Recessed lights, downlights, and troffers are direct luminaires.

Direct solar gain
Solar energy obtained directly through a window.

Direct sunlight
That portion of daylight arriving at a specified location directly from the sun, without diffusion.

Direct/indirect lighting
Lighting that is mixed from direct sources and indirect reflection. In daylighting, this means that some part of the light of the sky or the sun is bounced off some surface, while at least part of the sky is still visible from the point in question. In electrical lighting, it means that luminaires of different types are installed, or there are luminaires that emit light both up to the ceiling and down to the workspace. The advantages include a good balance between ambient illumination of the room and accent lighting and relatively good energy efficiency even in large spaces. The smaller direct component required makes it easier to control reflective glare on computer screens, and it renders three-dimensional objects well without harsh shadows. Disadvantages include relatively high installation and maintenance costs. Users often need instruction on how to use the system effectively.

Direct/indirect luminaire

Emits light upward and downward. Many suspended luminaires offer this feature and can be directly or indirectly balanced.

Direct-gain system

Type of passive solar heating system in which south-facing windows provide heat gain during the daytime, and high-mass, thermal-storage materials absorb and store that heat. At night, the stored heat radiates back out, warming the space. This is the simplest type of passive solar heating system, but careful design is required to prevent overheating.

Directional lighting

Produced mainly by adjustable sources, providing light from a particular direction.

Disassembly

Taking apart an assembled product. Design for disassembly in buildings allows building components to be readily reused and recycled.

Discharge lamp

A lamp that produces light by the discharge of electricity between electrodes in a glass-filled enclosure, such as a fluorescent lamp.

Discharging arch See Arch

Disconnect

A switch or circuit breaker adjacent to a piece of electrical equipment for the purpose of disconnecting power to the equipment for maintenance or servicing.

Discontinuous construction

Construction where there is no solid connection between the rooms of a building and the structure or between one section and another; used to prevent the transmission of sound along a solid path.

Disk

A flat, circular, raised ornament, carved as a series of disks adjacent to each other.

Dismantle

To take apart a structure piece by piece, often with the intention of reassembling it or moving it elsewhere for reconstruction.

Displacement ventilation

A method of space conditioning where conditioned air is supplied at or near the floor. Since the air is supplied at very low velocities, a cool layer of air collects in the occupied zone resulting in comfortable conditions for the occupants. Buoyant forces remove heat generated by occupants and equipment, as well as odors and pollutants, all of

Displacement ventilation

which stratify under the ceiling and are extracted from the space by return or exhaust fans. Displacement ventilation systems were originally used in industrial facilities and subsequently in office buildings, auditoria, performing arts centers, and spaces with large interior volumes. These systems are effective in improving indoor air quality as well as providing energy savings when compared with a conventional fully mixed system.

Display lighting

Illumination that is concealed by setup or placement and directed so as to illuminate objects on display in a case or display window. Needs an electrical source near the cabinet.

Distributed load

A load distributed over the surface; unless otherwise described, it is usually considered uniformly distributed.

Distribution duct

A raceway of various cross sections, placed within or just below the floor and from which the wires and cables serve a specific floor area.

Distribution lines

In electricity, the main feed line of a circuit to which branch circuits are connected.

Distribution panel

An insulated board or box containing circuit breakers, from which connections are made between the main feeder lines and branch lines.

District

A geographically definable urban or rural area with a concentration of sites, structures, or uses that are connected historically or esthetically by plan or development.

District energy system

District energy is an approach to supplying thermal energy in the form of steam, hot water, and cold water through a distribution system of pipe from a central plant to individual users. Users then extract the energy from the distribution system for their individual heating, cooling, and process requirements.

Distyle

Having two columns.

Disuse

That complete sequence and series of activities and actions that eliminate the building in its present form. There are basically two options: (1) demolition and return of the building, site, and all of its components to the natural environment, or (2) renovation. The renovation option essentially leads back to the beginning of the building life-cycle model or to some intermediate stage within that model.

Diurnal flux

Difference between day and night average temperatures.

Divided light
Window in which the glass is divided into several smaller panes.

Docomomo
International working party for documentation and conservation of buildings, sites and neighborhoods of the modern movement.

Document
A written, typed, or drawn record.

Documentary historic site
Documents an important historical event in the life of a person or family, usually in the form of a plaque.

Documentation
Any information that records the prior history of an existing building or site; such as maps, site plans, drawings or photographs, or written historic references as to its physical appearance.

Documentation standards
Those required to verify the significance of a particular property for listing on the National Register of Historic Places.

Dodecagon
A 12-sided regular polygon; the angle included between the 12 equal sides is 115 degrees.

DOE
U.S. Department of Energy.

Doesburg, Theo van (1883–1931)
Began his career as a painter and established the De Stijl movement with J.J.P. Oud. He later taught at the Bauhaus.

Dog tooth course See **Bond**

Dogleg stair See **Stair**

Dogtooth bond See **Bond**

Dogtooth ornament See **Ornament**

Dollman, George von (1830–1895)
A German Gothic Revival and Romantic architect, who designed Schloss Linderhoff in 1881 in an extravagant Neorococo style, and Schloss Neuschwanstein in 1886, two palaces for King Ludwig II of Bavaria.

Dollman, George von

Dolmen
Several large stones capped with a covering slab, as those erected in prehistoric times.

Dolomite See **Stone**

Dome
A curved roof structure that spans an area on a circular base, producing an equal thrust in all directions. A cross section of the dome can be semicircular, pointed or segmented.

bell-shaped dome
A dome in which the cross section is shaped in the form of a bell.

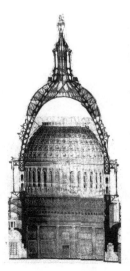

double dome
An outer dome built over an inner dome, with a space between them; used to provide a supporting structure for the outer dome, or a different shape or architectural treatment to each one individually.

elliptical dome
A dome with a cross section in the shape of an arc of an ellipse; may have a circular or an elliptical base.

geodesic dome
Consisting of a multiplicity of similar straight linear elements, arranged in triangles or pentagons; the members in tension have a minimal cross section and make up a spherical surface usually in the shape of a dome.

hemispherical dome
A dome with a constant radius of curvature that comes vertically from its springing line; the horizontal component of the thrust is absorbed by a continuous ring or chain at the base of the dome.

imperial dome
A round roof in the shape of an onion with a flared skirt at the base; it is commonly found in Greek Orthodox churches; also known as an onion dome.

interdome
The space between the inner and outer shells of a dome.

lattice dome
A steel dome structure having members which follow the circles of latitude and two sets of diagonals replacing the lines of longitude and forming a series of isosceles triangles.

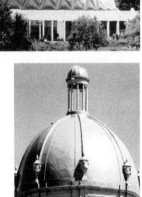

melon dome
A melon-like ribbed dome, either on the interior or on the exterior.

melon dome

onion dome

In Russian Orthodox church architecture, a bulbous dome which terminates in a point and serves as a roof structure over a cupola or tower.

radial dome

A dome built with steel or timber trusses arranged in a radial manner and connected by polygonal rings at various heights.

saucerdome

A dome whose rise is much smaller than its radius.

semicircular dome

A dome in the shape of a half sphere.

157

semidome
A dome equivalent to one-quarter of a hollow sphere, covering a semicircular area, such as an apse.

Turkish dome
An onion-shaped dome with a pointed top and a cylinder base; named for its use in Byzantine architecture.

Turkish dome

Dome light
Any window or opening in a dome.

Domenig, Gunther (1934–)
Austrian-born architect; a proponent of Organic Architecture, his A-Bank in Vienna (1979) is his best-known work, and features an undulating metal facade, and exposed mechanical and electrical systems in flexible pipe ducts on the interior.

Domenig, Gunther

Domestic architecture
Building types that relate to residential use, such as single-family homes, duplexes, townhouses, apartment buildings, and condominiums.

Domestic hardwood
Deciduous trees that grow in the United States; this is the only type of wood in the United States where on a general scale the growth of new trees easily exceeds the removal rate.

Domical vault See Vault

Dominance
Occupying a preeminent or influential position; architectural forms that exercise the most influence or governing control.

Dominick, Peter (1941–2009)
Designed contextual structures in the Denver area. He also worked with Disney and designed the Animal Kingdom Lodge at Disney World in Ortlando, FL 2008. He also designed Disney's Wilderness Lodge and Grand Californian Hotel & Spa.

Door
A hinged, sliding, tilting, or folding panel for closing openings in a wall or at entrances to buildings. Doors must relate to the facade or wall in which they are placed. They are an important element in setting the style of the exterior and are an important transitional element to the interior space.

accordion door
A hinged door consisting of a system of panels hung from an overhead track, folding back like the bellows of an accordion; when open, the panels close flat; when closed, the panels interlock with each other.

acoustical door
A door having a sound-deadening core, stops along the top and sides sealed by gaskets, and an automatic drop seal along the bottom; especially constructed to reduce noise transmission through it.

aluminum door
Used for storefront entrances, due to its capacity for high corrosion resistance.

automatic door
A power-operated door, which opens and closes automatically at the approach of a person or vehicle.

banded door
A wood door with a thin molded band applied to the outside edge of the face of each stile and the top and bottom rail.

batten door
A door formed by full height boards glued edge to edge with horizontal and vertical battens applied to give the appearance of paneling; a single batten door has battens on one side, and a double batten door has them on both sides.

bifold door
A folding door that divides into two parts, the inner leaf of each part being hung from an overhead track, and the outer leaf hinged at the jamb.

lank door
A recess in a wall, having the appearance of a door, usually used for symmetry of design; any door that has been sealed off but is still visible on the surface.

blind door
The representation of a door, inserted to complete a series of doors, or to give the appearance of symmetry.

bungalow door
Any of various front door designs featuring lights in the top portion of the door.

center-hung door
A door that is supported by and swings on a pivot that is recessed in the floor at a point located on the center line of the door's thickness; may be either single-swing or double-acting.

double-acting door
A door that opens in both directions, typically fitted with a double-acting hinge.

double door
A pair of swinging doors with hinges on each jamb, meeting in the middle.

double-faced door
A door with a different face detail on either side to match the decoration of the area in which each side faces. Normally constructed as two thin doors fixed back to back.

double-framed door
A door with stiles, rails, and panels set within a frame of stiles and rails.

Dutch door
A door consisting of two separate leaves one above the other; the leaves may operate independently or together.

fire door
A sheet-metal-clad door that automatically closes in the event of a fire; it is typically designed to slide across a doorway.

flush door
A smooth-surfaced door having faces in the same plane as the surface and which conceals its rails and stiles or other structural features.

flush-paneled door
A paneled door in which, on one or both faces, the panels are finished flush with the rails and stiles.

folding door
One of two or more doors which are hinged together so that they can open and fold in a confined space.

French door
A door having a top rail, bottom rail, and stiles, which has glass panes throughout its entire length; often used in pairs.

glass door
A door consisting of heat-strengthened or tempered glass, with or without rails or stiles; used primarily as an entrance door for retail stores.

half door
A short door installed in the frame with space left above and below the door.

hollow-core door
A wood flush door having a framework of stiles and rails encasing a honeycombed core of corrugated fiberboard or a grid of interlocking horizontal and vertical wood strips.

interior door
A hollow-core or solid-core door made to be soundproof.

jib door
A concealed door flush with the wall and usually decorated to match it.

louvered door
A door having a louvered opening, usually with horizontal blades, that allows for the passage or circulation of air while the door is closed

metal-clad door
A flush door having face sheets of light-gauge steel bonded to a steel channel frame; or a door having a structural wood core clad with galvanized sheet metal.

overhead door
A door of either the swing-up or the roll-up type constructed of one or several leaves; when open, it assumes a horizontal position above the door opening.

paneled door
A door having a framework of stiles, rails, and muntins which form one or more frames around thinner recessed panels.

pivoted door
A door hung on center or offset pivots as distinguished from one hung on hinges or a sliding mechanism.

pocket door
A door that slides in and out of a recess in a doorway wall requiring no room for the door swing.

revolving door
An exterior entrance door consisting of four leaves at right angles to each other, set in the form of a cross, which pivot about a vertical axis within a cylindrical-shaped vestibule

rolling door
A large door consisting of horizontal, interlocking metal slats guided by a track on either side, opening by coiling about an overhead drum at the head of the door opening.

roll-up door
A door made of small horizontal interlocking metal slats that are guided in a track; the configuration coils around an overhead drum which is housed at the head; may be manually or electrically operated.

self-closing door
A door for the control of fire or smoke, which closes by itself by the action of a spring which is held open by a fusible link that melts in a fire, causing the door to close.

sliding door
A door that is mounted on a track, which slides in a horizontal direction parallel to the wall on which it is mounted.

solid-core door
A wood flush door having a solid core of lumber or particleboard, or one consisting of mineral composition.

storm door
Auxiliary door installed in the same frame, as an entrance door to a house.

swinging door
A door that turns on hinges or pivots about a vertical edge when opened.

tempered glass door
Common application for commercial use.

Venetian door
A door having a long narrow window at each side similar in form to that of a venetian window, or a Palladian door.

wood door
Either solid core or hollow core with veneer; exterior doors are coated with waterproof adhesives.

Door buck
A metal or wood surface set in a wall, to which the finished frame is attached.

Door casing
The finished frame surrounding a door; the visible frame.

Door head
The uppermost member of a door frame; a horizontal projection above a door.

Door jamb
The vertical member located on each side of a door.

Door frame
An assembly built into a wall consisting of two upright members (jambs) and a head (lintel) over the doorway; encloses the doorway and provides support on which to hang the door.

Door handle
A device used to open or close a door and can refer to any fixed or lever-operated door latch device. Doorknob tends to refer to round operating mechanisms.

Door knocker
A knob, bar, or ring of metal, attached to the outside of an exterior door to enable a person to announce his or her presence, usually held by a hinge so that it can be lifted to strike a metal plate.

Door knocker

Door light
Glass area in a door.

Door louvers
Blades or slats in a door to permit ventilation while the door is closed; may or may not be adjustable.

Door mullion
The center vertical member of a double-door opening set between two single active leaves, usually the strike side of each leaf.

Door muntin
An intermediate vertical member used to divide the panels of a paneled door.

Door panel
A distinct section or division of a door, recessed below or raised above the general level, or one enclosed by a frame.

Door rail
A horizontal cross-member connecting the hinge stile to the lock stile, both at the top and bottom of the door and at intermediate locations, may be exposed as in panel doors, or concealed, as in flush doors.

Door sill
The horizontal member; usually consisting of a board covering the floor joint on the threshold of a door.

Door stile
One of the upright structural members of the frame that is located at the outer edge of a door.

Door stop
A strip against which a door shuts in its frame; a device placed on a wall behind a door or mounted on the floor to prevent the door from opening too wide.

Door surround
An ornamental border encircling the sides and top of a door frame.

Door threshold
A strip fastened to the floor beneath a door, usually required to cover the joint where two types of floor material meet; may provide weather protection at exterior doors.

Door transom
A crossbar separating a door from a light or window that is located above it.

Doorway
The framework in which the door hangs, or the entrance to a building; the key area of interest in a facade as a natural focal point and design element giving human scale and containing the street number.

163

Doorway

Doric capital See **Capital**

Doric order

The first and simplest of the orders, developed by the Dorian Greeks, consisting of relatively short shafts with flutes meeting with a sharp arris, simple undecorated capital, a square abacus, and no base. The entablature consists of a plain architrave, a frieze of triglyphs and metopes, and a cornice. The corona contained mutules in the soffit.

Dormer

A structure projecting from a sloping roof, usually housing a vertical window that is placed in a small gable, or containing a ventilating louver.

arched dormer
A dormer that has a semi-cylindrical-shaped roof; the head of the window in the dormer may be either rounded or flat.

eyebrow dormer
A low dormer on the slope of a roof. It has no sides; the roof is carried over it in a continuous wavy line.

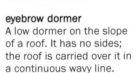

hipped dormer
A dormer whose roof has a miniature hipped appearance in front, dying into the main roof surface at the back.

inset dormer
A dormer that is partially set below a sloping roof, unlike the usual dormer that projects entirely above the sloping roof. Same as recessed dormer.

recessed dormer
A dormer with all or part of the window set back into the roof surface, resulting in the sill being lower than the roof.

shed dormer
A dormer whose eave line is parallel to the main eave line of the roof, and whose flat roof plane slopes downward in a direction away from the ridge line of the main roof.

wall dormer
A dormer whose face is integral with the face of the wall below.

Dormer cheek
The vertical sides of a dormer.

Dormer window See **Window**

Dormitory
A multiple-occupancy building which contains a series of sleeping rooms, bathrooms, and common areas.

Dosseret
A deep block sometimes placed above a Byzantine capital to support the wide voissoirs of an arch above; also, an isolated section of an entablature above a column.

Dot
A small square tile at the intersection of four larger tiles; usually placed at a 45-degree angle to the grid of the larger tiles.

Double dome See **Dome**

Double door See **Door**

Double glazing See **Glass**

Double header
A structural member made by nailing or bolting two or more timbers together for use where extra strength is required, as in the opening around stairs.

Double lancet window See **Window**

Double pane windows
Double or triple pane glass windows often contain argon, krypton, or other gases between panes to reduce heat flow and improve insulation.

Double tenon
Two tenons side by side, arranged within the thickness of a member; used for thick members to increase the glue line for added strength.

Double vault See **Vault**

Double wall
Construction system in which two layers of studs are used to provide a thicker-than-normal wall system to accommodate extra insulation. The two walls are often separated by several inches to reduce thermal bridging through the studs and to provide additional space for insulation.

Double window See **Window**

Double-acting door See **Door**

Double-decking
Insertion of a mezzanine or extra floor in a building with a high ceiling or a two-story space to increase the usable area.

Double-faced door See **Door**

Double-framed door See **Door**

Double-gable roof See **Roof**

Double-glazed window
A window with two panes of glass separated by an air space. Some double-glazed windows contain a gas between the panes that further increases the window's insulative value.

Double-hipped roof See **Roof**

Double-hung window See **Window**

Double-pitched roof See **Roof**

Double-stud wall

Construction system in which two layers of studs are used to provide a thicker-than-normal wall system so that a lot of insulation can be installed; the two walls are often separated by several inches to reduce thermal bridging through the studs and to provide additional space for insulation.

Double-sunk

Recessed or lowered in two steps, as when a panel is sunk below the surface of a larger panel.

Douglas fir See Wood

Dougong

Chinese term for a cluster of brackets cantilevered out from the top of a column to carry the rafters and overhanging eaves of a roof.

Dovetail joint See Joint

Dovetail molding See Molding

Dow, Alden (1904–1983)

An American architect; worked with Frank Lloyd Wright in 1933 before setting up his own practice in 1935. His later work represented his own style.

Dowel

A small pin inserted into two abutting pieces of wood; in stone or masonry construction, a wooden or metal pin placed between the different courses to prevent shifting.

Downcycling

Recycling of a waste stream to create a new material that has properties inferior to those of the original virgin materials. A good example is recycled plastic (HDPE) panels made of multicolored waste sources.

Downing, Andrew Jackson (1815–1852)

Served as consultant with Andrew J. Davies but did not practice on his own until 1850, when with Calvert Vaux designed the Daniel Parrish House in Newport, RI. He also planned an informal design for the Mall in Washington, DC, later replaced with the present one but still considered as one of the originators of the informal approach to landscape design.

Downlight

Luminaire that directs the majority of its flux downward, usually within a restricted cone.

Downlights

Used primarily for general illumination in a wide range of residential and commercial applications, lobbies,

Downlights

hallways, and corridors. Can be equipped with incandescent, halogen, low-voltage, compact fluorescent, or high-intensity discharge (HID) lamps. Types include open reflector, open baffle, lensed, diffuse, adjustable, and pull-down.

Downspout

A vertical pipe that carries water from the roof gutters to the ground or cistern.

Draft See Design drawing

Drafted margin

A narrow dressed border around the face of a stone, usually about the width of a chisel edge.

Dragon beam

A beam of a traditional timber-framed house that is set diagonally at the corner to support the corner post and floor joist when the building jetties on two different sides.

Dragon piece

A diagonal tie across the wall plate at the corner of a hipped-end roof for receiving the thrust of the hip rafter.

Drain

A channel, conduit, or pipe used to remove rain, wastewater, or sewage.

Drain tile

Pipe used at the bottom of foundation walls and footings to allow drainage of groundwater; typically made of terracotta with either a circular or hexagonal cross section.

Drainage

An assembly of pipes and fittings, in the ground, used for the removal of wastewater or rainwater from a building or site.

Drainage hole

A hole in a retaining wall; an open joint in masonry, to drain unwanted water.

Drainage plane

Path that water would take over the building envelope. Concealed drainage-plane materials, such as building paper or housewrap, are designed to shed water that penetrates the building's cladding. Drainage planes are installed to overlap in shingle fashion so that water flows downward and away from the building envelope.

Drainpipe
Any pipe used to convey drainage liquids; includes both soil pipes and waste pipes.

Drain-water heat-recovery system
A system that extracts heat from hot-water waste lines to preheat cold water before it reaches the water heater or tap; especially useful when paired with on-demand or solar hot-water systems. Also known as graywater heat-recovery system.

Drawbridge See **Bridge**

Drawing
A sketch, design, or other representation by lines.

Drawing room
A large formal space for entertaining.

Dress
To prepare or finish a wood member by planning, or cutting a stone piece by chipping away at the irregularities.

Dressed lumber
The lumber having one or more of its faces planed smooth.

Dressed stone
A stone that has been worked to a desired shape; the faces to be exposed are smooth, usually ready for installation.

Dressing
Masonry and moldings of better quality than the facing materials, used around openings or at corners of buildings.

Drip cap
A horizontal molding fixed to a door or window frame to divert the water from the top rail, causing it to drip beyond the outside of the frame.

Drip edge
Installed lip that keeps shingles up off the deck at edges and extends shingles out over eaves and gutters to prevent water from wicking up and under the shingles.

Drip molding See **Molding**

Dripstone
A hood mold on the outside of a wall, often used in Gothic architecture.

Dripstone cap
A continuous horizontal cap containing a drip molding on a masonry wall.

Dripstone course
A continuous horizontal drip molding in a masonry wall.

Drop
Any one of the guttae attached to the underside of the mutules or triglyphs of a Doric entablature.

Drop arch See **Arch**

Drop molding See **Molding**

Drop panel
The portion of a flat slab or flat plate that is thickened throughout the area surrounding the top of the column, to reduce the magnitude of shear stress.

Drought tolerance
The capacity of a landscape plant to function well in drought conditions.

Drought-tolerant plants
Species of plants, shrubs, and vines that generally do not require additional watering in order to thrive in their native habitats. Landscapes with drought-tolerant plants usually require little or no watering.

Drum

One of the cylinders of stone that form a column; a cylindrical or polygonal wall below a dome, often pierced with windows.

Dry bulb temperature

Air temperature as measured by an ordinary thermometer.

Dry masonry

A masonry wall laid up with out mortar.

Dry rot

A fungi that feeds on, and destroys, damp rather than wet timber. Most often found in damp, poorly ventilated under-floor spaces and roof areas. Causes timber to lose strength, develop cracks, and finally become so dry and powdery that it is easily crumbled.

Dry seam

A fracture in stone that, if left untreated, may expand and eventually result in structural failure.

Dry well

Underground structure that captures, then slowly releases stormwater runoff so that it can be absorbed by the soil.

Dry-pipe system

A system for fire protection that uses pipes containing air under pressure instead of water; controlled by a valve that allows water to enter the pipes in the event of a fire.

Dry-pressed brick

Clay bricks made with only a small amount of water, and compressed under high pressure which results in a very uniform brick with sharp edges and a surface sheen. They are often used as face bricks, although their fragile edges are subject to erosion due to the high clay-to-water ratio.

Drywall

An interior wall constructed with a material such as gypsum board or plywood; usually supplied in large sheets or panels, which do not require water to apply.

Drywall clips

Metal or plastic stops that are attached to framing at inside corners. The clips replace framing, thus leaving more room for insulation. Because such a corner floats, acting as a stop, the clip allows the first sheet of drywall to be trapped by the second, perpendicularly installed sheet, and cracking of the drywall joint is less common.

Dual-flush toilets

Toilets with two buttons for two flush options, one for liquid and another for solid waste. The button for liquid waste uses less water per flush.

Duany, Andrés (1949–)

American architect and urban planner, who grew up in Cuba until 1960. In 1977, Duany was cofounder of the Miami firm Arquitectonica, with his wife, Elizabeth Plater-Zyberk; Bernardo Fort-Brescia; Laurinda Hope Spear; and Hervin Romney. Duany and Plater-Zyberk founded Duany Plater Zyberk & Company (DPZ), headquartered in Miami. The firm first received international recognition in the 1980s as the designer of Seaside, Florida, and Kentlands, Maryland, and has completed designs and codes for over 200 new towns, regional plans, and community revitalization projects. Duany co-authored three books: *Suburban Nation: The Rise of Sprawl and the Decline of the American Dream* (2000), *The New Civic Art: Elements of town Planning* (2003), and *The Smart growth Manual* (2009). In 2001, he and Plater-Zyberk were awarded the Vincent Scully Prize by the National Building Museum in recognition of their contributions to the American built environment.

Duct

A nonmetallic or metallic tube for housing wires or cables, may be underground or embedded in concrete floor slabs; a duct usually fabricated of metal, used to transfer air from one location to another.

Duct blaster
Calibrated air-flow measurement system developed to test the air-tightness of forced-air duct systems. All outlets for the duct system, except for the one attached to the duct blaster, are sealed off, and the system is either pressurized or depressurized; the work needed by the fan to maintain a given pressure difference provides a measure of duct leakage.

Ductless mini-split
A small-capacity heat pump with a closely associated outside compressor and inside evaporating coil, often through-the-wall in design. These heat pumps often come with variable-speed compressors and blowers, giving them excellent modulation for thermal comfort. They are also well-suited for ultra-high-performance, small-volume homes.

Ductwork
An assembly of ducts and duct fittings, usually of sheet metal, arranged for the distribution of air in air-conditioning and mechanical ventilation systems.

Dudok, Willem (1884–1974)
A Dutch architect based in Hilversum who designed the Snellius School in Hilversum, The Netherlands, in 1922, as compositions of asymmetrical rectangular blocks, predating the International style, and the Town Hall, Hilversum, in 1931, in a distinctive version of the International style.

Dumbbell tenement
A five- to seven-story multiple dwelling unit in urban areas, characterized by a long, narrow plan with an indentation on each side, forming a shaft for light and air; hence its resemblance in plan to a dumbbell.

Duplex
A house having a separate apartment for two families, especially a two-story house having two separate entrances and a complete apartment on each floor; an apartment with rooms on two connected floors.

Durability
A factor that affects the life-cycle performance of a material or assembly. All other factors being equal, the more durable item is environmentally preferable, because it means less frequent replacement. However, durability is rendered moot as a factor if the material is replaced for aesthetic reasons prior to it actually wearing out.

Durability plan and inspection checklist
An effective durability management program will consider the important interrelationships among energy, moisture, and air quality, and how those dynamics are addressed. Other important areas of durability include exterior water strategies, interior moisture control, air infiltration, interstitial condensation, heat loss, and radiation control. Green rating programs can serve as guides and produce durability strategies.

Dutch bond See Bond

Dutch Colonial style (1650–1700)
A style adopted by the Dutch settlers in New York and New Jersey, characterized by the use of brick and stone walls with gambrel or double-pitched roofs and flared lower eaves which extend beyond the front and rear walls, forming a deep overhang.

Dutch door See Door

Dutch drain
Open drain that carries stormwater runoff from the bottom of a house wall away from the house.

Dutch gable See Gable

Dutch gambrel See Roof

Dutch roof See Roof

Dwarf order
A miniature range of columns or pilasters; for instance, those found on an attic story.

Dwarf order

Dwelling unit
An apartment, dormitory, or house used as a residence for one or more people.

Dymaxion house
A proposal for a unique circular house, conceived by R. Buckminster Fuller in 1928, that was to be mass-produced at the Beech Aircraft Company; a full-sized prototype featured a central shaft containing all the building's services, such as electrical wiring and all the waste disposal facilities within it.

Dynamic environmental chamber
Well-controlled system including temperature, relative humidity (RH), and air quality/purity that uses realistic air flows for the assessment of chemical emissions.

E e

Eagle See **Ornament: animal forms**

Early Christian architecture (200–1025)
The final phase of Roman architecture was influenced by the adoption of Christianity as the state religion and the rise of the Byzantine style. The Roman basilican form was adopted as the ground plan for most early Christian churches. These simple rectangular plans consisted of a nave with two side aisles and a longitudinal and horizontal emphasis.

Early English architecture (1200–1250)
The first English Gothic style to follow the Norman style featured moldings, consisting of rounds and deep hollows, which produced a strong effect of light and shadow. Arches were lancet-shaped; doorways were deeply recessed with numerous moldings in the arch and jambs. Windows were long, narrow, and almost always pointed. Pillars consisted of small shafts arranged around a larger central pier.

Early Gothic revival (1830–1880)
A favored style for secular buildings, it later became a favorite for classical elements used for ecclesiastical buildings as well. The style was characterized by verticality; pointed arches; steep, complex gables; roofs with finials; and medieval decorative motifs.

Ears
Projections on the sides of the upper part of door and window surrounds. Also called shoulders.

Earth floor
A hard-packed clay floor used in early dwellings.

Earth tube
Ventilation air intake tube, usually measuring 8 or more inches in diameter and buried 5 or more feet below grade. Earth tubes take advantage of relatively constant subterranean temperatures to preheat air in winter and precool it in summer. In humid climates, some earth tubes develop significant amounts of condensation during the summer, potentially contributing to indoor air quality problems.

Earth's thermal energy
A short distance below the surface, the Earth maintains a mostly constant temperature very close to the human comfort range. This can be used advantageously for geothermal heating systems.

Earth-sheltered design
Home design that is partially or totally below ground, either by digging into existing topography or filling over parts of the structure. Earth-sheltered design uses the constant temperature of the soil to improve energy efficiency and can be beneficial on hilly sites by decreasing maintenance and environmental impact.

Earthwork
Any construction that involves moving, forming, cutting, or filling earth.

Easement
A deed restriction on a piece of property granting rights to others to use the property; may include restrictions for use or development on the property.

Eastern Stick style (1855–1900)
An American residential style characterized by exposed framing overlaid on clapboard in horizontal, vertical, or diagonal patterns to suggest the frame structure underneath. Steeply pitched gable roofs, cross gables, towers and pointed dormers, and porches and verandas are also characteristic. Oversized corner posts, purlins, brackets, and railings complement decorative woodwork produced by the stickwork.

Eastlake style (1870–1880)
A style characterized by a massive quality, in which posts, railings and balusters were turned on a mechanical lathe. Large curved brackets, scrolls, and other stylized elements are placed at every corner or projection along the facade. Perforated gables, carved panels, and a profusion of spindles and latticework along porch eaves are typical. Lighter elements are combined with oversized members to exaggerate the three-dimensional facade.

Eave
The projecting overhang at the lower edge of a roof that sheds rain water.

Eaves channel
A channel or small gutter along the top of a wall; it conveys the roof rainwater to downspouts, or discharges it through gargoyles.

Eaves lath
A wooden strip placed under the last bottom course of roofing tile to raise it to conform to the same slope as the rest of the courses on the roof.

Ebony See **Wood**

Eccentric
Not having the same center or center line; departing or deviating from the conventional or established norm.

Echinus
The convex projecting molding of eccentric curve supporting the abacus of a Doric capital.

Eclectic style
The selection of elements from diverse styles for architectural decorative designs; particularly during the late nineteenth century in Europe and America.

Eclecticism
The practice of selecting from various sources, sometimes to form a new style.

Eco-city
Ecological cities are balanced with human society and balanced between humans and nature.

Eco-conscious
Marked by or showing concern for the environment.

Eco-design
A design process that considers the environmental impacts associated with a product throughout its entire life: from acquisition of raw materials through production/manufacturing and use to end of life. Eco-design seeks to improve the aesthetic and functional aspects of the product with due consideration to social and ethical needs while simultaneously reducing environmental impacts.

Eco-friendly
A widely used term for "environmentally friendly," something that is developed with a minimum impact on the environment.

Eco-indicator
A measure of the environmental impacts of manufacturing products or providing services, as evaluated according to the widely accepted life-cycle assessment procedure.

École des Beaux-Arts
A school founded in 1648 in Paris to teach painting and sculpture, literally the "School of Fine Arts"; architecture was added to the studies in 1819, emphasizing the study of Classical Greek and Roman buildings; the students were grouped in ateliers supervised by a master. Richard Morris Hunt was one of the first Americans to study at the school, followed by many other late nineteenth-century and early twentieth-century architects.

Ecological architecture
A concern with how ecological properties impact the building, its occupants, and the environment. The ecological elements are selected from natural or minimally processed earth resources—biodegradable, renewable, and clean elements with low-embodied energy. Elements consist of the soil and landscape, site selection, water resources, and waste management.

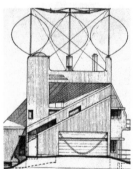

Ecological design
The meshing of human purposes with the larger patterns and flows of the natural world and the study of those patterns and flows to inform human purposes.

Ecological footprint
Measure of the resource use by a population within a defined area of land, including imported resources.

Ecological footprint

Assessment of the ecological footprints of nation-states or other defined geographic areas reveals the true environmental impact of those states and their ability to survive on their own resources in the long term. It has emerged as the world's premier measure of humanity's demand on nature. It measures how much land and water area a human population requires to produce the resource it consumes and to absorb its wastes, using prevailing technology.

Ecological impact
The effect that a human-caused or natural activity has on living organisms and their nonliving environment.

Ecological rucksack
Total weight of material flow carried by an item of consumption in the course of its life cycle. Like the ecological footprint, the ecological rucksack concept deals with displaced environmental impacts, but has a more technical focus.

Ecological technologies
Facilitate a reduction in polluting emissions from building maintenance and emissions not made from toxic chemicals or that do not deplete stratospheric ozone. Additional ecological technologies include storm-water and wastewater systems that reduce surface water and groundwater pollution.

Ecologically sustainable design
The use of design principles and strategies that help reduce the ecological impact of buildings, e.g., by reducing the consumption of energy and resources or by minimizing disturbances to existing vegetation.

Ecologically sustainable development
Any development that does not compromise the ability of future generations to enjoy similar levels of development. This is done by minimizing the effect of development on the environment. Also defined as using, conserving, and enhancing the community's resources so that ecological processes on which life depends are maintained, and the total quality of life now and in the future can be increased.

Ecology
The interrelationship of living things to one another and their environment and the study of such interrelationships. Also refers to the study of the detrimental effects of modern civilization on the environment, with a view toward prevention or reversal through conservation. Derived from the Greek word *oikus*, meaning home, and by extension the whole inhabited earth.

Economizer controls
HVAC system controls that operate mixed air dampers to mix return and outdoor air to obtain air of a temperature appropriate for free cooling. Economizer controls are used during periods when outdoor air requires less cooling energy input than return air.

Economizers
A collection of mechanical devices that decide how much outside air to bring into a building to save energy versus using refrigeration equipment to cool recirculated air.

Eco-preferred
Ecologically preferred products, materials, buildings, or services are those that embody one or more unique environmental attributes or qualities as a result of deliberately eliminating or reducing potential environmental impacts across their life cycle compared to other products in their category, such as photovoltaic panels, as a preference to fossil-fuel-powered electrical grid energy.

Ecosystem
A natural unit consisting of all plants, animals, and microorganisms functioning in an area, along with the nonliving factors of the built environment. Factors that affect an ecosystem are temperature, light, moisture, and air currents. There are overlapping and interdependent ecosystems from the entire globe, to a small city or individual building.

Edge beam See **Beam**

Edge laid
Paving or flooring that is installed edge-to-edge lengthwise.

Edge to edge
Two surfaces that butt together but do not overlap each other.

Edging
The finishing operation of rounding off the edges of a slab to prevent chipping or damage.

Edifice
A building, especially one of imposing appearance or size.

Effective leakage area
Calculation in square inches equal to the total area of all air leaks in a building envelope. Defined by the Lawrence Berkeley National Laboratory, the ELA is the area of a nozzle-shaped hole that would leak the same amount of air as the building does when pressurized to 4 pascals. Estimating leakage in this way enables one to visualize the cumulative impact many tiny holes can have on the air-tightness of a house.

Efficacy
In lighting design, a measure of the luminous efficiency of a specified light source, expressed in lumens per watt. For daylighting, this is the quotient of visible light incident on a surface to the total light energy on that surface. For electric sources, this is the quotient of the total luminous flux emitted by the total lamp power input.

Efflorescence
A deposit, usually white, formed on the surface of a brick, block, or concrete wall; it consists of salts leached from the surface of the wall.

Eggshelling
Chip-cracked plaster, either the base or finish coat.

Egress requirements
Building code regulations regarding the number, location and size of exit doors, corridors and stairs, for fire safety.

Egyptian architecture (3000 B.C.–200 A.D.)
An ancient architecture along the Nile River from Neolithic times, built of reed huts with inward sloping walls and thick bases to resist the annual inundation. The decorative "bundling" of reeds later influenced stone construction of fluted columns and capitals. Massive funerary monuments and temples were built of stone using post-and-lintel construction, with closely spaced columns carrying the stone lintels supporting a flat roof. A hypostyle hall, which was crowded with columns, received light from clerestories above. Walls were carved in ornamental hieroglyphs in low relief. There were many varieties of columns, often used side by side, and their capitals were distinctly ornate, based on the lotus, papyrus, or palm.

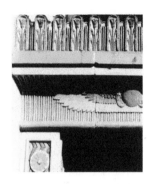

Egyptian cornice
The characteristic cornice of most Egyptian buildings, consisting of a large cavetto decorated with vertical leaves and a roll molding below.

Egyptian Revival style (1830–1850)
A revival style distinguished by distinctive columns and capitals and a smooth monolithic exterior finish. Characteristic battered walls are edged with roll moldings, tall straight-headed windows with splayed jambs, and a deep cavetto or gorge-and-roll cornice. Roofs are flat, and the smooth wall surface lends a monumental appearance reminiscent of pylons or gateways to Egyptian temples.

Ehrankrantz, Ezra (1932–2001)
American historic preservation architect in New York City. Projects include the Alexander Hamilton U.S. Customs House, NYC, originally designed by Cass Gilbert in 1900, and restored in 1994, now a branch of the Smithsonian Institution's National Museum of the American Indian, Washington, DC.

Eidlitz, Leopold (1823–1908)
American architect in New York City; a proponent of Gothic designs and structuralism. Designed the New York State Capitol in 1885, Albany, NY, with H. H. Richardson and Frederick Law Olmstead, and the County "Tweed" Courthouse, southern wing, in 1878, New York City.

Eiffel, Gustave (1832–1923)
French engineer who designed the Eiffel Tower, Paris, which was built of exposed steel for the 1887 Paris Exhibition and was the tallest structure in the world for 40 years.

Eisenman, John (1851–1924)
Best known for his design of the Cleveland Arcade, Cleveland, OH in 1882, with its two 9-story round arched blocks flanking a galleried iron-and-glass facade.

Eisenman, Peter (1932–)
Founded the Institute of Architecture and Urban Studies in New York City, and was associated with the New York Five. He is a Deconstructivist. His well-known Wexner Center for the Visual Arts, Columbus, OH (1989), is set on an angle to the existing campus buildings, as is the Convention Center, Columbus, OH (1993). Other works: Aronoff Center for Design and Art, University of Cincinnati, Cincinnati, OH (1996), City of Culture of Galicia, Santiago de Compostela, Galicia, Spain (1999), Memorial to the Murdered Jews of Europe, Berlin, Germany (2005), University of Phoenix Stadium, Glendale, AZ (2006).

El Tajin style (200–900 A.D.)
A style of Mesoamerican architecture as seen at the Pyramid of Niches, El Tajin, the Totonac capital in Veracruz, Mexico, characterized by elaborately carved recessed niches with geometric ornamentation.

Elastic deformation
The deformation that occurs instantly when a load is applied and is instantly and fully recovered when the load is removed.

Elastic limit
The limit of stress beyond which the strain cannot fully recover.

Elasticity
The ability of a material to instantly deform under load and to recover its original shape when the load is removed.

Elastomeric
Any material having the properties of being able to return to its original shape after being stressed, such as a roofing material that can expand and contract without rupture.

Elbasani, Barry (1941–2010)
A founder of ELS Architecture and Urban Design. His firm worked with the Rouse Company and designed the Grand Avenue in Milwaukee, WI; the mixed-use Pioneer Place in downtown Portland, OR; the Shops at Arizona Center in Phoenix, AZ; and the Village of Merrick Park in Coral Gables, FL. The firm also created the master plan for Summerlin, NV, and designed the Denver Pavilions in Colorado.

Elbow
Sharp corner in a pipe or conduit, as opposed to a bend, which has a larger radius of curvature.

Electric box
A metal box, open on one face with round knock-out openings on the ends for conduits to be attached to pull wires to the box. The three types are outlet boxes to connect fixtures, a switch box to control the flow of current, and a pull box to connect conductors at bends and long runs.

Electric current
The amount of electricity flowing in a wire or other conductor measured in amperes. Alternating current reverses its direction alternating at 60 cycles per second, abbreviated AC. Direct current flows in one direction only through a circuit from a power source, such as a battery.

Electric service

The means by which an electric utility company supplies power to a facility, either by overhead or underground wires and connected to a service meter to measure the usage.

Electric thermostat timer

Programmable thermostats that save energy by permitting occupants to set temperatures according to whether the house is occupied. These thermostats can automatically store and repeat settings daily with allowance for manual override. By eliminating manual setback, they allow the setting of more comfortable temperatures in the morning before occupants wake. Temperature setback can be adjusted for both heating and cooling seasons. These thermostats typically have a digital interface that allows more precise temperature control and a wider range of options or features.

Electrical engineer

An engineering discipline that deals with the study and/or application of electricity, electronics, and electromagnetism and covers a range of substudies including those that deal with power, electronics, control systems, signal processing, and telecommunications.

Electrical lighting

Covers all technical methods to illuminate spaces inside and outside of buildings in

Electrical system

The entire apparatus for supplying and distributing electricity; including transformers, meters, cables, circuit breakers, wires, switches, fixtures and outlets.

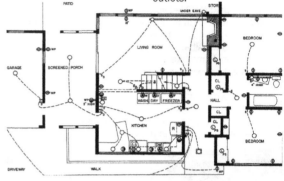

Electrical wiring

Copper wire covered with insulation that carries an electric current from the main panel box through metal conduit throughout the structure to fixtures and outlets.

Electrician

The skilled and licensed person or company who install and maintain electrical systems.

Electrostatic air cleaner

A device that has an electrical charge to trap particles traveling in the airstream.

Element

An integral part of the sub- or superstructure having its own functional requirements; such as foundations, walls, floors, roofs, stairs and structural framework.

Elephantine columns See Column

Elevation

A drawing showing the vertical elements of a building, either interior or exterior, as a direct projection to a vertical plane.

Elevator

A platform or enclosure that can be raised or lowered in a vertical shaft that transports people or freight. The hoisting or lowering mechanism which serves two or more floors, that is equipped with a cab or platform which moves in vertical guiderails for stability. Vintage elevators had exposed mechanisms, which are being used in buildings with a high-tech appearance.

Elgin Marbles

A collection of sculptures, taken from the Parthenon in Athens by Lord Elgin; preserved in the British Museum since 1816. The finest surviving work of Greek sculptural decoration of the Classical age; the collection includes a number of metopes, fragments of pediment statues, and an extended series of blocks carved in low relief of the cella frieze.

Elizabethan style (1558–1603)

A transitional style between the Gothic and Renaissance in England named after the queen, consisting mostly of designs for country houses, characterized by large windows and by strap iron ornamentation.

Ellipse

A closed loop obtained by cutting a right circular cone by a plane.

Elliptical

A plane figure resembling an ellipse, whose radius of curvature is continually changing; a three-centered arch is an example of one constructed to an elliptical shape.

Elliptical arch See **Arch**

Elliptical dome See **Dome**

Elliptical stair See **Stair**

Ellis, Harvey (1852–1904)

American architect whose drawings were published in the *American Architect and Building News*. He produced illustrations for L. S. Buffington that helped pioneer steel-framed skyscraper design.

Elm See **Wood**

Elmslie, George (1871–1952)

Partnered with William Greg Purcell (1880–1965). They were best known for their houses of the Prairie School style. Elmslie worked briefly with Louis Sullivan on the Carson Pirie Scott store in Chicago, IL.

Embankment
A sloped area of earth abutting a structure.

Embattlements
Having battlements; a crenellated molding.

Embedded column See **Column**

Embedded organization See **Organization**

Embellishment
Ornamentation; adornment with decorative elements.

Embodied energy
The energy consumed by all of the processes associated with the production of a building, landscape, or site, from the acquisition of natural resources to product delivery. This includes the mining and manufacturing of materials and equipment, the transport of the materials, and the administrative functions. Embodied energy is a significant component of the life-cycle impact of a home, building, landscape, or site. Included are the product's raw material extraction, transport, manufacturing, assembly, and installation, as well as capital and other costs of a specific

Embodied energy
material, in addition to disassembly, deconstruction, and/or decomposition.

Embodied water
The amount of water required to manufacture products, including the extracting the raw materials, transporting those materials, and processing them into the final product.

Emboss
To raise or indent a pattern on the surface of a material; sometimes produced by the use of patterned rollers.

Embrasure
The crenels or spaces between the merlons of a battlement; an enlargement of a door or window opening at the inside face of a wall by means of splayed sides.

Emergency generators
A gas- or diesel-driven device that generates electrical power when normal power fails. Critical in hospitals and data and communications centers.

Emerson, William R. (1833–1917)
American architect who designed in the Shingle and Stick styles, such as the Forbes House, Wilton, MA, in 1876; the C. J. Morrill House, Bar Harbor, ME, in 1879 in the Stick style; and the Loring House, Pride's Crossing, MA, in 1881, which was his most notable Shingle style work.

Emission factor
Quantity of a substance or substances released from a given area or mass of a material at a set point in time such as milligrams per square meter per hour.

Emissions
The release of gases, liquids, and/or solids from any process or industry. Liquid emissions are commonly referred to as effluents.

Emissivity
Ratio of radiation a surface gives off. Using products with low emissivity—for example, white roofs rather than black roofs—reduces the heat island effect, whereby urban landscapes become significantly warmer than surrounding rural landscapes.

Emphasis
A special importance or significance placed upon or imparted to an element or form by means of contrast or counterpoint; a sharpness or vividness of outline.

Empire style (1800–1830)
The elaborate Neoclassic style of the French Empire in the wake of Napoleon, characterized by the use of delicate but elaborate ornamentation, imitated from Greek and Roman examples, and by the use of military and Egyptian motifs.

Emulsion paint
A paint consisting of small particles of synthetic resin and pigments suspended in water. When the water evaporates, the resin particles form a film that binds the pigments.

Enamel
A high-gloss paint.

Encapsulation
The treatment of asbestos-containing material with a liquid that covers the surface with a protective coating or embeds the fibers in an adhesive matrix to prevent their release into the air.

Encarpus
A sculptural festoon of fruit and flowers.

Encased beam See **Beam**

Encaustic tile See **Tile**

End grain
The face of a piece of timber exposed when the fibers are cut transversely; exposure of this surface to the elements eventually causes deterioration.

End-lap joint See **Joint**

Energy
A resource such as oil, gas or coal from which usable energy can be produced; there are many alternative forms of energy that can be used to produce power.

blue energy
Called osmotic energy, it is generated from a chemical reaction between freshwater and seawater. It can either be installed near a saltwater resource or operated independently using stored water.

fuel cell energy
A kind of battery that produces electricity from the reaction between an externally supplied fuel and an oxidant, in the presence of an electrolyte. Fuel cells generate but do not store energy, and they can provide nonstop, continuous operation as long as the required energy flow is maintained. For home use they provide clean, renewable energy and do not need distribution lines. Small-size home fuel cell units produce heat as a by-product, which can then be used as a cogeneration system for domestic hot water and space heating. Fuel cells can be used in conjunction with power grid systems or used independently as an off-the-grid system in remote areas.

geothermal energy
Produced by using the heat below the earth's surface. Geothermal heat pumps are necessary for residential use; for commercial use, power plants are built on the surface to convert geothermal energy into electricity.

hybrid systems
A method that uses two or more distinct power sources to run a device, such as an on-board rechargeable energy storage system with a fueled power source internal combustion engine or fuel cell, air and internal combustion engines, and photovoltaic modules and wind turbines with electric power.

hydroenergy
Generated from the force of moving water, but a renewable energy resource that usually requires large installations, such as dammed water, to drive a water turbine and generator. There are small, mobile hydrogenerators available for individual home use.

solar photovoltaic
A nonmechanical device that converts sunlight into electricity. Consists of photovoltaic cells, mounting hardware, electrical connections, power-conditioning equipment, and an energy storage device. Individual cells can be grouped into modules to form larger collectors, which can in turn be further grouped into photovoltaic arrays, necessary for industrial-level electricity production. The number and size of the modules can vary depending on the availability and intensity of the sunlight, the geographical location of the modules, and the user's needs.

solar thermal
Uses sunlight to create heat, but needs water to operate. Once solar thermal devices receive sunlight, they concentrate the light and generate heat. The generated heat warms the water, which can either be circulated and used directly, or can be used to drive a turbine that generates electricity. There are three types of solar thermal systems: parabolic troughs, parabolic dishes, and power towers.

wind energy
Wind energy turns a windmill's blades on a rotor that is connected to a main shaft. The main shaft spins a generator, producing energy. The amount of energy generated depends on the speed and direction of the wind. One drawback is proximity of the wind generation facilities to the distribution centers and/or to the homes served, which affects the quality and cost of the energy.

Energy analysis
The analysis of the energy use of a structure.

Energy and atmosphere
LEED (Leadership in Energy and Environmental Design) Rating System category. Prerequisites and credits in this category focus on the individual aspects of energy efficiency, lighting, HVAC, and appliances and equipment.

Energy and Environmental Building Alliance (EEBA)
Provides education and resources to transform the residential design, development, construction, and remodeling industries to profitably deliver energy-efficient and environmentally responsible buildings and communities.

Energy assessment
A written report prepared by a qualified party evaluating energy usage, highlighting weak points in energy efficiency, and identifying cost-saving measures. This is a less rigorous process than an energy audit.

Energy audit
A special inspection performed to determine where there are energy inefficiencies in a home or existing building, including orientation and exposure, building materials and components, and existing mechanical systems; usually performed as a means to determine ways to reduce energy costs. A qualified tester uses methods and measurements that comply with industry standards and involves collection of detailed data and an engineering analysis. A written report includes recommendations and a detailed cost and savings analysis.

Energy conservation
Decreasing the demand for use of energy.

Energy efficiency
Measures advanced industrial processes and high-efficiency motors, lighting, and appliances that have the potential to provide significant reductions in electricity use.

Energy efficient
Products and systems that use less energy to perform as well or better than standard products. While some have higher upfront costs, energy-efficient products cost less to operate over their lifetime.

Energy factor
Overall efficiency of a water heater or other appliance. The amount of hot water produced per unit of gas or electricity purchased. The higher the energy factor number, the more efficient the water heater.

Energy guide label
An appliance label that provides an estimate of how much energy the appliance uses, compares energy use of similar products, and lists approximate annual operating costs. Required by the U.S. Department of Energy.

Energy heel
The point at which typical roof rafters connect with the top plate of the exterior wall leaves little room for the full depth of attic insulation, creating compressed insulation and

Energy heel
reduced performance. An energy heel raises the truss at least 6 inches to allow the insulation enough room to produce its full R-value at this critical location.

Energy management system
A control system capable of monitoring environmental and system loads and adjusting HVAC operations accordingly in order to conserve energy while maintaining comfort.

Energy modeling
Process to determine the energy use of a building based on software analysis. Also called building energy simulation. Common simulation software are DOE-2 and Energy Plus. This is typically a computer model that analyzes the building's energy-related features in order to project energy consumption of a given design.

Energy or water conservation
Using less energy or water. Conservation can imply a lifestyle change or a reduced level of service. Lowering thermostat settings or installing a shower flow restrictor are examples of energy conservation.

Energy or water efficiency
Using less water or energy to perform the same tasks. A device is energy efficient if it provides comparable or better quality service while using less energy than conventional technology. Building weatherization or high-efficiency showerheads are efficiency technologies.

Energy performance contracting
A turnkey service for the implementation of energy-cost-saving measures in buildings where the savings are guaranteed against some measure of performance; these two aspects, turnkey service and guaranteed performance, differentiate performance contracting from other traditional design and construction services.

Energy recovery
A process that creates electricity and heat, such as combustion, from waste or materials that would have otherwise been disposed in a landfill.

Energy recovery ventilator
A mechanical device that draws stale air from the house and transfers the heat or coolness in that air to the air being pulled into the house. This can help reduce energy costs and dilute indoor pollutants. Energy recovery units use the heat in the building to preheat in the winter or cool air in the building during summer to chill the incoming outside air. Ultimately, they use less fossil fuel by saving electricity to warm or cool the air; therefore, the air is cleaner because there is less fossil fuel burned to make the electricity. And the incoming outside air promotes better indoor air quality.

Energy recovery ventilator
An air-to-air heat exchanger or preconditioner designed to exchange temperature and moisture properties from one airstream to another. It captures the cooling or heating energy from the exhaust air before it leaves the building.

Energy retention

The amount of energy that can be retained, rather than dissipated, depends on several architectural factors. These include the size, shape, style, orientation, and construction of the building, as well as the insulation, heating/cooling, and envelope systems within the building. Nonarchitectural factors include the location, local climate, and users' preferences.

Energy smart

Meeting energy needs cost effectively and with the least impact on the environment.

Energy Star

A U.S. government program to promote energy-efficient consumer products. It began as a voluntary labeling program designed to identify and promote energy-efficient products, and computer products were the first to be labeled. It has since expanded to major appliances, office equipment, lighting, home electronics, new homes, and commercial/industrial buildings.

Energy Star appliances and lighting

The Energy Star program is a partnership between the EPA (U.S. Environmental Protection Agency) and the DOE (U.S. Department of Energy) that focuses on saving energy as a means to help preserve the environment and reduce the cost of energy. Lighting and appliances that have met the standards set by the Energy Star program are labeled as such. Energy-efficient choices can save up to one-third of monthly energy costs in addition to reducing the environmental impact of construction.

Energy Star homes

The Energy Star Home program is a joint program of the EPA (U.S. Environmental Protection Agency) and the DOE (U.S. Department of Energy). It helps the public save money and protect the environment through energy-efficient products and practices. New homes can earn the Energy Star, as well as many household products, including large appliances. Tax credits are also available to consumers through the program for certain cars, solar energy systems, and fuel cells. The program also provides tax incentives to residential and commercial builders for incorporating energy saving products and practices into their projects.

Energy Star indoor air package

An EPA (U.S. Environmental Protection Agency) program to promote the construction of new homes that are less likely to have indoor air problems than conventional homes. Homes complying must meet Energy Star Homes requirements as well as additional specifications addressing indoor air quality. Among the program's requirements are details designed to limit water entry into basements, requirements for radon mitigation components in geographical areas with potential radon problems, a requirement for sill pan flashing at window rough openings, mandatory roof gutters except in dry climates, a mandatory layer of rubberized asphalt under roof valleys, a duct-sealing specification, a prohibition against locating furnaces in garages, and a requirement for a mechanical ventilation system that complies with ASHRAE (American Society of Heating, Refrigerating and Air-Conditioning Engineers) 62.2.

Energy Star products

Rated products meet the energy efficiency guidelines specified by the EPA (U.S. Environmental Protection Agency) and the DOE (U.S. Department of Energy). Energy Star products range from computers and office equipment to refrigerators and air conditioners.

Energy Star rating

The label given by the EPA and the U.S. Department of Energy (DOE) to appliances and products that exceed federal energy efficiency standards.

Energy Star window standard

A standard for window performance established by the U.S. Department of Energy (DOE). Energy Star window criteria are climate-zone dependent; the DOE divides the country into four climate zones. In the northern zone, the U-factor of an Energy Star window must be no higher than 0.35. Energy Star windows sold in the southern zone must have a maximum U-factor of 0.75 and a maximum solar heat gain coefficient (SHGC) of 0.40. The criteria for the two central zones fall within the ranges established by the coldest and warmest zones.

Energy truss

Framing method in which roof trusses are raised at the point where the rafters connect to the top plate of the exterior wall to allow room for effective insulation. Also called high-heeled truss, raised-heel truss, energy heel, or Arkansas truss.

Energy-efficiency rating

Also referred to as energy-efficiency ratio, it is the operating efficiency of a room air conditioner, measured in Btu of cooling output divided by the power consumption in watt-hours; the higher the rating, the greater the efficiency.

Energy-efficient appliances

Products that use less energy than conventional models. The Energy Star label is a credible third-party certification of a product's energy efficiency. Consumers can also refer to the FTC's (Federal Trade Commission's) Energy Guide label, a yellow label affixed to most appliances today. Clothes washers, dishwashers, refrigerators, freezers, water heaters, window air conditioners, central air conditioners, furnaces, boilers, heat pumps, and pool heaters can get the label. Televisions, ranges, ovens, clothes dryers, humidifiers, and dehumidifiers do not receive such labels.

Energy-efficient light fixtures

The fixture or the type of bulbs used in a fixture. Compact fluorescent lights (CFLs) and light-emitting diodes (LEDs) are becoming more common in homes and buildings. They are more efficient and last longer than incandescent bulbs.

EnergyGuide

Label from the FTC (Federal Trade Commission) that lists the expected energy consumption of an appliance, heating system, or cooling system and compares consumption with other products in that category. The energy performance is based on specified operating conditions and average energy costs; actual performance may vary.

Energy-plus building
A building that over a typical year produces more energy from onsite renewable energy sources than it consumes.

Engaged
Applied or partially buried in a wall, such as a column with half or more of its shaft visible.

Engaged column See Column

Engineer
A person trained and experienced in the profession of engineering; one licensed to practice the profession by the authority in the area.

Engineered header
Framing member made of engineered lumber and used to carry a wall or roof load above a window or door.

Engineered wood
A range of derivative wood products made from fibers, chips, strands, or veneers bonded with adhesives to form composite materials. Engineered wood products include structural members such as I-beams, floor and ceiling joists, framing studs, and sheet products such as fiberboard or particleboard. Engineered wood products use smaller, younger trees and parts of the tree that were not previously used to reach the same or better structural characteristics of solid dimensional lumber. This makes better use of the resource and avoids the use of larger, older trees. The use of engineered wood products also eliminates the waste associated with warped, twisted, or otherwise unusable lumber.

Engineering
The design of the civil, structural, and electrical portions of a construction project.

English bond See Bond

English cross bond See Bond

Engraving
A design is incised in the reverse on a copper plate; this is then coated with printer's ink, which remains in the incised lines when the plate is wiped off. Damp paper is put on the plate and the two are put into a press; the paper soaks up the ink and produces a print of the original.

Enhanced air filtration
Superior media filters, such as high-level HEPA or even MERV filters on HVAC equipment.

Enriched
Having embellishments.

Enrichment
Any ornamentation on moldings, such as the egg-and-dart on ovolos.

Entablature
The superstructure composed of an architrave immediately above the columns, central frieze, and upper projecting cornice, consisting of a series of moldings. The proportions and detailing are different for each order, and they are strictly prescribed.

Entablature

Entasis

Intentional slight curvature given to the vertical profile of a tapered column to correct the optical illusion that it appears thinner in the middle if the sides are left straight.

Enterprise Green Communities

A nonprofit organization that provides resources and expertise to enable developers to build and rehabilitate homes that are healthier, more energy efficient, and better for the environment, yet still affordable. Green Communities is the first national green building program developed for affordable housing.

Enthalpy

A measure of the total heat content within a given sample of air. It is typically used to determine the amount of fresh outside air that can be added to recirculated air for the lowest heating/cooling cost.

Entrance

Any passage that affords entry into a building; an exterior door, vestibule or lobby.

Entrance

Envelope

The boundary that separates a building's conditioned and unconditioned spaces. The term usually refers to heat and air transfer, such as through walls, windows, and the roof. All of these are part of the building's envelope. Also, the imaginary shape of a building indicating its maximum volume; used primarily to check the plan, setback, and other restrictions regarding zoning regulations.

Envelope forms

With curtain-wall systems, the actual construction and arrangement of the surfaces enclosing the building are totally independent of the bearing system.

Environment

The combination of all external conditions which may influence, modify, or affect the actions of a person, piece of equipment, or any system.

Environmental architectural impact

An effect that is holistic, far-reaching and long-lasting. It includes the all resources expended to construct a building, such as technology, materials, energy, water, and all aspects of transportation. It also includes the waste from construction, the resources of the occupants over the lifetime of the building, and the waste from the demolition or recycling of the structure.

Environmental aspect

The way a manufacturer's activities or products can relate positively or negatively with the environment.

Environmental conditions

Natural elements that act upon or influence the condition of a building, such as earthquakes, wind, rain, sunlight, and humidity.

Environmental conservation

Concern for the maintenance of the environment in its entirety; involves maintaining the cherished aspects of heritage, such as buildings, trees, open spaces, landmarks, and the atmosphere.

Environmental design

The professions collectively responsible for the design of humanity's physical environment, including architecture, engineering, landscape architecture, urban planning and similar environment-related professions.

Environmental footprint

For an industrial setting, this is a company's environmental impact determined by the amount of depletable raw materials and nonrenewable resources it consumes to make its products and the quantity of wastes and emissions that are generated in the process. Traditionally, for a company to grow, the footprint had to get larger. Today, finding ways to reduce the environmental footprint is a priority for leading companies. An environmental footprint can be determined for a building, city, or nation and gives an indication of the sustainability of the unit.

Environmental impact

Any change to the environment, whether adverse or benefi-

Environmental impact

cial, wholly or partially resulting from human activity, industry, or natural disasters. Includes all the social and physical effects of a development or government policy on the natural and built environment.

Environmental impact statement

A document required of federal agencies by the National Environmental Policy Act for major projects or legislative proposals significantly affecting the environment. A tool for decision making, it describes the positive and negative effects of undertaking a project and an agency's alternative actions.

Environmental management plan

A management and communication process to enable the range of environmental aspects and impacts of a given activity, whether a design process, a project, or consultancy or corporate operation, to be systematically considered, addressed, monitored, and reviewed to ensure compliance with regulatory, project, and organizational environmental requirements, policies, and objectives. It is a tool that is used to manage environmental risks and issues on a project. The effectiveness of the plan is dependent on the commitment of management to follow the plan and on how well it is written and understood by construction personnel.

Environmental management policy

An environmental management policy is a statement of an organization's aims in relation to its impact on the environment. It sets overall goals and aspirations for the organization, as well as principles for action.

Environmental management system

A tool for managing an organization's impact on the environment. It provides a structured approach to planning and implementing environmental protection measures.

Environmental planning

The person or professional whose job it is to determine the future physical arrangement and conditions of a community; involves an appraisal of current conditions, a forecast of future requirements, a plan, and proposals to implement the plan.

Environmental preference

To revise product specifications, policies, and/or purchasing contract terms to request or give preference to products or services that minimize impacts on the environment throughout the process of manufacture, distribution, use, reuse and recycling, and disposal.

Environmental Protection Agency (EPA)

Established in 1970 to consolidate the U.S. federal government's environmental regulatory activities under the jurisdiction of a single agency, the mission of the EPA is to protect human health and to safeguard the natural environment. The EPA ensures that federal environmental laws are enforced fairly and effectively.

Environmental rating system

Design and management systems that provide a framework for identifying, setting, and reporting environmental design

Environmental rating system

and performance objectives for buildings; this system provides industry with mechanisms for industry-wide improvement and internal benchmarking while also providing a rating system through which to communicate the environmental performance characteristics of buildings to the community.

Environmental restoration

The act of repairing damage to a site caused by human activity, industry, or natural disasters. The ideal environmental restoration, though rarely achieved, is to restore the site as closely as possible to its natural condition before it was disturbed.

Environmental sustainability

Long-term maintenance of ecosystem components and functions for future generations.

Environmental testing verification

An EPA (U.S. Environmental Protection Agency) program that develops testing protocols and verifies the performance of new technologies.

Environments for living

A green building program that focuses on building science to improve home energy efficiency and comfort.

Epoxy

A plastic material that can be used as a filler to replace missing building components.

Epoxy resin

A group of thermosetting plastics. The uncured resin and hardener are kept separately and mixed just before use. They adhere strongly to metals, glass, concrete, stone, and rubber, and are resistant to abrasion, weather, acids and alkalis, and heat. Useful for repairing damaged concrete and for joining new concrete to old; for this purpose, a mixture of epoxy resin mortar can be used.

Equilateral arch See Arch

Equilateral triangle

A triangle where all three sides are of equal length, with three equal inside angles.

Equilibrium

The state of a body in which the forces acting on it are equally balanced.

Erection

The hoisting and installing in place of the structural components of a building, using a crane, hoist, or any other power system.

Erection bracing

A bracing which is installed during erection to hold the framework in a safe condition until sufficient permanent construction is in place to provide full stability.

Erection stresses

The stresses caused by construction loads and by the weight of components while they are being lifted into position.

Erickson, Arthur (1925–2009)

Erickson's body of work includes the San Diego Convention Center and Fresno City Hall, both in California; Napp Laboratories in Cambridge, England; the Kuwait Oil Sector Complex in Kuwait City; and the Kunlun Apartment Hotel Development in Beijing. He promoted environmentalism and corporate responsibility. He also designed the Canadian embassy in Washington, DC, and the Museum of Anthropology at the University of British Columbia.

Erosion

The wearing away of the land surface by rain or irrigation water, wind, ice, or other natural or anthropogenic agents that abrade, detach, and remove geologic parent material or soil from one point on the Earth's surface and deposit it elsewhere, including such processes as gravitational creep and so-called tillage erosion.

Ersatz style

A German word meaning "substitute" or "replacement," used by architectural critic Charles Jencks to describe architecture (1973–1975) with forms that are borrowed indiscriminately from various sources. This is partly the result of modern technology, which is capable of producing architecture in any style. It can also be considered as any "pastiche" that captures the essence of the original design.

Escalator

A moving stairway consisting of steps attached to an inclined continuously moving belt for transporting passengers up or down between the floors in a structure.

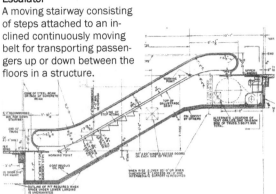

Escutcheon

A protective plate that surrounds the keyhole of a door or door handle.

Escutcheon pin

A small nail, usually brass, for fixing an escutcheon, often ornamental.

Esherick, Joseph (1914–1998)

San Francisco architect most noted for the renovation and redevelopment of the Cannery in 1968, a retail complex on Fisherman's Wharf, San Francisco, and the Monterey Bay Aquarium, Monterey, CA in 1984.

Esplanade

A flat open space used as a walkway.

Esquisse See Design drawing

Estate

A sizable piece of land, usually containing a large house. Because these properties represent an investment opportunity for public conservation and recreation, as well as other forms of private and commercial development, community organizations and local governments are becoming increasingly involved in the preservation of endangered estates.

Esthetic

The distinctive vocabulary of a given style.

Ethylbenzene

A component of paint formations and associated with some carpeting, ethylbenzene off-gases in the home, in office furniture products, in office buildings, and in a subject's breath. Ethylbenzene is considered a chronic toxin.

Etruscan architecture (700–280 B.C.)

A style which flourished in western central Italy until the Roman conquest; it is largely lost, except for underground tombs and city walls, but the characteristic true stone arch influenced later Roman construction methods. Examples that have survived show forms that were rich in ornamentation.

Eurhythmy

Harmony, orderliness, and elegance of proportions.

Evacuated tube solar collector

Solar collector consisting of a series of glass vacuum tubes in which an inner tube containing fluid, or in some types, a metal plate, absorbs heat energy and transfers it for practical use, usually heating water.

Evaporative cooler

Energy-efficient cooling system in which a fine mist of water is evaporated, lowering the air temperature. Evaporative coolers are most appropriate in dry climates, because they add humidity. Also known as a swamp cooler.

Evolutionary architecture

This style is defined by its major proponent, Eugene Tsui, as design that grows and develops based on climatic and ecological elements, as well as advances in science and technology. The design is approached as a living organism as if natural forces had shaped the structure.

Evolutionary architecture

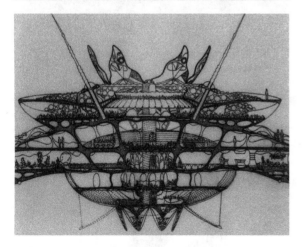

E-waste
Electrical or electronic equipment that is redundant, broken, or unwanted. The equipment can become a hazard, as many components contain toxic materials that some parts of the industry are working to phase out.

Excavation
The removal of earth from its natural position; the cavity that results from the removal of earth.

Exchange rate
The rate at which outside air replaces indoor air in a given space.

Executive mansion
A large stately house; a manor house.

Exedra
A large niche or recess, either roofed or unroofed; semi-circular or rectangular in plan, usually including a bench or seats.

Exfiltration
Uncontrolled outward air leakage from conditioned spaces through unintentional openings in ceilings, floors, and walls to unconditioned spaces or the outdoors caused by pressure differences across these openings due to wind, inside-outside temperature differences (stack effect), and imbalances between supply and exhaust airflow rates.

Exfoliation
Action caused by weathering or salt decay, resulting in the flaking off of surface layers of stone.

Exhaust air
The air extracted from a space and discharged outside.

Exhaust fan
A fan that withdraws air that is not returned to the central air-treatment center and is exhausted to the outside.

Exhaust-only ventilation
Mechanical ventilation system in which one or more fans are used to exhaust air from a house, and makeup air is supplied passively. Exhaust-only ventilation creates slight depressurization of the home; its impact on vented gas appliances should be considered.

Exoskeleton
The system of supports in a French Gothic church, including the ribbed vaults, flying buttresses, and pier buttresses. Also, the external structural skeleton of a building consisting of a framework of attached members, or a poured-in-place concrete framework.

Exotic species
A nonnative species that is artificially introduced to an area.

Expanded metal
The metal network formed from sheet metal by cutting a pattern of slits, followed by pulling the metal into a diamond pattern; used as metal lath for plaster, as reinforcing in concrete, and for making screens.

Expanded polystyrene
Type of rigid foam insulation that, unlike extruded polystyrene (XPS), does not contain ozone-depleting HCFCs. EPS frequently has a high recycled content. Its vapor permeability is higher and its R-value lower than XPS insulation. EPS insulation is classified by type: Type I is lowest in density and strength and Type X is highest.

Expansion anchor
Bolt that is inserted into a hole drilled into masonry or other material, with a device on the end that expands, prohibiting the bolt from being withdrawn.

Expansion joint See Joint

Experimental architecture
Architecture that is committed to experimentation with form, materials, new technology, and construction methodologies. It was the title of a book by Peter Cook (1971), which identified architects who were involved in experimental architecture, such as Friedman, Goff, Soleri, Otto, and Tange, and groups like Archigram and the Metabolists.

Exploded view See Projection drawing

Exploratory demolition
Action prior to remodeling, restoration, or renovation to uncover materials and areas sufficiently to verify the existing conditions of unexposed areas.

Exposed aggregate
A decorative finish for concrete; formed by removing the outer surface of cement mortar before it has hardened or by sprinkling aggregate on the wet concrete after placing.

Exposed brick
The stripping of plaster from interior walls to expose the brick to make the building appear old. A common practice in the adaptive use of old buildings as restaurants and boutiques, oftentimes changing the building's original character by so doing.

Exposure
The area on any roofing material that is left exposed to the elements.

Expressionism (1903–1925)
A northern European style that did not treat buildings only as purely functional structures, but as sculptural objects in their own right. Works typical of this style were Rudolph Steiner in Austria, Antonio Gaudí in Spain, P. W. Jensen Klint in Denmark, and Eric Mendelsohn and Hans Poelzig in Germany.

Extant
Still in existence; not destroyed, lost, or extinct.

Extant drawings
Same as measured drawings.

Extended use
Any process that increases the useful life of an old building through intervention, such as restoration, renovation, or adaptive use.

Extensive garden
Extensive gardens have thinner soil depths and require less management and less structural support than intensive gardens. They do not require artificial irrigation. Plants chosen for these gardens are low-maintenance, hardy species that do not have demanding habitat requirements. The goal of an extensive planting design is to have a self-sustaining plant community.

Exterior finish
The outside finish which is intended primarily to serve as a protection for the interior of the building and for ornamentation; consists of cornice trim, gutters, roof covering, wall material, door and window frames, water tables, corner boards, belt courses and other ornamentation.

Exterior grade plywood
Uses phenol formaldehyde (a volatile organic compound [VOC]) as an adhesive rather than the urea formaldehyde used in interior grade plywood and particleboard.

Exterior insulation finish system
A building product that provides exterior walls with an insulated finished surface and waterproofing in an integrated composite material.

Exterior wall See **Wall**

Extrados
The exterior curve or boundary on the visible face of the arch.

Extrados

Extradosed arch See **Arch**

Extruded polystyrene
Type of rigid foam insulation that is widely used below grade, such as underneath concrete floor slab

Eye
The convex disk at the vortex of an Ionicvolute spiral.

Eye of the dome
An aperture or skylight at the top of a dome; also known as an oculus.

Eyeball downlight
An adjustable recessed fixture that resembles an eyeball.

Eyebar
A metal tension member having an enlarged end, and containing a hole used to make a pinned connection to another part of the structure.

Eyebrow
A curved molding over the top of a window or door, often referred to as a hood.

Eyebrow dormer See **Dormer**

Eyebrow eave
An eave that is carried over a door entry or window in a wavy line.

Eyebrow lintel
A lintel above a window and carried in a wavy line.

Eyebrow window See **Window**

Eyelid wall washers
Eyelid-shaped shield on the room side of the fixture, to direct light onto a wall and shield the source from view.

Eyre, Wilson (1858–1944)
Italian-born American architect, celebrated mostly for his Shingle style residential designs in Philadelphia. He designed the Free Museum at the University of Pennsylvania in 1926, and was associated with Frank Miles Day and Cope & Stewardson.

F f

Fabric
An underlying framework or structure consisting of similar connected parts.

Facade
The main exterior face of a building, particularly one of its main sides, almost always containing an entrance and characterized by an elaboration of stylistic details.

Facadism
Preserving the facade of a building while demolishing or altering the rest of the building.

Face
The front facade of a building or the finished surface of an exposed member.

Face brick See Brick

Face side or edge
The side or edge of a piece of timber that has the best appearance. The face is also the first to be prepared and serves as a datum from which the piece may be brought to a final width and thickness.

Face string See String

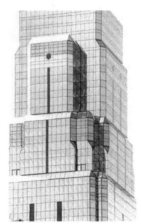

Faceted
Shapes that resemble any of the flat, angular surfaces that are similar to those cut on a gemstone.

Facilities management
A profession that encompasses multiple disciplines to ensure functionality of the built environment by integrating people, place, process, and technology.

Facilities management system
The planning, control, and management of buildings for optimizing the use of real estate, interior environment, energy usage, mechanical infrastructure, communication networks, and maintenance. It considers the life cycle of the building, i.e., purchase, construction, operation, relocation, renovation, demolition, or sale. In the HVAC industry, the term is sometimes used in a more limited way that is synonymous with a building automation system (BAS).

Facility
A physical environment containing open space, such as parks and gardens; infrastructure, such as roads, utilities; and built structures, such as homes, schools, hospitals, and factories.

Facing
A veneer of nonstructural material forming part of a wall and used as a finishing surface of a rougher or less attractive material, such as stone, terra-cotta, metal, stucco, plaster, and wood.

Factor of safety
A factor used in structural design to provide a margin of safety against collapse or serious structural damage. It allows for any inaccurate assumptions in the loading conditions, inadequate control over quality of workmanship, and imperfections in the materials, but not mathematical errors.

Factory
A building or group of buildings containing machinery and other facilities for the manufacture of goods.

Fading
The effect caused by excessive exposure to sunlight or weathering, usually in terms of color or luster.

Failure
A condition when a structure or material ceases to fulfill its required purpose. The failure of a structural member may be caused by elastic deformation, fracture, or excessive deflection. The nonstructural failure of a material may be due to weathering, abrasion, or chemical action.

False
Nonfunctional architectural element; such as a false arch, false attic, false front, false window.

False arch See Arch

False attic
An architectural construction concealing a roof, built without windows or enclosing rooms, and located above the main cornice.

False ceiling See Ceiling

False front
A front wall which extends beyond the side walls and above the roof of a building to create a more imposing facade.

False window See Window

False-face preservation
The retention of only the facade of a historic building during conversion, while the remainder is severely altered or demolished to accept the new use. Also called facadism.

Fan coil
Electric or hydronic heating or cooling element installed in a duct. In a highly energy-efficient home, fan coils in ventilation ducting can be used for heating or cooling the living space.

Fan tracery See Tracery

Fan vault See Vault

Fanlight
A semicircular window, usually over a door with radiating bars suggesting an open fan.

Fantastic architecture
Imaginative architecture by those defying traditional logic or considerations of use. See Visionary architecture.

Farrell, Terry (1938–)
British Postmodern architect, whose most well-known work includes Kowloon Station (1987), and the Victoria Peak Building (1998), both in Hong Kong.

Fascia
A broad horizontal member or molding with nominal thickness, projecting from the wall.

Fashionable
Stylish; conforming to the latest mode of fashion.

Fastener
A mechanical device, weld, bolt, pin or rivet for holding two or more parts, pieces, or members together.

Fast-track
An accelerated method of design and construction management of new construction or rehabilitation, under which sequential contracts for specialized trade work are awarded as plans and specifications are ready, while design is continuing on the other work. The object is to shorten the time between the beginning of design and the completion of construction.

Faucets and showerheads
New low-flow and dual-flow faucets and showerheads can save water, energy, and money. Faucet aerators increase the spray velocity, producing a water stream similar to conventional faucets, but using significantly less water. Dual flow gives the user the options of a wide full-force multiple stream spray, if needed, or a splash-free bubble stream that helps conserve water usage.

Faville, William B. (1866–1937)
American architect who designed the St. Francis Hotel in San Francisco, and the Oakland Hotel, Oakland, CA.

Feasibility evaluation
A detailed evaluation of a project's potential, which usually includes four key areas: (1) market support and economic evaluation, (2) site and location evaluation, (3) structural considerations, and (4) architectural and historical aspects.

Feasibility study
An analysis of the possibilities for a project or structure; which typically includes factors regarding zoning, alternative uses, building codes, financial, environmental, design and historic significance.

Federal style (1780–1820)
Low-pitched roofs, a smooth facade and large glass areas characterize this style. Geometric forms accentuate the rhythm of the exterior wall, which is elegant and intentionally austere. Although it rejected Georgian decoration, it retained its symmetry, pilaster-framed entrance, fanlight and sidelights. Windows were simply framed, and quoins were abandoned.

Federal style

Fehn, Sverre (1924–2009)
Norwegian Architect Sverre Fehn was a Modernist, yet he was inspired by primitive shapes and Scandinavian tradition. Fehn's works were widely praised for integrating innovative new designs with the natural world, such as the Norwegian Pavilion at the 1958 Brussels world exhibition. He was awarded the Pritzker Prize in 1997.

Feilden, Bernard (1919–2008)
British conservation architect who designed inventive building repairs, which he applied to the cracked stone spire of Norwich cathedral, St. Paul's cathedral, the Taj Mahal, and York Minster, in addition to numerous other historical structures.

Feline See Ornament: animal forms

Femur
The vertical surfaces of the triglyph left between the glyphs.

Fence
A structural barrier of wood or iron used to define, separate, or enclose areas like fields, yards, and gardens.

Fenestra
A loophole in the walls of a fortress or castle, from which missiles were discharged: the ancient equivalent of a window.

Fenestration
The design and place-
ment of windows, and
other exterior openings
in a building.

Fenwick, James (1818–1895)
Designed St. Patrick's Catherdral, New York City, a vast
Gothic church.

Ferriss, Hugh (1889–1962)
American architectural delineator and visionary. His images
of futuristic buildings were published in the *Metropolis of
Tomorrow* in 1929, which impacted architecture in the
1930s.

Ferro cement
Mix of Portland cement, sand, and water that is sprayed
on a steel mesh, creating a thin, strong material; often
used for cisterns.

Ferrocement
Concrete made with several layers of finely divided reinforcement, instead of the conventional larger bars.

Ferrous metal See Metal

Ferrule
A metal cap placed at the end of a post for strength and
protection against weathering of the end grain.

Festoon
Hanging clusters of fruit,
tied in a bunch with leaves
and flowers; used as
decoration on pilasters
and panels, usually hung
between rosettes and
skulls of animals.

Festoon

FHA
Federal Housing Administration, a U.S. governmental agency regulating federal housing policy.

Fiber cement
A siding that is more durable than wood and is termite resistant, water resistant, noncombustible, and warranted to last 50 years. It is composed of cement, sand, and cellulose fiber that has been cured with pressurized steam to increase its strength and dimensional stability. The fiber is added as reinforcement to prevent cracking.

Fiber cement siding
Cladding made from a mixture of Portland cement, cellulose or wood fiber material, sand, and other components.

Fiber optics
Light is transmitted from a tungsten-halogen or metal-halide light source through flexible bundles of optical fibers that carry the light to the end of the bundles.

Fiber stress
A term used to denote the direct longitudinal stresses in a beam, such as tension and compression.

Fiberboard
A generic term for building board made from felted wood or other fibers, and a soluble binder; includes particleboard and insulating board.

Fiberglass See Plastics

Field photography
Photographic documentation of existing conditions, construction progress, or completed works in the field, using small format or digital cameras.

Field records
Information gathered from a site or existing building survey, including sketches and measurement notations, and small format or digital pictures.

Fieldstone See Stone

Figure
Sculptural representation of a person or animal.

Filament
A very thin tungsten wire inside an incandescent light. When heated, it glows and emits light.

Filigree
Ornamental openwork of delicate or intricate design.

Filled insulation
A loose insulating material poured from bags or blown by machine into walls.

Filler
Any substance in paste form, used to fill cracks and imperfections in wood or marble.

Fillet molding See **Molding**

Fillet weld
A weld joining two surfaces at right angles to each other.

Final completion
A term denoting that the work is complete and all contract requirements have been fulfilled by the contractor.

Final inspection
A final review of the project by the architect before the issuance of the final certificate for payment.

Final payment
A payment made by the owner to the contractor upon issuance by the architect of the final certificate for payment of the entire unpaid balance of the contract sum as adjusted by any change orders.

Finger joint See **Joint**

Finger-jointed
High-quality lumber formed by joining small pieces of wood glued end to end, so named because the joint looks like interlocked fingers.

Finial
A small, sometimes foliated ornament at the top of a spire, pinnacle or gable which acts as a terminal.

Finish
The texture, color, and other properties of a surface that may affect its appearance.

Finish carpentry
A term applied to practically all finish work performed by a carpenter, except that classed as rough finish. It includes installing casings, laying finish flooring, stairs, fitting and hanging doors, fitting and setting windows, and installing baseboards and other finish material.

Finish flooring
The final floor covering, such as terra-cotta tile, linoleum, parquet flooring, marble or other masonry such as slate or encaustic tile.

Finish grade
The level above a datum when all site work is completed.

Finish hardware
Hardware such as hinges, door handles, and locks that has a finished appearance.

Finished floor level
The final level or position of the finished floor, including any tiles, as opposed to the level of the concrete or wood subfloor surface or floor joists.

Fink truss See **Truss**

Finsterlin, Hermann (1887–1973)
German designer of visionary buildings, many of which were published in books by Bruno Taut, but he did not construct any permanent structures.

Fir See **Wood**

Fire alarm
A device signaling the presence of a fire.

Fire barrier
Any element in a building so constructed as to delay the spread of fire from one part of a building to another; includes fire-resisting doors, enclosed stairways, and other similar constructions.

196

Fire barrier

Fire blocking

In wood frame construction, solid timbers that are of the same dimension, placed between the floor joists and studs.

Fire brick See Brick

Fire damper

A louver that closes automatically in an air duct in the presence of smoke or fire.

Fire detector

An automatic device that signals the presence of heat or flame in a structure.

Fire detector

An automatic device that signals the presence of heat or flame in a structure.

Fire door See Door

Fire escape

A continuous, unobstructed path of egress from a building in case of a fire.

Fire escape

Fire hazard

Any source of risk or danger from fire, such as the improper installation of wires for an electrical system; the use of combustible materials around fireplaces, and open spaces between floor joists that may create a draft and allow fire to spread; also anything that may obstruct or delay the operations of the fire department, or the egress of occupants in the event of a fire.

Fire indicator panel

A control panel that indicates a signal from an alarm and the zone where the signal was initiated.

Fire resistance

The capacity of a material or construction to withstand fire or give protection from it; characterized by its ability to confine fire or to continue to perform a structural function.

Fire resistance grading

The grading of building components according to the minutes or hours of resistance in a standard fire test.

Fire resistive

Applies to materials of construction that are not combustible in the temperatures of ordinary fires and that will withstand such fires without serious impairment of their usefulness for at least one hour, called a one-hour rating.

Fire stop

Obstructions across air passages in buildings which prevent the spread of hot gases and flames, such as the solid blocking between studs and floor and ceiling joists.

Fire wall

An interior or exterior wall having sufficiently high fire resistance and structural stability under conditions of fire to restrict its spread to adjoining areas or adjacent buildings.

Fire wall See Wall

Fire zone

An area of a city or municipality that has a certain level of fire risk associated with it, as defined in the building code; the relative risk is predicated on density, land use, and existing type of construction in the area.

Fireback

The back wall of a fireplace, constructed of heat-resistant masonry or ornamental cast or wrought metal, which radiates heat into the room.

Firebrick

Brick made of a ceramic material which will resist high temperatures; used to line furnaces, fireplaces and chimneys.

Fire-cut

An angular cut at the end of a joist which is anchored in a masonry wall. In the event of a fire, the joist will collapse without forcing the wall to fall outward.

Firehouse

A building for fire equipment and the housing of firefighters. Firehouses are characterized by a combination of features, primarily, a garage and living quarters. Their particular size and shape sets them apart from other buildings.

Fireplace

An opening at the base of a chimney, usually an open recess in a wall, in which a fire may be built.

Fireplace arch

The arch over the front of a fireplace that supports the chimney breast above.

Fireplace back

The rear wall of the firebox; usually constructed of firebrick and sloped forward to reflect heat back into the room, and provide room for a smoke shelf above it.

Fireplace cheeks

The splayed sides of a fireplace.

Fireplace jamb

The interior sides of the firebox, frequently splayed to reflect heat back into the room.

Fireproof construction

A method of building that employs noncombustible materials, such as a steel frame covered with a fire-resistant material.

Fireproofing

Any material that increases the resistance to fire, such as brick, stone, drywall, and sprayed asbestos.

Fire-resistance classification

A standard rating of fire resistance and protective characteristics of a building construction or assembly.

Fire-resistance rating

Refers to the time in hours that a material or construction will withstand exposure to fire.

Fire-retardant chemical

Chemicals used to reduce flammability or retard the spread of flames.

Fire-retardant paint

A paint based on silicone, casein, borax, polyvinyl chloride, or some other similar substance; a thin coating reduces the rate of flame spread of a combustible material.

First cost

The sum of the initial expenditures involved in capitalizing a property or building a project; includes items such as transportation, installation, preparation for service, as well as other related costs. In the context of a building, first cost includes land acquisition costs in addition to the cost of construction.

Fish scale

An overlapping semicircular pattern in woodwork that resembles the scales of fish.

Fish tape

An electrician's device to aid in pulling wires through a conduit. It is a thin steel wire that can be pushed through a system of conduits; then the actual wires are hooked onto the end of it and pulled back through the length of the conduit.

Fitch, James Marston (1909–2000)
Avowed preservationist and associate editor of the *Architectural Record* (1936–1941); then moved on to become editor of *House Beautiful.* He was also a preservation associate with the firm of Beyer Blinder Belle, New York City.

Fixed sash
A fixed window frame that does not open.

Fixed-price contract
A contract in which the work will be carried out for a fixed lump sum without rise and fall cost adjustments, usually used for newer and short-duration projects.

Fixture
Any item that is fixed permanently to a building, such as lighting and sanitary fixtures.

Flagg, Ernest (1857–1947)
American-born architect; New York City. Practitioner of the Beaux-Arts style. Designed the Singer Building in 1899 and Singer Tower in 1908, New York City; and the U.S. Naval Academy in 1908, Annapolis, MD.

Flagpole
A pole on which a flag, banner, or emblem may be raised and displayed; may be self-supporting or attached to a building.

Flagstone See **Stone**

Flamboyant style
The last phase of French Gothic architecture (1450–1500), characterized by flame-like tracery and the profuse use of ornamentation.

Flammable
Describes a material that burns with a flame.

Flange
A projecting collar, edge, rib, rim, or ring on a pipe, shaft, or beam.

Flank
The side of a building.

Flanking window A window adjacent to an external door and showing a common sill line; a side light.

Flansburgh, Earl (1932–2009)
Boston architect who specialized in educational facilities. One of his favorites was the Cornell (University) Store, a facility that he sunk largely underground to preserve the beauty of a landscaped quadrangle.

Flared eaves
An eave that projects beyond the surface of the wall and curves upward toward its outer edge.

Flashing
A thin impervious material placed in construction to prevent water penetration or provide water drainage between a roof and vertical walls and over exterior doors and windows.

Flat arch See **Arch**

Flat arris
The flat edge on a column where a fillet joins the concave surface of a flute

Flat keystone arch See **Arch**

Flat plate collector
A box-shaped solar collector that uses a dark-colored metal plate to absorb radiant heat and transfer it to a circulating liquid or gas that can be used immediately or stored for later use. Typically used for domestic hot water or space heating.

Flat plate slab
A reinforced-concrete floor slab of uniform thickness.

Flat roof See **Roof**

Flat slab
A concrete slab reinforced in two or more directions and supported directly on column capitals; also called mushroom slab because of the shape of the enlarged capitals.

Fleche
A comparatively small and slender spire, usually located above the ridge of a roof, especially one rising from the intersection of the nave and transept roofs of Gothic churches.

Flemish bond See **Bond**

Flemish diagonal bond See **Bond**

Flemish gable See **Gable**

Flemish Mannerism
North European mutation of Flamboyant Gothic and Mannerist styles, exploited cartouches, caryatids, grotesque ornaments, hermes, banded pilasters, obelisks, and other elaborate details; many examples are found in the Guild houses in the Grand Place, Brussels.

Fleur-de-lis ornament See **Ornament**

Fleuron
The small flower-like shape at the center of each side of the Corinthian capital abacus.

Flex duct
Flexible ductwork made with an interior liner, a layer of insulation, and an outer covering of plastic.

Flexible arm
An arm on a lamp that can be adjusted by swivels in a variety of horizontal or vertical directions.

Flicker
Momentary loss of light due to the fluctuation or loss of AC power.

Flight See **Stair**

Flitch
A portion of a log sawed on two or more sides and intended for remanufacture into lumber or sliced or sawed veneer. Also, a complete bundle of veneers after cutting, laid together in the sequence that they were sliced or sawed.

Flitch beam
A built-up beam formed by two wood members with a metal plate sandwiched in between, and bolted together for additional strength.

Float finish
A rather rough concrete finish, obtained by finishing with a wooden float.

Float glass See **Glass**

Floating floor See **Floor**

Floating foundation
A special type of foundation made to carry the weight of a superstructure which is to be erected on unstable soil. Such a foundation consists of a large raft-like slab composed of concrete and reinforced with steel rods.

Floating House
A flat-bottomed boat with walls and a roof, outfitted to function as a floating dwelling.

Floodlight
A spotlight that produces a wide beam of light.

Floor

The lowest surface of a room or structure, which can be a division between one story and another made up of other elements, or a homogeneous material. It is the base plane of any room or structure, and is usually characterized by a flat surface or a series of flat surfaces at different levels.

floating floor

The floor is separated from the rest of the building by supporting it on sleepers or a built-up structural system, to provide sound insulation or space for high-tech flexible electrical service, independent of wall locations.

hollow-tile floor

A reinforced-concrete floor that is cast over a formwork of hollow clay tile blocks, the concrete filling the voids between the tiles.

laminated floor

A structural floor that is constructed using a continuous series of lumber set on edge and nailed together.

raised floor

A false floor, which provides a space for cables or ducts above the structural floor, floor sections are usually supported on short, adjustable peg columns.

Floor cavity ratio

A number indicating floor cavity proportions calculated from length, width, and height. The floor cavity is formed by the workplane, the floor, and the wall surfaces between them.

Floor drain
A fixture providing an opening in the floor to drain water into a plumbing system.

Floor furnace
A heating unit specially adapted for small houses which have no basement or cellar; it is installed directly underneath the floor of a room.

Floor joist See **Joist**

Floor level
A designation for the height above a prescribed or established datum; or the location of the floor within the building itself; such as basement or first floor.

Floor load
The total weight on a floor, including dead weight of the floor itself, and any line or transient load; permissible loading may be stated in pounds per square foot.

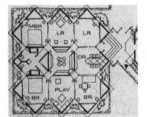

Floor plan
A drawing representing a horizontal section taken above a floor to show, diagrammatically, the enclosing walls of a building, its doors and windows, and the arrangement of its interior spaces.

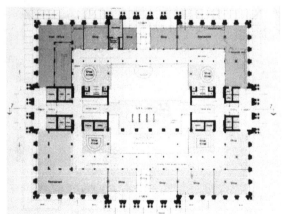

Floor sander
A heavy-duty machine for surfacing floors with sandpaper.

Floor tile See **Tile**

Floor-area ratio
The ratio between the total floor area of all floors of a building which is permitted by code to the area of the lot on which the building is constructed.

Floorboards
The close boarding fixed to the floors joists to provide a surface; normally either timber, plywood, or chipboard.

Flooring

Any material used for the surface of a floor, such as boards, bricks, planks, tile, or marble. Nontoxic green flooring is made from environmentally friendly and renewable resources. Those that are FSC-certified or approved by international associations dedicated to socially responsible trade practices, include bamboo, cork, and natural linoleums made of linseed, cork, tree rosin, limestone, and jute.

Flooring nail
A steel nail with a mechanically deformed shank, often threaded helically, having a countersunk head and a blunt diamond point.

Florentine arch See **Arch**

Florentine mosaic See **Mosaic**

Floriated
Decorated with floral patterns.

Floriated

Florid
Highly ornate; extremely rich to the point of being overly decorated.

Flue
An incombustible and heat-resistant enclosed passage in a chimney to control and carry away products of combustion from a fireplace to the outside air.

Flue lining
Fire clay or terra-cotta pipe, either round or square, usually in 2-foot lengths of varying size, used for the inner lining of chimneys.

Fluorescent bulb
Introduced in 1930, the bulb produces light by passing an electrical current through a gas-filled tube. Available in either warm or cool light, with long bulbs that fit into a ballast, which can be mounted onto a ceiling or under cabinets.

Fluorescent lamp
A discharge lamp of low intensity. Light is produced by an electric arc, producing ultraviolet light that energizes the phosphors coating the inside of the tube. This action converts the low-voltage electrical charge into visible light. They are very efficient and produce very little heat, but they require a ballast to start the lamp and regulate the current through the lamp.

Fluorescent lighting
Type of energy-efficient lighting introduced in the 1930s in which electric discharge within a sealed glass tube energizes mercury vapor, producing ultraviolet (UV) light, which is absorbed by a phosphor coating, which in turn fluoresces, generating visible light.

Fluorescent luminaire
A lighting fixture containing a series of fluorescent lamps in a reflector, either suspended from the ceiling or housed in the surface of a ceiling, with diffusing grilles that can be lowered when the lamps need replacement.

Fluorinated gas
A greenhouse gas typically associated with refrigerants and aerosols.

Fluorocarbons

Carbon-fluorine compounds that often contain other elements such as hydrogen, chlorine, or bromine. Fluorocarbons are used as refrigerants and propellants in aerosol products, among other uses. Common fluorocarbons include chlorofluorocarbons (CFCs), hydrochlorofluorocarbons (HCFCs), hydrofluorocarbons (HFCs), and perfluorocarbons (PFCs). In 1978 the United States began phasing out the use of CFCs because of their ozone-depleting effect.

Flush

Signifying that the adjoining surfaces in a building or in a wall are even, level, or arranged so that their edges are close together and on the same plane.

Flush bead molding See Molding

Flush door See Door

Flush eaves

The eaves fascia is set against the wall surface and is attached directly to it.

Flush joint See Joint

Flush molding See Molding

Flush mortar joint See Joint

Flush panel

A panel whose surface is in the same plane as the face of the surrounding frame.

Flush paneled door See Door

Flush-out

A period after work is finished and prior to occupation that allows the building's materials to cure and release volatile compounds and other toxins. A building flush-out procedure is normally carried out, with specified time periods, ventilation rate, and other criteria.

Flute

A groove or channel that is usually semicircular or semielliptical in section; especially one of many such parallel grooves that is used decoratively, as along the shaft of a column.

Fluted glass See Glass

Fluting

The hollows or parallel channels cut vertically on the shaft of columns, pilasters and piers, separated by a sharp edge or arris, or by a small fillet.

Fly ash

A fine glass powder recovered from the gases of burning coal during the production of electricity. It is an ash residue from high-temperature combustion processes. Electric motor plants using western coal produce a nontoxic fly ash that because of its very high calcium content can be a substitute for Portland cement, the common bonding material in concrete, and these micron-sized Earth elements consist primarily of silica, alumina, and iron. When mixed with lime and water, fly ash forms a cementitious compound with properties very similar to that of Portland cement. Because of this similarity, fly ash can be used to replace a portion of cement in the concrete, providing some distinct quality advantages. The concrete is denser, resulting in a tighter, smoother surface with less bleeding. Fly ash concrete offers a distinct architectural benefit with improved textural consistency and sharper detail. Regulations vary from state to state; however, ASTM International suggests that fly ash must not contain more than 6 percent unburned carbon to be used for its cementitious qualities. Substitution of fly ash for Portland cement in concrete is considered a sustainable building strategy, as it reduces the amount of energy-intensive (and CO_2-producing) cement in the mix.

Flying buttress See Buttress.

Flying facade

The continuation of the facade wall above the roofline of a building.

Focal lighting
A method of creating brighter areas with the general lighting of a space. Used to light for specific tasks, such as reading, or areas, such as under kitchen cabinets.

Focus
A center of interest or activity drawing attention to the most important aspect of a design scheme, such as the main space, materials, scale, lighting, or orientation.

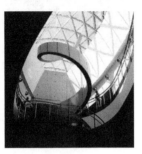

Fog
To fog a room or building is to use a fog machine during a blower door test, in order to reveal the locations of air leaks where the fog escapes. The fogging material is usually a glycol-based solution, completely nontoxic.

Foils
The foliation consists of the cusps, which are the projecting portions and the spaces between the cusps.

cinquefoil
A five-lobed pattern divided by cusps.

multifoil
Having more than five foils, lobes, or arcuate divisions.

quatrefoil
Having four foils, lobes, or acute divisions.

sexfoil
Having six-lobed pattern divided by cusps.

trefoil
A three-lobed cloverleaf pattern.

Folding casement window See **Window**

Folding door See **Door**

Folding partition
A floor-to-ceiling folding door used to subdivide a room into smaller spaces.

Foliated
Adorned with foils, as on tracery; decorated with a conventionalized representation of leafage, applied to capitals, friezes, panels, or other ornamental moldings.

Folk architecture
Simple structures usually intended to provide only basic shelter suitable for the surrounding terrain, without concern for following any architectural style; built of local materials and available tools by people who would inhabit them.

Folk Victorian style (1870–1915)
An architectural style characterized by overall simplicity of form. Decorative treatment is usually confined to porch trim, gable trim, and brackets under the eaves.

Folly
A functionally useless, whimsical or extravagant structure; often a fake ruin; sometimes built in a landscaped park to highlight a specific view, serve as a conversation piece, or to commemorate a person or event.

Foot
The lowest part of an object, such as the end of a rafter where it meets the top plate.

Footbridge See **Bridge**

Foot-candle
A unit of measurement for the average light on a surface; 1 foot-candle equals 1 lumen per square foot.

Footing

That portion of the foundation of a structure that transmits loads directly to the soil; may be enlarged to distribute the load over a greater area to prevent or to reduce settling.

continuous footing

Combined footing, which acts like a continuous beam on the foundation.

spread footing

A footing that is especially wide, usually constructed of reinforced concrete.

Foot-lambert

The unit of brightness of a light source or of an illuminated surface.

Footprint

The projected area of a building or piece of equipment on a horizontal surface.

Force

Anything that changes or tends to change the state of rest of a body; common forces in buildings are the weight of the materials from which they are built, the weight of the contents, and the forces due to wind, snow, and earthquakes.

Forced circulation

Movement of a fluid by pumping as opposed to gravity circulation.

Forced-air heating

Heat distribution system in which heat is delivered by forcing warm air through a network of ducts. A furnace or heat pump typically generates the warm air.

Forecourt
A court forming an entrance plaza for a single building or a group of several buildings.

Forest Stewardship Council (FSC)

A nonprofit organization that encourages the responsible management of the world's forests. FSC sets high standards to ensure that forestry is practiced in an environmentally responsible, socially beneficial, and economically viable way. Landowners and companies that sell timber or forest products seek certification as a way to verify to consumers that they have practiced forestry consistent with FSC standards. Independent, certification organizations are accredited by FSC to assess forest management and determine if

Forest Stewardship Council

standards have been met. Certifiers also verify that companies claiming to sell FSC-certified products have tracked their supply back to FSC-certified sources.

Form

The contour and structure of an object as distinguished from the matter composing it; a distinctive appearance as determined by its visible lines, figure, outline, shape, contour, configuration and profile.

Form lining

The lining of concrete formwork can be used to impart a smooth or patterned finish to the concrete surface; to absorb moisture in order to obtain a drier surface; or to apply a set-retarding chemical, which allows the aggregate to be exposed.

Form lumber

In building forms, any lumber or boards used for shaping and holding green concrete until it has set and thoroughly dried.

Formal balance See Balance

Formal garden
A garden that has plantings, walks, pools and fountains that follow a definite recognizable plan, and are frequently symmetrical and emphasize geometrical forms.

Formaldehyde

A gas used widely in production of adhesives, plastics, preservatives, and fabric treatments and commonly emitted by indoor materials that are made with its compounds. It is highly irritating if inhaled and is now listed as a probable human carcinogen.

Formalism

A style representing a new classicism in American architecture (1950–1965), manifested in buildings designed by Mies van der Rohe, Phillip Johnson, Paul Rudolph and Minuro Yamasaki.

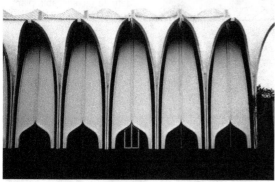

Formalism

Formwork

A temporary framework of wood to contain concrete in the desired shape until it sets.

Fort

A fortified place or position stationed with troops; a bastion, fortification. The first buildings erected in towns had protective walls.

Fort-Brescia, Bernado (1951–)

Peruvian-born American architect who with his wife, the U.S. born Lauinda Hope-Spear (1950–) established the architectural firm Arquitectonica (1977). Among their best-known work is the Atlantis Condominium, Miami (1985) and multipurpose hotel and entertainment developent in the Times Square area of New York City (2001).

Fortification

A structure designed to defend the town or area in which it was built; every element had a military defensive purpose.

Fortress

A fortification of massive scale, generally of monumental character; sometimes including an urban core as a protected place of refuge.

Forum

A Roman public square surrounded by monumental buildings, usually including a basilica and a temple; the center of civic life was often purely commercial.

Foster, Sir Norman Robert (1935–)

English architect; and one of the most distinguished practitioners of the High-Tech style. Among his largest commissions were the Hong Kong and Shanghai Bank, Hong Kong (1979), Willis Faber U. Dumas Building, Ipswich (1974), Century Tower, Tokoyo (1987), the Chek Lap Kok International Airport, Hong Kong, the largest enclosed space in the world (1998), the Millennium Tower project, London (1997), and the City Hall (illus,), London,

Fouihoux, Jacques-Andre (1879–1945)

Paris-born American architect who as partner with Raymond Hood designed the McGraw-Hill Building, New York City, in 1932. The partners were partly responsible for the design of Rockefeller Center, New York City; when Hood died, Fouihoux joined Wallace K. Harrison to complete it. He also contributed to the design of the New York World's Fair in 1939.

Found space

A space within an existing building that was not utilized prior to rehabilitation or adaptive use after having been abandoned; such as converting an attic, basement, or constructing new mezzanine levels.

Foundation

The lowest division of a building that serves to transmit and anchor the loads from the superstructure directly to the earth or rock, usually below ground level.

Foundation wall See Wall

Fountain

An architectural setting incorporating a continuous or artificial water supply, fed by a system of pipes and nozzles through which water is forced under pressure to produce a stream of ornamental jets.

Fountain

Fowler, Orson Squire (1809-1887)
Not an architect, but a leading phrenologist, who published the book *A Home for All* in 1848, which established the octagonal plan as a viable building type. It was constructed in many different materials all over America, including the Armour-Stiner House in 1859, a multistyled structure in Irvington, NY, and the McElroy House in San Francisco, CA, in 1861.

Foyer
An entranceway or transitional space from the exterior of a building to the interior spaces.

Fractable
A coping on the gable end of a building when carried above the roof, and broken into steps or curves forming an ornamental outline

Frame
The timber work which encloses and supports the structural components of a building

Frame building
A building built with a wood frame rather than masonry; includes balloon frame, half-timbered, and timber frame.

Frame construction
Any type of construction in which the building is supported mainly by a frame, and not mainly by load-bearing walls. Balloon-framed lumber houses, steel-framed buildings, and reinforced concrete frame buildings all belong in this category.

Framework
Composed of individual parts that are fitted and joined together as skeletal structures designed to produce a specific shape, or to provide temporary or permanent support.

Framing
A system of rough timber structural woodwork that is joined together in order to support or enclose, such as partitions, flooring and roofing.

Francois I (Premier) style
The culmination of the early phase of French Renaissance architecture (1515-1547), named after Francis I, merged Gothic elements with the full use of Italian decoration. Fontainbleau and the chateaux of the Loire, among them Chambord, are outstanding examples.

Franta, Greg (1951-2009)
He founded the Ensar Group, a sustainable design-consulting firm, in Boulder, Colorado, in 1981 and worked on more than 800 projects worldwide. He was a founder of the AIA Committee on the Environment (COTE) and is considered to be a father of the green building movement.

Franzen, Ulrich (1921-)
Designed the Alley Theater, Houston (1968) and the Multi-categorical Animal Lab at Cornell University, Ithaca, New York (1974).

Fraternity house
A building used for social and residential purposes by an association of male students called a fraternity.

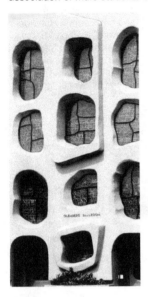

Free-form
Shapes that are characterized by a free-flowing rather than a geometric structure, resembling forms found in nature.

Free-form style (1965–1973)
A style relating to organic and biomorphic forms, such as kidney and boomerang shapes as opposed to rectangular or circular shapes produced by the compass. These forms were also popular in applied arts and the design of furniture.

Freestanding
A structural element that is fixed by its foundation at its lower end but not otherwise constrained throughout its vertical height.

Freight elevator
In a building, an elevator with an extra heavy platform, reinforced side braces to keep the platform from tipping, extra rigid car frames, and additional guide-rail brackets to anchor the guide rails more securely to the building structure; used for transferring freight from one floor level to another.

French arch See **Arch**

French Colonial architecture
A style developed by the French colonists in America, particularly in New Orleans from 1700 onward, which featured a symmetrical facade with a porch reached by steps and a projecting roof across the entire front and sometimes around the sides, and wrought-iron balconies extended over the sidewalk. They typically had high steeply pitched roofs, decorated with ornamental finials at each end of the roof ridge.

French door See **Door**

French Second Empire style (1860–1875)
Called "mansard" for its characteristic roof, similar to the Louvre in Paris; its height was emphasized by elaborate chimneys, dormer windows, and circular windows protruding from the roof. In residences, frequently of wood, the style was asymmetrical and included porches and towers.

French window See **Window**

Fresco
A mural painted into freshly spread moist lime plaster; in such work, groundwater-based pigments unite with the plaster base; retouching is done after it has dried.

Fresco

Fresh air
Outside air drawn into a space or HVAC system.

Fresh air ventilation
A mechanical ventilation component of the HVAC system that draws in fresh air rather than recirculating and filtering air within a home.

Fret See **Ornament**

Fretting
Decoration produced by cutting away the background of a pattern in stone or wood, leaving the rest as a grating.

Fretwork
A rectangular motif used in early Greek border ornament or pattern, rarely as an isolated ornamental device: an angular counterpart of the spiral or wave.

Friable
Ability of a solid material to break down or disintegrate. A friable insulation material may lose its effectiveness; some friable materials release hazardous dust into a house.

Frieze
An elevated horizontal continuous band or panel that is usually located below the cornice, and often decorated with sculpture in low relief.

Frieze-band window See **Window**

Frigidarium
The cold water swimming bath in a Roman bath.

Frilled
An ornamental edging such as a scroll which has added decorative carving along its projecting edges.

Fritted glass
A special type of glass that uses ceramic-enamel coatings in a visible pattern such as dots or lines to control solar heat gain. The pattern is created by opaque or transparent glass fused to the substrate glass material under high temperatures. The substrate is heat-strengthened or tempered to prevent breakage due to thermal stresses.

Frontage
The extent of land along a public highway, stream, or city lot; also, the extent of a building along a public street, road, or highway.

Front-gabled See Gable

Frontispiece
The decorated front wall or bay of a building; a part or feature of a facade, often treated as a separate element, and ornamented highly; an ornamental porch or main pediment.

Frontispiece

Frosted
Rusticated, with formalized stalactites or icicles; given an even, granular surface to avoid shine; closely reticulated or matted to avoid transparency.

Frosted glass
Glass which has been surface-treated to scatter light or to simulate frost.

Frost-protected shallow foundation
A foundation system in which foam insulation is placed around the perimeter of a foundation to reduce heat loss through the slab and/or below-grade walls, subsequently raising the frost depth of a building and allowing foundations to be as shallow as 16 inches below grade.

Fry, Edwin Maxwell (1899–1987)
English modernist architect, known for his buildings in Britain, Africa, and India. Originally trained in the Neoclassical style of architecture, Fry grew to favor the new Modernist style and practiced with eminent colleagues including Walter Gropius, Le Corbusier, and Pierre Jeanneret. In the 1950s he and his wife, the architect Jane Drew, worked on a development to create a new capital city of Punjab at Chandigarh, India. Fry's works in Britain range from railway stations to private houses to large corporate headquarters.

Fuel cell
An electrochemical device in which hydrogen is combined with oxygen to produce electricity with heat and water vapor as by-products. Natural gas is often used as the source of hydrogen, and air as the source of oxygen. Since electricity is produced by a chemical reaction and not by combustion, fuel cells are considered to be green power producers. Fuel cell technology is quite old, dating back to the early days of the space program. Commercial use of fuel cells has been sporadic; however, the use of fuel cells in buildings is expected to increase in the next decade.

Full round
Sculpture in full and completely rounded form.

Full-cost accounting
An accounting system in which environmental costs are built
directly into the prices of products and services.

Fuller, Millard (1935–2009)
Habitat for Humanity founder Millard Fuller applied "the
theology of the hammer" to more than 300,000 homes
in more than 100 countries, Fuller's construction initiative
transformed the lives of more than a million low-income
people. Habitat homeowners pay no interest and live in
homes built with donated money and materials and volun-
teer labor; homes are sold without making a profit. A
second generation of Habitat called the Fuller Center for
Housing is active in 24 states and 14 countries.

Fuller, R. Buckminster (1895–1983)
Developed the geodesic dome, protecting an interior space,
suitable for any arrangement, by using a vast "space frame."
The most well known of these are the American Pavilion,
Expo 67, Montreal, in 1967; and Epcot Center at Disney
World in Florida in 1982.

Full-spectrum lights
The lights that imitate the natural light spectrum and are
therefore considered healthier.

Functional performance test
A range of checks and tests carried out to determine
whether all components, subsystems, systems, and interfaces between systems function in accordance with the
contract documents. In this context, function includes all
modes and sequences of control operation, all interlocks,
conditional control responses, and all specified responses
during design day and emergency conditions. Functional
tests are performed after construction checklists have been
completed.

Functional performance testing
The process of determining the ability of the commissioned
systems to perform in accordance with the owner's project
requirements, basis of design, and construction documents.

Functionalism
A design movement (1920–1940) evolved from several
previous movements in Europe, advocating the design of
buildings and furnishings as a direct fulfillment of functional requirements. The construction, materials, and
purpose was clearly expressed, with the aesthetic effect
derived chiefly from proportions and finish, to the exclusion
or subordination of purely decorative effects.

Fundamental building systems commissioning
Verification by an outside source that the fundamental
building elements and systems are designed, installed,
and calibrated to operate as intended.

Fungicide
Substance that kills fungi, including mold and mildew,
and yeasts.

Fungus
Molds, mildews, yeasts, mushrooms and puffballs; a
group of organisms that are lacking in chlorophyll and
usually nonmobile, filamentous, and multicellular. Some
grow in soil; others attach themselves to decaying trees
and other plants to obtain nutrients. Some are pathogens;
others stabilize sewage and digest composted waste.

Funk architecture (1969–1979)
An alternative form of architecture, using makeshift structures erected from waste materials, developed by members
of rural communes.

Furnace
System used to heat air for a forced-air heating system.
Furnaces can be gas-fired, oil-fired, wood-fired, or electric.

Furness, Frank (1839–1912)
American architect who designed the Provident Life and
Trust Company building, Philadelphia, in 1876, in the high
Victorian Gothic style, and the Academy of Fine Arts, (illus.),
and the Universsity of Pennslyvania ibrary (illus.), now the
Fisher Fine Arts Library, both located in Philadelphia. In
1876. Louis Sullivan worked in the office of Frank Furness
before moving to Chicago.

Furness, Frank

Furnishing and outfitting
That complete sequence or series of activities and actions that begin with the structure and result in the completed building.

Furring
A series of parallel wood strips used to support and level plaster lathing, drywall, or sheathing, and to form an air space; typically used on walls, beams, columns, and ceilings.

Furring brick
A hollow brick that is grooved on one face for furring the inside face of a wall; usually the size of an ordinary brick, and grooved or scored on the face to afford a key for plastering.

Furring channel
Small metal channels used to hold the metal lath in a plaster ceiling.

Furring strip
A thin piece of wood used to build up areas to align panels.

Fuse
An electrical safety device inserted in a circuit to prevent overload; excessive current melts a wire inside a fuse, which interrupts the flow. It is no longer functional once this happens, unlike a circuit breaker, which can be reset.

Future-proofing
Attempting to anticipate the future in order to minimize possible negative consequences. In the building sector, future-proofing is demonstrated by sustainable buildings that use less water and energy than conventional buildings, thereby providing a buffer against future increases in wa- ter- and energy-service costs and protecting against service shortages.

Futurist style (1914–1916)
A movement that began with a publication by two young architects, Antonio Sant'Elia and Mario Chiattone. They presented a series of designs for a city of the future. Their manifesto proclaimed that architecture was breaking free from tradition, starting from scratch. It had a preference for what is light and practical. None of these designs were ever constructed.

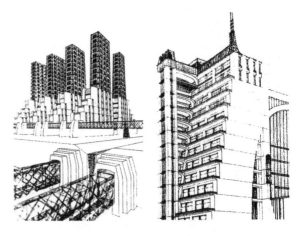

G g

Gable

The entire triangular end of a wall, above the level of the eaves, the top of which conforms to the slope of the roof which abuts against it, sometimes stepped and sometimes curved in a scroll shape.

bell gable

A gable-end parapet with an opening that supports a bell; it is found in Spanish Colonial architecture.

broken gable

A vertical surface at the end of a building having a broken-pitch roof; extending from the level of the cornice to the ridge of the roof.

crowstep gable

A masonry gable extended above the roof with a series of setbacks; often found in European medieval architecture, especially Dutch architecture.

Dutch gable

A gable, each side of which is multicurved and surmounted by a pediment.

Flemish gable
A masonry gable extended above the roof with set-back stages that may be stepped or curved profiles in any of a wide variety of combinations.

front-gabled
A term describing a building having a gable on its main facade.

hanging gable
A small extension of the roof structure at the gable end of a barn or house.

hipped gable
The end of a roof formed into a shape intermediate between a gable and a hip; the gable rises about half-way to the ridge, resulting in a truncated shape, the roof being inclined backward from this level.

mouse-tooth gable
Dutch term referring to the infilling in the steps of a crow-step gable. Brick is laid at an angle that is perpendicular to the slope of the gable within the steps, and the gable is finished off with a smooth stone coping.

multicurved gable
A gable having an outline containing two or more curves on each side of a central ridge, as in a Flemish gable.

paired gables
A facade having two gables

parapeted gable
A gable end-wall that projects above a roof; typical shapes include boltel, fractable, and square.

side gable
A gable whose face is on one side of a house, perpendicular to the main facade.

stepped gable
A gable with a stepped profile; usually constructed of brick; also called a corbiestep gable or a crowfoot gable.

straight-line gable
A gable that rises above the roof line with a straight incline following the roof below it.

wall gable
A portion of a wall that projects above the roof line in the form of a gable.

Gable front
Building with a gable roof where the main entrance is located in one of the gable ends, also known as front-gabled.

Gable ornamentation
Any type of decorative element, such as spindlework or scrollwork at the apex of a gable.

Gable post
A post directly under the ridge of one end of a gable roof; supports the intersection of the barge-board at the roof edge.

Gable roof See **Roof**

Gable wall See **Wall**

Gable window See **Window**

Gabled tower
A tower that is finished with a gable on two or all sides, instead of terminating in a spire.

Gablet
A small gable; usually found over a dormer window or at the apex of a gable wall.

Gabriel, Jacques (1698–1782)
One of the greatest eighteenth-century Neoclassical architects in France; architect to King Louis XV, with additions to the Royal Palace at Fontainbleau in 1748 and the Place de la Concorde, Paris, in 1753.

Gaine
A decorative pedestal, taking the place of a column, tapered downward and rectangular in cross section, forming the lower part of a herm, on which a human bust is mounted; often with a capital above.

Gaine

Galleria
A long interior passageway lit by a continuous skylight and lined with retail stores; also called an arcade.

Gallery
A long covered area acting as a corridor inside or on the exterior of a building or between buildings. A room, often top-lit, used for the display of artwork.

Gallier, James, Sr. (1798–1866)
American architect in New Orleans; partner of Minard Lafever, Charles B. Dakin, and James Dakin. Leading Greek Revival architect; he designed the Government Street Presbyterian Church, Mobile, AL, in 1835; Saint Charles Hotel, New Orleans, in 1836; and the City Hall in New Orleans in 1850.

Gallons per flush
Measurement of water use in toilets. Since 1992, toilets sold in the United States have been restricted to 1.6 gallons per flush or less. The standard for high-efficiency toilets is 1.28 gallons per flush.

Gallons per minute
Measure of liquid flow.

Garage
A building or portion of a residence where motor vehicles are kept; a place for repairing and maintaining vehicles. Also see Parking garage.

Garden
A piece of ground, open or enclosed, appropriated to plants, trees, shrubs, or other landscape features.

Garden apartment
A ground-floor apartment with access to a garden or other adjacent outdoor space: two- or three-story apartment buildings with communal gardens.

Garden city
An idea put forward by Ebenezer Howard (1902) which suggested building medium-size communities of about 32,000 people, each complete with its own facilities and industries, and a planned layout surrounded by open countryside. Houses were grouped so they opened onto their own gardens. This concept can be seen in variations throughout history, including Broadacre City, designed by Frank Lloyd Wright.

Garden house
A summer house in a garden or garden-like situation.

Garden, Hugh (1873–1961)
American architect in partnership with Richard E. Schmidt.

Gargoyle
A spout carrying water from the roofs above, frequently carved with grotesque figures or animals with open mouths, from which water is discharged away from the building's walls

Garland

An ornament in the form of a bank, wreath, or festoon of leaves, fruits, or flowers.

bay leaf garland

A stylized laurel leaf used in the form of a garland to decorate torus moldings.

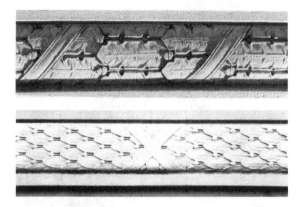

Garnier, Charles (1825–1898)

A French architect who developed the luxury flat style in the rebuilding of Paris. He also designed the grandiose Neobaroque Opera House in Paris in 1861, a triumph of rich color, ornate decoration, and highly disciplined control of mass and space. Other work includes the casino at Monte Carlo in 1878.

Garnier, Tony (1869–1948)

While a student in Rome, published *Une Cité Industrielle* in 1918, which constituted a revolutionary plan for a model town of 35,000. This publication influenced Le Corbusier and other Modernists.

Garret

A room or space located just beneath the roof of a house usually with sloping ceilings; sometimes called an attic.

Garret window

A skylight so arranged that the glazing lies along the slope of the roof.

Gas

A state of matter, including natural gas and propane, used as a fuel to produce energy, generally for lighting and heating.

Gas bracket

A wall-mounted ornamental bracket that carries a gas-lighting fixture.

Gas fitting

Providing gas piping, fixtures and appliances and installing them within a building.

Gas fixture

A gas-operated device used for lighting interiors, and for exterior lamps prior to electrification.

Gas heater

A gas-fired space heater, installed in a wall, or one used to heat water in a storage tank.

Gas station

A retail establishment at which vehicles are serviced, especially with gasoline, oil, air, water, and antifreeze or coolant. They were the first structures built in response to the automobile, and are undoubtedly the most widespread type of commercial building in America.

Gas-fired absorption chiller

Mechanical equipment that is used to generate chilled water for cooling of buildings. Conventional chillers use electricity as the energy source, whereas gas-fired absorption chillers use clean-burning natural gas. While conventional chillers have a compressor and use refrigerants to produce cooling, absorption chillers contain an absorber, generator, pump, and heat exchanger, and they do not use ozone-depleting substances. The absorption cycle uses environmentally friendly working fluids, namely, water (refrigerant) and lithium bromide (absorbent).Some absorption chillers use ammonia as the refrigerant and water as the absorbent.

Gate

A passageway in a fence, wall, or other barrier which slides, lowers, or swings shut, and is sometime of open construction.

gopuram

In Indian architecture, a monumental gateway tower to a Hindu temple, usually highly decorative.

moon gate

A circular opening in a wall, in traditional Chinese architecture.

pai-lou

A monumental Chinese arch or gateway with one, three or five openings, erected at the entrance to a palace, tomb, or processional way; they are usually built of stone in imitation of earlier wood construction.

pai-lou

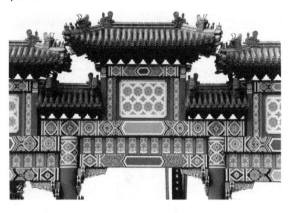

torana

An elaborately carved ceremonial gateway in Indian, Buddhist and Hindu architecture, with two or three lintels between two posts.

tori

A monumental, freestanding gateway to a Shinto shrine, consisting of two pillars with a straight cross piece at the top and a lintel above it, usually curving upward.

Gate to gate

A term used to describe the product boundary encompassing the fabrication and assembly of business and institutional furniture. For purposes of the assessment, the entry gate is the receiving dock of the first facility where basic materials used in the manufacture of the furniture, e.g., steel, particleboard, fabric, and laminate, begin the conver-

Gate-to-gate

sion to furniture components. The end gate is the shipping dock where the ready-to-install furniture is transported for distribution to the end user. The gate-to-gate assessment will include transportation of intermediate materials and components between facilities where more than one physical location is included in the manufacturing process.

Gate tower
A tower containing a gate to a fortress.

Gatehouse
A building, enclosing or accompanying a gateway for a castle, manor house, or similar building of importance.

Gateway
A passageway through a fence or wall; the structures at an entrance or gate designed for ornament or defense.

Gaudí, Antonio (1852–1926)

One of the most original architectural talents, inspired by Islamic and Gothic sources, whose work is mainly found in Barcelona. Casa Vicens, 1878, was his first work, a suburban house decorated with polychrome tiles and sinuous ironwork. The commission for La Sagrada Familia Church, 1884, with completed transept facades, with its extraordinary ceramic-covered spires, is his most fantastic work. The Palacio Guell, 1885, is dominated by a pair of parabolic arches at the entrance and topped by chimneys encrusted with colored tiles. Casa Batlo, 1906, has a unique tile roof and tile facade, and Casa Mila, 1910, has an undulating facade and huge ceramic-covered chimney pots on the roof. Park Guell, 1914, is a playful open space surrounded by undulating and tile-encrusted benches along with fanciful gatehouses with imaginative use of rock and tile.

Gauge
To shape a brick by rubbing or molding it into a particular size.

Gauged arch See **Arch**

Gauged brick See **Brick**

Gayle, Margot (1908–2008)
Centenarian and American historic preservationist and author Margot Gayle and her supporters established the SoHo Cast Iron Historic District in New York City, which encompasses 26 square blocks of a prior industrial district called Hell's Hundred Acres. She also helped SoHo maintain its unique low-rise character and brought back features like cobblestone streets. Gayle is also credited with starting the Victorian Society in America and the Friends of Cast Iron Architecture.

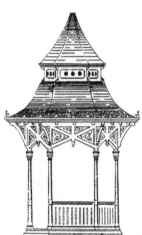

Gazebo
A fanciful small structure, used as a summer house; usually octagonal in plan with a steeply pitched roof topped by a finial. The sides are usually open, or latticed between the supports.

Geddes, Norman Bel (1893–1958)
American designer identified with the Streamlined style. Designed the General Motors Pavilion at the New York World's Fair in 1939. He produced many interiors and developed a scheme for prefabricated housing.

Gehry, Frank O. (1929–)
Canadian-born American architect, who has been identified with Deconstructivism. He settled in Santa Monica, CA, where he built several single-family houses from which most traditional forms were eliminated. Examples are the Wosk House, Beverly Hills, CA (1984), and the Gehry House, Santa Monica, CA (1988). In the latter, the rectangular form of the house is distorted, a tilted cube emerges from the façade, and layers of the house are peeled away to reveal the structure. Other works include the California Aerospace Museum, Santa Monica, CA (1984); the University of Minnesota Art Museum (illus.), Minneapolis, MN (1990); the Guggenheim Museum, Bilbao, Spain (1977); and the Experience Music Project, Seattle, WA (2000). More recent work includes the steel-clad Performing Arts Center (illus.), Annandale-on-Hudson, NY (2002); the Rasin (Fred and Ginger) Building (illus.), Prague, Czech Republic (2002); the Walt Disney Concert Hall, Los Angeles, CA (2003); and the 76-story undulating facade of 80 Spruce Street in NYC.

226

Frank O. Gehry

General Services Administration (GSA)
Owns or manages most U.S. federal buildings except those of the military, the Department of Veteran's Affairs (VA), and the U.S. Postal Service (USPS); transfers surplus historic public buildings to state and local governments for reuse; and is required to locate federal offices in buildings of historical, architectural, or cultural significance when possible. They also commission works of art for federal buildings,

Gentrification
English term for the process by which young professionals (gentry) buy into inner-city areas as part of a neighborhood preservation trend.

Genus loci
The prevailing spirit of a place, implying the conscious act of developing the character of a given place rather than imposing a foreign character on it.

Geodesic dome See **Dome**

Geometric style (1200–1250)
The early development of the decorative age of English architecture characterized by the geometrical forms of its window tracery.

Geometric tracery See **Tracery**

Geometrical
Refers to forms that can be generated into three-dimensional plane figures and divided into three groups: continuous, such as those used in bands; enclosed panels; or flat patterns on walls.

Geminated
Coupled, as with columns and capitals.

Geminated capital See **Capital**

General color-rendering index
A measure of the average appearance of eight standardized colors chosen to be of intermediate saturation and spread throughout the range of hues.

General contractor
The building contractor who has legal responsibility for the entire construction project and who coordinates the work of all the subcontractors.

Georgian architecture (1714–1776)
A formal arrangement of parts within a symmetrical composition and enriched Classical detail characterize this style. The simple facade is often emphasized by a projecting pediment, with colossal pilasters and a Palladian window. It often includes dormers, and the entrances ornately decorated with transoms or fanlights over the doors. The style was transmitted through architectural pattern books.

Geotextiles
Cloth or clothlike materials intended for use in the soil, usually for filtering or containing soil water. Some types are used to prevent or control erosion.

Geothermal
The heat derived from the Earth. It is the thermal energy contained in the rock and fluid in the Earth's crust.

Geothermal energy
Hot water or steam extracted from reservoirs beneath the Earth's surface; can be used for heat pumps, water heating, or electricity generation. The term may also mean the use of near-constant underground temperatures by ground-source heat pumps to provide heating and cooling.

Geothermal heat
A technology that uses the warmth from subsurface water to heat buildings; it also extracts this heat to put back into the ground for cooling.

Geothermal heat pumps
A system that consists of indoor pump equipment, a ground loop, and a flow center to connect the indoor and outdoor equipment. The ground loop uses the temperature of the earth or water, several feet underground, to heat or cool the dwelling. A pump circulates a temperature-sensitive fluid through this ground loop buried below the freezing line where the temperature stays constant at approximately 60 degrees F. The heat pumps transfer stable temperatures into houses in reverse order according to the season. In the winter, warm fluid carries heat into the house, and in the summer, cool fluid draws heat out of the house.

Geothermal heat system (closed loop)
Geothermal heat pumps use the constant temperature of the Earth to provide cooling and heating for a home. A loop of piping is buried in the ground, and fluid circulates through the loop. In the summer, the fluid uses the cooler

Geothermal heat system
temperature of the ground to provide indoor cooling. During colder months, the geothermal heat pump uses the below-ground temperature, which is significantly warmer than the outside air, to warm a home.

Geothermal heat-exchange technology
During the cold months, a geothermal heat-exchange system uses the natural heat of the earth subsurface, around 55 degrees, to heat water, which is then used to heat buildings. During warmer months, geothermal heat exchange extracts heat from the building into subsurface water for cooling.

Geothermal/ground-source heat pump
Underground coils that transfer heat from the ground to the inside of a building. This type of heat pump can realize substantial energy savings over conventional heat pumps, by using the naturally more stable temperature of the Earth as its heat source.

Gesso
A mixture of gypsum plaster, glue, and whiting; applied as a base coat for decorative painting.

Ghost mark
An outline that shows earlier construction that was removed; includes outlines created by missing plaster and patched holes showing the parts of the building that were demolished.

Ghost town
A town, especially a boom town of the American old West, that has been completely abandoned. Once-living places like Madrid, NM, or Bodie, CA, were abandoned as quickly as they were built, leaving only an empty shell, oftentimes with the contents still intact.

Gibbs surround
The surrounding trim of a doorway or window, consisting of alternating large and small blocks of stone, like quoins. These are often connected with a narrow raised band along the face of the door, window, or arch.

Gibbs, James (1682-1754)
English Neoclassical architect who created a restrained version of the Baroque style in Britain.

Gideon, Sigfried (1888-1968)
Swiss art historian who became a powerful advocate of the Modern movement and with Le Corbusier, was a founder in CIAM. His book, *Space, Time, and Architecture*, written in 1941, was fashionable in schools of architecture. He also wrote several other books on architecture.

Gilbert, Cass (1859-1934)
Designed the Woolworth Building, New York City, which was the tallest building in America for 17 years. It was ornamented with Gothic details. He also won the commission for the Minnesota State Capitol in Minneapolis, MN in 1903 and for the U.S. Customs House, New York City, in 1907.

Gilding
Gold leaf, gold flakes, or brass, applied as a surface finish.

Gill, Gordon (1964-)
Canadian American architect and an associate partner in the Chicago office of Skidmore, Owings & Merrill, where he led design teams on award-winning projects such as the Virginia Beach Convention Center, Virginia Beach, VA and the Pearl River Tower, Guangzhou, China. After working with Adrian Smith and Robert Forest at SOM, the three opened their own firm in 2006 to pursue the design of high-performance, energy-efficient, and sustainable architecture, such as the Masdar City Headquarters Building, Abu Dhabi, UAE.

Gill, Irving (1870-1936)
Designed the Dodge House, Los Angeles, CA a composition of white-painted blocks reminiscent of contemporary European houses, similar to the Cubistic style of Adolf Loos.

Giller, Norman (1918-2008)
A leading designer of the Miami Modern style. Commonly known as MiMo, Miami's version of postwar modernism featured dramatic sweeping roof lines, biomorphic forms, bold colors, and liberal use of glass. In Florida, Giller's projects included the Carillon Hotel in Miami Beach, the Thunderbird Motel in Sunny Isles, the Singapore Hotel in

Giller, Norman
Bal Harbor, and the original Diplomat Hotel and Country Club in Hollywood. Giller was instrumental in the creation of Miami Beach's Design Review Board, and he later served as its chairman.

Gillham, Oliver (1949-2008)
Land-use expert, author, and architect, his 2002 book, *The Limitless City,* explored the history and development of urban sprawl in the United States. His projects included work on Terminal A at Logan Airport in Boston, MA land use around Bangor International Airport in Maine, the redevelopment of Australian docklands in Melbourne, Australia and the planning of new towns in India.

Gilman, Arthur, Jr. (1821-1882)
American architect and designer of the Arlington Street Church in 1861, Boston, MA and later in partnership with Gridley James Fox Bryant worked on the Boston City Hall, Boston, MA in 1865. He designed the Equitable Life Assurance Company Building, New York City, in 1867, which was one of the world's first skyscrapers.

Gingerbread
The highly decorative and often superfluous woodwork applied to a Victorian style house or commercial structure.

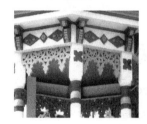

Gingerbread style (1830–1880)
A rich and highly decorated style featuring the ornate wood-working of American buildings, particularly in vogue during the Victorian era.

Giotto di bondone (c. 1266-1337)
Italian architect; designed the Florence Cathedral campanile (1334), which combines Romanesque, Classical, and Gothic elements.

Girder
A large or principal beam used to support concentrated loads at isolated points along its length.

Girdle
A horizontal band ringing the shaft of a column.

Girt
In a braced frame, a horizontal member at an intermediate level between the columns, studs, or posts; a heavy beam, framed into the studs, which supports the floor joists.

Glare

A state that reduces the ability to perceive the visual information needed for a particular activity. It arises when some parts of the visual field are much brighter than their surroundings.

blinding glare

So intense that for an appreciable length of time after it has been removed, no visual perception is possible.

discomfort glare

Glare that is distracting or uncomfortable, interfering with the perception of visual information required to satisfy biological needs, it does not significantly reduce the ability to see information needed for activities.

direct glare

Results from high luminances directly visible from a viewer's position.

disability glare

Reduces the ability to perceive the visual information needed for a particular activity.

reflected glare

The reflection of incident light that partially or totally obscures the surface details by reducing the contrast on a surface.

Glare index

A value for predicting the presence of glare as a result of daylight entering an area. The glare index is affected by the size and relative position of fenestration, orientation to the sun, sky luminance, and interior luminance. The glare index is similar to the index of sensation and the discomfort glare rating, which are used for electric-lighting applications.

Glass

A hard, brittle, usually transparent or translucent substance, produced by melting a mixture of silica oxides; while molten, it may be easily blown, drawn, rolled, pressed, or cast to a variety of shapes. It can be transparent, translucent, or mirrored; and made nonglare, pigmented, or tinted. It can be shaped by casting, rolling, pressing, or baking. It can also be bonded to metal for use as an exterior cladding.

art glass

A type of decorative leaded glass window in which scenes or patterns are produced by using colored rather than stained glass; it is common in works of the Art Nouveau style. Also works of blown glass.

art glass

colored glass

Originated over 2000 years ago when pieces of colored glass were embedded in heavy matrices of stone or plaster.

corrugated glass

A glass sheet manufactured by pressing molten glass in a mold, with a cross section in the form of a wave.

crown glass

The glass made by blowing a mass of molten material, which is then flattened into a disk and spun into a cular sheet.

decorative glass

Embossing and sandblasting techniques create a subtle form of ornamentation. Etching and beveling are also used to create ornamentation in glass.

double glazing

Insulating glass that is composed of an inner and outer pane, with a sealed air space between them.

float glass

Sheets of glass made by floating molten glass on a surface of molten metal, which produces a polished surface.

fluted glass

Glass whose solar transmittance is reduced by adding varius coloring agents to the molten glass; the most common colors are bronze, grey and green.

heat-absorbing glass

A glass whose solar transmittance is reduced by adding various coloring agents to the molten glass; the most common colors are bronze, gray, and green.

insulating glass

The glass that has insulating qualities, made by sandwiching two layers of glass separated by a vacuum-sealed edge.

laminated glass

Two or more plies of flat glass bonded under heat and pressure to inner layers of plastic to form a shatter-resisting assembly that retains the fragments if the glass is broken; it is called safety glass.

leaded glass

Dates from the Middle Ages, where glass was set into malleable lead frames.

leaded glass

low-emissivity glass

A glass that transmits visible light while selectively reflecting the longer wavelengths of radiant heat; made by a coating either the glass itself or the transparent plastic film in the sealed air space of insulating glass.

luster glass

An iridescent glass, of the type made by Tiffany.

obscure glass

A glass that has one or both faces acid-etched or sandblasted to obscure vision.

opal glass

The glass that contains calcium phosphate, which is derived from bone ash, and which renders the glass white and opaque.

opalescent glass

A type of iridescent glass showing many colors; first used by Louis Comfort Tiffany in the late nineteenth century, and now called Tiffany glass.

painted glass

A type of stained glass formed by painting a plain piece of glass with enamel, then baking or firing it in a kiln at a high temperature.

patterned glass

A glass that has an irregular surface pattern formed in the rolling process to obscure vision or to diffuse light; usually on one side only, the other side is left smooth.

plate glass

A high-quality float glass sheet, formed by rolling molten glass into a plate that is subsequently ground and polished on both sides after cooling.

prismatic glass

Rolled glass that has parallel prisms on one face. These refract the transmitted light and thus change its direction.

reflective glass

Window glass having a thin, translucent metallic coating bonded to the exterior or interior surface to reflect a portion of the light and radiant heat and light that strikes it.

reflective glass

rolled glass

Molten glass from a furnace is passed through rollers to produce a pattern on one or both surfaces of the glass.

safety glass

A glass containing thin wire mesh reinforcement; glass laminated with transparent plastic; glass toughened by heat treatment, causing it to break into small fragments without splintering.

sheet glass
A float glass fabricated by drawing the molten glass from a furnace; the surfaces are not perfectly parallel, resulting in some distortion of vision. Used for ordinary window glass.

sound-insulating glass
A glass consisting of two lights in resilient mountings, separated by spacers, and sealed so as to leave an air space between them; the air space contains a desiccant to assure dehydration of the trapped air.

spandrel glass
An opaque glass used in curtain walls to conceal spandrel beams, columns, or other internal structural construction.

stained glass
A glass given a desired color in its molten state or by firing a stain into the surface of the glass after forming; used for decorative windows or transparent mosaics.

structural glass
A glass which is cast in the form of cubes, rectangular blocks, tile, or large rectangular plates; used widely for the surfacing of walls.

tempered glass
Annealed glass reheated to just below the softening point and then rapidly cooled with water. When fractured, it breaks into relatively harmless pieces.

tinted glass
A glass that has a chemical admixture to absorb a portion of the radiant heat and visible light that strikes it to filter out infrared solar energy, thereby reducing the solar heat gain.

vision-proof glass
A glass that has been given a pattern during its manufacture, so that it is not transparent.

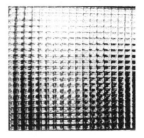

wire glass
Flat or patterned glass having a square or a diamond wire mesh embedded within the two faces to prevent shattering in the event of breakage or excessive heat. Wire glass is considered a safety glazing material.

Glass block
Composed of two sheets of plate glass with an air space between them, formed into a sealed modular hollow block; laid up with mortar, similar to masonry blocks as a modular material; comes in several distinct styles, patterns and degrees of transparency and translucency.

Glass block

Glass door See Door

Glass mullion system

A glazing system in which sheets of tempered glass are suspended from special clamps, stabilized by perpendicular stiffeners of tempered glass, and joined by a structural silicone sealant or by metal patch plates.

Glass wool insulation

A kind of insulating material composed of glass fibers which are formed into lightweight blankets of uniform thickness, fastened securely to heavy paper or vapor barrier; manufactured in standard widths to go between the studs in a wall, or joists in the ceiling.

Glaze

A ceramic coating, usually thin, glossy, and glass-like, formed on the surface of pottery earthenware; the material from which the ceramic coating is made.

Glazed brick See Brick

Glazed tile See Tile

Glazed work

Brickwork built with enameled brick or glazed brick.

Glazing

Transparent or translucent layer of window or door that transmits light. High-performance glazing that may include multiple layers of glass, plastics or acrylics, low-emissivity coatings, and low-conductivity gas fill.

Glazing bars

The bars of wood or metal that hold panes of glass in place; also called astragals.

Glazing factor

The ratio of interior illuminance at a given point on a given plane (usually the work plane) to the exterior illuminance under known overcast sky conditions. LEED (Leadership in Energy and Environmental Design) uses a simplified approach for its credit compliance calculations. The variables used to determine the daylight factor include the floor area, window area, window geometry, visible transmittance, and window height.

Global warming

An increase in the near surface temperature of the Earth. Global warming has occurred in the distant past as the result of natural influences, but the term is most often used to refer to the warming predicted to occur as a result of increased emissions of greenhouse gases.

Globe

A spherical ornament, fabricated out of solid wood or a metal shell, usually found on the top of steeples and cupolas.

Gloss

A property of paint finish that determines its reflective quality; either shiny, semireflective, soft finish, or flat.

Gloss enamel

A finishing material that forms a hard coating with maximum smoothness and a high shine.

Glue

An adhesive substance; types include liquid glue, casein glue, animal glue, epoxy resin, vegetable glue, synthetic resin, cellulose cement and rubber compounds.

Glue block

A wood block usually triangular in cross section, glued to an angular joint between two members at right angles to stiffen the joint.

Glue injector

A metal syringe with a small pointed nozzle for injecting the glue into a joint or through a very small hole.

Glue line

The line of glue visible on the edge of a plywood panel; also applies to the layer of glue itself.

Glue-laminated arch See Arch

Glulam

Glue laminated lumber, normally a structural timber member, glued and laid up from fairly small sections and lengths to either large cross sections, long lengths, curved shapes, or a combination of these.

Gmelin, Paul (1859–1937)

German-born American architect who studied skyscraper design; designed the original New York Times Building, New York City, and did presentation drawings for the Boston Public Library project, Boston, MA.

Gneiss See Stone

Godefroy, Maximillien (1765–1840)

French architect who introduced the latest Neoclassicism to the United States when he settled here. He designed the Battle Monument, Baltimore, MD in 1827, the first civic monument to be erected in the United States.

Goff, Bruce (1904–1982)

One of the most creative and idiosyncratic architects in America; began his career with Rush, Endicott and Rush in Tulsa, OK. He designed the Boston Avenue Methodist Church in Tulsa, in 1929. In his residential design he used found materials such as coal, rope, and glass cullets. His best-known works are the Ford House in Aurora, IL, in 1948; the Bavinger House in Norman, OK, in 1950; and the Price House in Bartlesville, OK, in 1956, all of which were designed with great individuality based on the different needs of the clients.

Goff, Bruce

Gold leaf

Very thin sheets of beaten or rolled gold, used for gilding and inscribing on glass; usually contains a very small percentage of copper and silver. Heavy gold leaf can be classified as gold foil.

Goldberg, Bertrand (1913-1997)

Chicago architect most noted for the Hilliard Center apartments (1966) and Marina City Apartments (1967), both in Chicago, IL.

Golden mean

A proportional relationship devised by the Greeks that expresses the ideal relationship of unequal parts. It is obtained by dividing a line so that the shorter part is to the longer part as the longer part is to the whole line. It can be stated thus: a is to b as b is to $a + b$. If we assign the value of 1 to a, and solve as a quadratic equation, then $b = 1.618034$. Therefore, the golden mean is 1:1.618.

Goodue, Bertram (1869-1924)
American architect; in partnership with Ralph Cram from 1892 to 1913. On his own, he designed St. Bartholomew's Church, New York City, in a Byzantine Romanesque style. His most well-known work is the Nebraska State Capitol in Lincoln, NE in 1920.

Goodwin, Philip (1885-1958)
American architect who worked for McKim Mead and White. He was a partner with Edward Durrell Stone on the design for the Museum of Modern Art, New York City.

Goodwin, Michael Kemper (1939–2011)
Architect in the Phoenix area who also served two terms in the Arizona House of Representatives in the 1970s. His father, Kemper Goodwin, was also an architect (with two buildings on the National Register of Historic Places in Arizona) with whom Michael would later work. Works include the Tempe Municipal Building (1970), an upside-down pyramid designed to shade and cool itself; Marcos de Niza High School (1971), considered a revolution in open space campus design; and Corona del Sol High School (1976), a very early design using solar technology.

Goody, Joan (1936-2009)
A founding partner of Boston's Goody, Clancy & Associates with a long and varied project list including Boston's Harbor Point. Goody also cited the restoration of H. H. Richardson's Trinity Church at Copley Square in Boston; a federal courthouse in Wheeling, WV; the Salomon Center for Teaching at Brown University; Heaton Court, an affordable housing project in Stockbridge, MA; and the transformation of St. Elizabeth's Hospital in Washington, DC, into the new headquarters for the Department of Homeland Security. Goody helped found an advocacy group called Women Architects, Landscape Architects and Planners (WALAP) in 1970.

Gopuram See **Gate**

Gorgoneion
In Classical decoration the mask of a gorgon, a woman with snakes for hair believed to avert evil influences.

Gothic arch See **Arch**

Gothic architecture (1050–1530)

A revolutionary style of construction of the High Middle Ages in western Europe which emerged from Romanesque and Byzantine forms. The term "Gothic" was originally applied as one of reproach and contempt. The style was characterized by a delicate balance between the lateral thrust from loads and the force of gravity. It was most often found in cathedrals employing the rib vault, pointed arches, flying buttresses and the gradual reduction of the walls to a system of richly decorated fenestration. The style's features were height and light, achieved through a mixture of skeletal structures and increasing use of windows. Walls were no longer necessary to support the roof and could be replaced with large, tall windows of stained glass. One of the finest and oldest examples of French Gothic architecture is Notre Dame in Paris.

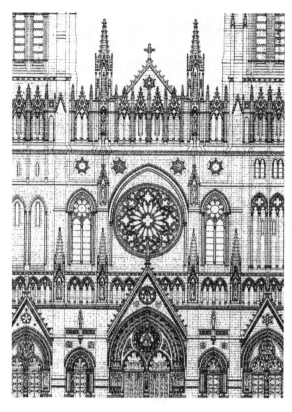

Gothic Revival style (1830–1860)

A romantic style distinguished by vertically pointed arches, steeply pitched complex gable roofs, finials, and medieval decorative motifs. Country houses featured wide verandas and octagonal towers or turrets. Windows in dormers had hood molds with gingerbread trim running along the eaves and gable ends. Variety was the standard of the style.

Gouache

A method of painting using opaque pigments pulverized in water and mixed with gum.

Grade

The designation of the quality of a manufactured piece of wood; also the level of the ground in relation to the building.

Grade beam See **Beam**

Grade level
The level of the ground around a building.

Gradient
The slant of a road, piping, or the ground; expressed as a percentage of slope from true horizontal.

Grading
The modification of the ground surface by cuts and/or fills.

Graffiti
The designs resulting from painting words or symbols on a building, wall or other object.

Graffito
A decorative pattern produced by scratching the top layer of a two-color stucco finish.

Graham, Bruce (1926–2010)
Designer of Chicago's John Hancock Center and Sears Tower (now Willis Tower), he led the Chicago office of Skidmore, Owings & Merrill from the 1960s through the 1980s with a staff that included Gordon Bunshaft, Walter Netsch, and structural engineer Faziur Khan. Graham is responsible for projects in London, Barcelona, Egypt, Korea, Guatemala, Houston, Los Angeles, Boston, Atlanta, Kansas City, Milwaukee, Wichita, Nashville, Tulsa, Madison, and many other locales. Their revolutionary design, engineered by Khan, placed the structural supporting elements on the outside, greatly reducing construction costs and providing vast open interiors. His 1957 Inland Steel building, Chicago, IL featured the use of structural columns on the building's perimeter and a separate adjoining tower for elevators and other services to open up the interiors. With William Hartmann, another partner at SOM, Graham developed the Chicago 21 Plan of 1973 for the Chicago Central Area Committee, which called for alterations and additions to Chicago's lakefront as the city entered the 21st century, including rebuilding the dilapidated Navy Pier into a tourist destination and the creation of the museum campus.

Graham, Ernest R. (1866–1936)
American architect, partner to Daniel H. Burnham in Chicago, IL.

Graham, John Jr. (1908-1991)
Designed the Space Needle at the Science Center, and the revolving "eye of the needle" restaurant in Seattle, WA, in 1962.

Grain
The pattern of fibers found on the cut surface of wood.

Graining
The painting of a surface to imitate the grain of wood or the veining of marble.

Grandstand
The viewing platform with seats at a racecourse, parade, or sports event, typically covered with a roof.

Granite See **Stone**

Granules
Crushed rock that is coated with a ceramic coating and fired; used as a top surface on shingles.

Grapevine ornament See **Ornament**

Grate
A frame that consists of parallel metal bars, attached by cross bars at regular intervals; used as a grille or security device.

Gravel
Small pieces of stone that are of varying sizes; used as an ingredient in concrete.

Gravel fill
Crushed rock, pebbles, or gravel deposited at the bottom of an excavated wall, or footing, to ensure adequate drainage of water.

Gravel roof
A roof built up with a layer of pebbles embedded in the hot tar undercoating; the pebbles protect the tar and break up raindrops.

Gravel stop
A continuous band of bent sheet metal with a vertical or sloped projection that prevents gravel on a built-up roof from falling off.

Michael Graves (1934–)
One of the most controversial American architects, he was identified as one of the New York Five, and first became known for a series of private houses based on reworked themes of Le Corbusier. His work was Postmodern. The Public Services Building (illus.), Portland, OR (1980), and the Humana Tower, Louisville, KY (1982), the Whitney Museum of Art addition, NYC (1985), and the Walt Disney World, Dolphin and Swan Hotels (illus.), Orlando, FL (1987) are celebrated examples of his work.

Gravity retaining wall
A retaining wall that relies on the weight of the masonry or concrete for its stability.

Gravity-flush toilet
A toilet whose flush is powered solely by the force of falling water.

Gray, Eileen (1878–1976)
A furniture designer and architect, Gray's contributions were overlooked for many years, but she is now considered one of the most influential designers of modern times. Many Art Deco and Bauhaus architects and designers found inspiration in her unique style.

Grayfield
A site, such as a mall or commercial facility, which has been abandoned, leaving behind a large, developed, but empty area.

Grayfield development
The development of noncontaminated retail areas such as old malls, strip malls, or institutional areas into complete, livable communities.

Graywater
Appendix G of the Uniform Plumbing Code (UPC) defines graywater as "untreated household wastewater which has not come into contact with toilet waste. Graywater includes water from bathtubs, showers, bathroom wash basins, and

Graywater

water from clothes-washer and laundry tubs. It shall not include wastewater from kitchen sinks or dishwashers." The International Plumbing Code (IPC) defines graywater in its Appendix C as "wastewater discharged from lavatories, bathtubs, showers, clothes washers, and laundry sinks." Some states and local authorities allow kitchen sink wastewater to be included in graywater. Other differences with the UPC and IPC definitions can be found in state and local codes. Project teams should comply with the graywater definitions as established by the authority having jurisdiction in their areas.

Graywater recycling
Graywater is water that is discharged from household appliances and fixtures in the laundry, kitchen or bathroom. Graywater recycling is water taken from the wastewater stream that is treated and reused. The level of treatment and quality of resultant water depends on the intended use.

Graywater reuse
A strategy for reducing wastewater outputs from a building by diverting the graywater into productive uses such as subsurface irrigation or onsite treatment and use for nonpotable functions such as toilet flushing.

Graywater system
Graywater can be recycled for irrigation, toilets, and exterior washing, and such recycling conserves water. Incorporating plumbing systems that separate graywater from blackwater can result in water cost savings.

Grazing light
Lights located close to the lit surface that strengthen highlights and shadows and emphasize surface textures. Also used for inspection to detect surface blemishes and errors in workmanship.

Greek architecture (800–300 B.C.)
The first manifestation of this style was a wooden structure of upright posts supporting beams and sloping rafters. The style was later translated into stone elements with a wood roof. It was a "kit of parts" characterized by austerity and free of ornate carvings. The decorative column orders were an integral part of this style: the Doric, which is the simplest and sturdiest, the Ionic, which was more slender, and the Corinthian, which had a very elaborate capital. Greek ornament is refined in character. The materials were limestone and marble, and were prepared with the highest standards of masonry, including sophisticated optical corrections for perspective (entasis).

Greek architecture

Greek cross plan
A cruciform plan with four arms of equal length.

Greek Revival style (1750–1860)
The Greek contribution to Neoclassical architecture stood for a purity and simplicity of structure and form. The buildings are square or rectangular, proportions are broad, details are simple, facades are symmetrical and silhouettes are bold. Freestanding columns support a pedimented gable. Many government and civic buildings are designed in this style, which is more suited to these building types than to smaller domestic buildings.

Green
An abstract concept that includes the terms sustainability, ecology, and performance. The level of greenness is determined by the extent of interaction of these three categories. It is now commonly used to describe something or someone that is environmentally conscious or friendly or sustainable, or has positive environmental attributes, effects, or objectives.

Green Advantage®
A green building Environmental Certification program that brings consumers together with certified building prac-titioners who have proven knowledge about green building techniques and approaches.

Green belt
Areas of green land around urban areas to prevent further expansion; they are kept open by severs and normally permanent planning restrictions.

Green building
Also known as sustainable building or environmental building, green building is the practice of increasing the efficiency with which buildings and their sites use and harvest energy, water, and materials, and reducing the impacts of buildings on human health and the environment, through better site planning, design, construction, operation, maintenance, and removal—the complete building life cycle.

Green Building Certification Institute (GBCI)
Established as a separately incorporated entity with the support of the U.S. Green Building Council, GBCI administers credentialing programs related to green building practice such as the development and implementation of a maintenance program for LEED AP (Leadership in Energy and Environmental Design Accredited Professional) credential holders.

Green building rating systems
Designed in response to growing concerns in the building industry in topics such as sustainability, building performance, environmental impact, energy, cost efficiency, and maintenance. The rating systems proposed quantifiable tools to evaluate and measure the level of a building's environmental performance, such as energy, design, construction, site, technologies, and materials performance.

Green code
A requirement for new construction, it is intended to serve as a model, enforceable building code in contrast to voluntary rating systems such as LEED (Leadership in Energy and Environmental Design). The International Codes Commission (ICC) is the primary organization behind the code. I-Codes of various kinds have been adopted in all 50 states and many federal agencies. It is a baseline mandatory building code and will affect the architectural profession.

Green design
A design, usually architectural, conforming to environmentally sound principles of building; site selection and design; building design; selection of appropriate materials; improved indoor environmental quality; and reduced materials, water, and energy use.

Green development
A sustainable approach to real estate development that incorporates environmental issues such as efficient and appropriate use of land, energy, water, and other resources; protection of significant habitats, endangered species, archeological treasures, and cultural resources; and integration of work, habitat, and agriculture. A development approach that integrates environmental responsiveness, benefits the surrounding environment, provides resource efficiency in the construction, development and operations of buildings and/or communities in ways that are not wasteful, and sensitivity to the existing culture of the community. Green development supports human and natural communities and cultural development while remaining economically viable for owners and tenants.

Green electricity
The electricity generated from renewable energy sources, such as photovoltaic solar power, wind power, geothermal, hydropower, and biomass (i.e., wood and animal waste), and landfill mass.

Green Globes™
A green building guidance and assessment program that integrates a comprehensive environmental assessment protocol, software tools, qualified assessors with green building expertise, and a rating/certification system.

Green Guide
Guidelines published by the Federal Trade Commission (FTC) providing standards for advertising claims that a product is green.

Green guide for healthcare
Voluntary sustainable design toolkit for the healthcare sector integrating enhanced environmental and health principles and practices into the planning, design, construction, operations, and maintenance of their facilities.

Green home guide
A guide for homeowners and green residential construction including many links to websites.

Green infrastructure
A strategically planned and managed network of wilderness, parks, greenways, conservation easements, and working lands with conservation value that supports native species, maintains natural ecological processes, sustains air and water resources, and contributes to the health and quality of life for communities. More specifically, the use of soil, trees, vegetation, wetlands, and open space, either preserved or created, in urban areas to capture rain while enhancing wastewater and stormwater treatment. Also includes nonliving complementary solutions such as porous pavement or rain barrels. Green infrastructure can be used in lieu of or in conjunction with traditional hard infrastructure approaches such as pipes, retention basins, and treatment facilities.

Green label
A certification program by the Carpet and Rug Institute for carpet and adhesives meeting specified criteria for release of volatile compounds.

Green lease
A lease that sets out mutual obligations for building owners, managers, and tenants regarding environmental performance.

Green lumber
Boards cut from green or unseasoned logs; lumber that has not been dried or seasoned with a moisture content over 20 percent.

Green mortgage
Type of mortgage in which the lending institution raises the allowable loan amount for an applicant's earning level because the applicant's green home has lower monthly operating costs and may even reduce the applicant's transportation costs. See energy-efficient mortgage.

Green movement
Emerged as a combination of all previous environmental movements, such as Greenpeace (1971) and other global green organizations. See Greenpeace.

Green neighborhood
A neighborhood that is typically moderately dense, includes a range of uses, and is designed for people and pedestrians first—including a dense network of paths and streets, human-scaled buildings, and pedestrian-oriented street design. It has green elements, including a network of green spaces and corridors, street trees, and significant private landscaping including possibly green roofs. Buildings have excellent environmental performance. Green infrastructure is commonplace, from low-impact stormwater management to district energy systems.

Green philosophy
A way of living that involves a holistic approach to preservation and conservation of natural resources. It aims to provide a better understanding of the balance between human action and natural environmental resources and improve health and well-being. It also entails creating a better understanding of social responsibility and what effect choices made by people and business have on the environment.

Green Playbook
Provides local governments with guidance and resources to rapidly advance green buildings, neighborhoods, and infrastructure.

Green roof
A roof system in which living plants are maintained in a growing medium using a membrane and drainage system. Green roofs can reduce stormwater runoff, moderate temperatures in and around the building by providing insulation and reducing the heat island effect, as well as provide a habitat for wildlife and recreational space for humans. When properly constructed, green roofs increase roof durability, because the roof assembly's air and water barriers are buffered from temperature fluctuations and UV exposure. Green roofs are considered a sustainable building strategy in that they have the capacity to reduce stormwater runoff from a site, modulate temperatures in and around the building, have thermal insulating properties,

Green roof
can provide habitat for wildlife and open space for humans, and contain other benefits as well. Green roofs help to invest in the protection of the environment by diminishing developmental impact on communities while providing a fresh approach with visually appealing organic architecture, a potentially ideal union of aesthetics, economics, and ecology.

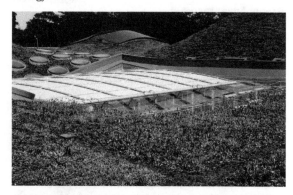

Green screen
A welded trellis system that has a captive growing space that is 3 inches deep to allow plants to intertwine and grow within the panel. Made from recycled steel, it is an adaptable clip system that can attach to a building facade and span openings between floors or horizontally between posts.

Green Seal
An independent, nonprofit organization that strives to achieve a healthier and cleaner environment by identifying and promoting products and services that cause less toxic pollution and waste, conserve resources and habitats, and minimize global warming and ozone depletion. It is also a certification for construction products such as windows, paints, and adhesives, attesting that the products were manufactured and can be used with minimal impact on the environment.

Green Seal certification
Products are awarded the Green Seal if they meet the Green Seal standard through test data gathered for the environmental and performance requirements, as well as the quality-control procedures of the manufacturing facility.

Green Star
A national voluntary environmental rating scheme that evaluates the environmental design and achievements of buildings against eight environmental impact categories, plus innovation.

Green technologies
Environmentally friendly technologies including those that promote sustainability via efficiency improvements, reuse/recycling, and substitution.

Green/living roof
A roof surface covered by a waterproofing membrane, a drainage plane, a water-retention medium, and plantings of drought-resistant species. The benefits of a green roof include control of stormwater runoff, which can reduce urban water pollution; absorption of airborne toxins and an increase in oxygen in the air; and a reduction of surface temperature of the roof, called the heat island effect. A living roof can also increase the lifespan of the roof system and provide building and noise insulation. Green roofs are most common in multifamily or other large urban buildings.

Greenbelt zones
The zones or areas in or around a city where the removal of native vegetation is prohibited; e.g., parks and other open, undeveloped, and vegetated space is protected.

Greene, Charles Sumner (1868–1957)
American-born architect; studied at MIT and set up practice with his brother Henry in Pasadena, CA. They used projecting roofs, flat gables, and timber construction. The most well known is the Gambel House in Pasadena built in 1908.

Greene, Henry Mather (1870–1954)
Partner with Charles Sumner Greene, his brother. They both studied at MIT before setting up practice in California.

Greene, Lane (1935–2009)
Greene is credited with the preservation and/or restoration of numerous structures in Georgia, including the First A.M.E. Church and Morton Theater in Athens; the Crawford W. Long House and Madison County Courthouse in Danielsville; and the Wren's Nest, home of *Uncle Remus* author Joel Chandler Harris, in Atlanta.

Greenfield
Semirural real property that is undeveloped (except for possible agricultural use) and is being considered as a site for development.

Greenfield site
Land on which no urban development has previously taken place; usually understood to be on the periphery of an existing built-up area.

Greenguard
Established performance-based standards to define goods such as building materials, interior furnishings, furniture, cleaning and maintenance products, electronic equipment, and personal care products with low chemical and particle emissions for use indoors. The standards establish certification procedures including test methods, allowable emissions levels, product sample collection and handling, testing type and frequency, and program-application processes and acceptance.

Greenguard certification
A third-party product certification program for low-emitting interior building materials, furnishings, and finish systems through evaluation of more than 75,000 chemicals, including volatile organic compounds(VOCs), carcinogens, and reproductive toxins.

Greenhouse
A glass-enclosed, heated structure for growing plants and out-of-season fruits and vegetables under regulated, protected conditions.

Greenhouse effect
The steady, gradual rise in the temperature of the earth's atmosphere due to the heat that is retained by layers of ozone, water vapor and carbon dioxide.

Greenmail

When companies threaten to close or relocate (often to another country) if they are forced to comply with environmental laws.

Greenpeace

A nonprofit organization, with a presence in 40 countries across Europe, the Americas, Asia, and the Pacific. As a global organization, Greenpeace focuses on the most crucial worldwide threats to the Earth's biodiversity and environment.

GreenSmart

A voluntary Housing Industry Association (HIA) program that educates builders, designers, product manufacturers and consumers about the benefits of an environmentally responsible approach to housing and development. HIA GreenSmart aims to increase energy efficiency, reduce water consumption, minimize waste, and encourage better environmental management at each stage in the construction of today's housing.

GreenSpec

Product information service from BuildingGreen that contains detailed listings for more than 1,700 green building products with environmental data, manufacturer information, and links to additional resources.

Greenwalls

A vertical planting system, vertical garden, plant wall, and/or vegetated wall, created by providing a planting substrate into a modular structural wall system. Most commonly used are tropical plants, vines, and ferns rooted into a modular panel system, with an integrated irrigation drip system and control, a catch basin to control the water runoff, and a structural support system. When air is circulated through a greenwall, it becomes an active bio-wall air-filtration system, breaking down harmful VOCs (volatile organic compounds) and creating clean oxygen. Greenwalls keep buildings cooler in the summer and create a sound barrier.

Greenwash

The practice of trying to convince others that a product or service is providing something good for the environment, while hiding involvement in activities that are damaging to the environment.

Greenwashing

Misleading information given by an organization so as to present an environmentally responsible public image that conceals its true environmental impact.

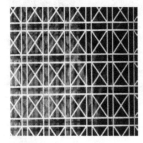

Grid

A framework of parallel, crisscrossed lines or bars forming a pattern of uniform size; sets of intersecting members on a square or triangular matrix, which make up a three-dimensional structural system.

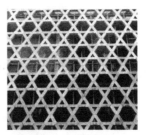
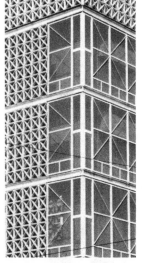

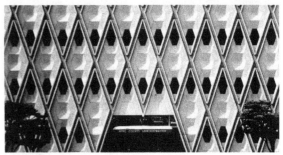

power grid

A network of power transmission and distribution facilities used to provide electricity to users such as homes, businesses, and industry. Large power plants, wind-power-generating facilities, and small power producers such as photovoltaic farms feed electrical power into the grid for distribution to users.

Grid systems

A method of laying out an area in a grid, using squares, or any other geometric shapes to facilitate the arrangement of objects within the space.

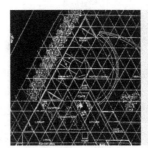

Grid-based organization See **Organization**

Grid-connected power system
Electricity generation system that usually relies on photovoltaics or wind power that is hooked up to the utility company's electric grid through a net-metering arrangement, so that electricity can be obtained when the locally generated power is not sufficient.

Gridiron plan
A town or city street layout based on a geometrical grid pattern, similar to the plan of old Savannah, GA.

Griffin See Ornament: animal forms

Griffin, Marion Mahony (1871–1961)
Worked for Frank Lloyd Wright in Oak Park, IL, and was responsible for many of the drawings of Wright's work published in the *Wasmuth Portfolio* of 1910. She was married to architect Walter Burley Griffin (1876–1937).

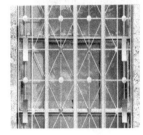

Griffin, Walter Burley (1876–1937)
Worked in Chicago with Frank Lloyd Wright before being appointed director for the design and construction of the Federal Capitol at Canberra, Australia, in 1913. Here he also designed many major works with his wife, Marion Lucy Mahoney (1871–1961).

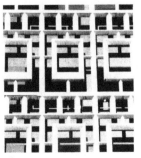

Grille
An ornamental arrangement of bars to form a screen or partition, usually of metal, wood, stone, or concrete, to cover, conceal, decorate, or protect an opening.

Grillwork
Materials arranged with voids to function as, or with the appearance of, a grille.

Groin
The curved area formed by the intersection of two vaults.

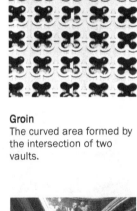

Groin arch See Arch

Groined rib See Rib

Groined vault See Vault

Groove
A continuous recess formed centrally along the edge of a timber board in order to accommodate a tongue.

Gropius, Walter (1883–1969)
German and American architect. His design for the Bauhaus, Dessau, Germany, in 1925, was the first example of the new International-style architecture to be built. He left Germany in 1928. He designed the Graduate Center, Harvard University, Cambridge, MA, in 1949, while professor of architecture at Harvard.

Gross floor area
The area within the perimeter of the outside walls of a building as measured from the inside surface of the exterior walls without reduction for hallways, stairs, closets, and thickness of the walls.

Grotesque
Sculptured or painted ornament involving fanciful distortions of human and animal forms, sometimes combined with plant motifs, especially those without a counterpart in nature.

Grotto
A natural or artificial cave, often decorated with shells or stones and incorporating waterfalls or fountains.

Ground
A piece of wood or metal embedded in and flush with the plastering of walls to which moldings, skirting, and other joiners work is attached. Also used to stop plastering around door and window openings. Also, an electrical conductor in contact with the earth, such as a steel pipe containing a cold water line that runs underground and also the continuation of a circuit back to the supply source, which is grounded at some point in its run. Its purpose is to conduct electricity to a harmless location should a fault in the circuit occur, causing possible lethal harm.

Ground cover
Low-growing plants often grown to keep soil from eroding and to discourage weeds.

Ground floor
The floor or story of a building at or slightly above grade level; excluding the basement but including all of the construction up to the floor above.

Ground glass
Glass whose surface has been roughened with an abrasive, diffusing the light passing through it; produced by sand blasting or etching with acid.

Ground joint See Joint

Ground light
Visible radiation from the sun and sky, reflected by exterior surfaces below the plane of the horizon.

Ground line See Perspective projection

Ground plane See Perspective projection

Ground story
The part of a building between the ground and first floor.

Groundsill
In a framed structure, the sill which is nearest the ground or on the ground; used to distribute the concentrated loads to the foundation.

Ground-source heat pump
Home heating and cooling system that relies on the mass of the Earth as the heat source and heat sink. Temperatures underground are relatively constant. Using a ground-source heat pump, heat from fluid circulated through an underground loop is transferred to and/or from the home through a heat exchanger. The energy performance of ground-source heat pumps is usually better than that of air-source heat pumps; ground-source heat pumps also perform better over a wider range of above-ground temperatures.

Groundwater
Freshwater found beneath the Earth's surface, usually in aquifers, that supply wells and springs.

Groundwork
Site work that takes place on the ground, such as excavations, drain laying, pathways and driveways.

Grouped columns See Columns

Grouped pilasters See Pilasters

Grouping
Arrangement of the major architectural forms of a building; such as a main structure plus its wings, or the arrangement of elements on the facade.

Grout
Mortar containing a considerable amount of water so that it has the consistency of a viscous liquid, permitting it to be poured or pumped into joints, spaces, and cracks within masonry walls and floors.

Grout injection
Stabilizing or waterproofing the ground by injecting a liquid cement slurry under pressure. Grout-injected piers are made by drilling into sandy or unstable soil and inserting the grout through the hollow drill stem as the auger is withdrawn.

Grouted masonry
Concrete masonry construction composed of hollow units when the hollow cells are filled with grout.

Gruen, Victor (1903-1980)
Viennese architect who settled in the United States in 1938, starting his own firm in Los Angeles, specializing in shopping centers, such as Northland Center in Detroit in 1954 and Southdale Center, Minneapolis, MN.

Grunsfeld III, Ernest Alton (1929-2011)
In 1956 Grunsfeld joined Wallace Yerkes, formerly a partner of his father, noted Chicago architect Ernest Alton Grunsfeld, Jr., in a small residential practice in Chicago. Before long, Yerkes and Grunsfeld became the architects of choice for many prominent clients in Chicago's North

Grunsfeld III, Ernest Alton
Shore communities. Grunsfeld continued in private practice, designing private houses, apartments, and commercial projects across the United States. Working in a modern aesthetic, Grunsfeld is highly regarded for his sensitivity to site and landscape, architectural massing, and building materials.

Guaranteed maximum cost
The amount established in an agreement between owner and contractor as the maximum cost for performing specified work on the basis of the cost of labor and materials plus overhead expenses and profit.

Guarini, Guarino (1624-1683)
Italian architect. Trained as a mathematician, he used this knowledge to create highly intricate compositions of building forms. Most were built in Turin, Italy.

Guest house
A separate residence for guests, or a small secondary house on a private estate.

Guestimate
The educated guess or estimate of an experienced architect, such as the approximation of total costs without having undertaken detailed calculations.

Guilloche molding See Molding

Guimard, Hector (1867-1942)
Designed the metro stations of Paris using an Art Nouveau motif, and other notable structures.

Gum See Wood

Gunite
A construction material composed of cement, sand or crushed slag, and water mixed together and forced through a cement gun by pneumatic pressure; sold under the trademark Gunite.

Gusset
A plate, usually triangular in shape, used to connect two or more members, or to add strength to a framework at its joints.

Gut rehab
Building renovation in which the walls are gutted and reduced to the wall framing and sometimes sheathing, then the building is insulated, resheathed, and finished.

Gutta
A small conical-shaped ornament resembling a droplet used in groups under the triglyph or the cornice found in Classical architecture.

Gutter
A shallow channel of metal or wood at the edge of a roof eave to catch and drain water into a downspout.

Gutter hook
A light metal strap used to secure or support a metal gutter.

Guy rope
A rope that secures or steadies a derrick or a temporary structure.

Gwathmey, Charles (1938–2009)
A founding partner of Gwathmey Siegel and Associates. With Peter Eisenman, Michael Graves, John Hejduk, and Richard Meier, Gwathmey was one of the New York Five—self-proclaimed rethinkers of Corbusian Modernism—who produced the 1972 publication *Five Architects*. Gwathmey's residential client list included Steven Spielberg, David Geffen, Jerry Seinfeld, and Jeffrey Katzenberg, and his design style was labeled High Modernist. The firm's many prominent large projects included the International Center of Photography in NYC, the Museum of the Moving Image in Queens, and a 1992 addition to the Guggenheim Museum, in NYC.

Gymnasium
In Greek and Roman architecture, a large open court for exercise, surrounded by colonnades and rooms for massages and lectures.

Gypsum board
A wallboard having a noncombustible gypsum core; covered on each side with a paper surface.

Gypsum
A mineral used in drywall. Mining of gypsum disrupts habitats, uses energy, and causes pollution. Synthetic gypsum is made with fly ash, a by-product of manufacturing and energy-generating processes, which reduces the need for mined material. Recycled gypsum drywall is made with reclaimed drywall and waste from drywall manufacturing.

Gypsum plaster
A hard, fire-resistant, quick-setting plaster with ground gypsum mixed with a retardant as the main ingredient.

Gypsum roof deck
A structural roof deck of gypsum cement.

H h

Habitat
The sum of the environmental conditions that determine the existence of a community in a specific place where humans, animals, plants, and microorganisms live and its surroundings, both living and nonliving.

Habitat fragmentation
Habitat disruption where natural habitat is broken into small, relatively isolated sections.

Habitation site
An archeological site where people lived or worked over a period of time.

Hacienda
A large estate or ranch in areas once under Spanish influence; now the main house on such an estate.

Hadid, Zaha (1950–)
Iraqi-British architect who established her own London-based practice in 1980. A winner of many international competitions. In 2002 Hadid won the international design competition to design Singapore's One-North master plan; in 2004 she became the first female recipient of the Pritzker Architecture Prize; in 2005 her design won the competition for the new city casino of Basel, Switzerland. Other projects include the Science Center, Wolfsburg, Austria; the Contemporary Art Museum, Cincinnati, OH; the Center for Contemporary Arts, Rome, Italy; and the Guangzhou Opera House, Guangzhou, China.

Half arch See **Arch**

Half baluster
A baluster that projects from the surface to which it is attached, by about one-half its diameter.

Half door See **Door**

Half landing See **Stair**

Half-column See **Column**

Half-pitched roof See **Roof**

Half-round molding See **Molding**

Half-space landing See **Landing**

Half-timbered
Descriptive of buildings of the 16th and 17th centuries, which were built with strong timber foundations, supports, knees, and studs, and whose walls were filled in between with plaster or masonry materials.

Half-timbered

Half-timbered wall See **Wall**

Hall
A large room or building used for the transaction of public business and the holding of courts of justice; used also for public meetings and assemblies and other entertainment.

Hall chamber
A room directly above the hall on the upper floor or one opening directly off the hall on the ground floor.

Hall church
A church with side aisles as high, or nearly as high, as the central nave.

Hallet, Etienne-Sulpice (1760–1825)
French architect who settled in the United States and supervised the building of the U.S. Capitol, Washington, DC, based on William Thornton's designs.

Hallway
A corridor or a passageway in a house, hotel, office, institutional or commercial building.

Halocarbon
Class of man-made chemicals, including chlorofluorocarbons (CFCs), hydrochlorofluorocarbons (HCFCs), and hydrofluorocarbons (HFCs), whose heat-trapping properties are among the most damaging of the greenhouse gases. This, coupled with their tendency to remain in the atmosphere for hundreds of years, has resulted in limits on their use. Halocarbons are most commonly used in refrigeration, air conditioning, and electrical systems, as well as blowing agents in some foam insulation products.

Halogen
A type of incandescent lamp with a higher energy efficiency than standard ones.

Halogen bulb
A bulb containing a filament filled with a halogen gas, which burns very brightly and has a long life. Used for direct lighting and spot lighting.

Halon
Bromine-containing compounds with long atmospheric lifetimes whose breakdown in the stratosphere causes depletion of ozone. Halons are used in fire suppression systems and fire extinguishers.

Halprin, Lawrence (1916–2009)
A landscape architect and environmental planner based in the San Francisco Bay area, Halprin designed Ghirardelli Square in San Francisco and the Franklin Delano Roosevelt Memorial (illus.), in Washington, DC. He was married to a dancer, the former Anna Schuman, and described his blend of modernism, environmentalism, and movement as a choreography. Before committing to a design, he often held a workshop with different constituencies—artists, community activists, clients, and developers—to gauge emotional responses to different places. His many projects included the plazas and grand fountains of Portland, OR; Seattle, WA Freeway Park; Nicollet Mall in Minneapolis, MN; and the Approach to Yosemite Falls in Yosemite National Park, CA dedicated in 2005. One of his most prominent projects that required great environmental sensitivity was Sea Ranch in Sonoma County, CA, a development of 1,500 homes on 5,000 acres. Halprin preserved and enhanced the natural features of the site, planting more than a half-million trees. It was the subject of his 1995 book *The Sea Ranch: Diary of an Idea.* Halprin wrote nearly a dozen books during the course of his career on projects and the process of design.

Halprin, Lawrence

Hammell, Robert (1945–2008)
Washington, DC–based architect Hammell worked on 12 projects involving Category One National Landmarks, including the Jefferson Memorial, the Treasury Department Building, and the Pentagon. In 2001 Hammell cofounded the National Landmark Institute to study and teach design, technical, and management skills for high-profile public projects.

Hammer beam
One of a pair of short horizontal members, attached to the foot of a principal rafter, located within a roof structure in place of a tie beam.

Hammer brace
A bracket under a hammer beam to support it.

Hammer-beam roof
A roof without a tie beam at the top of the wall.

Hand saw
Any saw operated by hand instead of by electricity; may be either a cross-cut saw, ripsaw, or keyhole saw used by carpenters, or a hacksaw used by plumbers.

Hand screw
A clamp with two parallel jaws and two screws used by woodworkers; the clamping action is provided by means of the screws, one operating through each jaw.

Hand-hewn
Wood beams that have been trimmed with hand tools, such as an adze; typical of early barn timbers.

Handicapped access
Efforts to ensure that facilities and programs in buildings are made available to those with limited mobility, vision impairment, and other handicapped conditions. This may not necessarily apply to converted historic buildings and museums if it cannot be done without compromising the building's historical and architectural integrity.

Handover
The handover is the formal acceptance by the building owner that construction of the building is substantially complete and that it is ready for occupancy; that all systems have been installed correctly and are operational; and that a handover commissioning record has been completed and is available for the owner.

Handrail

A rail providing a handhold and serving as a support at the side of a stair or elevated platform.

Hand-sawn

Wood that has been sawn by hand, as opposed to sawn in a mill.

Hanger

A strap or rod attached to an overhead structure to support a pipe, conduit, or the framework of a suspended ceiling; a stirrup bracket used to support the end of a beam or joist at a masonry wall or girder.

Hanging gable See Gable

Hanging scaffold

A suspended scaffold supported by cables at each end, and raised or lowered by a small electric motor located on the cable.

Hanging stair See Stair

Hankar, Paul (1859–1901)

Belgian architect who became one of the proponents of Art Nouveau, as shown in his designs for the Hotel Hankar (1893) and the Hotel Ciamberlani, both in Brussels, Belgium. He was influenced by Japanese art, and his work in turn influenced Otto Wagner.

Hannover principles

Design guidelines written by Bill McDonough and Michael Braungart for the 2000 World's Fair that were issued at the World Urban Forum of the Earth Summit in 1992.

Hardboard See Wood

Hardouin-Mansart, Jules (1646–1708)

French architect who designed the Church of the Invalides, Paris (1680), the most Baroque of Parisian churches, with a dome derived from St. Peter's, Rome, Italy.

Hardpan

An extremely hard soil containing gravel and boulders.

Hardwood See Wood

Hardy, Hugh (1932–)

Spanish-born American architect; formed a partnership with Malcolm Holzman and Norman Pfeiffer in New York City. The firm designed many theaters and became known for its use of disparate parts, giving an impression of incompleteness. He restored the glory of radio City Music Hall (1999), and revived the New Victory Theater (illus.), and the New Amsterdam theater, all in New York City.

Harmika

A square enclosure on top of the dome of a stupa.

Harmon, Arthur Loomis (1878–1958)

American architect who was associated with the firms of Carrere and Hastings and McKim Mead and White. He was a partner in the firm of Shreve, Lamb and Harmon, designers of the Empire State Building, New York City.

Harmonic proportions

Relates the consonances of the musical harmonic scale to those of architectural design, particularly to those theories of proportion.

Harmony

The pleasing interaction or appropriate orderly combination of the elements in a composition.

Harp

A metal device fitted into the socket of a lamp that holds a lampshade.

Harris, Harwell H. (1903–)
American architect who worked with Richard Neutra on the city planning scheme "Rush City Reformed" and other CIAM projects.

Harris, John A. (1920–2008)
A British architect instrumental in the development of modern Dubai. He drew the first master plan for the city in 1960, when Dubai had no paved roads and no utility networks, water was carted into town, and communication with the outside world was difficult. Following the discovery of oil there in 1966, a building boom of housing, hospitals, schools, and commercial buildings ensued, and Harris wove together his plan for the old city with its future needs and designed many of these new structures. His firm designed the Dubai World Trade Center—which opened in 1979 and at 40 stories was the tallest building in the region for more than 20 years.

Harrison, Peter (1716–1775)
American architect who worked in Boston, Newport, RI, and Cambridge, MA. One of the first colonial architects to use the Palladian style. Designed Kings Chapel, Boston, adapted from Gibbs's style. Also designed the Redwood Library (1750), the Touro Synagogue (1763), and the Brick Market (1773), all located in Newport, RI.

Harrison, Wallace K. (1895–1981)
Formed one of the most successful practices in the United States with Raymond Hood. He worked on Rockefeller Center, New York City, and designed the United Nations building, New York City, with Le Corbusier; the Phoenix Mutual Life Insurance building, Hartford, CT; Lincoln Center, New York City; U.S. Steel Building (illus.), Pittsburgh, PA, and the gigantic South Mall Government Complex in Albany, NY.

Harvested rainwater
The rain that falls on a roof and is channeled by gutters to a storage tank or cistern. The uses of this water are limited by any pollutants that may be picked up from the roof surface and atmospheric pollutants.

Hastings, Thomas (1860–1929)
American architect who graduated from the École des Beaux-Arts and began a career with McKim Mead and White; later formed a partnership with John Carrere and designed enormous hotel complexes in Florida, such as the Ponce de Leon Hotel (1889), St. Augustine.

Hatch
Opening in a floor or roof with a removable cover.

Haunch
The middle part of an arch, between the springing point and the crown.

Haunch arch See Arch

Haunched beam
A beam whose cross section is thicker at the supports than in the middle of the span.

Haviland, John (1792–1852)

English-born American architect who designed in a severe Greek Revival style and later incorporated Egyptian Revivalist elements. He became known as the greatest Egyptian Revivalist. His design for the Philadelphia Arcade was unmatched. He also designed the Eastern State Penitentiary (1821), which was a prototype for the radial plan prison, and the Franklin Institute (1825), both of which are in Philadelphia, PA.

Hayden, Sophia (1869–1953)

The first woman to graduate from MIT's four-year program in 1890; then at age 22 she won the coveted competition to design the Woman's Building at the World's Columbian Exposition, Chicago.

Hazard

A material or condition that may cause damage, injury, or other harm, frequently established through standardized assays performed on biological systems or organisms. The confluence of hazard and exposure create a risk.

Hazardous material

Any material used in building construction that can be harmful to the construction workers or future occupants of the building; such as asbestos.

Hazardous use units

Any building designed or used for the purpose of occupancy that has contents which are liable to burn rapidly or cause explosions.

Hazardous waste

By-products of society with physical, chemical, or infectious characteristics that pose hazards to the environment and human health when improperly handled, specifically characterized by one or more of the following properties: ignitable, corrosive, reactive, or toxic.

Head

In general, the uppermost member of any structure. The upper horizontal cross-member between jambs, which forms the top of a door or window frame; may provide structural support for construction above.

Head

Head contractor

Main contractor engaged to be responsible for the majority of work on a building site, including subcontract work, materials, and labor supplies.

Head mortar joint See Mortar joint

Header

A masonry unit laid so that its short end is parallel to the face, overlapping two adjacent widths of masonry; a framing member supporting the ends of joists, transferring the weight of the latter to parallel joists and rafters.

Header bond See Bond

Headroom

The clear vertical distance between a floor or stair tread and the ceiling or overhead obstruction.

Healthy Building Network (HBN)

A national network of green building professionals, environmental and health activists, socially responsible community advocates and others who are interested in promoting healthier building materials as a means of improving public health and preserving the global environment.

Hearth

That part of the floor directly in front of the fireplace, and the floor inside the fireplace on which the fire is built, made of fire-resistant masonry.

Hearthstone

A single large stone forming the floor of a fireplace; materials such as firebrick and fireclay products, used to form a hearth.

Heartwood See Wood

Heat distribution
System for delivering heat throughout a house.

Heat exchanger
Device that transfers heat from one material or medium to another. An air-to-air heat exchanger, or heat-recovery ventilator, transfers heat from one airstream to another. A copper-pipe heat exchanger in a solar water-heater tank transfers heat from the heat-transfer fluid circulating through a solar collector to the potable water in the storage tank.

Heat gain
Increase in the amount of heat in a space, including heat transferred from outside in the form of solar radiation, and heat generated within by people, lights, computers, copiers, mechanical systems, and other sources. This additional migration of heat into a space is accomplished by conduction, radiation, or the natural exchange of air. This increase in the amount of heat in a given space must be mitigated by air-conditioning.

Heat island effect
A phenomenon that occurs in developed areas where the replacement of natural land cover with paving, buildings, roads, and parking lots results in an increase in outdoor temperatures. The heat island effect can be mitigated by vegetation, green roofs, and light-colored materials that reflect heat. Urban heat islands can be as much as 10 degrees Fahrenheit hotter than the surrounding undeveloped areas.

Heat islands
A phenomenon that causes urban and suburban temperatures to be 2 to 10 degrees Fahrenheit hotter than nearby rural areas. Elevated temperatures can impact communities by increasing peak energy demand, air-conditioning costs, air pollution levels, and heat-related illness.

Heat loss
Migration of heat from a space: by conduction, radiation, or exchange of air.

Heat pipe
A device that can quickly transfer heat from one point to another. Heat pipes are often referred to as the "superconductors" of heat, as they possess an extraordinary heat-transfer capacity and rate of transfer with almost no heat loss.

Heat pump
An electric device with both heating and cooling capabilities. It extracts heat from one medium at a lower (the heat source) temperature and transfers it to another at a higher temperature (the heat sink), thereby cooling the first and warming the second. Because they move heat rather than generate heat, heat pumps can significantly reduce energy costs. Air-source, ductless mini-split, geothermal, and absorption are some of the varieties of heat pumps used in homes.

Heat sink
Where heat is dumped by an air conditioner or by a heat pump used in cooling mode; usually the outdoor air or ground.

Heat transfer
A generic term for thermal conduction, convection, and radiation.

Heat-absorbing glass See Glass

Heating
Raising the temperature of an interior space, either by a fire in a fireplace, hot air from a furnace, gas heater, electric heater, or radiator.

Heating load
Rate at which heat must be added to a space to maintain a desired temperature.

Heating, ventilation, and air conditioning (HVAC)
Controls the ambient temperature, humidity, air flow, and air filtering of a building. It must be planned for and operated along with other data center components such as computing hardware, cabling, data storage, fire protection, physical security systems, and power.

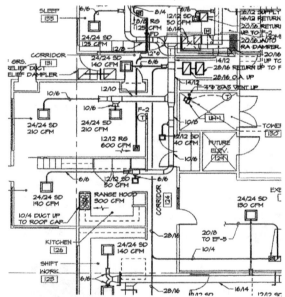

Heating/cooling systems
Methods for the delivery of hot or cold air depend on the type of unit served and distribution system. For example, forced air is used for heated air systems, whereas for fluid and electrical systems, convection and radiation are used to transfer heat. Central heating systems use a furnace, boiler, or heat pump to transfer heated air, water, or steam through ductwork or radiators. For smaller environments and/or temporary circumstances, heating may also be provided via local, nonmechanical means, such as

Heating/cooling systems

electric space heaters, baseboard and wall heaters, heat pumps, and passive systems such as green roofs and natural ventilation.

boilers

Used to heat water for domestic use and hydronic heating systems. Similar to furnaces, boilers also run on a variety of energy resources, including electric and biomass.

central heating systems

Used in mass dwellings in cold climates to simultaneously distribute a treated hot/cold, air/fluid to multiple spaces. Central systems use a furnace, boiler, or heat pump to transfer heated air, water, or steam through ductwork or radiators. For smaller environments and/or temporary circumstances, heating may also be provided via local, nonmechanical means, such as electric space heaters, baseboard and wall heaters, heat pumps, and passive systems.

electric space heaters, baseboard, and wall heaters

Units for small spaces with local heating requirements do not require a comprehensive duct-work or piping. Wall heaters are installed on the exterior wall; baseboard heaters are installed at the bottom of any interior wall; and space heaters are portable.

furnaces

Used to heat air only. These devices operate efficiently and can use a wide range of energy resources, such as natural gas, propane, biomass, fossil oil, or electricity.

green roofs

A natural heating/cooling alternative that provides thermal insulation and evaporative cooling due to water circulating from plants.

heat pumps

Used to heat air and water for both heating and cooling. These systems collect heat from a variety of sources including air, water, or specialized liquids and circulate the substance through pipes. Air-based heat pumps are used when cooling a structure, and ground-based pumps are used when heating one. Both use electricity to pump the air or liquid and to operate the compressor.

Heat-pump water heater

An appliance that uses an air-source heat pump to heat domestic hot water. Includes an insulated tank equipped with an electric resistance element to provide backup heat whenever hot-water demand exceeds the capacity of the heat pump. Since heat-pump water heaters extract heat from the air, they lower the temperature and humidity of the room in which they are installed.

Heat-recovery system

A process by which the exhaust air preheats the supply air when the outdoor air is cooler than the inside air. When the outside air is warmer than the inside air, the exhaust air will cool the supply air. This system saves energy depending on the season and only operates for sensible heat recovery.

Heat-recovery ventilation

To reduce energy costs, new homes have been built to be airtight when doors and windows are closed. But indoor air quality can be compromised by pollutants, moisture, and emissions without sufficient ventilation. Heat-recovery ventilators solve this problem by using a heat exchanger to transfer heat from outgoing heated indoor air to incoming fresh air without mixing the airstreams. The heat-exchange core of a typical HRV is made up of thin aluminum passages with incoming and outgoing airstreams flowing in alternate passages.

Heat-recovery ventilator

A system that reclaims the heat from warm exhaust air exiting a building and uses it to preheat entering fresh air.

Heavy-timber construction

Fire-resistant construction obtained by using wood structural members of specified minimum size; wood floors and roofs of specified minimum thickness; and exterior walls of noncombustible construction.

Heel

The lower end of an upright member, especially one resting on a support.

Hejduk, John Quentin (1929–2000)

Architect, artist, and educator noted for a profound interest in the fundamental issues of shape, organization, representation, and reciprocity. Hejduk worked in the firms of I. M. Pei and Partners and A. M. Kinney and Associates and established his own practice in New York in 1965.

Helical reinforcement

Small-diameter reinforcement wrapped around the main or longitudinal reinforcement of columns, restraining the lateral expansion of the concrete under compression, and thus strengthening the column.

Helical stair See **Stair**

Helices

Figures like the tendrils of a vine.

Helix

Any spiral form, particularly a small volute or twist under the abacus of the Corinthian capital.

Helix

Hellenic architecture (480–323 B.C.)
The style of architecture of the Classical Greek period up to death of Alexander the Great.

Hellenistic architecture (323–30 B.C.)
The style of Greek architecture after the death of Alexander the Great.

Helm
A bulbous termination to the top of a tower, found in central and eastern Europe.

Helm roof See Roof

Helmie, Frank (1869–1939)
American architect and partner in the firm of Helmie and Corbett.

Hemispherical
A rounded form resembling half of a sphere bounded by a circle.

Hemispherical dome See Dome

Hemispherical vault See Vault

Hemlock See Wood

Hemp
One of the oldest cultivated crops, its use dates back to the stone age. Hemp is one of the most environmentally friendly fibers in the world and requires no pesticides, herbicides, or fertilizers, and it uses very little water. Hemp exhibits eight times the strength of cotton and can be woven into a variety of textures. Although currently illegal to cultivate in the United States, hemp is a fast-growing, environmentally sound substitute for cotton and wood fibers that requires little or no chemicals to produce paper, textiles, and a variety of other products.

Henry II (Deux) style (1547–1559)
The second phase of the early French Renaissance, named after Henry II who succeeded Francis I. It was characterized by Italian classic motifs which supplanted Gothic elements. The west side of the Louvre in Paris (illus.) is the most characteristic example.

Henry IV (Quatre) style (1586–1610)
The early phase of the Classical period of French architecture preceding the architecture of Louis XIII and Louis XIV; the style was particularly strong in domestic architecture and town planning.

Heptagon
A seven-sided regular polygon; the angle included between the seven sides is 128.6 degrees.

Herm
A rectangular post, usually of stone, tapering downward, surmounted by a bust of Hermes or other divinity or by a human head.

Herrera, Juan de (1530–1597)
Spanish architect who developed a Renaissance style of great purity and simplicity that reflected the severe taste of his patron, King Phillip II. From 1572 he was in charge of completing El Escorial, Toledo, Spain.

Herringbone
A way of assembling, in a diagonal zigzag fashion, brick or similar rectangular blocks for paving; also strips of wood or other materials having rectangular shapes for facing walls and ceilings or for use in parquetry.

Herringbone blocking
Solid blocking between joists or studs with alternate blocks offset to allow for end nailing.

Herringbone bond See Bond

Herringbone matched
The assembly of wood veneers from the same flitch, so that successive sheets are alternated face up and face down. Side by side sheets show a symmetrical mirror image at the joints.

Herron, Ronald (1930–1998)
London-born architect associated with Archigram in the 1960s. His vision for "Walking Cities" (illus.) was published widely and is exhibited internationally.

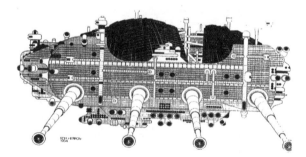

Herzog, Thomas (1941–)
German architect from Munich known for his focus on climate and energy use through the use of technologically advanced architectural skins. He established the firm Herzog + Partners in 1983. Perhaps best known for the conversion of the giant Bankside Power Station in London to the new home of the Tate Modern. In 2001, the firm of Herzog & de Meuron was awarded the Pritzker Prize.

Hexagonal
Refers to a plane geometric figure containing six equal sides and six equal angles; occurring in nature as minerals, snow crystals, and honeycombs.

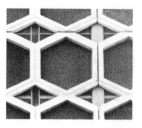

Hexastyle
A building having a row of six columns across a facade or portico.

Hickory See Wood

Hicks, Margaret (1858–1883)
The first woman to graduate from Cornell University in 1880 and the first woman architect to have her work published.

HID
High-intensity discharge (lighting).

Hierarchy
An arrangement or system of ranking one above the other or arranged in a graded series or sequence such as size (large to small), shape (similar or dissimilar), and placement (emphasis or location).

Hieroglyph
A figure representing an idea and intended message; a word or root of a word; a sound that is part of a word, especially applied to the engraved marks and symbols found on the monuments of ancient Egypt.

Hieroglyph

High albedo
Reflective roof coatings that lower the absorption of solar energy and can reduce building air-conditioning energy use.

High chairs
A manufactured device used to hold up the welded wire frame at approximately one-half the thickness of the cement slab during placement of the concrete.

High efficiency
General term for technologies and processes that require less energy, water, or other inputs to operate. A goal in sustainable building is to achieve high efficiency in resource use when compared to conventional practice. Setting specific targets in efficiency for systems and designs help put this general goal of efficiency into practice.

High reflectance
The ability of materials to effectively reflect the sun's energy. Roof materials with high reflectance stay cooler in the sun, reduce energy costs, improve occupant comfort, and reduce the urban heat island effect.

High Renaissance
Refers to the culmination of the Italian Renaissance style in the late 16th century, characterized by the imitative use of the orders and Classical compositional arrangements.

High style
The more ornately detailed version of an architectural style; used in contrast to simpler examples, from different periods or of the same period.

High Victorian
The more ornate architecture and interior design work of the Victorian period, which combines a variety of stylistic sources with great freedom.

High Victorian Gothic
A style characterized by complex exteriors, often with bays, towers and turrets, typically with contrasting colors and textures of brick or stone, especially as horizontal bands and voussoirs in alternating colors.

High-efficiency furnace
Furnaces that have an annual fuel utilization efficiency (AFUE) of 85 percent (oil) and 90 percent (gas) or higher. In general, the higher the AFUE, the more efficient the furnace. The Energy Guide label on the equipment indicates the degree of efficiency.

High-efficiency hot-water heaters
Models with an Energy Star rating are considered highly efficient.

High-efficiency particulate air filter (HEPA)
An air filter that captures a high percentage of all particles, including very small particles not caught by other types of filters.

High-efficiency rooftop units
Air-conditioning units that have a higher AFUE (annual fuel utilization efficiency) rating.

High-efficiency toilet
A toilet that provides at least 20 percent water savings over the federal standard of 1.6 gpf (gallons per flush), and still meets the most rigorous standards for flush performance. HETs include pressure-assist toilets that use as little as 1.0 gpf, gravity-flush toilets that consume 1.28 gpf, and dual-flush toilets that offer two different flush volumes.

High-heeled truss
Roof truss design that allows space for insulation near the eaves. Conventional truss design limits the amount of insulation that can be applied in this area.

High-intensity discharge lamp
A lamp that produces light by passing an electrical current through a gas or vapor under high pressure. HID lamps have a long life and consume little energy to produce a great amount of light. However, they distort the color rendition of objects and are used mostly as ambient lighting for commercial interiors. Ballasts are required to start the lamp. Types include mercury vapor lamps, metal-halide lamps, and sodium vapor lamps. The different types contain different gases, which produce different colors of light.

High-light window See Window

High-mass construction
Passive building strategy of constructing buildings of massive, heat-retaining materials, such as masonry or adobe, to moderate diurnal temperature swings, especially in arid climates.

High-performance building
A green or sustainable building, often with an emphasis on the use of advanced technology, or "smart infrastructure," and its impact on the occupants' ability to control key building.

High-performance glazing
Generic term for glazing materials with increased thermal efficiency.

High-performance green building
Buildings that include design features that conserve water and energy; use space, materials, and resources efficiently; minimize construction waste; and create healthy indoor environments.

High-performance windows
Windows rated by U-factor and solar heat gain coefficient (SHGC). The U-factor is the measure of the speed with which a window transfers heat by conduction. The SHGC rates a window according to how much radiant solar energy it lets through.

High-pressure sodium lamp
Among the most efficient sources of light; used primarily for exterior street lighting. The color rendition is similar to incandescent lamps.

High-quality duct system
This option avoids the potential of significant heating and cooling losses, as well as the potential health threats caused by depressurizing or pressurizing a house. All ducts are sealed using a latex mastic and fiberglass tape. Inner and outer linings of the duct are both sealed. The air handler, support platform, and return plenum are sealed airtight at the joints. Duct tape is not used in any part of the system. Ductwork is not run inside the building envelope walls. The system can be performance-tested to ensure proper installation.

High-rise building
Describing a building having a comparatively large number of stories usually above 10–12 stories, and equipped with elevators.

High-tech architecture
A style of architecture where mechanical and electrical building services and structural elements are not only exposed but emphasized by using different colors to indicate their respective functions, which accentuates the machinelike aspects of the building.

High-tech architecture

Hildebrandt, Lucas von (1668–1745)
Leading Austrian architect; pupil of Fontana and admirer of Guarini. His Baroque facades are plain, while the interiors are dramatic and richly decorated. Works include the Residenz (illus.), Wurzburg, Germany, and the Belvedere (illus.), Vienna, Austria (1724).

Hildebrandt, Lucas von

Hinge
A movable joint used to attach, support, and turn a door about a pivot, consisting of two plates joined by a pin which supports the door and connects it to its frame, enabling it to swing open or closed.

Hip
The external angle at the junction of two sloping roofs or sides of a roof: the rafter at the angle where two sloping roofs or sides of a roof meet.

Hip and valley roof See **Roof**

Hip knob See **Ornament**

Hip rafter See **Rafter**

Hip roll
A rounded piece of tile, wood, or metal used to cover, finish, and sometimes add a decorative effect. Also called a ridge roll.

Hip roof See **Roof**

Hip tile
A convex-shaped tile used to cover a roof hip, installed overlapping along the edge of the hip.

Hipped dormer See **Dormer**

Hipped end
The sloping triangularly shaped end of a hipped roof.

Hipped gable See **Gable**

Hippodrome
In ancient Greece, a stadium for horse and chariot racing.

Historiated
Ornament incorporating human or animal figures or plants that have a narrative as distinct from a purely decorative function.

Historic
Having importance in, or influence on history; famous, renowned.

Historic American Buildings Survey (HABS)
A documentation program initiated in 1993 to collect measured drawings, photographs, and historical and architectural data with the help of teams of architects and architectural students, and deposit them for public use in the Library of Congress. To date it includes thousands of drawings, photographs and data that have been published in scores of catalogs.

Historic American Engineering Record (HAER)
A program of the National Park Service that has documented exemplary engineering works since 1969. The HAER has established a collection of the same nature as HABS on engineering landmarks, such as bridges, mills, factories, foundries, subways, and tunnels. Documentation includes photographs, measured drawings, and written historical information, and is archived at the Library of Congress.

Historic bridge foundation
A clearinghouse for information on preservation of endangered bridges, provides assistance to list bridges in the National Register of Historic Places, or to complete grant applications for preservation projects, and consults to devise reasonable alternatives to the demolition of historic bridges.

Historic building
A building that has been recognized and documented as having historic significance, or part of a historic district, especially those listed in a register or inventory of historic places.

Historic district
Definable geographical area containing a number of related historic sites, buildings, structures, features, or objects that are united by past events. It also includes those buildings linked esthetically by plan or physical development that have been designated on a local, state, or national register of historic places; may encompass a neighborhood or all of a small town; some districts may comprise individual elements separated geographically but linked by association or history.

Historic district ordinance
A municipal law designating and protecting an area of significant historical, architectural, cultural or other special character, or authorizing the general establishment of historic districts, typically establishing a review commission, procedures, and an appeals process. It sometimes grants the supervising agency power to restrict demolition or alteration within an area. It parallels the goals of zoning laws to regulate growth and to control the distribution of people and traffic. It protects environmental amenities and keeps them for the public enjoyment.

Historic district survey
Site and location research to determine the extent and nature of an existing or proposed historic district. Each building within the survey is evaluated as to its appropriateness to the district.

Historic house museums
Museums that depict tangible evidence of the original events which are identified with a particular house.

Historic image restoration
The renovation of a building or community that saves and restores the historic image of that place.

Historic interiors specialist
An architect or interior designer whose specialty involves the investigation, analysis, and documentation of interior finishes, furnishings, and fixtures of historic buildings.

Historic photographs
Photographic records of historic buildings or areas, for the purpose of placing in an archive, collection, or book.

Historic preservation
Encompasses a broad range of activities related to preservation and conservation of the built environment by physical and intellectual methods.

Historic preservation architect
A licensed architect who specializes in the conservation, preservation, restoration, renovation, rehabilitation or

Historic preservation architect

adaptive reuse of structures. May prepare historic structure reports, and produce construction drawings for all types of preservation work, as well as, provide other services on other building types.

Historic preservation certification
The application form required for certified historic structure designation if the property is not itself listed in the National Register.

Historic resource
A building, site, or district deemed to be of historic significance.

Historic significance
The importance of an element, building, or site owing to its involvement with a significant event, person, or time period, or because it is an important example of a past architectural style.

Historic site
A location where a historical event took place, as designated by a marker or plaque.

Historic structure
A structure famous because of its association with a historic event, or the history of the locality.

Historical archeology
The study of the cultural remains of literate societies, as distinct from prehistoric archeology. American historical archeology deals with excavated material as well as above-ground resources, such as buildings, pottery, weapons, tools, glassware, cutlery and textiles.

Historical commission
A municipal entity, usually without regulating authority; its responsibilities are often limited to identification of landmarks and districts, research and public education.

Historical frieze
A frieze decorated with bas-relief scenes illustrating a historical event.

Historical function
The original use of a building before any restoration or alteration.

Historical marker
A permanent, descriptive sign or plaque affixed or adjacent to a historic building or historic site.

Historical monument
Building or site having architectural or historical, municipal, state, or national significance.

Historical museum
A museum, which may or may not be housed in a historic building, whose primary purpose is the interpretation of history. Its collections may include objects, sites and structures related to persons and events significant in history.

Historical research
The study of documents: photographs, publications, and other data concerning a historic site, building, structure, or object.

Historicism
Purposely citing historical precedents in historic architectural styles of the past.

Hittite architecture (2000–1200 B.C.)
An architecture found in northern Syria and Asia Minor, characterized by fortifications constructed with stone masonry and gateways ornamented with sculpture.

Hittorf, Jacques-Ignace (1792–1867)
German-born architect and scholar who settled in Paris, France. Worked with François-Joseph Belanger on the ornamental iron-and-glass dome of the Halle au Blé (1813), Paris; most noted work was the Gare du Nord (1866), Paris.

Hoban, James (1762–1831)
American architect in Washington, DC; winner of the competition for the White House (1801). He was also the superintendent of construction of the U.S. Capitol (1802) in Washington, DC.

Hoffman, Frances Burrall, Jr. (1884-1980)
American architect, graduate of Harvard, who trained at the École des Beaux-Arts in Paris. Worked with Carrere and Hastings before setting up office with Harvey Creighton Ingalls. He designed Viscaya (1916), a winter home for industrialist John Deering, on Biscayne Bay near Miami. It was designed to look as if it had been there for 300 years.

Hoffman, Josef (1870-1956)
Designed the Stocklet House, Brussels (1905). The exterior was enhanced with marble and bronze, while the interior dining hall featured murals by Gustav Klimt.

Hoist
A projecting beam with block and tackle, used for lifting goods; often seen above openings in the upper stories of medieval houses. Also, a platform for lifting people and/or materials. The platform is lifted by cables and contained within an open frame that is supported by the building.

Holabird, John (1921-2009)
In 1880 Holabird's grandfather established the Chicago firm, Holabird & Root. He too eventually joined the firm, retiring as a partner in 1987. From the Chicago Board of Trade to the Marquette Building to the Art Deco Palmolive Building, the firm's effect on the Chicago skyline was profound.

Holabird, William (1854-1923)
Designed the Tacoma Building, Chicago, a 12-story building that established the Chicago School of Architects as leaders in skyscraper design.

Holistic
In holistic design, it expresses an integrative and comprehensive approach that considers the interrelatedness of a projects parts, components, systems, and subsystems, in order to optimize energy and environmental performance during the entire life of a project.

Holl, Elias (1573-1646)
Leading Renaissance architect in Germany; was influenced by Palladio and Mannerism but modified them by using typically German features such as high gables.

Holl, Steven (1947-)
American architect and watercolorist, perhaps best known for the 1998 Kiasma Contemporary Art Museum in Helsinki, Finland; the 2003 Simmons Hall at MIT in Cambridge, MA; the celebrated 2007 Bloch Building addition to the Nelson-Atkins Museum of Art, Kansas City, MO; and the 2009 Linked Hybrid mixed-use complex in Beijing, China.

Hollien, Hans (1934-)
Austrian-born architect, who established his reputation with small crafted shops, detailed with meticulous care, and in sharp contrast to the surrounding facades. He also designed the Museum of Modern Art, Frankfurt, Germany (1987), and the Haas House (illus.), Vienna, Austria.

Hollow block masonry See **Masonry**

Hollow block unit See **Masonry**

Hollow molding See **Molding**

Hollow square molding See **Molding**

Hollow tile See **Tile**

Hollow-core door See **Door**

Hollow-tile floor
A reinforced concrete floor that is cast over a formwork of hollow clay-tile blocks, with concrete filling the voids between the tiles.

Hollyhock ornament See **Ornament**

Home energy audit
A home energy audit provides information on the need for, cost, and associated savings of energy improvements in a home or building. A home energy audit is recommended prior to upgrading insulation or windows.

Home energy performance audit
Energy audit that also includes inspections and tests to assess moisture flow, combustion safety, thermal comfort, indoor air quality, and durability.

Home energy rating system
A collection of programs throughout the country that assign energy ratings based on predicted energy use of the house. Ratings are either on a scale of 1 to 100 points or 1 to 5-plus stars. Most houses built today without any special attention to energy efficiency typically earn an 80-point or three-star rating.

Home Performance with Energy Star
A residential weatherization program jointly administered by the U.S. Department of Energy and the U.S. Environmental Protection Agency. The program connects homeowners interested in improving the energy performance of their homes with contractors trained to assess home performance and perform energy retrofit work.

Home-run plumbing system
Water-distribution piping system in which individual plumbing lines extend from a central manifold to each plumbing fixture or water-using appliance; piping is typically cross-linked polyethylene (PEX). Depending on patterns of use, hot water may be delivered more quickly in such a system because the diameter of the tubing can be matched to the flow of the fixture or appliance.

Homestead
A piece of land, limited to 160 acres, deemed adequate for the support of one family, occupied by the owner as a home; in many states, such a property is protected by statute and exempt from seizure or attachment to satisfy a judgment rendered in favor of a creditor.

Homestead Act of 1862
U.S. federal legislation permitting settlers to occupy a homestead on designated public land in the western states and own it after five years.

Homesteading
The process of acquiring land by occupying and using it; may be done through the Homestead Act or without official sanction.

Homogeneous
Likeness in nature or kind: similar, congruous and uniform in composition or structure throughout.

Honeycomb
Any hexagonal structure or pattern, or one resembling such a structure or pattern.

Honeycomb brickwork
A brick bond characterized by the absence of certain bricks for decorative purposes, or to allow ventilation or provide a screened effect.

Honeycombed concrete
Concrete with voids caused by failure of the cement mortar to fill all the spaces between the particles of the coarse aggregate.

Honeysuckle ornament See **Ornament**

Honnecourt, Villard de (c. 1175–1235)
French architect whose notebook is an invaluable source of information on thirteenth-century building materials.

Hood
A projection above an opening, such as a door or window, serving as a screen or as protection against the weather.

Hood

Hood mold See **Molding**

Hood, Raymond (1881–1934)
Won the competition with John Mead Howells to design the Chicago Tribune Tower in 1922, a high point in Beaux-Arts eclecticism with a Gothic superstructure. He designed both the Daily News Building (1929) and the McGraw-Hill Building (illus.), in New York City, in the Art Deco style (1930).

Hopper window See **Window**

Horizon line See **Perspective projection**

Horizontal cornice See **Cornice**

Horizontal recycling
A recycling system that turns the majority of the original product back into a similar product as the original.

Horn
The diagonally projecting points of the Corinthian and Composite orders; usually chamfered to protect the sharp edges.

Horse See **Ornament: animal forms**

Horseshoe arch See **Arch**

Horta, Victor (1861–1947)
The leading architect in the Art Nouveau style (c. 1900). The Tassel House, Brussels (1892) was his first work in the Art Nouveau style. Both the L'Innovation department store (1901) and the Maison de Peuple (1899), Brussels (1901), had large metal and glass Art Nouveau facades (both now demolished).

Hotel
A building with rooms or suites for rent by the day; typically includes public facilities for dining.

Hot-rolled sections
Structural steel sections produced by passing a red-hot billet through a series of rollers, gradually forming it closer to the desired shape.

House
A structure serving as a dwelling for one or several families. A place of residence, an abode. The design of the house runs the gamut of nearly every design style that has ever existed. Although there are many reasons to preserve a particular house or group of houses, other external development factors usually dictate the ultimate fate of those in danger.

House drain
A horizontal piping system within a building, which receives waste from the soil stack and carries it to the sewer.

House museum
A museum whose structure itself is of historical or architectural significance and whose interpretations relate primarily to the building's architecture, furnishings, and history.

Houseboat
A barge-like boat equipped for use as a domicile. Houseboats have been around since the biblical times of Noah, and many people still live on boats in many parts of the world; in the East, the Middle East, and northern Europe. In the United States they were popular in the San Francisco Bay area (1890s) and are still in wide use today in Sausalito, CA.

Housing and Community Development Act of 1974
Encompasses urban and rural Community Planning through block grants from the U.S. Department of Housing and Urban Development, and puts the burden on local communities of satisfying federal protective laws and procedures, as a condition of receiving project funds.

Housing and Community Development Act of 1977
Established the Urban Development Action Grant Program.

Housing code
Regulations that define the livability of a housing unit; it usually requires minimum physical requirements for light, security, ventilation, room size, and fire egress.

Housner, George (1911–2008)
Acclaimed as the father of modern earthquake engineering and credited with developing an innovative mathematical system to analyze the effects of vibration throughout structures. After the catastrophic 1933 earthquake at Long Beach, he later helped design California's water storage and transportation systems, and his techniques helped strengthen dozens of dams and aqueducts across the state.

Hovel
A open-sided shed, covered overhead for shelter of livestock, produce, or people; a poorly constructed and ill-kept house.

Howard, John Galen (1864–1931)
American architect in San Francisco; apprenticed to H. H. Richardson and McKim Mead and White. Educator and designer in Classical styles; executed a plan for the University of California (1923), Berkeley, CA.

Howard, Sir Ebenezer (1850–1928)
English proponent of the Garden City movement; published *Garden Cities of Tomorrow* in 1902. Howard's ideas led to the new-town policy adopted in Britain after World War II.

Howe truss See Truss

Howe, George (1886–1973)
American architect; trained at the École des Beaux-Arts in Paris but abandoned the style in favor of modernism. With William Lescaze, he designed the Philadelphia Saving Fund Society office building, the paradigm of an International Modernist skyscraper. He later joined Louis Kahn, and in 1950 he became chair of the Department of Architecture at Yale University.

Howells, John Mead (1868–1959)
American architect who worked with McKim Mead and White before setting up an office with Isaac Newton Phelps (1867–1944) in New York. The firm designed the Madison Square Church Mission House (1989), NYC and Woodbridge Hall, Yale University (1901), New Haven, CT. Howells designed the Chicago Tribune Tower (1925), with Raymond Hood, the Daily News Building (1930), and Beekman Tower (1928), both in New York City, with Art Deco modeling. He was a sensitive restorer of early American architecture, and wrote *Lost Examples of Colonial Architecture* (1931) and *The Architectural Heritage of the Merrimack* (1941).

H-section
A rolled-steel member with an H-shaped cross section with parallel flanges and faces; used for structural columns and piles, due to its ability to withstand rotation; usually square in its outer dimensions.

Hue
Name of a color.

Human scale See Scale

Humidistat
A device for measuring and controlling relative humidity.

Hundertwasser, Friedensreich Regentag Dunkelbunt (1928–2000)
Austrian painter and architect, whose original and unruly artistic vision expressed itself in pictorial art, environmentalism, philosophy, and design of facades, postage stamps, flags, and clothing. The common themes in his work are bright colors, organic forms, a reconciliation of humans with nature, and a strong individualism, rejecting straight lines. His architectural work is comparable to Antonio Gaudí (1852–1926) in his use of biomorphic forms and tile. He was also inspired by the art of the Vienna Secession and by the Austrian painters Egon Schiele (1890–1918) and Gustav Klimt (1862–1918). The Hundertwasserhaus apartment block in Vienna (illus.), designed by Hundertwasser, has undulating floors, a roof covered with earth and grass, and large trees growing from inside the rooms, with limbs extending from windows.

Hundertwasser, Friedensreich Regentag Dunkelbunt

Hunt, Myron (1868–1952)
American architect who was one of the group in Steinway Hall in Chicago that included Frank Lloyd Wright; designed the Huntington Art Gallery and Ambassador Hotel in Los Angeles, CA.

Hunt, Richard Morris (1827–1895)
American architect. Trained in France, he produced buildings in the United States in a variety of styles, mainly grandiose pastiche, for millionaire clients.

Hut
A small, simple shelter or dwelling. Can be constructed out of a variety of building materials

HVAC
Abbreviation for heating, ventilation, and air conditioning, which all seek to provide thermal comfort, acceptable indoor air quality, and reasonable installation, operation, and maintenance costs.

HVAC (16 SEER+)
A rating system, seasonal energy efficiency ratio (SEER), used to measure the efficiency of central air conditioners and air-source heat pumps. The higher the rating, the more energy efficient it is. Air conditioners that are 14 or higher SEER meet Energy Star criteria.

HVAC systems
The equipment, distribution network, and terminals that provide, either collectively or individually, the processes of heating, ventilation, or air conditioning for a building.

Hybrid vehicles
A car that runs on both electric battery and fuel, making the gas mileage efficient and producing fewer emissions, both of which help control pollution in the environment.

Hydraulic elevator
An elevator powered by the energy of a liquid under pressure in a cylinder which acts on a piston or plunger to move the elevator car up and down on guide rails.

Hydrocarbon (HC)
Chemical compound consisting entirely of carbon and hydrogen.

Hydrochlorofluorocarbon (HCFC)
Compound commonly used as a refrigerant in compression-cycle mechanical equipment, such as refrigerators, air-conditioners, and heat pumps, or as a blowing agent to produce foam insulation. It is an alternative refrigerant type that has reduced ozone-depleting effects.

Hydroelectric energy
The energy produced by moving water that is harvested for conversion into usable electrical energy.

Hydroelectric power
The use of artificial or natural waterfalls to generate electricity.

Hydroelectricity
Electricity that is produced when falling water turns generators. It is a renewable energy source derived from gravity and rain. Very small generation facilities, producing up to 50 kilowatts, are called micro-hydro.

Hydrofluorcarbon
A greenhouse gas.

Hydrology
The science of water relating to the occurrence, properties, distribution, circulation, and transport of water.

Hydronic heating
Heat distribution system in which hot water produced by a boiler is circulated through pipes and baseboard radiators or tubing in a radiant floor. Also called baseboard hot-water heating.

Hygrometer
A device that measures relative humidity of air.

Hygrothermal
A term used to characterize the temperature (thermal) and moisture (hygro) conditions, particularly with respect to climate, both indoors and out.

Hyperbola
The section of a right circular cone by a plane that intersects the cone on both sides of the apex.

Hyperbolic paraboloid
A shell in the form of a hyperbolic paraboloid, that is, saddle shaped. Generated by two systems of straight lines,

Hyperbolic paraboloid

its formwork is more readily constructed from straight pieces of timber than that of a dome.

Hyperthyrum

A latticed window constructed over the door of an ancient building. Also, a frieze and cornice arranged and decorated in various ways for the lintel of a door.

Hypostyle hall

A structure whose roofing is supported within the perimeter by groups of columns or piers of more than one height; clerestory lights are sometimes introduced.

I i

I beam See **Beam**

I joist
A manufactured wood product so named because its section looks like an uppercase letter I. The top and bottom chord are lumber or laminated wood, and the vertical web is plywood or oriented strand board.

Ice dam
A ridge of ice that forms along the lower edge of a roof, possibly leading to roof leaks, caused by heat leaking from the attic, which melts snow on the upper parts of the roof. The water then refreezes along the colder eaves and works its way back up the roof and under shingles, thus causing leakage.

Icon
An image of sacred personages that are objects of veneration; found on buildings.

Icosahedron
A regular polygon bounded by 20 equilateral triangles, with 12 vertices and 20 edges.

Ictinus (c. 500 B.C.)
Greek architect and designer of the Parthenon, with Callicrates, and the Temple of Apollo Epicurius, Bassae, which had a Doric Order outside, an Ionic Order inside, and a Corinthian Order at the end.

Icynene®
Open-cell, low-density spray foam insulation that can be used in wall, floor, and roof assemblies. It has an R-value of about 3.6 per inch and a vapor permeability of about 10 perms at 5 inches thick.

Igloo
An Inuit house constructed of snow blocks or various other materials such as wood, sod, poles and skins; when of snow, a domed structure is employed.

Illuminance
Commonly called light level, illuminance refers to the light intensity arriving on a surface, measured in foot-candles (fc). It is the standard international unit that is used to measure the amount of light per unit of surface area, also known as lux (lx). Measurements of illuminance are used to select lighting fixtures and to evaluate a lighting design. The photometric data to be considered include the luminous-intensity distribution curve (LIDC), the coefficient of utilization (CU), and the light loss factor (LLF).

Illuminance values
Standards for lighting and illuminance in the United States are established by the Illuminating Engineering Society of North America (IESNA) and are summarized in its handbook, which shows categories and ranges of illuminance.

Illumination Engineering Society of North America (IESNA)
The professional society of lighting engineers, including those from manufacturing companies and others professionally involved in lighting. It produces standard practice documents for applying proper lighting techniques to indoor and outdoor spaces.

Illumination quality
Standards set by the IESNA (Illuminating Engineering Society of North America) generally measured in foot-candles 30 meters above the floor.

Illustration
The representation by artistic means of an idea or scene.

Image See **Design drawing**

Imbrex
A curved tile, typically a half-cylinder shape; used to cover joints between roofing tiles.

Imbrication

Overlapping rows of shaped tiles or shingles that resemble overlapping fish scales; it is also called fish-scale pattern.

Imhotep (c. 2600 B.C.)

History's first named architect, and counselor to King Zoser of Egypt. He created the huge funerary complex at Saqqara; it's stepped pyramid design, and sophisticated stonework and use of columns, set the pattern for Egyptian monuments for 2,500 years.

Imitation

The representation of one material using another, generally implies copying the color and surface appearance of another material; one of the most common form is wood graining and marbleizing.

Impact analysis

The second stage of life-cycle assessment, in which the environmental impacts of a process, product, or facility are determined.

Impact resistance

Capacity of a material to resist loads that are suddenly applied.

Imperial dome See Dome

Impermeable

Resistance to penetration by fluids or vapor through the material.

Impervious

Resistant to water.

Impervious surface

A surface that sheds the precipitation falling on it, rather than infiltrating. They contribute to stormwater runoff and the heat island effect.

Impervious surfaces

Constructed surfaces that are impenetrable by water. Impervious surfaces can lead to excessive stormwater runoff and limit the amount of stormwater that remains onsite or recharges local aquifers.Pervious or porous surfaces allow some water infiltration, thereby reducing runoff.

Impluvium

A pool for receiving water draining from the roof in an ancient Roman atrium.

Imposed load

The weight of any movable load in a building or structure, such as wind, rain and snow, occupants, furniture and other belongings.

Impost

The horizontal molding or capital on top of a pilaster, pier, or corbel which receives and distributes the thrust at the end of an arch.

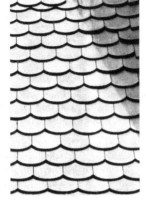

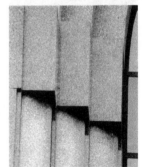

Impregnation of timber
The process of saturating timber with a preservative, such as creosote oil.

Improved land
Building sites which have water and sewage connections available and access to streets and telephone, gas, and other electrical services.

Improvement
Any changes made in a project which increases its value, such as the addition to a building or a home. An improvement may also be made by drainage, removing unsightly objects or growths; any act which serves to add utility, beauty, or otherwise increases the value of the property.

Improvement analysis
The third stage of life-cycle assessment (LCA), in which design for environmental techniques are used with the results of the first and second LCA stages to improve the environmental plan of a process, product, or facility.

Improvement on land
Any addition to a site which tends to increase its value or utility, such as the erection of buildings and retaining walls, also building fences and driveways.

Improvements to land
Buildings or facilities which, though not embraced within the boundaries of a property, add to its value and usefulness, such as water mains, sewers, street lighting, sidewalks and curbs.

In cavetto See Relief

In situ
In position; a term applied to building elements and improvements that are assembled or cast in their permanent position on site rather than prefabricated elsewhere; for example, cast concrete rather than precast concrete.

In situ pile
A concrete pile cast in its final location, with or without a casing, as distinct from a pile that is precast and subsequently driven.

In the clear
The uninterrupted linear measurement of a space; not including the structural parts themselves.

Inca architecture (1200–1400)
The last of the Pre-Columbian cultures buildings were characterized by megalithic masonry, as exemplified in the ceremonial buildings of the mountain city Machu Picchu, the last fortress to resist the Spanish invaders.

Incandescent lamp
A lamp consisting of metal elements that are heated within a glass enclosure until they glow. The bulb is filled with an inert gas so the filament will not burn up. These lamps are available in various wattages and have a low efficacy rating. A very low percentage of wattage goes to the production of light, and the rest goes to heat.

Incandescent light
The light produced when electric current heats a tiny coiled filament to glowing. Such light bulbs convert about 90 percent of the electricity into heat and only 10 percent into light.

Incentive
A discount, contribution, or amenity offered to a lessee of a property or facility.

Incident
Subordinate to the whole scheme, but used to give points of reference along the way, and temporarily create interest.

Incise
To cut a shallow mark into a material.

Inclinator
A platform, open or enclosed, running on an inclined rail for moving passengers to a higher level at an inclined angle, rather than as a vertical-lift elevator.

Incombustible
Material that does not burn in a standard test in a furnace.

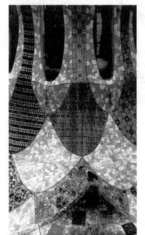

Incrustation
The covering of wall surfaces with some precious material, such as tile or jewels.

Indented
A gap left by the omission of stone, brick, or block units in a course of masonry, used for bonding future masonry.

Indented joint See **Joint**

Indented molding See **Molding**

Index of Fame
In doing research for her book *American Architects and the Mechanics of Fame*, (1991), author Roxanne Kuter Williamson studied more than 600 American architects who have earned a name in history. Her criteria for "fame" were geared toward those deemed important enough to be included in history textbooks. The Index of Fame was compiled from the indexes of two dozen major architectural histories and four encyclopedias. The number of times the name appeared was put into a matrix, and each architect was then given a specific place in the index.

Indian (Buddhist) architecture (300 B.C.–320 A.D.)
The earliest surviving buildings are of timber and mud-brick construction, of which the stupa is the most characteristic; it is a hemispherical mound with a processional path around the perimeter and elaborately carved gateways. The most typical is the stupa at Sanchi. In rock-cut Buddhist temples, the main forms and details follow early wooden prototypes, with elaborately carved stone shrines in which the exterior is more important than the interior.

Indian (Hindu) architecture (600–1750)
All types of temples in this style consist of a small unlit shrine crowned by a spire and preceded by one or more porch-like halls, used for religious dancing and music. The stone was laid up rough-cut and carved in place by Hindu sculptors who treated every element on every surface as unique, using the repetition of sculptural forms to achieve a unifying context. There was no attempt to evolve a style or to perfect any particular pillar or column.

Indian (Hindu-Buddhist) architecture (1113–1150)
The Hindu and Buddhist religions had a strong influence on Far East temple architecture. One of the most well known and representative sites is Angkor Wat located in Cambodia, a temple complex of shrines that was intended as a funerary monument. It is perhaps one of the world's largest religious structures and was conceived as a "temple mountain" within an enormous enclosure and surrounded by a wide moat. A monumental causeway, framed by giant mythical serpents, leads to the entrance gate. The temple is built on a series of stepped terraces, surrounded by towers at each corner. Vaulted galleries receive light from an open colonnade illuminating the continuous relief friezes which adorn the inner walls. The central sanctuary is a large pagoda-like tower on top of a stepped pyramid. It is joined by passageways to towers at each of the four corners at the base.

274

Indian (Hindu-Buddhist) architecture

Indian (Jain) architecture (1000–1300)
An architecture in which temples are enclosed shrines preceded by an open porch, which is often elaborately carved. They have a lighter appearance and are more elegant than Hindu temples.

Indigenous American architecture (500 B.C.–1500 A.D.)
The native styles include the wigwams and long houses of forested areas, teepees of the Plains Indians, igloos of the Inuits, and sophisticated communal pueblo cities carved out of mountain sides or built out of adobe in the Southwest.

Indigenous materials
Materials that are produced near a construction site. This reduces building costs and helps to boost local economies.

Indigenous planting
Landscaping strategy that uses native plants. Provided that the native plants are placed in the proper growing conditions, such plantings can have low or zero supplemental water needs.

Indirect light
The light provided by reflection usually from wall or ceiling surfaces. In daylighting, this means that the light coming from the sky or the sun is reflected on a surface of high reflectivity like a wall, a windowsill, or a special redirecting device. In electrical lighting, the luminaires are suspended from the ceiling or wall mounted and distribute light mainly upward, so it gets reflected off the ceiling or the walls. Creates a soft, undisturbing environment suitable for concentrated work or viewing paintings or drawings. Reflective glare on computer monitors can be controlled more easily. Can be installed without disturbing the ceiling surface, as in historical buildings or a painted ceiling.

Indirect lighting
The lighting reflected from the ceiling or other surface, not received directly from a luminaire.

Indirect water heater
Water heater that draws heat from a boiler used for space heating; a separate zone from the boiler heats potable water in a separate, insulated tank via a water-to-water heat exchanger.

Indoor adhesive
An adhesive or sealant product applied onsite, inside a building's weatherproofing system.

Indoor air pollution
Chemical, physical, or biological contaminants in indoor air.

Indoor air quality

According to the U.S. Environmental Protection Agency (EPA) and the National Institute of Occupational Safety and Health, good indoor air quality includes the introduction and distribution of adequate ventilation air, control of airborne contaminants, and maintenance of acceptable temperature and relative humidity. According to ASHRAE (American Society of Heating, Refrigerating and Air-Conditioning Engineers) Standard 62-1989, indoor air quality is air in which there are no known contaminants at harmful concentrations as determined by cognizant authorities, and that a substantial majority (80 percent or more) of the people exposed do not express dissatisfaction.

Indoor air quality procedures

One of two procedures listed in ASHRAE (American Society of Heating, Refrigerating and Air-Conditioning Engineers) Standard 62-1989 to determine appropriate ventilation rates for buildings. The IAQ procedure provides a method of measuring and controlling outdoor airflow in order to keep harmful substances diluted to acceptable levels. It is inherently a more rigorous strategy than the ventilation rate procedure because it considers all contaminants.

Indoor environmental quality

A LEED (Leadership in Energy and Environmental Design) Rating System category. Prerequisites and credits in this category focus on the strategies and systems that result in a healthy indoor environment for building occupants. Takes into consideration all impacts of the indoor environment on human health and performance, including indoor air quality, daylighting, views, and visual and thermal comfort.

Indus Valley architecture (1500–1200 B.C.)

Cities that flourished in the Indus Valley were carefully planned on a grid system, with main boulevards forming rectangular blocks; they were mostly of mud-brick construction.

Industrial archeology

Studying the remains and saving landmark examples of American engineering, such as factories, bridges, railroads, steel plants, and mills.

Industrial design

Utilizing the resources of technology to create and improve products and systems which serve humans, taking into account factors such as safety, economy, and efficiency in production, distribution, and use.

Industrial lighting fixture

A fixture with a utilitarian or functional appearance. Strip lights in open spaces with simple reflectors that are surface mounted or hung by chains or rods. HID (high-intensity discharge) industrials include high- and low-bay lighting systems.

Industrial Revolution style (1750–1890)

The evolution of this style was based on the production of iron and steel in quantities that could be used as a primary building material. The first iron frame structures were industrial buildings, which evolved into the steel-frame skyscrapers of modern times. The few pioneers of this new style

Industrial Revolution style

were engineers and not architects. Walls were still made of masonry over a steel skeleton, and the use of large glass skylights was widespread.

Industrial structure

A building derived from or used for industry or the commercial manufacture of goods. No single building type exists in a greater divergence of scales, styles, shapes, materials and other variables, than industrial structures. From plants no larger than houses, to massive amorphously styled steel mills.

Infill

A smart growth term referring to using available properties within the community, as opposed to expanding outward or using existing greenfields within the community.

Infill site

Building site sandwiched between existing buildings in a developed area. The use of infill sites reduces pressure on greenfield sites and often provides residents with better access to public transportation. For the purposes of LEED (Leadership in Energy and Environmental Design) for Homes credits, an infill site is defined as having at least 75 percent of its perimeter bordering land that has been previously developed.

Infilling

Material used to fill the space in a framework; also the process of placing additional buildings between existing ones.

Infiltration

In stormwater management, entry of runoff into the soil. Also, air that leaks into a building through cracks, joints, or damaged areas in a building.

Infiltration rate

The quantity of water that can enter the soil in a specified time interval.

Infiltration trench

An excavation backfilled with coarse aggregate stone that gradually infiltrates runoff into the surrounding soil.

Inflow

Entry of extraneous rainwater into a sewer system from sources other than infiltration, such as basement drains, manholes, storm drains, and street washing.

Informal balance See Balance

Infrared thermometer

A digital thermometer capable of measuring the temperature of a surface from a distance ranging from a few inches to a few feet. Most handheld infrared thermometers include a laser to help aim the device; the laser plays no role in temperature measurement. Used as an inexpensive substitute for a thermal imaging camera, an infrared thermometer can detect hot or cold spots on walls, ceilings, and duct systems.

Infrastructure

Any offsite utilities, services, and structures that serve a real estate development; may include gas, water, electric service, water and sanitary sewer systems, roads, railroads; sometimes used to refer to facilities within a site or building.

Inlay

A shaped piece of one material embedded in another, usually in the same plane, as part of a surface ornamentation.

Inn

A place which provides eating and drinking, but not lodging, for the public; a tavern.

Innovation and design process

A LEED (Leadership in Energy and Environmental Design) Rating System category. Prerequisites and credits in this category recognize projects for innovative building features and sustainable building knowledge.

Inorganic fiberboard

Board made from inorganic fibers, such as fiberglass, as opposed to organic fibers.

Inscription

Lettering that is carved or engraved in stone or wood, or on the surface of other materials, often of monumental scale, used primarily on exterior surfaces.

Inset dormer See Dormer

Inside finish

Includes all the fittings, door trims, window trims and shutters; most commonly used to denote woodwork, but is sometimes extended to the most elaborate and permanent work, such as marble, tile, and other items.

Insolation

Amalgamation of the words "incoming solar radiation" that means the total amount of solar energy that strikes a given surface in a given time. It is commonly expressed in kilowatt-hours per square meter per day.

Inspection

An onsite examination of construction to determine whether it is being performed according to the drawings, specifications, and building codes; may include examination of concrete reinforcement, repointing, steel connections, and waterproofing installations.

Inspection certificate

A certificate issued by a building inspector stating that the work is in accordance with the regulations in the appropriate building code.

Inspection list

A list of items of work to be completed or corrected by the contractor, commonly called a punch list.

Installation

The assembly of building components on site and affixing in position for use.

Installation inspection

The process of inspecting components of the commissioned systems to determine if they are installed properly and ready for systems performance testing.

Insulate
To reduce sound or heat transfer through an element of construction normally by the inclusion of lightweight porous material, dense material or discontinuous construction.

Insulated column See Column

Insulated concrete form (ICF)
Hollow insulated forms, usually made from expanded polystyrene used for building walls, both foundation and aboveground; after stacking and stabilizing the forms, the aligned cores are filled with concrete, which provides the wall structure.

Insulated glass window
A window consisting of two panes of glass separated by a space. The perimeter of the glass is sealed, allowing no movement of outside air into the space. The space, itself, can be filled with dehydrated air or with a special gas. The type of glass, spacer, and gas used in the space contribute to the overall insulating efficiency of the glass.

Insulating board
Any board suitable for insulating purposes, usually manufactured from vegetable fibers.

Insulating glass See Glass

Insulation
A material used to reduce transmission of sound or heat; types include batt insulation, loose-fill insulation, and urethane foam. Blown-in products or the installation of batts typically decreases the opportunity for air leakage. Typical insulation levels are R-15 in the walls, R-38 in the attic, and R-11 on basement walls.

Insulation board
Any type of building board used to prevent the passage of heat, cold or sound through walls and floors.

Insulation foamed in place
A product that acts as an air barrier and provides insulation and air sealing in one step. Most foam insulation products have a higher R-value per inch than fiberglass batt insulation. Using foam insulation increases energy efficiency.

Insulation systems
Systems that protect the building against temperature fluctuations in the environment, reducing unwanted cold or heat.

bio-based insulation
Made of renewable biomaterials that are more energy efficient, healthier, and more durable than traditional thermal insulation. They can be applied as foams, spray-foams, polymers, or biofibers, and are safe for the environment. Bio-based insulations are lightweight, contain no volatile organic compounds, provide a minimum R-value of 3.5 per inch, have excellent thermal and acoustical property, and meet U.S. government requirements for renewable resources. They have a Class 1 fire rating, have low levels of free-floating dust and allergens, and are not affected by moisture, mold, or insects.

glass-based insulation
Made of glass fibers combined with reinforcing agents. Fiberglass is the most widely used. Recycled glass insulation is a composite made from glass fibers and a polymer. Glass-based insulation systems are used for roofs, attics, walls, ducts, pipes, as well as other home appliances and equipment. Glass-based insulation products have relatively small deterioration rates and high R-values, ranging from R-11 to R-38. They can be blown into any void to any desired R-value. Since the base material is made of sand and recycled glass, it is naturally noncombustible, requiring no additional chemical treatment. Moreover, glass-based insulations are not absorbent and retain their R-value, even after they are exposed to moisture.

green insulation systems
Adaptations of traditional systems, offering enhanced performance with the added benefit of being a sustainable and ecological solution to energy retention problems. The primary green insulation systems are bio-based insulation, rigid panel insulation, glass-based insulation, wood-cement forms, natural fiber insulation, and radiant barriers.

insulating wood-cement forms
Systems that can be cast in place or installed as pre-assembled pieces that lock together to create structurally solid insulated forms. Made from a variety of different materials, such as cement-bonded wood fiber, polystyrene beads, and expanded polystyrene. They are highly efficient insulation systems, especially for low-rise residential and commercial constructions, because their structure minimizes air leaks and heat loss. They are also durable, with a high sound absorption rate, a long lifespan, and relatively low maintenance costs.

natural fiber insulation
Made from animal, plant, and mineral resources. They contain no VOCs (volatile organic compounds) or chemical irritants, have excellent thermal and sound blockage, have the ability to absorb moisture, and can breathe and react to climatic changes. Natural fibers are classified based on their origins. Plant sources are extracted from seeds or seed cases, such as hemp, kapok, and cotton fiber. Wood fibers are made directly from wood and are not inherently moisture resistant, and so must be coated with enhancers

natural fiber insulation

or other adhesive materials. Cellulose fiber is made from 80 percent recycled postconsumer paper. Typically it is blown or poured into hard-to-reach areas, such as attics and finished wall cavities. Animal sources contain hair extracted from animals, such as sheep, mohair, and alpaca. Mineral sources either occur naturally or are modified from fibers extracted from minerals. Their visual and performance properties closely resemble fiberglass. They are primarily applied as loose-fill and batting in hard-to-reach areas, or as rigid boards in roofs and attics. Since they are highly fire-resistant, they are also often used for furnace and chimney insulation.

radiant barriers

Composite material insulation systems made of highly reflective materials or coatings that stabilize a structure's thermal balance by reducing its heat gain and/or heat loss. Unlike conventional insulation systems that trap heat within the material, radiant barriers reflect heat before it enters the building. Often applied in attics or underneath the roofing layers in residential, commercial, and industrial buildings.

rigid panel insulation

Mostly used as roof and wall coverings, which contribute to the building's overall structural strength. Rigid foam panels can be made fireproof when shielded with additional fire-resistant materials. The panels can be made from various types of base materials, including cellular glass, wood fiber, gypsum board, and composites.

Intaglio
Incised carving in which the forms are hollowed out of the surface; the relief in reverse, often used as a mold.

Intaglio

Intarsia

Italian term for the flat decorations made from pieces of variously colored woods, inlaid to form ornamental patterns, architectural perspectives, or figurative scenes

Integral collector storage solar water heater

Solar water heater in which potable water is heated where it is stored.

Integral porch
A porch whose floor is set within the main structure of the house, rather than being attached to the house.

Integrated ceiling

A suspended ceiling system in which acoustical, illumination, and air-handling components are combined as an integral part of a grid.

Integrated design

A holistic design process that mobilizes multidisciplinary design input and cooperation, ideally to maximize and integrate environmental and economic life-cycle benefits. It is a process that delivers value by understanding

Integrated design

impacts across a broad range of disciplines during design and can also describe a resulting solution or building-services system, which is the physical integration of services and building components. A developer or owner who commits to high performance and energy efficiency ensures that these commitments are recognized by each team member who undertakes whole-system analyses, considering interactions among systems, such as lighting, both electrical and natural, with mechanical systems, daylighting with envelope systems, water supply and waste, with heating and cooling, and windows with ventilation and lighting.

Integrated design team

All individuals involved in a project from very early in the design process, including the architect, engineers, landscape architect, and interior designer; the owner's representatives, investors, developers, building users, facility managers, and maintenance personnel; and the general contractor and subcontractors.

Integration

An essential concept in sustainable building. Viewing a building as a system allows the discovery of synergies and potential trade-offs or pitfalls with design choices. An integrated design approach helps maximize synergies and minimize unintended consequences.

Integrative design process

A collaborative design methodology emphasizing knowledge integration in the development of a holistic design. The basis for integrated design practices are in the "whole-building design" approach. By viewing a building system interdependently as opposed to its separate elements (site, structure, systems, and use), this approach facilitates sustainable design practices and can serve to mitigate additional costs that result from "greening" at a later time. The integrated design process requires multidisciplinary collaboration, including key stakeholders and design professionals, from conception to completion. Decision-making protocols and complementary design principles must be established early in the process in order to satisfy the goals of multiple stakeholders while achieving the overall project objectives.

Intelligent building

A building designed with extensive automated systems to detect, diagnose and control the response to varying any of the environmental requirements.

Intelligent materials

Materials that are able to adapt to their environment by altering their properties. Example of intelligent materials include liquid crystal glass, which changes from transparent to opaque upon application of a current, and thermochromic glazing, which changes transparency in response to ambient temperatures.

Intensive and extensive gardens

Intensive gardens have thicker soil depths and generally require more management and artificial irrigation systems.

Intensive and extensive gardens

The plants chosen for these gardens must thrive in the specific roof environment they inhabit. Intensive gardens are heavier than extensive gardens, requiring more structural support. Extensive gardens have thinner soil depths and require less management and structural support. They do not require artificial irrigation. Plants chosen for extensive gardens are low maintenance, hardy species that do not have demanding habitat requirements. The idea of an extensive planting design is to have a self-sustaining plant community.

Intent

A LEED (Leadership in Energy and Environmental Design) Green Building Rating System component. Identifies the primary goal of each prerequisite or credit.

Interaxial span

The distance from axis to axis of columns, which includes the width of one column; one module greater than the intercolumniation.

Intercolumniation

The clear distance between two columns, measured at the lower part of the shafts, according to a system of proportions in Classical architecture, based on the diameter of the column as the governing module.

Interdome See **Dome**

Interfenestration

The space between the windows and their decorations on a facade.

Interfenestration

Interior decoration
The surface treatments and furnishings of the inside of a building; typically refers to those elements applied after construction is complete.

Interior decorator
One who works primarily with furnishings, fixtures and wall coverings.

Interior designer
A profession in which creative and technical solutions are applied within a structure to achieve a built interior environment. An interior designer is qualified by means of education, experience, and examination to protect and enhance the life, health, safety, and welfare of the public.

Interior door See Door

Interior finish
The final appearance of all exposed interior surfaces; floors, walls, ceilings, and all finish materials, such as tile, marble, plastic laminate, wood, and paint.

Interior furnishings
Those temporary or semi-permanent systems and components that are generally required for the normal utilization of the building for its intended purpose. Examples include interior design elements, paint, furniture, flooring, ceilings, and wall fixtures.

Interior landmark
In New York City, any interior or portion more than 30 years old with special historical or aesthetic value that is customarily accessible to the public.

Interior lot
A property, or lot, which is bounded by a street or highway on only one side.

Interior trim
A general term for all the finish molding, casing, and baseboard.

Interior wall See Wall

Interlace
An ornament of bands or stalks elaborately intertwined, sometimes including fantastic images.

Interlaced
Intermixed forms that cross over each other with alternation as if woven together.

Interlaced arcade See Arcade

Interlaced arches
Arches, usually circular, constructed so that their forms intersect each other.

Interlaced arches

Interlaced ornament
A band of ornamental figures that are overlapped or intertwined to create resultant forms.

Interlocked
Two or more components, members, or items of equipment which are arranged mechanically or electrically to operate in some specific relationship with each other.

Interlocking
Forms that are united firmly or joined closely by hooking or dovetailing.

Interlocking joint See **Joint**

Interlocking organization See **Organization**

Intermediate rib See **Rib**

Intern architect
One pursuing a training program under the guidance of practicing architects with the objective of qualifying for registration as an architect.

International Code Council Evaluation Service (ICC-ES)
A nonprofit public benefit corporation that evaluates building products and issues final reports on code compliance of building products and materials. These reports are then made available at no charge to the building community at large.

International Council on Monuments and Sites (ICOMOS)
An international, nongovernmental, professional organization composed of 65 national committees and 14 International specialized committees for the study and conservation of historic buildings, districts and sites; founded in 1965; maintains a secretariat in Paris.

International Council on Museums (ICOM)
An organization devoted to the promotion and development of museums and the museum profession.

International Energy Agency (IEA)
An organization committed to energy policies around the world. The goal is to ensure a cost-effective and dependable renewable energy system for a country's citizens through their commitment to energy security, economic development, and environmental protection.

International Organization for Standardization (ISO)
The body based in Geneva that coordinates International Standards. It works with national bodies such as the British Standards Institute (BSI), the American National Standards Institute (ANSI), and the European Committee for Standardization (CEN).

International Residential Code (IRC)
The one- and two-family dwelling model building code copyrighted by the International Code Council. The IRC is meant to be a stand-alone code compatible with the three national building codes—the Building Officials and Code Administrators (BOCA) National code, the Southern Building Code Congress International (SBCCI) code, and the International Conference of Building Officials (ICBO) code.

International style (1920–1945)
A style of architecture in both Europe and America pioneered by Le Corbusier, which spread to the Bauhaus, where it was the most influential. It was characterized by an emphasis on function combined with a rejection of traditional decorative motifs and regional characteristics. It was further characterized by flat roofs, smooth and uniform surfaces, large expanses of windows and projecting or cantilevered upper floors. The complete absence of ornamentation is typical, and cubistic shapes were fashionable. White was the preferred color. Horizontality was emphasized with windows that continued around corners. Roofs without eaves terminated flush with the plane of the wall. Wood and metal casement windows were set flush to the wall as well. Sliding windows were popular, and clerestory windows were also used extensively. There were fixed panes of glass from floor to ceiling, and curtain-like walls

International style

of glass were common. Popular building materials were reinforced concrete, steel frames, and an unprecedented use of prefabricated parts, since the style had its roots in industrial architecture. The resultant forms were much akin to cubist and abstract art.

Interpenetrate
A decorative feature, such as a molding, that enters another element, such as a column, and reappears on the other side; it was commonly found in the Gothic Revival style.

Interpretation
All the educational activities designed to explain the history and meanings inherent in historic sites, including tours, furnishings, displays, exhibits and related programs.

Interpretive restoration
Replacement of existing elements based on knowledge other than exact physical evidence or incomplete documentation.

Interrupted arch
A segmental pediment whose center has been omitted, often to accommodate an ornament.

Intersecting arcade see Arcade

Intersecting tracery See Tracery

Intersecting vault See Vault

Interstices
Spaces or intervals between parts of a structure or between components.

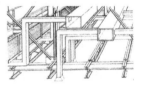

Interstitial
Forming a narrow or small space between parts of other elements, or between floors in a structure.

Interstitial condensation
The condensation that occurs within spaces inside the construction, as opposed to surface condensation.

Interweaving counterpoint See Counterpoint

Intonaco See Plaster

Intrados
The inner curve or face of an arch or vault forming the concave underside.

Invasive plants

Both indigenous and nonindigenous species or strains that are characteristically adaptable and aggressive, have a high reproductive capacity, and tend to overrun the ecosystems they inhabit. Collectively they are a threat to biodiversity and ecosystem stability.

Invasive vegetation

An exotic plant adapted to similar growing conditions as those found in the region to which it is imported. Because such a species usually has no natural enemies, such as pests, diseases, or grazers, it flourishes, disrupting the native ecosystem and forcing out native plant species, which results in habitat loss and water-table modification.

Inventory analysis

The first stage of life-cycle assessment, in which the inputs and outputs of materials and energy are determined for a process, product, or facility.

Inverted arch See Arch

Inverter

Device for converting direct current (DC) electricity into the alternating current (AC) form required for most home uses; necessary if home-generated electricity is to be fed into the electric grid through net-metering arrangements.

Investment architecture

Major comprehensive planning, redevelopment, and architectural schemes associated with large complex design problems in the early 1950s involving coordinating various talents to resolve complex design issues.

Investment property

Buildings or property purchased for the purpose of producing rental income, plus the appreciation in value gained over the life of the rental.

Invisible architecture

A form of architecture that would represent an expenditure of energy to create walls and furniture by the use of jets of air instead of conventional building materials, allowing for instant buildings.

Invitation to bid

A portion of the bidding requirements soliciting bids for a privately financed construction project.

Ionic order

An order of architecture invented by the Greeks, distinguished by an elegantly molded base; tall, slender shafts with flutes separated by fillets; and capitals, using a spiral volute that supports an architrave with three fascias; an ornamental frieze; and a cornice corbeled out on egg-and-dart and dentil moldings.

Ionic order

Iranian architecture (500–1000)

Decorative patterned brickwork, colored tile and molded stucco characterize this style. Other attributes are the use of stalactite vaults. The essential elements are richly decorated surfaces with brightly colored tiles and molded stucco. The minaret evolved into a standard form that had an influence on Indian architecture.

Iron Age (700 B.C.)
The period characterized by the introduction of iron metallurgy for tools and weapons.

Iron framing
A system of structural ironwork for buildings.

Iron See Metal

Ironwork
Objects made of cast iron or wrought iron; most often with utilitarian form in colonial America, but thereafter elaborate and ornamental.

Irradiance
The amount or density of light energy incident on a surface.

Irrigation
Supplying water to grass, trees, and other plants.

Irwin, Harriet (1828–1897)
Without having had architectural training, patented the first dwelling plan by an American woman (1869).

I-section See Beam

ISL bulb
A reflector type bulb with a silver-covered reflector.

Islamic architecture
Mesopotamian and Greco-Roman forms are the two main sources for this style (600–1500), which makes use of symbolic geometry, using pure forms such as the circle and square. The major sources of decorative design are floral motifs, geometric shapes and Arabic calligraphy. The major building types are the mosque and the palace. Mosque plans are based on strongly symmetrical layouts featuring a rectangular courtyard with a prayer hall. Forms are repetitive and geometrical; the surfaces are richly decorated with glazed tiles, carved stucco and patterned brickwork, or bands of colored stonework. Plaster made from gypsum was carved and highly polished to give it a marble-like finish.

Islamic architecture

ISO 14000

Set of generic standards developed by the International Organization for Standardization (ISO) to give business a method by which to measure environmental impacts.

Isolation

Reduction of vibration or sound; usually involving resilient surfaces or mountings or discontinuous construction.

Isolation joint See Joint

Isometric projection See Projection drawing

Isozaki, Arata (1931–)

Japanese architect who synthesized western and Japanese themes, concentrating on the clarity of geometry and pure forms as in the Gumma Prefecture Museum of Fine Arts, Takasiki, Japan (1974). His recent works include the Museum of Contemporary Art, Los Angeles, CA, (1981), and Team Disney Headquarters (illus.), Buena Vista, FL. (1990).

Italian Villa style (1830–1880)

The main feature of this style is the combination of a tall tower with a two-story L- or T-shaped floor plan. Gently pitched roofs resembling the pediment of Classical temples had wide projecting eaves. Windows are grouped into threes or placed within arcades. A smooth stucco finish highlighted Classic simplicity, while exuberant ornamentation recalls the Baroque. The overall massing is asymmetrical, intending to produce a picturesque quality, coupled with a low gable or hipped roof and wide eaves supported by decorative brackets, based on Italian farmhouses. Features also include square or octagonal towers, cupolas or glass belvederes, bay windows and ornamental brick.

Italianate style (1840–1880)

A style typified by a rectangular two- or three-story house with wide eaves supported by large brackets; tall, thin first-floor windows; and a low pitched roof topped with a cupola. There are pronounced moldings, details and rusticated quoins. Earmarks of the style are arched windows with decorative "eyebrows" and recessed entryways. The style appeared in cast-iron facades, whose mass-produced sections featured many stylized Classical ornaments.

Iwan

In Near Eastern architecture, a large porch or hall with a pointed barrel vault, as in the Sassanian palace at Ctesiphon and many later Islamic structures where it served as an entrance.

J j

Jack rafter See **Rafter**

Jacobean style
An English architectural and decorative style (1600–1625) adapting the Elizabethan style to continental Renaissance influence, named after James I.

Jacobsen, Hugh Newell (1929–)
Jacobsen briefly apprenticed to Philip Johnson, but later opened his practice in the Georgetown area of Washington, DC. Widely known for his modern pavilion-based residences, Jacobsen drew inspiration from the vernacular architecture of the American homestead, recalling the barns, detached kitchens, and smokehouses—the outbuildings—of rural America. Jacobsen designed the "1998 Life Dream House," a promotion by *Life* magazine where famed architects designed homes and plans were made publicly available. He also designed the home of Jacqueline Kennedy Onassis on Martha's Vineyard in the early 1980s.

Jahn, Helmut (1940–)
German-born American architect; joined the firm of C.F. Murphy in Chicago, in 1967. Studied with Mies van der Rohe, but his later work moved to a new richness of expression drawing on aspects of Art Deco for a unique expression. In Chicago, he designed O'Hare International Airport (1965); the Exhibition Building at McComick Place (1971); Xerox Center (1980), and the State of Illinois Center (1985). He began incorporating High-Tech and Art Deco inspired elements in his later high-rise office structures, such as 750 Lexington Avenue, New York (1989), and Messe Tower, Frankfurt, Germany (1991).

Jalousie window See **Window**

Jamb
The vertical member at the side of a window, door, chimney, or other exterior opening.

Jamb shaft

A small shaft having a capital and a base, placed against the jamb of a door or window.

Japanese architecture (500–1700)

An architecture based exclusively on timber construction, strongly influenced by Chinese design. Simple pavilion structures consist of a wooden framework of uprights and tie beams supported by a wooden platform. Nonbearing walls are constructed of plaster and wood, and sliding partitions of light translucent screens divide interior spaces, with doors and windows of lightweight material. Tiled hipped roofs project wide overhangs with upturned eaves as the result of elaborate bracket systems. Stone is used only for bases, platforms, and fortification walls. Great emphasis is placed on the integration of buildings with their surroundings, with open verandas providing the transition. There is a strictly modular approach to the layout, based on the tatami mat, which governs the entire design of the house. Carpenters became skilled in designing individual types of wooden joints.

Jefferson, Thomas (1743–1826)

American statesman and third president of the United States. He was also a gifted amateur architect. Inspired by Palladio, he designed Monticello, his own house (1769), and the University of Virginia (1826), Charlottesville, both in Virginia.

Jeffersonian

Neoclassical architecture based on the architecture of Thomas Jefferson, as expressed in the buildings on the campus of the University of Virginia in Charlottesville, VA.

Jenney, William Le Baron (1832–1907)

American architect who studied in Paris, France; set up an office in Chicago, IL. He was the first to use structural steel in a building for columns and girders. They were prototype skyscrapers. He taught Sullivan, Holabird, Roche, and Burnham in the practice of constructing tall buildings.

Jensen-Klint, Peder Vilhelm (1853–1930)

Danish architect, designer, painter, and architectural theorist, best known for designing Grundtvig's Church (illus.) in Copenhagen (1926), generally considered to be one of the most important Danish architectural works of the time. Its Expressionist style relies heavily on Scandinavian Gothic traditions. Built entirely of brick, it features a steep stepped-gabled facade resembling organ pipes. Jensen-Klint's son, architect Kaare Klint, assumed responsibility for completing work on Grundtvig's Church after his father's death.

Jerde, Jan (1940–)

American-born architect whose high-tech projects include the Fremont Street Experience redevelopment (illus.), Las Vegas, NV (1995), and Canal City, Fukuoka, Japan (1996).

Jerde, Jan

Jerkinhead

Gable end that slopes back at the top to form a small hipped roof end; also called a hipped gable.

Jerkinhead roof See **Roof**

Jerry-built

Traditionally applied to building speculators who built houses between the First and Second World Wars, using shoddy materials and shortcut methods to make a quick profit.

Jettied house

A building having a second story which overhangs the lower one.

Jetty

The upper story that juts out over the lower story of a timber-framed house.

Jib door See **Door**

Jigsaw
A power woodworking tool with a reciprocating thin blade suspended in a frame; used to cut complex curves in thin material.

Job captain
A member of the architect's staff normally responsible, on a given project, for the preparation of drawings and their coordination with other elements.

Jog
A change in direction in a joint or member.

Joggle joint See **Joint**

Joggle post
A post made of two or more pieces joggled together.

Jogglework
A stone keyed by joggles.

Johanson. John (1916–)
American architect who worked with Marcel Breuer and SOM. He is best known for the circular Chancellery for the U.S. Embassy, Dublin (1958), and the Mummer's Theater, Oklahoma City, OK (1970).

Johnson, Philip (1906–2005)
American architect. His glass house, New Canaan, CT (1949), saw the style of Mies Van der Rohe reach its ultimate development. It is a glazed box in which only the bathroom in enclosed. He designed the Seagram Building (with Mies Van der Rohe) in NYC (1958). Other works include the Museum of Modern Art additions and sculpture garden, the New York State Theater at Lincoln Center, NYC (1964), and the Kline Science Center (with Richard Foster) at Yale University, New Haven, CT (1965).

Johnson, Reginald (1882–1952)
American architect who designed in the Spanish Revival style for houses in California.

Johnston, William L. (1811–1849)
American architect who designed the Jane Building (1849), Philadelphia, in the vein of Louis Sullivan, with large expanses of glass that appeared later in Sullivan's work. Johnson died before its completion, which was taken over by Thomas Ustick Walter.

Joint
The space between the stones in masonry or between the bricks in brick work. In concrete work, joints control the shrinkage on large areas and isolate independent elements.

angle joint
Any joint formed by uniting two members at a corner which results in a change of direction.

bevel joint
Any joint in which the ends of the two abutting elements are cut at an angle, especially when not forming a right angle.

blind joint
A joint that is invisible.

bridle joint
A carpentry joint connecting a slotted end of one timber to the double-notched end of another timber; used to connect a rafter to a tie beam or two rafters at a ridge.

broken joint
A pattern where the elements are installed so that the adjacent butt joints between pieces are not aligned, such as in flooring and brickwork.

butt joint
A plain square joint between two members, when the contact surfaces are cut at right angles to the faces of the pieces; the two are filled squarely against each other rather than lapped.

cogged joint
A carpentry joint formed by two crossed structural members, each of which is notched at the place where they cross.

construction joint
A separation provided in a building that allows its component parts to move with respect to each other; a joint where two placements of concrete meet.

control joint
A joint that is premolded, tooled, or sawed, and installed to prevent shrinkage of large areas. It creates a deliberately weakened section to induce cracking at the chosen location rather than at random.

dovetail joint
A splayed tenon, shaped like a dove's tail, broader at its end than at its base; the joint is formed by such a tenon fitting into the recess of a corresponding mortise.

end-lap joint
A joint formed between the ends of two pieces of timber, normally at right angles; each piece is notched equal to the width of the other piece, to form a flush surface in the assembled joint.

expansion joint
Designed to permit the expansion or contraction due to temperature changes. It generally extends through the entire structure from the footings to the roof.

finger joint
An end joint made up of several meshing fingers of wood made with a machine and glued together.

flush joint
Any joint finished even or level with the surrounding surfaces.

indented joints
A joint used in joining timbers end to end; a notched fishplate is attached to one side of the joint to fit into two corresponding notches in the joined timbers; the entire assembly is fastened with bolts.

interlocking joint
A form of joggle in which a protrusion on one member complements a slot or routed groove in another; a joint formed between sheet-metal parts by joining their preformed edges to provide a continuous locked piece.

isolation joint
A joint that separates one concrete section from another so that each one can move independently; found in floors, at columns, and at junctions between the floor and walls.

joggle joint
A notch or projection in one piece of material, which is fitted to a projection or notch in a second piece, to prevent one piece from slipping past the other.

lap joint
A joint in which one member overlaps the edge of another and is connected.

miter joint
A joint between two members at an angle to each other; each member is cut at an angle equal to half the angle of the junction, usually at right angles to each other.

mortise and tenon
A joint between two members, formed by fitting a tenon at the end of one member into a mortise cut into the other.

rigid joint
A joint that is capable of transmitting the full extent of force at the end of the member to the other members framing into the joint.

scarf joint
A wood joint formed by two members cut diagonally to overlap and interlock; pegs, glue, straps, or other devices are used to attach the members.

semirigid joint
A joint in either steel or concrete that is designed to permit some rotation; also called a partially fixed joint.

spline joint
A joint formed by inserting a spline of long strips of wood or metal in a slot cut into the two butting members.

standing-seam joint
In metal roofing, a type of joint between the adjacent sheets of material, made by turning up the edges of two adjacent sheets and then folding them over.

straight joint
A line created by the meeting of two or more separate elements or pieces, often continuing in a straight line from one end to another.

tongue-and-groove joint
A joint formed by the insertion of the tongue of one member into the corresponding groove of another.

tooled joint
Any mortar joint finished with a tool, other than a trowel, that compresses and shapes the mortar; common types include a beaded joint, concave joint, and raked joint.

Joint reinforcement
Steel wire, bar or fabricated reinforcement which is placed in horizontal mortar joints.

Joint venture
A legal arrangement in which two or more parties undertake to share the risks and rewards of a project on an agreed upon basis.

Jointing
In masonry, the finishing of joints between courses of bricks or stones before the mortar has hardened.

Joist
One of a series of parallel timber beams used to support floor and ceiling loads, and supported in turn by larger beams, girders, or bearing walls; the widest dimension is placed in the vertical plane.

ceiling joist
Any joist which carries a ceiling; one of several small beams to which the ceiling of a room is attached. They are mortised into the sides of the main beams or suspended from them by strap hangers.

floor joist
Any joist or series of joists which supports a floor.

solid-web joist
A conventional joist with a solid web formed by a plate or rolled section, as opposed to an open-web joist.

trimming joist
A joist supporting one end of a header at the edge of an opening in a floor or roof frame, parallel to the other common joists.

Jones, Euine Fay (1921–2004)
American architect and designer Jones was an apprentice of Frank Lloyd Wright. He was also the only one of Wright's disciples to have received the AIA (American Institute of Architects) Gold Medal. Among his most famous buildings are the Thorncrown Chapel in Eureka Springs, AK; the Mildred B. Cooper Memorial Chapel in Bella Vista, AK; and

Jones, Euine Fay
the Pinecote Pavilion at the Crosby Arboretum in Picayune, MI. In 2009, the University of Arkansas School of Architecture was dedicated in his honor and is known as the Fay Jones School of Architecture.

Jones, Inigo (1573–1632)
London-born Royal Architect who introduced the Palladian style to Jacobean England and started the Palladian revival. He worked on Old St. Paul's Cathedral, London (1631–1671), adding Classical elements showing the power and scale of Roman architecture, which paved the way for Sir Christopher Wren when he began rebuilding the cathedral.

Jones, Stephanie Tubbs (1950–2008)
A Cleveland congresswoman and advocate of urban revival who championed legislation to boost federal preservation funding and a resident of Cleveland's historic Wade Park, OH. Congress passed a bill she cosponsored with the National Trust for Historic Preservation to enhance the federal rehabilitation tax credit for the first time in many years.

Journeyman
A craftsperson who has finished apprenticeship in a trade and qualifies for wages, but remains employed by others.

Jugendstil
A term meaning "youth style." The German version of Art Nouveau; after the journal *Die Jugend* that publicized the style; it was associated with the Sezession movement in Vienna, Munich, and Dresden; chief proponents were Endell, Hoffman, Olbrich, and Wagner.

Jullian de la Fuente, Guillermo (1931–2008)
A protégé of Le Corbusier, he managed the Paris office until Corbusier died. He worked on the architect's only U.S. project, Harvard University's Carpenter Center, Cambridge, MA. His Paris firm designed the fairgrounds in Valencia, Spain and also the French Embassy in Rabat, Morocco. Other projects included the home of astronomer Carl Sagan in Ithaca, NY.

Jump duct
Flexible duct that connects a room to a common space to balance pressure and to provide a pathway for return air in forced-air heating and cooling systems. Jump duct grilles are typically located in the ceiling.

Junction box
In electrical work, a box in a street distribution system, where one main is connected to another main; also a metal

Juxtaposition
The state or position of being placed close together or side by side, so as to allow comparison or to create contrast.

K k

Kahn, Albert (1869–1942)

German-born American architect who formed a practice with his brothers Julius and Moritz; designed industrial buildings for automobile manufacturers Packard, Ford, and Chrysler.

Kahn, Ely Jacques (1884–1972)

An American architect, trained in Paris, who designed many Art Deco skyscrapers in NYC in the 1920s and 1930s.

Kahn, Louis I. (1901–1974)

Born in Estonia, settled in the United States and became known for the monumentality, dignity and sculptural form of his buildings, include the Yale Art Gallery, New Haven, CT (1951). The Salk Institute Laboratories, San Diego, CA (1959) was an important work, as was the Kimbell Art Museum, Fort Worth, TX (1967), and the Phillips Exeter Academy Library, Exeter, NH (1967). Richards Laboratories in Philadelphia, PA (1957) has a bold silhouetting of towers. His last work was the National Assembly of Bangladesh, Dacca (1974).

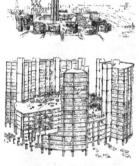
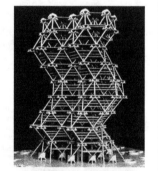

Kallman, Gerhard (1915–)

German-born American architect; trained in London before coming to the United States in 1948. Was partners with Noel McKinnell (1935–) and Edward Knowles (1929–) on the Boston City Hall competition, Boston, MA (1961). He had taught at Columbia University, and McKinnell was one of his students.

Kando

Main sanctuary of a Japanese Buddhist temple.

Kanner, Stephen (1956–2010)

A Los Angeles architect and cofounder of the Los Angeles Architecture and Design Museum. Among his award-winning designs are the Harvard Apartments in Koreatown, CA and the In-N-Out Burger in Westwood, CA, that is a tribute to 1950s jet-age architecture.

Kaplický, Jan (1937–2009)

Czech architect who was the leading force behind the innovative design office Future Systems. Best known for the futuristic Selfridges Building in Birmingham, England, which won seven awards including the RIBA Award for Architecture 2004, and the Media Centre at Lord's Cricket Ground in London, England. In 2007, he won the international architectural competition for the new building of the National Library of the Czech Republic in Prague, a project that was subsequently cancelled due to criticism of the design. With the office of Renzo Piano and Richard Rogers (1971–1973), London, he developed the competition-winning design for the Centre Georges Pompidou (1977) in Paris. France. He developed an architectural style that combined organic forms with high-tech futurism. Among the drawings he made were structures orbiting the earth built by robots, weekend houses resembling survival capsules that could be transported by helicopter, and home interiors that could be manipulated.

Keep

Inner tower of a castle.

Keichline, Anna (1889–1943)

The first woman to become a registered architect of Penn-

Keichline, Anna

sylvania, but best known for inventing the hollow, fireproof "K Brick," which was a precursor to the modern concrete block.

Kellogg, Kendrick Bangs (1934–)

American architect. An innovator of organic architecture, Kellogg built a wide assortment of distinctive buildings, including the Yen House, Wingsweep, the Joshua Tree house, and the Onion House. Public buildings include the Hoshino Wedding Chapel in Japan and the Charthouse restaurants. Kellogg is related to Frederick Law Olmsted, the "Father of Landscape Architecture" (of the 1800s), who was a cousin to Kellogg's grandfather. Olmsted's landscape designs were curvilinear and irregular, a significant break from the formal symmetrical patterns of the time, a practice that Kellogg continued. Kellogg's architecture does not fit neatly into the same category as Frank Lloyd Wright, Bruce Goff, Bart Prince, or other organic architects. The Onion House (1962) is landmark of organic architecture; it was constructed in Kona, Hawaii, for Elizabeth von Beck. With no outside walls, the division between interior and exterior consists of screen or stained glass. The stained glass and the mosaic dining table were built by artist James Hubbell. The curvilinear architecture of the three-dome complex is reminiscent of the fanciful creations of Antonio Gaudí in Spain, while the central fireplaces and geometric lines were influenced by teachings of Frank Lloyd Wright. Each dome is ventilated at the top to take advantage of the natural ocean breezes.

Kelvin

Standard unit for color temperature, abbreviated K. A Kelvin unit is the basis of all temperature measurement, starting with 0 K at absolute zero.

Kent, William (1684–1748)

English architect and landscape designer, whose revolutionary informal gardens created a new relationship between a building and its natural setting.

Kerf

A series of grooves, or kerfs, cut down to about two-thirds of the thickness of the wooden piece, so that it can be bent around curves.

Key

A tapered or wedge-shaped piece that locks pieces of timber together.

Key console

A console that acts as the keystone of an arch.

Key course

A continuous course of keystones in an arch, used in a deep archway where a single keystone will not suffice; a course of keystones used in the crown of a barrel vault.

Keyhole saw

A small handsaw with a thin tapering blade designed for cutting small rounded openings, such as a keyhole. They are also used to enlarge holes and notch structural members for cables and conduit.

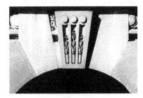

Keystone

The central stone or voissoir at the top of the arch, the last part to be put into position to lock the arch in place, often embellished.

Khalili, Nader (1936–2008)
Khalili founded the Cal-Earth Institute near Hesperia, CA, which teaches students how to construct low-cost adobe homes. Among his many innovations is the "super adobe" Earthbag construction system, developed in the 1980s for NASA in response to a call for designs for human settlements on the Moon and Mars. He also developed the Gelaftan Earth-and-Fire system, a technique for ceramicizing adobe to provide a low-cost building material for homes in earthquake-and storm-prone areas. He was the author of five books, including *Ceramic Houses and Earth Architecture: How to Build Your Own*.

Khan, Ali Mardan (1630–1653)
One of many credited for the design for the Taj Mahal, Agra, India; the most well-known Islamic tomb monument.

Kick switch
A switch located in the base of a lamp.

Kickplate
A metal plate placed along the lower portion of the door to prevent marring the door by showing marks.

Kiesler, Federick (1890–1965)
Vienna-born American visionary architect; his "endless house" encapsulated his organic ideas of curves and continuous wall and ceiling planes as a contrast to the rectangular

Kikutake, Kiyonori (1929–)
Japanese architect and leading light in Metabolism. His Sky House, Tokyo, Japan (1958), extending cities into the sea, was partially realized at Aquapolis, Okinawa (1975).

Kiln
A large oven used for the artificial seasoning of lumber, for the firing of brick, and for the burning of lime.

Kiln dried
Wood that has been seasoned by the heat of a kiln, rather than dried in the air.

Kilowatt-hour (kWh)
Measure of electricity consumption; a 100-watt light bulb burning for 10 hours consumes 1 kWh.

Kinetic architecture (1971–1985)
A style depicted by forms that are dynamic, adaptable and responsive to the changing demands of the users. This broad category includes a number of other concepts, such as mobile architecture, which would not necessarily be constantly moving, only capable of being moved if required.

King closure See Brick

King post
A vertical member extending from the apex of the inclined rafters to the tie beam between the rafters at the lower ends of a truss, as well as in a roof.

King truss See Truss

Kiosk
A small ornamented pavilion or gazebo, usually open for the sale of merchandise, or to provide cover or shelter to travelers.

Kitchen
Room where food is prepared and cooked.

Kitchen recycling center
A built-in section of kitchen cabinetry that allows convenient separation of recyclable materials.

Kitsch
Work that is pretentious, and in some opinions, in bad taste, It represents applying styles without understanding them. It is part of camp taste that values the outrageous or worst excesses of commercialism..

Klenze, Leo von (1784–1864)
Versatile German architect who created dignified and monumental public buildings, some in the Greek style, others in the Renaissance style. Work includes Walhalla (illus.), near Regensburg, Germany.

Klerk, Michael de (1884–1923)
Dutch architect and member of the Amsterdam School who designed projects in Amsterdam, the Netherland's and include Eigen Haard Housing, (1921), one of several housing estates, with his colleague Piet Kramer (1881–1961). It was built of brick and featured skillfully curved corners and details. He also designed the Navigation House (illus.), Amsterdam (1916) and the De Dageraad housing complex (1920), Amsterdam.

Klerk, Michael de

Kliment, Stephen (1930–2008)
Writer and architectural activist and editor of *Architecture and Engineering News* from 1960 to 1968, Kliment was an acquisitions editor at John Wiley & Sons before joining *Architectural Record*. During his tenure there, from 1990 to mid-1996, the magazine underwent a redesign by Vignelli Associates and celebrated its 100th anniversary. As an author, Kliment's books include *Architectural Sketching and Rendering: Techniques for Designers and Artists* and *Writing for Design Professionals*.

Kling, Vincent (1916-)
Philadelphia-based architect; he designed the Municipal Services Building (illus.), Philadelphia, PA (1965).

Kling, Vincent

Knee
A bent or curved element used to stiffen a joint where two members meet at an angle, such as a timber frame column and beam.

Knee brace See Brace

Knee wall
A low wall resulting from one-and-one-half-story masonry construction.

Kneestone
A stone which is sloped on top and flat on the bottom that supports inclined coping on the side of a gable, or a stone that breaks the horizontal joint pattern to begin the curve of an arch.

Knob
A protuberance, useful or ornamental, which forms the termination of an isolated member; a handle, more or less spherical, used for operating the mechanism for opening a door.

Knob and tube
In electrical work, a means of concealing wiring and one of the earliest methods to wire a house. Insulated wires are supported with porcelain knobs and pass through porcelain tubes in wood members of a floor or wall. Although it still meets the National Electric Code, it is not approved by many local organizations, and it is used only for temporary installations such as construction shacks.

Knob ornament See Ornament

Knot
Hard, cross-grained section in a piece of timber, where a branch had formed in the trunk of the tree.

Knot ornament See Ornament

Knotted column See Column

Knotwork ornament See Ornament

Kohn, A. Eugene (1930-)
Partner in the firm Kohn, Pederson, Fox, with William Pederson and Sheldon Fox. They designed the Procter and Gamble general offices, Cincinnati, OH (1985); 333 Wacker Drive, Chicago, IL (1983); One Logan Square, Philadelphia, PA; and the DG Bank, Frankfurt, Germany (1933), among many other office structures.

Knuckle
The part of a hinge containing the holes through which the pin passes.

Koch, Carl (1912-1998)
Noted American architect who believed that the American lifestyle would be best served by a housing system that could be easily assembled, disassembled, and reconfig-ured. This passion led him to pioneer prefabrication tech-nologies. His Techbuilt series of homes was designed to be built with prefabricated panels for the walls, floor, and roof. Notable projects include the Lewis Wharf Condominiums (1973), residential renovation of a Boston Waterfront granite warehouse that was built in 1840 on the site of Paul Revere's silversmith shop and John Hancock's ware-houses. *Progressive Architecture* magazine gave him the unofficial title "The Grandfather of Prefab" in 1994. He outlined his experiences on prefabrication in a book that he wrote with Andy Lewis titled *At Home with Tomorrow*.

Kocher, Lawrence (1885-1969)
American architect who was an early member of CIAM. He was the editor of *Architectural Record* from 1928 to 1938.

Koolhaas, Rem (1944-)
Dutch architect who formed the Office for Metropolitan Architecture (OMA) in 1975, producing a number of vision-ary and theoretical projects, including *Delirious New York*, later published as a book (1978). He was a publicist for *Deconstructivism*.

Korsunsky, Yaroslaw (1926-2009)
A native Ukrainian who practiced architecture though he spoke no English. His firm helped design the Mall of America in Minneapolis, MN along with many nursing homes, schools, and community centers. In Chicago, IL he designed the Saints Volodymyr and Olha Ukrainian Catholic Cathedral, a local landmark with a gold dome in the Byzan-tine Ukrainian style.

Kramer, Piet (1881-1961)
Dutch architect of the Amsterdam School; collaborated with de Klerk on the Navigation House (illus), in 1916 and the

Kramer, Piet

De Dageraad housing complex (illus.), in 1925, both in Amsterdam the Netherlands; both regarded as superb examples of Expressionism.

Krier, Leon (1946–)

Luxembourg-born architect and theorist: who championed Rational Architecture. His view of the city as a document of intelligence, memory, and pleasure is the antithesis of the concept of the disposable, adaptable, Plug-in City of Archigram and other advocates. He was critical of Post-modernism.

Krier, Rob (1938–)

Luxembourg-born Austrian architect, brother of Leon Krier, who built many housing projects in Berlin (1970s). He has been dubbed a devotee of Neo-Rationalism.

Krypton

A colorless, odorless inert gas, often used with argon in fluorescent lighting and sometimes used as a gas fill in high-performance glazing.

Kump, Ernest (1911–1991)

His most notable project was Foothill College, Los Altos Hills, CA (1961), consisting of 40 buildings on a 122-acre campus, all designed and built as a unit.

Kurokawa, Kisho (1934–2007)

A student of Kenzo Tange, Kurokawa cofounded the Metabolism Movement in 1960. Kurokawa was cast onto the world stage when three of his buildings were constructed for the Osaka World Expo of 1970, and a year later he was a finalist for the design of Pompidou Center in Paris. In 1972 his Nakagin Capsule Tower was constructed in Tokyo. It featured 140 pods that were prefabricated offsite and attached with four bolts onto a central tower. Among his many other notable projects are the National Ethnological Museum in Osaka, the Hiroshima City Museum of Contemporary Art, a new wing of Amsterdam's Van Gogh Museum, and Tokyo's National Arts Center. For the 1998 Kuala Lumpur International Airport, Kurokawa transplanted a tropical rain forest into a design based on Islamic domes.

Kyoto Protocol

A legally binding agreement adopted by the countries in attendance at the December 1997 United Nations Framework Convention on Climate Change in Kyoto, Japan. Delegates from the 160 industrialized nations present agreed to reduce their greenhouse gas emissions by an average of 5.2 percent below 1990 emissions levels by 2010. The United States pledged a 7 percent reduction, although the U.S. Congress did not ratify the agreement.

L l

L'Enfant, Pierre Charles (1754–1825)
Designed the city plan for Washington, DC.

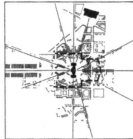

L'Orme, Philibert de (1514–1570)
Original and inventive, he was instrumental in creating a distinctive version of Renaissance classicism that drew on French traditions as well as Italian models. Works include Tuileries Palace, now destroyed.

Label molding See **Molding**

Label stop
The termination of a hood mold or arched dripstone in which the lower ends are turned in a horizontal direction away from the door or window opening.

Labrouste, Pierre Henri (1801–1875)
French architect whose reputation rests on the Bibliothèque Ste. Genevieve, Paris (1838), in which an iron structure was slotted into a masonry cage. The reading room of the Bibliothèque Nationale, Paris, employed the same exposed iron-and-glass interior (1854).

Labyrinth
A maze of twisting passageways; a garden feature of convoluted paths outlined by hedges, often with a garden house at the center; in medieval cathedrals, the representation of a maze inlaid in the floor.

Lacework
Architectural patterns or decorations resembling lace.

Lacing course
A course of brick or tile inserted in a rough stone or rubble course as a bond course.

Ladder
A wooden object consisting of two side pieces connected to each other at regular intervals by rungs; used for climbing up or down during construction, or used as a temporary stair.

Lafever, Minard (1798–1854)
American architect in New York; author of popular builder's guides that disseminated the Greek Revival nationwide. Also designed numerous churches: Sailors' Snug Harbor (1833), Staten Island, NY; New Dutch South Reformed Church (1840), New York; and Church of the Holy Trinity (1847), Brooklyn, NY.

Lally column See **Column**

Lamb, Thomas White (1871–1942)
Lamb achieved recognition as one of the leading architects of the boom in movie theater construction of the 1910s and 1920s. Particularly associated with the Fox Theatres, Loew's Theatres, and Keith-Albee chains of vaudeville and film theaters, known as "movie palaces," as showcases for the films of the emerging Hollywood studios. His first theater design was the City Theatre, built in NYC in 1909 for film mogul William Fox. The 1914 Mark Strand Theatre, the 1916 Rialto Theatre, and the 1917 Rivoli Theatre, all in New York's Times Square, set the template for what would become the American movie palace. Among his most notable are the 1929 Fox Theatre in San Francisco and the 1919 Capitol Theatre in New York, both now demolished. Among his most noted designs that

Lamb, Thomas White

have been preserved and restored are the B.F. Keith
Memorial Theatre in Boston, MA (1928), now the Boston
Opera House; Warner's Hollywood Theatre (1930) in NYC,
now a church; and the Loew's Ohio Theatre (1928) in
Columbus, OH. Lamb designed, with Joseph Urban, New
York's Ziegfeld Theatre, as well as the third Madison
Square Garden (1925) and the Paramount Hotel in mid-
town Manhattan.

Lamb's tongue
The end of a handrail which is turned out or down from
the rail and curved to resemble a tongue.

Lamella roof See Roof

Laminated
Any construction built up out of thin sheets or plates
which are fastened together with glue, cement, or other
similar adhesive.

Laminated arch
A structural arch built up from multiple layers of smaller
timbers.

Laminated beam See Beam

Laminated floor See Floor

Laminated glass See Glass

Laminated shingles
Asphalt-based shingles made from two separate pieces
that are laminated together.

Laminated strand lumber
Engineered wood product developed in the 1980s in which
wood strands are glued together and pressed into forms
using steam injection. The strength and stability of LSL
falls between that of conventional lumber and laminated
veneer lumber (LVL).

Laminated timber See Wood

Laminated veneer lumber
Engineered wood product in which wood veneers are
glued together in thick sections for use as beams or other
structural members. LVL is stronger, straighter, and less
prone to warping or shrinkage than conventional lumber
and does not require the destruction of mature trees.

Laminated wood
An assembly of pieces of wood with the fibers or grain
in each piece parallel to the fibers of the other pieces.
The wood is built up in plies or laminations and joined
with glue or mechanical fasteners. They are held together
under pressure until the glue sets.

Laminating
The process of bonding laminations, or thin plates of
material together with an adhesive.

Lamp
The electric bulb or tube within a luminaire, used for illumination. The three most common types are incandescent, fluorescent, and high-intensity discharge lamps.

Lamp life
Varies from source to source.

Lampshade
Used to mute the direct glare of a bulb. Fabric shades
are stretched over a metal wire frame. Clip-on lamp-
shades attach directly onto the bulb where harps are not
available.

Lanai
A living room or lounge area which is entirely, or in part,
open to the outdoors.

Lancet arch See Arch

Lancet window See Window

Land development and management practices
All land-related practices involved in the development
and management of a site, including design, construction,
operations, and ongoing maintenance practices.

Land subdivision
A general area which has been divided into blocks or
plots of ground, such as residential, commercial, industrial, or agricultural sites, with suitable street, roadways,
and open spaces.

Land survey
A survey made to determine the lengths and directions
of boundary lines, and the area of the tract bounded by
these lines.

Land Trust Alliance
An association of hundreds of local land trusts, dedicated
to preserving open space and natural habitat in the
United States.

Land use
The manner in which a particular piece of property or
district is permitted to be used; typical usage includes
residential, commercial, industrial, institutional, and
agricultural.

Land use regulation
Restrictions imposed on development by governing
agencies, such as zoning, architectural review boards,
or public participation in the planning or review process.

Landfills
Sanitary landfills are disposal sites for nonhazardous
solid wastes spread in layers, compacted to the smallest
practical volume, and covered by material applied at the
end of each operating day.

Landing
The horizontal platform at the end of a stair flight or
between two flights of stairs.

half-space landing

A stair landing at the junction of two flights which reverses direction, making a turn of 180 degrees. Such a landing includes the width of both flights, plus the well.

quarter-space landing

A square landing connecting two flights of stairs that continue in a straight line.

Landmark

Any building structure or place that has a special character or special historic or aesthetic interest or value as part of the heritage or cultural characteristics of a city, state, or nation.

Landmark protection ordinances

Typically deal with the protection of isolated landmarks located outside the confines of a historic district and provide essentially for a stage of demolition for periods varying from 90 days to one year.

Landmarks Commission ordinances

Grant landmark jurisdiction over an entire area such as a city, but their power generally applies only to single buildings within that area. Some have authority over historic districts.

Landmarks Register

A listing of building and districts designated for historical, architectural or other special significance that may carry a variety of forms of protection for listed properties; local registers are often maintained by a preservation commission and modeled after the National Register of Historic Places.

Landscape

The visible features of an area of land, including physical elements such as landforms, living elements of flora and fauna, abstract elements such as lighting and weather conditions, and human elements (e.g., human activity or the built environment).

Landscape archeology

A branch of archeology that concentrates on negative evidence and other signs of human occupancy, such as boundary lines, earth mounds, buildings and their siting and the outlines of paths and roads as indicators of buried features, and clues to original land use.

Landscape architect

A person trained and experienced in the design and development of landscape and gardens; a designation reserved for a person professionally licensed to perform landscape architectural services.

Landscape architecture

The person and/or professional whose job it is to design, arrange, or modify the features of a landscape, for aesthetic or practical purposes.

Landscape area

The area of the site equal to the total site area less the building footprint, paved surfaces, water bodies, patios, and similar elements.

Landscape design

The art, planning, design, management, preservation, and rehabilitation of the land and the design of large-scale man-made constructs. The scope of the profession includes architectural design, site planning, estate development, environmental restoration, town or urban planning, park and recreation planning, regional planning, and historic preservation.

Landscape waste

All accumulations of grass or shrubbery cuttings, leaves, tree limbs, and other materials accumulated as the result of the care of vegetation such as lawns, shrubbery, vines, and trees.

Landscape window See **Window**

Lang, Roger P. (1944–2008)

Lang had been the director of community programs and services for the New York Landmarks Conservancy. He joined the Boston firm of Perry Dean Stahl and Rogers in 1968 and worked on some of that city's premier restoration projects, including Quincy Market and Faneuil Hall. Among the many landmarks he helped save were the abandoned buildings located on the south side of Ellis Island, NY.

Langdon, Geoffrey Moore (1955–2008)

Author, teacher, and CADD (computer-aided design and drafting) expert Geoffrey Langdon was the founder of Boston's Architectural CADD Consultants and taught classes on design, solar energy, and CADD at several area colleges. A sought-after lecturer and speaker, Langdon wrote scores of articles for industry journals and magazines and was the author of several books. These included *Architectural CADD: A Resource Guide to Design and Production Software Appropriate for Architects, CADD and the Small Firm* (with Evan Shu), and *ArchiCAD for AutoCAD Users* (with Ralph Grabowski).

Lantern

A tower or small turret with windows, crowning a dome or cupola. often glazed; provides light or ventilation to the space below.

Lantern light

A decorative frame with glass panels, originally used to protect an open flame. Used for centuries to light hallways and entrances. Besides hanging from a ceiling, it can be mounted on a wall.

Lap

The length by which one piece of material overlaps another.

Lap joint See Joint

Lap siding

The siding that looks like individual boards, typically 8–12 feet long. Each piece of siding is lapped over the piece below it to provide a waterproof covering for the house.

Lap splice See Splice

Lapis lazuli

A decorative variety of calcite, stained a deep blue by trace minerals; used as a stone veneer, and in powdered form as the original ultramarine pigment.

Larch See Wood

Large-format photography

Pictures taken with a format of 4 by 5 inches or larger in cameras with tilts and swings for perspective controlled documentation of existing buildings.

Larsen, Henning (1925–)

Danish architect, internationally known for the Ministry of Foreign Affairs building in Riyadh, Saudi Arabia and the Copenhagen Opera House, Denmark. He is the founder of the company that bears his name, Henning Larsen Architects.

Latent load

Cooling load that results when moisture in the air changes from a vapor to a liquid through condensation. A latent load puts additional demand on cooling systems in hot, humid climates.

Lateral brace See Brace

Latex paint

A water-based paint using vinyl-based binders.

Lath

Originally a rough-sawn strip of wood fixed to timber framing with small gaps between adjacent laths as a ground for plaster. Now applied to other materials used for the same purposes, such as metal lath.

Lathe

A machine for turning metal or wood. The piece to be turned is mounted between two chucks that rotate and the cutting tool can slide back and forth on a track, and be moved closer or farther away from the wood, allowing a variable profile to be cut

Latin cross

A cross which has an upright much longer than the crossbeam; three arms are the same length, and the fourth lower arm is much longer.

Latrobe, Benjamin (1764–1820)

Trained in Europe but emigrated to the United States, where he met George Washington. He built a number of public buildings, including work on the U.S. Capitol, Washington, DC (1803–1814), and advised Thomas Jefferson on the design of the University of Virginia campus (1817).

Lattice
A network of bars, straps, rods, or laths crossing over and under one another; the result is a rectangular or diagonal checkered pattern, which may be varied by the width of the bands and the spacing of the members

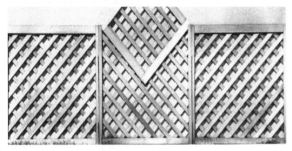

Lattice dome See **Dome**

Lattice molding See **Molding**

Lattice structure
An open-web joist, column, cylindrical shell, dome, or other structural type, built up from members that intersect diagonally to form a lattice.

Lattice truss See **Truss**

Latticed window See **Window**

Latticework
Reticulated or netlike work formed by the crossing of laths or thin strips of wood or iron, usually in a diagonal pattern.

Lautner, John (1911–1994)
American architect who studied with Frank Lloyd Wright and set up practice in Los Angeles, CA designing original private homes. The most well known are the Malin House, now the Kuhn house (illus.), Los Angeles, CA (1960), where the entire structure is carried on one pier, and the Elrod House, Palm Springs, CA (1968), with massive roof frames and wedge-shaped windows.

Lavatory
A wash basin, or by extension a room containing a wash basin and a water closet.

Le Corbusier (1887-1965)
Pseudonym of Charles-Edouard Jenneret; the most influential of 20th century architects. The Villa Savoy, Poissy, France (1931), is typical of his residential designs, freely planned with geometric shapes and using modern construction techniques. He called such houses "machines for living." The Pavilion Suisse, Paris, France (1932), was built on stilts, and featured curved walls, cubist blocks, and a random rubble wall to contrast with the white concrete. The Unite d'Habitation, Marseilles, France (1946), shows emphasis on mass and on the untreated concrete. The proportions for his buildings were worked out on his "modular system." He designed Notre-Dame-du-Haut (illus.), Ronchamps, France (1954), which is molded in concrete, to create a huge sculptural form penetrated by windows placed at random. Chandigarh (illus.), East Punjab, India (1956), was planned as a new state capital, but only three buildings were completed. Late work includes the Carpenter Center, Harvard University, Cambridge, MA (1963).

Le Corbusier

Leaching
A process by which chemicals can escape from certain materials in the environment. For example, arsenic can leach out of older pressure-treated wood.

Leaching cesspool
A cylindrical chamber in the ground lined with stones without mortar that allows liquids to seep out into the soil.

Leaching field
A system of underground tile or pipes arranged in trenches filled with gravel on the downhill side of a septic tank. The waste is discharged into the tile bed and seeps into the soil along the lengths of the feeders.

Lead
Toxic heavy metal often found in paints made or applied before 1978. When renovating, proper lead-abatement procedures are required to avoid lead poisoning.

Lead paint
Usually associated with oil-based paint, or those containing white lead. It is now considered a long-term health hazard, and must be removed if found among habitable spaces.

Lead poisoning
A condition arising from breathing the fumes of lead compounds, or ingesting lead paint and water contaminated by lead pipes.

Lead See Metal

Lead ventilation
The ventilation of an unoccupied building space immediately prior to its occupancy. Lead ventilation is performed to dilute contaminants from building and HVAC sources to acceptable levels by the time occupants arrive.

Leaded glass See Glass

Leaded light
A window having small diamond-shaped or rectangular panes of glass set in lead canes.

Leader
A vertical pipe that carries rainwater from the roof gutter to the ground or drain system.

Leadership in Energy and Environmental Design (LEED)
A third-party certification program and the nationally accepted benchmark for the design, construction, and operation of high-performance green buildings. LEED gives building owners and operators the tools they need to have an immediate and measurable impact on their buildings' performance. LEED promotes a whole-building approach to sustainability, factoring in community resources and public transit, site characteristics, indoor environmental quality, awareness and education, and innovation by recognizing performance in five key areas of human and environmental health: sustainable site development, water savings, energy efficiency, materials selection, and indoor air quality. There are LEED ratings (project certification) for commercial and residential construction, as well as for specific applications such as neighborhoods, and specific applications such as retail, multiple buildings/campuses, schools, healthcare, laboratories, and lodging. This program was developed and is operated by the U.S. Green Building Council. Buildings evaluated by LEED are rated as certified, silver, gold, or platinum. There are a total of 69 LEED credits available in the six categories: 26 credits are required to attain the most basic level of LEED certification; 33 to 38 credits are needed for silver; 39 to 51 credits for gold; 52 to 69 credits for the platinum rating.

Lead-free paint
A paint that does not contain any lead compounds.

Lean-to
A shed or building having a single pitched roof, with its highest end against an adjoining wall or building.

Lease
An agreement granting permission of a property for a specified time period, usually for a specific amount of rent, without transferring ownership.

Leaseback
A transaction that occurs when a property owner sells property to another, who subsequently leases back possession of the property to the original owner.

Leaves
Hinged or sliding components, as in a door.

LeBrun, Napoleon (1821–1901)
Both his major works won in competitions: the Roman Catholic Cathedral (1865) and the Academy of Music (1855), both in Philadelphia, PA.

Ledger
A wood strip nailed to the lower side of a girder to provide a bearing surface for joists.

Ledoux, Claude-Nicholas (1736–1806)
French architect. His Neoclassical buildings combined simple shapes and austere treatments.

Lee, Wanchul (1926–2008)
A designer of embassies and public buildings, Lee completed more than 80 projects for the U.S. State Department across the world, renovating embassies and building new ones. Sometimes they were in politically unstable environments, such as U.S. embassies in Beirut, Lebanon and Tel Aviv, Isreal during times of unrest, and a new embassy in Tanzania, Africa following the destruction of the previous building by terrorists in 1998.

LEED AP (Accredited Professional)
LEED Professional Accreditation distinguishes building professionals with the knowledge and skills to successfully steward the LEED certification process. LEED Accredited Professionals (LEED APs) have demonstrated a thorough understanding of green building practices and principles and the LEED Rating System.

LEED Certification
Developed by the U.S. Green Building Council, LEED is the nationally accepted benchmark for the design, building, and operation of green buildings. There is a LEED certification system for homes.

LEED for Healthcare
Promotes sustainable planning, design, and construction for high-performance healthcare facilities.

LEED for Retail
Recognizes the unique nature of retail design and construction projects and addresses the specific needs of retail spaces.

LEED for Schools
Recognizes the unique nature of the design and construction of K–12 schools and addresses the specific needs of school spaces.

LEED Green Building Rating System (LEED GBRS)
A voluntary, consensus-based, market-driven, building rating system based on existing proven technology. It

LEED Green Building Rating System (LEED GBRS)
represents the U.S. Green Building Council's (USGBC's) effort to provide a national standard for what constitutes a "green building." Through its use as a design guideline and third-party certification tool, it aims to improve occupant well-being, environmental performance, and economic returns of building using established and innovative practices, standards, and technologies.

LEED Rating System
LEED (Leadership in Energy and Environmental Design) is a self-assessing system designed for rating new and existing commercial, institutional, and high-rise residential buildings. It evaluates environmental performance from a "whole-building" perspective over a building's life cycle, providing a definitive standard for what constitutes a green building.

LEED See Leadership in Energy and Environmental Design

LEED Steering Committee (LEED SC)
Oversight committee of the USGBC (U.S. Green Building Council) responsible for direction and decisions for the LEED program.

LEED Technical and Scientific Advisory Committee (LEED TSAC)
The Technical and Scientific Advisory Committee is a standing LEED committee composed of six to eight individuals representing a diversity of building community perspectives and technical areas of competency. The committee provides support for each of the LEED products and advice on topics as assigned by the LEED Steering Committee and the USGBC (U.S. Green Building Council) Board of Directors.

LEED-CI
LEED for Commercial Interiors. One of the six LEED Green Building Rating Systems. It is a benchmark for the tenant improvement market that gives the power to make sustainable choices to tenants and designers.

LEED-CI Reference Guide
A supporting document to the LEED-CI Green Building Rating System. The Guide is intended to assist project teams understand LEED-CI criteria and the benefits of compliance with the criteria.

LEED-CS
LEED for Core and Shell. One of the six LEED Rating Systems. It focuses on buildings being developed where the developer is responsible for the core and shell of the structure and has no responsibility for the design and decisions concerning the interior space fit-outs.

LEED-EB
LEED for Existing Buildings. One of the six LEED Green Building Rating Systems. It establishes a set of performance standards for the sustainable upgrades and operation of existing buildings.

LEED-H
LEED for Homes. One of the six LEED Green Building Rating Systems. It addresses single-family homes, both detached and attached, and multifamily residential buildings with up to three stories, developed on a single lot.

LEED-NC
LEED for New Construction. One of the six LEED Green Building Rating Systems. It focuses on the design and construction process for new construction and major reconstruction of buildings. It is designed to guide and distinguish high-performance commercial and institutional projects.

LEED-NC Reference Guide
The LEED-NC Reference Guide is a supporting document to the LEED-NC Green Building Rating System. The Guide is intended to assist project teams understand LEED-NC criteria and the benefits of compliance with the criteria.

LEED-ND
LEED for Neighborhood Developments. One of the six LEED Green Building Rating Systems. It addresses the design and location of new, multilot residential, commercial, or mixed-use developments. Integrates the principles of smart growth, urbanism, and green building into the first national program for neighborhood design.

Leeward
The side of the structure that is sheltered from the wind.

Legal description
A written documentation of the location and boundaries of a parcel of land. A legal description may be based on a metes and bounds survey, the rectangular system of survey, or it may make reference to a recorded plot of land.

Legorreta, Ricardo (1931-)
Mexican-born architect whose abstract forms and bold colors evoke images of his heritage. His work includes projects in both Mexico and Texas.

Lensed troffers
Fixtures that use an acrylic or polycarbonate lens to refract light and distribute it.

Lenses
Glass or plastic devices used in electrical fixtures to alter the effect of light coming from the fixture, such as frosted, lenticular, and the like.

Leonardo da Vinci (1452-1519)
He built nothing, but he produced a number of influential architectural schemes and designs.

Lescaze, William (1896-1969)
American architect, born and trained in Switzerland. His Philadelphia Saving Fund Society Building, Philadelphia, was an early skyscraper designed in the International style.

Lesche
In ancient Greece, a public portico or clubhouse for conversation or hearing the news. Buildings were numerous in cities, and their walls were decorated by celebrated painters.

Lescot, Pierre (1510-1578)
French architect responsible for rebuilding the Louvre in Paris.

Lesene
Vertical strips resembling a pilaster, but without a base or capital; used to subdivide wall surfaces and domes into framed panels.

Lethbridge, Francis Donald (1921-2008)
A Washington, DC, architect who led the preservation movement in the nation's capital. While his own works were unabashedly modern, Lethbridge designed the 1989 Visitors Center at Arlington National Cemetery as well as multifamily housing, churches, institutional buildings, and the U.S. embassy in Lima, Peru. Lethbridge compiled the first list of

Lethbridge, Francis Donald

DC-area buildings and sites the group wanted to save. Among the 292 entries were Union Station, the Old Executive Office Building, and the boundary stones placed at one-mile intervals around the city in 1792.

Level spreaders
A stormwater management device installed parallel to a slope that changes concentrated flow to sheet flow.

Leverage
The use of fixed-cost funds to acquire property that is expected to produce a higher rate of return either by way of income or through appreciation.

Liability insurance
Insurance that protects the insured against liability because of injury to the person or property of another.

Libeskind, Daniel (1946–)
American architect, artist, and set designer of Polish-Jewish descent. Libeskind founded Studio Daniel Libeskind in 1989 with his wife, Nina. His buildings include the Jewish Museum in Berlin. Germany (1999); the extension to the Denver Art Museum (2005) Denver, CO; the Grand Canal Theatre in Dublin, Ireland; the Imperial War Museum North in Salford Quays, England; the Michael Lee-Chin Crystal at the Royal Ontario Museum in Toronto, Canada; the Felix Nussbaum Haus in Osnabrück, Germany; the Danish Jewish Museum in Copenhagen, Denmark (2003); and the Wohl Centre at the Bar-Ilan University in Ramat-Gan, Israel. In 2003, Libeskind won the competition to be the master plan architect for the reconstruction of the World Trade Center site in Lower Manhattan, NY (2002), which was destroyed in the September 11, 2001, attacks. The studio's completed projects include the Contemporary Jewish Museum (illus.) in San Francisco, CA (2005); The Ascent at Roebling's Bridge in Covington, KY; and the Royal Ontario Museum in Toronto, Canada.

Library
A repository for literature and artistic materials such as books, kept for reading or reference by the public. Libraries

Library
have been in existence since around 200 B.C. and were evident in Colonial America. Andrew Carnegie stimulated development through his funding of 1,649 public library buildings in 1,412 communities as well as 108 academic library buildings (1897–1917).

Library of Congress
Serves as the national library of the United States; maintains collections of manuscripts, photographs, maps and related historical documents, preserves its own building, and produces publications and exhibits. It is the repository for HABS and HAER documentation.

Lien
A legal charge against property which is made securely for the payment of a debt or for the performance of an obligation.

Lienau, Detlef (1818–1887)
German-born American architect who designed in the eclectic style. He designed several large blocks of townhouses in New York City, including the Rebecca Jones group (1870), which were very influential in the 1870s and 1880s.

Lierne
Short rib that runs from one main rib of a vault to another rib.

Life cycle
The consecutive, interlinked stages of a product or structure and its fixtures, beginning with raw materials acquisition and manufacture and continuing with its fabrication, manufacture, construction, and use, and concluding with any of a variety of recovery, recycling, or waste management options.

Life cycle of buildings
The entire process involved in construction of a building. It involves mining for resources; manufacturing the building product, transportation of the materials and technology, building life-cycle maintenance, demolition, and recycling. Every stage has its own activities, environmental impacts, and energy consumption that take place during the building's life-cycle phase.

Life safety
Items in general building codes that deal with aspects of life safety, such as fire protection.

Life-cycle analysis
The assessment of a product's full environmental costs, from raw material to final disposal, in terms of its extraction costs, transportation, manufacturing, consumption, use, and disposal throughout its lifetime. Life-cycle analysis is used as a tool for evaluating the relative performance of building materials and technologies. The life-cycle cost is the amortized annual cost of a product, including capital costs, installation costs, operating costs, maintenance costs, and disposal costs discounted over the lifetime of a product.

Life-cycle assessment

Examination of environmental and health impacts of a product or material over its life cycle; provides a mechanism for comparing different products and materials for green building.

Life-cycle assessment in sustainable architecture

A software program that is a streamlined LCA decision support tool for construction to help identify key environmental issues in construction, give designers an easy to use tool for evaluating the environmental aspects of building design, and enable designers and specifiers to make informed choices based on environmental considerations, such as life-cycle analysis.

Life-cycle cost

The total cost of acquiring, owning, operating, and disposing of a building or building system over its entire useful life. It includes the cost of land acquisition, construction costs, energy costs, the cost to maintain, service and repair the building and its systems, costs of system replacement, financing costs, and residual or salvage value at the end of the building's useful life.

Life-cycle cost method

A technique of economic evaluation that sums over a given study period the costs of initial investment less resale value, replacements, operations including energy use, and maintenance and repair of an investment decision, expressed in present or annual value.

Life-cycle cost of material

The costs accruing throughout the service life of a material. Life-cycle costs address the capital costs involved in production, maintenance, and disposal, and can also include other environmentally related capital costs and societal costs.

Life-cycle costing

A technique that enables a comparative cost assessment of a building structure or infrastructure to be made for various investment alternatives, over a specified extendede period of time, taking into account all relevant factors, both in terms of initial capital costs and future estimated cost. This analysis should include the initial cost, maintenance costs, and replacement costs based on the life expectancy of its components. The technique usually compares alternative components and alternative systems. The objective is to identify the most economic overall choice over the life of the building or infrastructure.

Life-cycle inventory

An accounting of the energy and waste associated with the creation of a new product through use and disposal.

Lift slab

A reinforced prestressed concrete flat plate cast at ground level.

Lift-slab construction

Casting floor and roof slabs one upon another, then jacking or hoisting them into final position, saving on formwork for cast-in-place floors in a multistory structure.

Light

An opening through which daylight is admitted to the interior space of a building; a pane of glass, window, or compartment of a window.

Light adaptation

The process by which the retina becomes adapted to a luminance greater than about 1.0 foot-lambert.

Light fixture

A luminaire secured in place or attached as a permanent appendage or appliance. It consists of a lighting unit with lamps and components to protect the electrical circuits from the weather, and other devices to spread the light in a prescribed pattern.

Light fixture

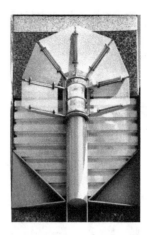

Light fixtures

An assembly that is part of a buildings electrical system that produces and controls light. The lampholder secures and energizes the lamp connected to the power supply; the reflector controls how the light is distributed; ridged baffles at the bottom of the fixture reduce the intensity of the light; and the diffuser controls the quality of the light. Louvers redirect the emitted light and shield the light source from view at certain angles. Parabolic reflectors spread, focus, or otherwise redirect the light. Lenses are used to focus or disperse the emitted light. Types include recessed, semi-recessed, surface-mounted, pendant-mounted, and track-mounted. Fixtures within a building consist of three basic types: recessed, when lamps are housed above the ceiling, and the lens is flush with the surface; surface-mounted, when fixtures are attached and project from the wall or ceiling; and pendant fixtures, which are suspended from the ceiling by a rod or chain, such as a chandelier.

Light loss factor

Also known as maintenance factor, LLF is used to calculate the illumination of a lighting system at a specific point in time under general conditions; dirt accumulation, lamp-output depreciation, maintenance procedures, and atmospheric conditions.

Light pole

A street light, lamppost, street lamp, light standard, or lamp standard is a raised source of light on the edge of a road or walkway, which is turned on or lit at a certain time every night.

Light pole

Light pollution

The wasted light from building and landscape designs that produces glare, is directed upward to the sky, or is directed off the site.

Light properties

Visible electromagnetic wave energy in the spectrum range from 3,800 to 7,200 angstrom units, a measurement of light waves. Longer wavelength is infrared light, and that of shorter wavelength is ultraviolet light. It is the natural agent that stimulates the sense of sight and is measured by candle, foot-candle, and lumen. It spreads equally in all directions at once over large areas, and diminishes in intensity according to the square of the distance from the source. Light is usually measured in terms of luminous flux. When light hits a surface, it will either be absorbed, reflected, transmitted, or refracted. Also, an opening through which daylight is admitted to the interior space of a building; an individual pane of glass, window, or compartment of a window.

Light shelf

A horizontal device usually positioned above eye level to reflect daylight to penetrate further into a building by reflecting some of the light onto the ceiling and to shield direct sunlight from the area immediately adjacent to the window. The light shelf may project into the room, beyond the exterior wall plane, or both. Exterior light shelves may also function as sunshades. The upper surface of the shelf may be specular or nonspecular but should be highly reflective, that is, having 80 percent or greater reflectance.

Light sources

Light can emanate from three different sources: point, line, or area sources. Point sources are bare incandescent amps, recessed incandescent, or high-intensity discharge lamps with small apertures where specular reflection can be precisely controlled. Line sources consist of bare fluorescent tubes and linear fluorescent fixtures. They can be controlled in their transverse axis, but not longitudinally. This makes them useful for lighting larger areas, where repetitive rows

Light sources

of fixtures are suitable. Area sources includes windows, skylights, and diffused elements with little or no directional control.

Light types

There are many different types of lights and lamps, such as incandescent, tungsten halogen, fluorescent, compact fluorescent, neon, deluxe mercury, metal halide, high-pressure sodium, white sodium, ceramic metal halide, and low-voltage halogen.

Light well

A small courtyard commonly placed in large buildings to admit daylight into interior areas not exposed to an open view.

Light-emitting diode (LED)

Illumination technology that produces light by running electrical current through a semiconductor diode. LED lamps are much longer lasting and much more energy efficient than incandescent lamps; unlike fluorescent lamps, LED lamps do not contain mercury and can be readily dimmed.

Light-emitting diode (LED) lamp

Typically very low light output from a very tiny lamp. They are commonly used for decorative lighting and theater step lighting. Different semiconductors create different colors of light. LEDs are very long lasting; unlike fluorescent lamps, they do not contain mercury.

Lighthouse

A tower or other structure supporting one or more lights to assist in the navigation of ships into harbors or to warn of dangerous shoals; often, quarters for the lighthouse keeper may be within or adjacent to the structure.

Lighting design

The process of integrating lighting systems within the structure of the building, the shape of the interior space,

Lighting design

the function of the space, and the designer's concept for lighting the space.

Lighting efficiency
A factor crucial to a light source's acceptability for energy conservation. Incandescent and tungsten halogen are not efficient. Most HID lamps are very efficient, but have less desirable attributes.

Lighting layout
Lighting is an integral part of any architectural scheme and should be coordinated with the interior features. The primary elements of a lighting design consist of point sources, linear sources, planar sources, or volumetric sources. Volumetric sources provide ambient light; linear and planar sources provide general illumination, and point sources provide specific task lighting. A lighting layout coordinates the fixture locations with diffusers, return grilles, speakers, sprinkler heads, and other structural elements.

Lighting levels
Established by building, fire, life safety, and health codes. Too much light will lead to excessive energy usage and failure to meet energy code limits.

Lighting system
The lighting fixtures in a building, sometimes including portable lights, subdivided into smaller spaces.

Lightning conductor
A metal device that leads an electrical charge from a strike by lightning safely to the ground; typically a braided copper wire connected to a lightning rod on the roof and grounded to a water pipe or other underground metal conductor.

Lightweight concrete
The concrete that is composed of any lightweight aggregate, such as sand.

Limba
A straight-grained, fine textured wood used for paneling.

Lime stucco
A stucco composed of lime, sand, and aggregate; may also contain pigments and waterproofing materials.

Limestone See Stone

Limited partnership
A form of ownership in which partners are divided into two classes; the general partners who actively manage the operations and bear full responsibility; and the limited partners, whose exposure is normally limited in amount of obligation and have no control over the affairs of the partnership.

Lin, Maya (1959–)
Trained as an artist and an architect, she is best known for her large, minimalist sculptures and monuments. At the age of 21 and still a student, Lin created the winning design for the Vietnam Veterans Memorial in Washington, DC.

Lin, Maya

Lindsey, Gail (1945–2009)
A pillar of the green building movement, Lindsey was the founder of the Wake Forest, NC, environmental consulting firm Design Harmony. She had been chair of the American Institute of Architects (AIA) Committee on the Environment (COTE) and led more than 200 workshops and programs on green building, including the Greening of the White House, the Greening of the Pentagon, Sustainable Design Initiatives for the National Park Service, the U.S. Department of Energy's International Green Building Challenge, the Sustainable Design Training Program of the U.S. Department of Defense, and the U.S. Environmental Protection Agency's Energy Star program. A cocreator of the LEED Rating System, she was one of the first 12 LEED trainers for the U.S. Green Building Council. She also helped create the Army's SPiRiT rating system, which contains checklists and strategies for maintaining sustainable facilities.

Lineal foot
A running foot, as distinct from a square foot or a cubic foot.

Lineal measure
One-dimensional measurement of material or object.

Lineal organization See Organization

Linear
Forms that describe a line, are related to a line, or are defined by being arranged in a line.

Linear diffuser
An air-conditioning device that distributes air to a space from a linear slot, often at the terminals of the duct, but also along the run.

Linear grouping
An arrangements of units along one side of a gallery, or along both sides of a corridor.

Linear wall grazing
Sometimes called wall slots; vertical lighting troughs to illuminate walls in lobbies and corridors.

Linen-scroll ornament See **Ornament**

Lining
Material which covers any interior surface, such as a framework around a door or window or boarding that covers interior surfaces of a building.

Link
The part of a building addition that connects to an existing building; often recessed from the facade and sometimes constructed of different materials, when the style of the addition is not similar to the existing building.

Linked organization See **Organization**

Linoleum
A resilient flooring product developed in the 1800s, manufactured from cork flour, linseed oil, oak dust, and jute. Linoleum's durability, renewable inputs, antistatic properties, and easy-to-clean surface often make it classified as a "green" building material.

Lintel
The horizontal beam that forms the upper structural member of an opening for a window or door and supports part of the structure above it.

Lintel course
In stone masonry, a course set at the level of a lintel; it is commonly differentiated from the wall by its greater projection, its finish, or its thickness, which often matches that of the lintel.

Little, Arthur (1852–1925)
Shingle style architect and author. Wrote the first major book on early American architecture (1878). He also performed the restoration of Bulfinch's Harrison-Gray-Otis House in Boston, MA.

Live load
A load that is not permanently applied to a structure, as compared with a dead load representing the building component's permanent weight.

Living Building Challenge
A program launched and operated by the Cascadia Region, OR, Green Building Council, one of the first three chapters of the U.S. Green Building Council (USGBC), with standards that go beyond the USGBC's LEED standard. Buildings, neighborhoods, renovations, or infrastructure (non-conditioned space) that meet this rigorous green building standard must achieve all corresponding prerequisites or imperatives.

Living history
Interpretive programs, generally in conjunction with historic sites, that emphasize participating activities involving visitors and re-creations of historical events and techniques.

Living machine
Ecological wastewater treatment system that relies on biological systems, such as microorganisms, plants, and animals, to purify wastewater; usually used in municipal-scale treatment systems.

Load
The demand upon the operating resources of a system. In the case of energy loads in buildings, the word generally refers to heating, cooling, and electrical loads.

Load controller
An outdoor computer installed next to a breaker panel and connected to 220-V appliances like air-conditioning units, clothes dryers, water heaters, and electric spa heaters. It measures the usage of power in the home and controls peak demand energy usage by defaulting to a preset level. As demand increases, the load controller shuts off lower priority appliances to maintain a user preset demand level.

Load-bearing wall See **Wall**

Load-dominated building
A building whose energy use is driven by internal loads like lighting, plug loads, and heat from people. Air conditioning that runs throughout the year in a cool or cold climate is probably load-dominated.

Loading dock
The area of a building accessible from the street, and convenient to the transportation systems within the building, that provides for the loading and unloading of commercial vehicles.

Lobby
A space at the entrance to a building, theater, hotel, or other structure.

Local buckling
Crinkling of a strut or of the compression flange of a beam because it is too thin; particularly liable to occur in thin-walled sections.

Local or regional materials
Building products manufactured and/or extracted within a defined radius of the building site. For example, the U.S. Green Building Council defines local materials as those that are manufactured, processed, and/or extracted within a 500-mile radius of the site. Use of regional materials is considered a sustainable building strategy because these materials require less transport, which reduces transportation-related environmental impacts. Additionally, regional materials support local economies, supporting the community goal of sustainable building.

Lock
A device which fastens a door, gate, or window in position; may be opened or closed by a key or a dead bolt.

Locust See **Wood**

Loft
An open space beneath a roof often used for storage; one of the upper floors of a warehouse or factory, typically unobstructed except for columns, with high ceilings; the upper space in a church, choir or organ loft.

Loft building
A former commercial or industrial building containing large open floor areas; used currently for conversion into a number of residential units.

Log building
A structure built from cut logs rather than dressed lumber; one of the earlier forms of dwellings in many parts of the world.

Log cabin siding
Wood siding with a bulging, rounded profile, designed to resemble the logs of a log cabin.

Log house
A house built of logs that are horizontally laid and notched.

Loggia
An arcaded or colonnaded structure, open on one or more sides, sometimes with an upper story; an arcaded or colonnaded porch or gallery attached to a much larger structure.

Lombard architecture (600–700)
A northern Italian pre-Romanesque architecture during the rule of the Lombards, based on early Christian and Roman forms, and characterized by the development of the ribbed vault shaft.

Long column See **Column**

Long house
A Native American dwelling, used by the Iroquois and other eastern tribes, constructed of poles set in the ground and lashed together to form an arched or triangular-shaped frame, connected with horizontal poles; approximately 16 feet wide and 50 to 100 feet long, and divided internally into compartments 6 to 8 feet wide opening onto a con-tinuous aisle.

Longhena, Baldassare (1595–1682)
Italian architect who designed St. Maria della Salute, Venice, Italy (1630–1687), sited at the head of the Grand Canal.

Longhena, Baldassare

Longitudinal section See **Projection drawing**

Lookout
Rafter, bracket, or joist at the ridge of a roof that projects beyond an end-wall of a building: may support an overhanging portion of the roof or cornice.

Loop window See **Window**

Loophole Any opening in a parapet or wall to allow for vision, light, or air.

Loos, Adolf (1870–1933)
Viennese architect, primarily a designer of houses; disclaimed all forms of ornament. Designed the Kartner Bar and the Gustav Sheu House, both in Vienna, Austria with typically boxlike exteriors and rectangular windows of various sizes.

Loose-fill insulation
A material that can be blown into an area, such as an attic space, which provides insulating qualities by the air pockets created within the material.

L'Orme, Philibert de (1514–1570)
Original and inventive, was instrumental in creating a distinctive version of Renaissance classicism that drew on French traditions as well as Italian models. Works include Tuileries Palace, Paris, France now destroyed.

Lot
One of the smaller portions of land into which a village, town, or city block is divided or laid out; also, a parcel of a subdivision, described by reference to a recorded plot or by definite boundaries. Also a portion of land in one ownership; if two or more lots are occupied by a single building unit, then such a plot is considered a single lot.

Lot line
The line which bounds a plot of ground described as a lot in the title to the property. The line is represented on paper, and may be expressed on the property by boundary markers at corners, a fence, stone wall, or other means of delineation.

Lotus
A fan-shaped decorative motif formed by symmetrically arranging lotus petals with a spreading curvature.

Lotus capital See **Capital**

Louis XIV–XVI style (1643–1792)
A high Classical style typified in the architecture, decoration and furniture of France, culminating in the building of Versailles. It developed into the ornate Rococo style.

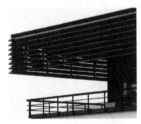

Louver
A window opening made up of overlapping boards, blades or slats, either fixed or adjustable, designed to allow ventilation in varying degrees without letting the sun or rain come in.

Longhena, Baldassare

 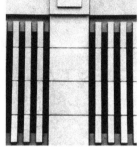

Louvered door See **Door**

Low biocide
Paint that does not contain toxic additives, such as fungicide or pesticide.

Low toxic
Generic term for products with lower levels of hazard than conventional products. Specific criteria need to be applied to this term to make it meaningful in the selection of sustainable building materials.

Low VOC
Building materials and finishes that exhibit low levels of "off-gassing," the process by which VOCs (volatile organic compounds) are released from the material, impacting health and comfort indoors and producing smog outdoors. Low or zero VOC is an attribute to look for in an environmentally preferable building material or finish.

Low-conductivity gas fill
Transparent gas installed between two or more panes of glass in a sealed, insulated window that resists the conduction of heat more effectively than air. The fill boosts a window's R-value and reduces its U-factor.

Low-emissivity glass See **Glass**

Low-emissivity windows
The window technology that lowers the amount of energy loss through windows by inhibiting the transmission of radiant heat while still allowing sufficient light to pass through. Low-E (low-emissivity) windows reflect heat, not light, and keep spaces warmer in winter and cooler in summer.

Low-flow fixture
A faucet with an aerator installed to reduce the flow of water but not reduce water pressure.

Low-flow fixtures
Plumbing fixtures that use less water than average fixtures. Low-flow toilets use no more than 1.6 gallons per flush compared to the traditional 3.5 gallons per flush.

Low-flow showerhead
The showerhead that restricts water flow to less than the 2.5 gallons per minute limit, mandated by the EPA (U.S. Environmental Protection Agency).

Low-flow toilet
A toilet that combines efficiency and high performance. Design advances enable these toilets to save water with no trade-off in flushing power. Such toilets often have the EPA's (Environmental Protection Agency's) WaterSense label.

Low-impact development
An innovative stormwater management approach with a basic principle that is modeled after nature, using design techniques that infiltrate, filter, store, evaporate, and detain runoff close to its source. Maximizing open space and use of rain barrels, rain gardens, cisterns, pervious concrete, and bioswales are just some of the methods used to retain rainwater onsite, thereby preventing pollution of nearby streams from excessive runoff.

Low-maintenance landscaping
Identification of areas that require a more manicured look and noting areas that are not as important to heavily maintain.

Low-rise
A relatively short multistory building, often described in building codes as not more than 75 feet tall, as opposed to a high-rise.

Low-voltage lighting
The lighting operated at a lower voltage for safety reasons; must be used in conjunction with a transformer. Primarily used for lighting artwork and landscaping.

Lozenge
An equilateral four-sided figure with pairs of equal angles, two acute and two obtuse; a rhombic or diamond-shaped figure.

Lozenge window See **Window**

Lubetkin, Berthold (1901-1990)
Russian-born architect, who emigrated to England and founded Tecton, a practice producing International style modern buildings notable for their simplicity and clean lines.

Lucarne window See **Window**

Lucite See **Plastic**

Luckman, Charles (1909-1999)
Partner with William Periera in the firm of Periera and Luckman. They designed the space-age restaurant (illus.) at the Los Angeles International Airport. Their most notable buildings are the Prudential Center, Boston, MA (1965), and the Transamerica Building (illus), in San Francisco, CA.

Luckman, Charles

Lumber
Timber that is sawn or split in the form of beams, boards, joists, planks, or shingles; refers especially to pieces smaller than heavy timber.

Lumen
A unit of light from one source. Light falling on 1 square foot of surface of an imaginary sphere having a 1-foot radius around 1 candle.

Lumen maintenance control
Allows lighting to be dimmed automatically when it is new and through photoelectric sensing, to be increased gradually over time as lamps age and luminaires get dirty. It is similar to equipment that controls daylighting, and most controls can be programmed to do both.

Lumen method (daylighting)
A method of estimating the interior illuminance from window daylighting at three locations within a room, based on empirical studies.

Luminaire
A complete lighting unit, consisting of one or more lamps together with housing, the optical components to distribute the light from the lamps, and the electrical components' ballast and starters, which are necessary to operate the lamps.

Luminance
The physical measure of brightness, or luminous intensity per unit of projected area of any surface, as measured from a specific direction. It is the amount of visible light.

Luminance
leaving a point on a surface in a given direction. This "surface" can be a physical surface or an imaginary plane, and the light leaving the surface can be due to reflection, transmission, and/or emission. The standard unit of luminance is candela per square meter (cd/m2).

Luminous
Emitting light, especially emitting self-generated light as opposed to reflected light.

Luminous ceiling
A lighting system in which the whole ceiling is translucent with lamps that are installed above and suspended from a structural ceiling.

Luminous flux
The rate of flow of light, analogous to the rate of flow of a fluid.

Luminous intensity
The luminous flux emitted from a point per unit solid angle in a particular direction. The luminous intensity is the official base unit for light.

Luminous-intensity distribution curve (LIDC)
A representation of the pattern of light produced by a lamp or light fixture from the center of the source. This information is usually supplied by the manufacturer of the fixture.

Lump sum bid
A bid of a set amount to cover all labor, equipment, materials, overhead, and profit necessary for construction of an improvement to real estate,

Lundquist, Oliver Lincoln (1917-2009)
A New York architect who designed the United Nations logo, Lundquist joined Raymond Loewy's industrial design firm in 1937, working on displays for the 1939 World's Fair. He also designed track lighting for Lightolier.

Lundy, Victor (1923-)
He trained under Walter Gropius and worked with Marcel Breuer. His best known work was the I. Miller Shoe Salon, New York City (1961), employing timber ribs and mirrors.

Lunette
A semicircular window or wall panel framed by an arch.

Luster glass See Glass

Lutyens, Sir Edwin Landseer (1869-1944)
Important English architect; designed the plan for the city of New Delhi, India (1912). His later work includes many commercial bank buildings in London, England (1920).

Lux
Lumens measured by the square yard or meter.

Lyndon, Donlyn (1936-)
Partner in the firm of Moore, Lyndon, Turnbull in San Francisco, CA.

M m

M roof See Roof

Machicolation
Openings formed by setting the parapets out on corbels so as to project beyond the face of the wall. Some parapets set out on corbels have a similar appearance, even if there are no openings.

Machine room
A space that houses machinery and equipment, such as elevator equipment, generators, boilers, or air-conditioning equipment.

Machu Picchu
The most celebrated Inca citadel, on a promontory 2,000 feet above the valley in the Andes in Peru. The site includes buildings which surround an oblong plaza. The houses were built around courts, with stairs, windows, interior niches, narrow doorways and thatch-covered gable roofs. Some houses were carved out of the rock; some connecting stairs were hewn out of the mountain.

Macintosh, Charles Rennie (1868-1928)
A highly original architect and designer who created his own version of Art Nouveau, combining logical planning and expressive ornament. His first major work was the Cranston Tea Room in Edinburgh. He also designed the Glasgow School of Art, a highly original Art Nouveau design.

Macintosh, Charles Rennie

Maderno, Carlo (1556-1679)
Italian architect who designed St. Susannah facade, Rome (1603), a Baroque elevation crowded with orders and set with niches rather than windows. One of his last works, the Palazzo Barberini, Rome (1628), was completed by Bernini.

Maginnis, Charles (1867-1955)
American architect in partnership with Timothy F. Walsh (1868-1934).

Maher, George Washington (1864-1926)
American architect of the Prairie School, influenced by Arts and Crafts; designed mostly domestic architecture reminiscent of Greene and Greene, and Frank Lloyd Wright.

Mahogany See Wood

Maillart, Robert (1872-1940)
Structural engineer; built his first of 40 reinforced-concrete bridges in 1901. He contributed to the design of mushroom slab construction in high structures, in which the columns, beams, and floors are integrated.

Main
In electrical work, the current from which all other branch circuits are taken.

Main beam
In floor construction, one of the principal beams which transmits loads directly to the columns.

Main rafter
A roof member extending at right angles from the plate to the ridge.

Main Street Program
Encourages economic revitalization of central business districts in small to medium-sized communities through preservation.

Main tie
The lower tension member of a roof truss which connects the feet of the principal rafters.

Maintenance

Providing upkeep, repair and care for a building's structural integrity and appearance after acquisition or after restoration, at an acceptable level to enable it to fulfill its function over its life cycle, and to prevent deterioration. Green products and structures feature low- or no-maintenance materials and designs that reduce the resources required for their continued use, as well as reduce the likelihood that replacement will be needed.

Makeup air

Outside air supplied to replace household air that was used in a combustion appliance or exhausted through a ventilation system.

Maki, Fumihiko (1928–)

Major figure in Japanese architecture since the late 1950s. A member of the Metabolist movement, a group that advocated the embrace of new technology with a belief in architecture's organic, humanistic qualities. He was awarded the Pritzker Prize in 1993. U.S. projects include the Center for the Arts, Yerba Buena Gardens, in San Francisco, CA and numerous projects in his native Tokyo.

Makovecz, Imre (1935–)

Hungarian architect influenced by the work of Rudolph Steiner. He designed the Hungarian Pavilion for Expo '92, Seville, Spain (1992), with seven churchlike spires rising through the roof. He also produced set designs for several dramatic performances.

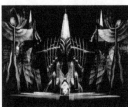

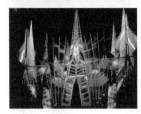

Malling of America

Outdoor strip shopping centers were built as early as the 1920s in California and became common everywhere by the late 1950s. The two-level mall, developed by the Rouse company in Baltimore, was the first fully enclosed temperature-controlled mall. Another first was the Southdale Center near Minneapolis, MN designed (1956) by Victor Gruen Associates, which revolutionized shopping everywhere.

Maltese cross

A cross that is formed by four equal triangles or arrowheads joined at their points; the outer edge of each arm is indented with an angle.

Manastaba

In Indian architecture, a freestanding upright pillar in front of a temple.

Mandapa

In Indian architecture, a large porch-like hall leading to a Hindu temple and used for religious dancing and music.

Manifold

Component that distributes the water in a home-run plumbing system. It has one inlet and many outlets, each of which feeds one fixture or appliance.

Manigault, Gabriel (1758–1809)

Trained in London as a lawyer, and designed as a hobby. Designed early churches and houses, which became popular objects of affection; also designed the City Hall (1800), Charleston, SC.

Mannerism (1530–1600)

A style of Italian architecture which was a reaction against the classical perfection of High Renaissance architecture, either responding with a rigorous application of classical rules and motifs or flaunting Classical convention in terms of shape and scale. It was a relaxed nonconformist style, using unnatural proportion and stylistic contradictions.

Manor house
The house occupied by the lord of a manor; the most important house in a country or village neighborhood.

Mansard roof See Roof

Mansart, Francois (1598–1666)
French architect who designed in the French Classical style.

Mansion
A large and imposing dwelling; a large apartment in a building.

Mantel
The frame and shelf surrounding the fireplace; often used to denote just the shelf.

Mantelpiece
The fittings and decorative elements of a mantel, including a cornice and shelf carried above the fireplace.

Manueline architecture (1495–1521)
The last phase of Gothic architecture in Portugal, so named after King Manuel.

Map
A vertical graphic description of a geographic area; including a site map, land-use map, subdivision map, topographic survey map, and National Geodetic Survey map.

Maple See Wood

Marble See Stone

Marbleized
Painted in imitation of the surface color and pattern of stone, especially veined marble; base materials may include slate, plain marble, cast iron, and plaster.

Marbling
The process of painting a wood surface so that it will resemble marble.

Margin draft
A narrow dressed border along the edge of a squared stone, usually the width of a chisel, as a border surrounding the rough central portion.

Margin light
A narrow pane of glass at the edge of a sash window or door.

Marine frieze
A frieze decorated with symbols related to the sea, such as seahorses, tritons, fish and shells.

Maritime site
Buildings located on or near the sea and connected with shipping or navigation. Many have been converted to museum or commercial use.

Market
An open place or building where goods are offered for sale; a store or shop that sells a particular type of merchandise. Many such buildings have been restored and are still functioning in the same capacity, while others have been adapted to similar uses.

Market analysis
An economic study of the potential market for a proposed use for a site or building; may include a demographic analysis of potential users.

Market evaluation
Part of a feasibility study which includes community market dynamics, social and demographic characteristics, recent development trends, specific markets, the competition, and geographic factors.

Market rent
The current rental value of a property, which varies according to local or national market conditions.

Market valuation
A valuation of the current market rent of a property, obtained by comparison with equivalent properties.

Marmoset
An antic figure, usually grotesque, introduced into architectural decoration in the 13th century.

Marmoset

Marquee

A permanent projecting roof-like shelter over an entrance to a building, often displaying information about performances.

Marquetry

Inlaid pieces of a material, such as wood or ivory, fitted together and glued to a common background.

Mascaron

The representation of a face, a human, or partly human head, more or less caricatured, and used as an architectural element.

Mask

A corbel, the shadow of which bears a close resemblance to that of a human face. It was a favorite ornament under the parapet of a chancel.

320

Mason
A craftsperson skilled in shaping and joining pieces of stone or brick together to form walls and other parts of buildings and structures.

Masonite See Wood products

Masonry
Includes all stone products, all brick products and concrete block units, including decorative and customized blocks.

ashlar masonry
Smooth square stones laid with mortar in horizontal courses.

broken rangework masonry
Stone masonry laid in horizontal courses of different heights, any one course of which may be broken into two or more courses.

cavity wall masonry
An exterior wall of masonry, consisting of an outer and inner course separated by a continuous air space, connected together by wire or sheet-metal ties; the dead air space provides improved thermal insulation.

concrete masonry
Construction consisting of concrete masonry units laid up in mortar or grout.

coursed masonry
A masonry construction in which the stones are laid in regular courses, not irregularly as in rough or random stonework.

coursed rubble masonry
A masonry construction in which roughly dressed stones of random size are used, as they occur, to build up courses; the spaces between them are filled with smaller pieces or mortar.

cyclopean masonry
Often found in ancient cultures, characterized by huge irregular stones laid without mortar and without coursing.

diamondwork masonry
A masonry construction in which pieces are set to form diamond-shaped patterns on the face of the wall.

hollow block masonry
Extruded block of concrete or burnt clay, which consists of voids and consequently is a good insulator. It is used for walls and as a backing for brick.

hollow masonry unit
A brick or concrete block that is less than 75 percent solid in the plane that rests in the mortar bed.

patterned block masonry
Concrete block with a recessed decorative pattern on the front face.

pebble wall masonry
A wall built of pebbles set in mortar, or one faced with pebbles embedded in a mortar coating on the exposed surface, either at random or in a pattern.

pitch-faced masonry
In masonry, a surface in which all arrises are cut true and in the same plane, but the face beyond the arris edges is left comparatively rough, dressed with a chisel.

polygonal masonry
A masonry constructed of stones having smooth polygonal surfaces.

Masonry

quarry-faced masonry
Squared blocks with rough surfaces that look as if they just came out of the ground.

random ashlar masonry
Ashlar masonry in which regular stones are set without continuous joints and appear to be laid without a drawn pattern, although the pattern may be repeated.

random broken coursed ashlar
An ashlar masonry bond pattern with random-sized stone blocks laid in short, discontinuous courses, with stones of varied sizes within each course.

random rubble
A rubble masonry consisting of stones of irregular size and shape with roughly flat faces, set randomly in a wall.

rubble masonry
Very irregular stones, used primarily in the construction of walls where the irregular quality is desirable.

rubblework
Stone masonry built entirely of rubble.

rustic stone masonry
Any rough, broken stone suitable for rustic masonry, most commonly limestone or sandstone; usually set with the longest dimension exposed horizontally.

rusticated masonry
Coursed stone masonry where each unit is separated by deep joints; the surface of each unit is usually very rough.

square rubble masonry
Wall construction in which squared stones of various sizes are combined in patterns that make up courses at every third or fourth stone.

vermiculated masonry
A form of masonry surface, incised with discontinuous wandering grooves resembling worm tracks; a type of ornamental winding frets or knots on mosaic pavements, resembling the tracks of worms.

Masonry field
In brickwork, the expanse of wall between openings, composed principally of stretchers.

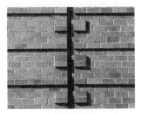

Masonry saw
A portable electrically powered handsaw similar to a circular saw with a variety of blades, such as diamond tipped.

Masonry wall See Wall

Mass effect
Describes the effect of a high-mass material on heating or cooling requirements. High-mass materials such as concrete, used in floors and walls, can absorb and store significant amounts of heat, which is later released. In climates with lots of sunshine, low humidity, and large daily temperature fluctuations, high-mass materials can mean a reduction in cooling and heating requirements by delaying the time at which the heat is released into the house.

Mass
The physical volume or bulk of a solid body; or a grouping of individual parts or elements that compose a body of unspecified size.

Mass transit vehicles
Vehicles typically capable of serving 10 or more occupants, such as buses, trolleys, and light rail.

Massing
The overall composition of the exterior of the major volumes of a building, especially when a structure has major and minor elements.

Mastaba
A freestanding tomb found in ancient Egypt, consisting of a rectangular superstructure with inclined sides, from which a shaft leads to underground burial chambers.

Master builder
An individual of broad experience and training, who is distinguished in the craft of building.

Master plan
A planning document designed to guide the future development on an entity; may be employed for the construction or remodeling of a building, site planning or access, or for a political jurisdiction indicating proposed land uses.

Mastic
Any heavy-bodied, dough-like adhesive compound; a sealant with putty-like properties used for applying tiles to a surface or for weatherproofing joints.

Matched lumber
Any lumber which has been edge-dressed and shaped to make a tongue-and-groove joint on the edges when laid edge to edge.

Matched veneer
Wood or stone veneer pieces installed with the grain of adjoining pieces aligned at the edges; types include herring-bone-matched and book-matched.

Matching
Used to describe the arrangement of timber veneers, such as book-matching, slip-matching and quarter-matching.

Material safety data sheets (MSDSs)
Informational fact sheets that identify hazardous chemicals and health and physical hazards, including exposure limits and precautions for workers who may come into contact with these chemicals. Green design professionals review product MSDSs when specifying materials and require their submittal during the shop drawing phase.

Matte surface
A surface from which reflection is predominantly diffuse, with or without a negligible specular component.

Matte surface
A surface that redistributes the incident light uniformly in all directions, so that the luminance is the same and without sheen, even when viewed from an oblique angle.

Mausoleum
A large and stately tomb, or a building housing such a tomb or tombs; originally the tomb for King Mausolos of Carla, about 350 B.C.

Mayan arch See Arch

Mayan architecture (600–900)
Sites such as Tikal in Guatemala, Copan in Honduras, and Palenque in Mexico represent the highest development of this style It is characterized by monumental constructions, including soaring temple pyramids, palaces with sculptural facades, ritual ball courts, plazas and interconnecting quadrangles. Buildings were erected on platforms, often with a roof structure. The lower section contained a continuous frieze carrying intricate decoration of masks, human figures, and geometric forms. Decorative elements formed open parapets. Exterior surfaces were covered with a lime stucco and painted in bright colors, and interior walls were massive and decorated. The sites were totally rebuilt periodically, leaving previous structures completely covered and intact. One of the most notable examples is Chichen Itza in Yucatan, the largest center of the Mayan civilization.

Maybeck, Bernard (1862–1937)

Designed the Christian Science Church (illus.), Berkeley, CA, in a mixture of styles. He also designed the Palace of Fine Arts (illus.), San Francisco, for the Pan Pacific International Exposition of 1915. The Exposition buildings were demolished, but the Palace remains, rebuilt out of permanent materials, an exact replica of the original structure.

Mayne, Thom (1944–)

Los Angeles–based architect Mayne helped found the Southern California Institute of Architecture (SCI-Arc) in 1972, where he is a trustee. He is principal of Morphosis, an architectural firm in Santa Monica, CA. Mayne received the Pritzker Architecture Prize in March 2005. Under the Design Excellence program of the U.S. government's General Services Administration, Mayne has become a primary architect for federal projects. Recent commissions include new academic building for the Cooper Union, NYC (2009); Federal Building (illus.), San Francisco, CA (2006); University of Cincinnati Student Recreation Center, OH (2006); Wayne L. Morse U.S. Courthouse in Eugene, OR (2006); Science Center School in Los Angeles, CA (2004); graduate housing at the University of Toronto, Canada (2000); and the Diamond Ranch High School in Pomona, CA (1999). Visually, the firm's architecture includes sculptural forms.

Mayne, Thom

Maze

A confusing and intricate plan of hedges in a garden, usually above eye level, forming a labyrinth.

McArthur, John, Jr. (1823–1890)

Scottish-born American architect; designer of the Philadelphia City Hall, PA (1894).

324

McComb, John, Jr. (1763–1853)
American architect who won the competition for the New York City Hall (1812), with his partner, Joseph Francis Mangin.

McDonough, William Andrews (1951–)
American architect, founding principal of William McDonough + Partners in 1981, cofounder of McDonough Braungart Design Chemistry (MBDC) with German chemist Michael Braungart. McDonough's first major commission was the 1984 Environmental Defense Fund Headquarters, whose requirement for good indoor air quality in the structure exposed him to the need for sustainable development. In 1996 McDonough became the first and only individual recipient of the Presidential Award for Sustainable Development. In 2002 he wrote (with Michael Braungart) *Cradle to Cradle: Remaking the Way We Make Things*. The Ford Motor Company's legendary River Rouge Plant in Dearborn, MI (2002), includes the world's largest "living roof," covered with more than 10 acres of sedum, a low-growing ground cover. The Bernheim Arboretum Visitor Center in Clermont, KY (2005), blurs the line between outdoor and indoor space, draws heavily on the biophilia hypothesis—the study of the human desire and physiological need for contact with nature. The NASA Ames Research Center's Sustainability Base—designed to be a net energy positive building—produces more energy than it consumes. The firm's duplex design for the "Make It Right Foundation" will offer residents natural ventilation, roof-mounted PV (photovoltaic) panels, water cisterns to harvest rainwater runoff and rain gardens to absorb storm runoff.

McIntire, Samuel (1757–1811)
A carpenter-builder in Salem, MA, who worked with Bullfinch, and excelled in ornamental woodwork. Projects include the Darby Summer House (1794), Danvers, MA; and John Gardner House (1805) and South Congregational Church (1805), both in Salem, MA.

McKim, Charles F. (1847–1909)
American architect, in partnership with William Mead and Stanford White, the largest architectural practice of its time in the United States. His bold design was expressed best in Pennsylvania Station (illus.), NYC (1911), now demolished. Another important work is the layout and design of the Columbia University campus (1901) and the Morgan Library (1903), both in NYC.

McLaughlin, Donal (1907–2009)
Working as the head of graphics for the Office of Strategic Services (OSS, precursor of the CIA), McLaughlin, an architect and graphic designer, led the team charged with putting together the printed elements for the 1945 United Nations Conference on International Organization in San Francisco, CA including the emblem that eventually became the UN's logo. He collaborated with Walter Dorwin Teague and Raymond Loewy on exhibits for the 1939 World's Fair, Queens, NY and the interiors of Tiffany & Company's Fifth Avenue store, NYC before joining the OSS to develop visual material for the military.

Mead, William R. (1846–1928)
Partner in the New York City architectural firm of McKim Mead and White.

Meander
A running ornament consisting of a fret design with many involved turnings and an intricate variety of designs.

Measured drawing
An architectural drawing of an existing building, object, site, or detail that is accurately drawn to scale on the basis of field measurements.

Measured drawings
A set of drawings that accurately record an existing site, building, structure, or object, that is based on field measurements and accurately drawn to scale; which typically includes plans, elevations, sections, and details.

Measurement and verification
The process of monitoring building systems to ensure optimal performance.

Measuring line See **Perspective projection**

Mechanical code
Building code that governs the safe design of heating, ventilating, plumbing, and air-conditioning systems for a structure.

Mechanical engineer
A person trained, skilled or professionally engaged in a branch of engineering related to mechanical equipment, particularly heating and air-conditioning systems.

Mechanical engineering
An engineering discipline that involves the application of principles of physics for analysis, design, manufacturing, and maintenance of mechanical systems. Mechanical engineers require a solid understanding of key concepts including mechanics, kinematics, thermodynamics, and energy, and they use these principles in the design and analysis of automobiles, aircraft, heating and cooling systems, manufacturing plants, industrial equipment and machinery, medical devices, and more.

Mechanical room
A room devoted to mechanical equipment and controls, such as a boiler, furnace, ductwork, plumbing, and water heater.

Mechanical services
Building services such as heating, ventilation, air-conditioning, and gas installations.

Mechanical systems
Construction related to furnishing heating, ventilating, and air-conditioning, plumbing, and fire-suppression systems.

Mechanical ventilation
The use of fans and intake and exhaust vents to mechanically distribute ventilation and other conditioned air.

Mechanics lien
A lien filed by a contractor or tradesman for unpaid work performed on a project, which becomes an encumbrance on the property until released by agreement, paid, or dismissed by action in the courts.

Medallion
An ornamental plaque, usually round or oval in shape, inscribed with an object in low relief, such as a head, flower or figure, and applied to a wall or frieze.

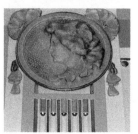

Medallion molding See Molding

Medieval architecture (400–1400)
The architecture of the European Middle Ages where the use of Byzantine, Romanesque and Gothic elements spanned a millennium. It was an age of the fortified castle, where bishop's palaces rivaled cathedrals in splendor and served public and private functions. As the population grew, smaller houses nestled around castle walls, creating medieval towns. As urban land grew more valuable, tall narrow houses with upper stories were common.

Medium-density fiberboard (MDF)
An engineered wood product made of softwood fibers, wax, and resin and formed into panels with the use of heat and pressure. Similar to particleboard, MDF has finer texture, offering more precise finishing. Most MDF is made with formaldehyde-emitting urea-formaldehyde binders, which can contribute to poor indoor air quality.

Medlon dome See Dome

Medusa
In Greek mythology, the mortal one of the three Gorgons, who had snakes for hair and whose head was cut off by Perseus to present to Athena as an ornament for her shield.

Meeting house
A house of worship, especially that of the Society of Friends, or Quakers, and the Mormons.

Meeting rail
One of the horizontal rails of a double-hung sash.

Megalithic
Built of very large stones, used as found in nature or roughly hewn, especially as used in ancient construction.

Megalopolis
A single vast urban area formed by the expansion and merging of adjacent cities and their suburbs.

Megastructure
A type of structure, popular during 1964 to 1976, in which individual buildings become merely components or lose their individuality altogether. Vast new structures were proposed to replace existing cities. Their overall purpose is to provide a total environment for work and leisure.

Megastructure

Megeron
A rectangular hall, fronted by an open, two-columned porch, traditionally used in Greece since Mycenaean times.

Meier, Richard (1934–)
Worked with Marcel Breuer and SOM, and was the most prolific of the New York Five. He persisted in using white in his buildings, such as the Saltzman House in East Hampton, NY (1967), and the Douglas House, Harbor Springs, MI (1971). His later works include the High Museum of Art (illus.), Atlanta, GA (1984), the Barcelona Art Museum, Spain (1996), and the Getty Center Museum (illus.), Brentwood, CA (1997).

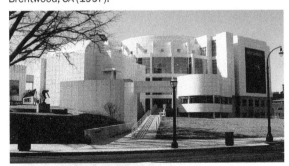

Meier, Richard

Melnikov, Konstandin (1890–1974)
Russian architect whose work anticipated certain aspects of Deconstructivism, which gained him popularity among the avant-garde. He is also associated with Constructivism.

Member
Any individual element of a building, such as a framing member; also one of the individual shapes that make up a molding, such as a cornice or a water table.

Membrane
A thin, flexible surface such as a net or form with a fabric surface, supported by tension cables or an air system.

Membrane structure
A roof of flexible membranes of canvas or plastic, supported by cables or ropes.

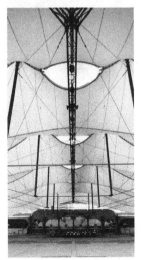

Memento mori

An image meant to serve as a reminder of death, usually a human skull or skeleton.

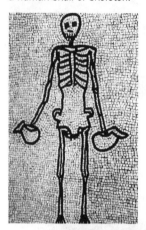

Memorial

An object designed to establish or serve as a remembrance of a person or an event, commemorated by a monument or plaque.

Memorial arch

An arch commemorating a person or event, popular during the Roman Empire and again at the time of Napoleon.

Memorial arch

Memorial plaque

A flat inscribed stone that commemorates a special event or to serve as a memorial; set into or fixed to the surface of a wall.

Mendelsohn, Eric (1887–1953)

German architect who designed the Einstein Tower in Potsdam (1921), a highly plastic building. Its design is an outcome of expressionist demands. He also designed the Luckenwalde Hat factory (1923) and the Schoken Department Stores in Stuttgart (1928), using bold new forms to express a new function.

Mendes da Rocha, Paulo (1928–)
Pritzker Prize–winning, modernist architect Mendes da Rocha is known for bold simplicity and innovative use of concrete and steel.

Mengoni, Guiseppe (1829–1877)
Italian architect and designer of the Galleria Vittorio Emanuelle, Milan (1871), the largest shopping arcade of its type in Europe, and the architect's most famous project.

Merlon
One of the solid alternates between two crenels, or open spaces, in a battlement parapet.

Merrill, John (1896–1975)
Partner in the firm of Skidmore, Owings and Merrill (SOM).

Mesh reinforcement
Welded wire fabric used as reinforcement for concrete, especially used inslabs.

Meshrebeeyeh
An elaborately turned or carved wood screen or wood lattice that encloses a balcony window, as found in Islamic countries.

Mesoamerican architecture (1300–500 B.C.)
A characteristic feature of this architecture is the great temple pyramids of pre-Columbian America, which are equivalent in complexity to those of ancient Egypt and the Middle East. The main centers in Mexico and Peru are divided into four main cultures: Mayan, Toltec, Aztec, and Inca. All four of these civilizations conceived of their architecture in monumental terms characterized by strong grid plans, huge walled enclosures, and vast stone cities.

Mesolithic era
The cultural period between the Paleolithic and Neolithic eras, marked by the appearance of cutting tools.

Mesopotamian architecture (3000–500 B.C.)
A massive architecture constructed of mud-bricks set with clay mortar; producing heavy walls articulated by pilasters and recesses and faced with glazed brick. Columns were seldom used, and openings were infrequent and small.

Metal
Any of a class of elementary substances which are crystalline when solid and characterized by opacity, ductility and conductivity; mined in a form called "ore" and manufactured to specific applications.

aluminum
A lightweight metal which is malleable and nonmagnetic and has good conductivity; it is a good reflector of heat and light and is resistant to oxidation; it is often anodized for better corrosion resistance, color and surface hardness.

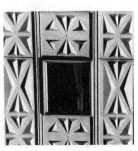

brass
Any copper alloy having zinc as the principal alloying element, but often with small quantities of other elements.

bronze
An alloy of copper and tin, bronze in color, having a substantial admixture of copper to modify the properties of the principal element, such as aluminum bronze and magnesium bronze.

cast iron
A hard, nonmalleable iron alloy containing carbon and silicon, which is poured into a sand mold and then machined to a desired architectural shape.

copper
A metal with good electrical conductivity, used for roofing, flashing, hardware and plumbing applications; when exposed to air, copper oxidizes and develops a greenish "patina" that halts corrosion.

ferrous metal
A metal in which iron is the principal element.

iron
A metallic element found in the earth's crust, consisting of a malleable, ductile, magnetic substance from which pig iron and steel are manufactured.

lead
A soft, malleable, heavy metal that has a low melting point and a high coefficient of thermal expansion; very easy to cut and work.

stainless steel
A high-strength, tough steel alloy; contains chromium with nickel as an additional alloying element and is highly resistant to corrosion and rust.

steel
A hard and malleable metal when heated; produced by melting and refining it according to the carbon content; used for structural shapes due to its alloy of iron and carbon which has a malleable high tensile strength.

steel

tin
A lustrous white, soft, and malleable metal having a low melting point; relatively unaffected by exposure to air; used for making alloys and solder, and in coating sheet metal.

weathering steel
A high-strength, low-alloy steel that forms an oxide coating when exposed to rain or moisture, which adheres to the base metal and protects it from further corrosion.

wrought iron
A commercially pure iron of fibrous nature, valued for its corrosion resistance and ductility; used for water pipes, water tank plates, rivets, and other forged work.

zinc
A hard bluish-white metal, brittle at normal temperatures, very malleable and ductile when heated; not subject to corrosion; used for galvanizing sheet steel and iron in various metal alloys.

Metal gutters
A type of attached gutter, prefabricated of sheet metal; usually obtainable in two styles; either half round or given an ornamental profile resembling a cornice molding.

Metal roofing See **Roof**

Metal valley
A V-shaped valley or gutter between two roof slopes, lined with pieces of lead, zinc, copper, or sheet metal to prevent the leaking of water.

Metal wall ties
Strips of corrugated metal used to tie a brick veneer wall to a framework.

Metal-clad door See **Door**

Metal-halide lamp
Similar to mercury-vapor lamps, except that metal halide is added to provide better color; can be used for interior lighting.

Metes and bounds
A term used to define boundary lines when describing the location of land in terms of directions and distances from one or more points of reference.

Methane (CH_4)
A greenhouse gas that is emitted during the production and transport of coal, natural gas, and oil or from the decomposition of organic wastes in municipal solid waste landfills and the raising of livestock. Although CO_2 is more prevalent in the atmosphere, methane is over 20 times more effective in trapping heat than CO_2 over a 100-year period.

Methyl diisocyanate binder
Nonformaldehyde binder used in some medium-density fiberboard and particleboard products, including straw-based particleboard.

Metope
A panel, either plain or decorated with carvings, between the triglyphs in a Doric frieze.

Metope medallion
A raised, round rosette in the otherwise blank surface of a metope on a Doric frieze.

Mews
An alley or court in which stables are or once were located or have been converted into residences.

Mezzanine
A low-ceilinged story located between two main stories, it is usually constructed directly above the ground floor, often projecting over it as a balcony.

Mezzo-relievo See **Relief**

Michelangelo Buonarroti (1475–1564)
Architect, sculptor, painter, and poet; represents the Italian Renaissance at its height. In 1546, he was appointed architect of St. Peters, Rome.

Microclimate
Localized climate conditions within an urban area or building.

Microturbines
Small rotary engines, usually fueled by natural gas, that provide onsite electricity generation.

Mid-wall column See Column

Miesian
Designed in the style of Mies van der Rohe, German-born architect who designed many International Style buildings; typical elements include exposed structural supports, reveals at joints between materials, and a total lack of applied ornament. See Van der Rohe, Ludwig Mies.

Mihrab
A niche in the mosque of any religious Muslim building that indicates the direction of prayer toward Mecca.

Mill
A building equipped with machinery for grinding grain into flour. The evolution of industry in America left the country with a great legacy of mills, ranging from small flour-grinding gristmills, or village-scaled New England textile companies, to vast Midwestern grain milling operations. Many still standing are characterized by structural strength and by large open interiors which make them highly adaptable for new uses. This potential for reuse is fortunately joined by an awareness of the mill's value as a physical record of America's historical, technical, and social development.

Milling
In stonework, the processing of quarry blocks, through sawing, planning, turning, and cutting, to produce finished stone.

Mills, Robert (1781–1855)
Designed the Washington Monument, the tallest obelisk that epitomized the romantic Classical ideals.

Millwork
Wood products, such as cabinets, moldings, door and window frames, panels, and stair components that are manufactured by machines.

Minaret
The tall slender tower of a mosque with stairs leading up to one or more balconies from which followers are called to prayer.

Mineral fibers
Very fine insulation fibers made from glassy minerals that have been melted and spun. The fibers are hazardous to inhale.

Minoan architecture (1800–1300 B.C.)
A Bronze Age civilization that flourished in Crete, whose gate buildings with porches provided access to unfortified compounds. Foundation walls, piers and lintels were stone with upper walls framed in timber. Rubble wall masonry was faced with stucco and decorated with colorful wall frescoes. Ceilings were wood, as were the many columns with balloon capitals, and featured a distinct downward tapering shaft, as in the Palace of King Minos at Knossos in Crete.

Miralles, Enric Moya (1955–2000)

Spanish Catalan architect who in 1993 formed a new practice with his wife, the Italian architect Benedetta Tagliabue, under the name EMBT Architects. Their most important project was the Scottish Parliament Building in Edinburgh. The independent architectural language of Miralles can be difficult to classify in terms of contemporary architecture. It is influenced by Spanish architects, such as Alejandro de la Sota, José Antonio Coderch and Josep Maria Jujol, and also by international greats such as Le Corbusier, Louis Kahn, and Alvar Aalto, as well as the Russian Constructivist movement of the early 20th century.

Mirror

A surface capable of reflecting light without appreciable diffusion.

Mirror angle

In reference to a viewer and an observed surface, the angle equal and opposite to the viewing angle.

Mismatched

A bad fit at a joint, poor grain; or imperfect color matching of veneers.

Mission

A diplomatic office in a foreign country; a small church or monastic order.

Mission architecture

The church and monastery architecture of the Spanish religious orders in Mexico and California in the 18th century.

Mission parapet

A low freestanding wall at the edge of a roof, frequently curved; as found many in Spanish missions of the southwestern United States.

Mission style (1890–1920)

A characteristic of this style is its simplicity of form. Round arches supported by piers form openings in the thick stucco walls, with roof eaves that extend beyond the wall surface. Towers, curvilinear gables and small balconies were used on large buildings. The only ornamentation is a plain stringcourse that outlines arches, gables or balconies.

Mission tile See Tile

Mitchell, Ehrman (1924–)

American architect and partner with Romaldo Guirgola.

Mitchell, William (1945–2010)

An architect and urban theorist who founded and led the Smart Cities research group at the MIT Media Lab and oversaw a building campaign on the campus. The project included five buildings: Frank Gehry's Stata Center, Kevin Roche's Sports and Fitness Center, Steven Holl's Simmons Hall, Charles Correa's Brain and Cognitive Sciences Complex, and Fumihiko Maki's Media Lab Complex. His many books include *Reinventing the Automobile* (2010), with Christopher Borroni-Bird and Lawrence Burns; *World's Greatest Architect: Making, Meaning and Network Culture* (2008); and his seminal work, *Computer-Aided Architectural Design* (1977).

Miter

The line formed by the meeting of moldings or other surfaces that intersect each other at an angle; each member is cut at exactly half the angle of the junction

Miter arch See Arch

Miter box

A device used for guiding a handsaw at the proper angle for cutting a miter joint in a piece of wood. Can be made of wood, using precut slots on the sides to hold the saw, or metal, which has an adjustable carriage to change the angle of the miter.

Miter joint

A joint between two members at an angle to each other; each member is cut at an angle equal to half the angle of the junction, usually at right angles to each other.

Miter saw

A deep saw with a stiffening piece along the upper edge or back; often used in adjustable miter boxes.

Mixed air

The mixture of outdoor air and return air in an HVAC system. When filtered and conditioned, mixed air becomes supply air.

Mixed mode ventilation

A ventilation strategy that combines natural ventilation with mechanical ventilation, allowing a building to be ventilated either mechanically or naturally, and at times both.

Mixed use

A variety of authorized activities in an area or a building, as distinguished from the isolated uses and planned separation outlined by many zoning ordinances.

Mixed-use development

A large-scale real estate project that may incorporate old buildings and is characterized by three or more significant uses (retail, office, residential) with functional and physical integration of the project components, and development under a coherent plan.

Mixtec architecture (700–1000)
An architecture characterized by great mass, use of interior stone columns, and emphasis on horizontal lines, developed in Oaxaca, Mexico. The minutely detailed fretwork of the interior and exterior paneled friezes was produced by assembling thousands of small decorative elements, and setting them into clay. At Mitla there are freestanding buildings surrounding large courts oriented toward cardinal points of the compass.

Mizner, Addison (1872–1933)
Apprenticed to Willis Polk in San Francisco before moving to New York City to become a social architect like Stanford White. He began a commission in Palm Beach, FL, to design a hospital, which was later completed as the Everglades Club (1918). It was designed in a Mediterranean Revival style, which influenced many buildings in the area. He later became a developer, noted for his Spanish Revival buildings and Florida resort architecture. Mizner was the visionary behind the development of Boca Raton, but the Florida land boom ended, and his career never attained the potential.

Moat
A broad, deep trench, filled with water, surrounding the ramparts of a town or fortress.

Mobile
A type of sculpture made of movable parts that can be set in motion by the movement of air currents.

Mockbee, Samuel (1944–2001)
American architect who, along with D. K. Ruth, founded the Rural Studio program at Auburn University. In many cases, buildings designed and built by students incorporated novel materials that otherwise would have been considered waste. The buildings often consisted of a combination of vernacular architecture with modernist forms. In 1993, Mockbee was awarded a grant from the Graham Foundation for Advanced Studies in the Fine Arts to work toward the publication of his book, *The Nurturing of Culture in the Rural South; An Architectonic Documentary*. Mockbee was nominated posthumously for the American Institute of Architects (AIA) Gold Medal in 2003.

Mock-up See **Design drawing**

Model
A three-dimensional representation of a building or part of a building, executed at a small scale for the purposes of studying the massing or details of a proposed project.

Model energy code
A building code that requires houses to meet certain energy efficiency standards such as insulation levels or energy consumption. Like most building codes, it is adopted on either a state or local basis, if at all, and may be amended.

Modeling
The shaping of three-dimensional forms in a soft material, such as clay; also, the gradations of light and shade reflected from the surfaces of materials.

Modelscope
Device that attaches to the lens of a camera; allowing users to view a model from the level of a pedestrian.

Modern architecture
Building design in the currently fashionable architectural style; the term was originally used to describe a movement that combined functionalism with ideals that rejected historical design concepts and forms; it included styles such as Art Deco, International Style, Organic Architecture, and Prairie Style.

Modernism (1960–1975)

A term meaning "just now." The Modern Movement exemplified a conscious attempt to find an architecture tailored to modern life, and one that made use of new materials. It rejected the concept of applied style and the use of any ornament. It used concrete, steel and glass to help evolve an architecture directly related to construction methods. Exterior and interior forms were conceived and expressed as a single entity.

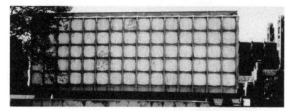

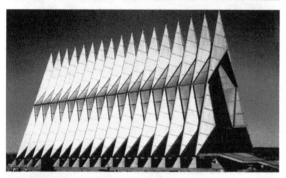

Modernistic (1920–1940)

Known through the years as Art Deco, Art Moderne, and Depression Modern, its ornamentation around doors and windows stressed the verticality of the skyscrapers for which the style was popular.

Modernistic style (1920–1940)

A style characterized by a mode of ornamentation combining rectilinear patterns and zigzags with geometrical curves. One of the distinctive forms consisted of polychrome low-relief frames. Ornamentation around doors and windows and on panels stresses the verticality in skyscraper designs. Stepped setbacks are also common, reflecting local urban zoning ordinances.

Modernization

Redesign of an existing structure to make it look new, or contemporary in style, as opposed to restoration.

Modifier

A method to influence color, transparency, surface normals, or light emission, depending on the type of modifier and the type of material. Modifiers can be combined, so that each modifier in the chain influences a specific property of the material.

Modillion

A horizontal bracket or console, in the form of a scroll with acanthus, supporting the corona under a cornice.

block modillion

A modillion in the form of a plain block.

Modillion cornice

A cornice supported by a series of modillions, often found in Composite and Corinthian orders.

Modular brick See Brick

Modular building

Technique that uses standardized components as a building practice. Modularization can occur as preconstructed components or predefined rules for construction, such as the spacing of structural elements. This technique is good for reducing waste and making the building process more efficient, saving time and money.

Modular construction

A type of construction in which the size of all building materials is based on a common unit of measure.

Modular grid

Reference grid in which the grid lines are spaced at exact multiples of the module.

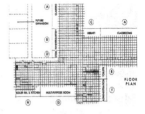

Modular system

A method of designing or constructing buildings and equipment in which standardized modules are widely used.

Modulation

To measure, to adjust to, or regulate by a certain proportion; to temper or to soften in passing from one element, form, or material to another.

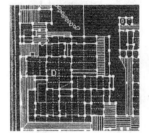

Modulor

A system of proportion developed by Le Corbusier in 1942. It was based on the theories of early civilizations and on the human form, and was related to the golden section.

Module

A simple ratio by which all parts of a building are related as part of an ordered system.

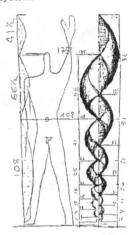

Mogul architecture

The later phase of Indian Islamic architecture, named after the Mogul dynasty, typified by monumental palaces and mosques and detailed decorative work. The Taj Mahal is the most famous example.

336

Moisture barrier

Any material, such as specially treated paper or plastic sheeting, which is impervious to water; used in walls and other areas to stop moisture from entering and thus prevent condensation. Also called a vapor barrier.

Mold

A mold is a fungus that typically grows in a filamentous cobweb-like mass under damp conditions and is capable of producing a staggering numbers of reproductive spores in as little as a few days. Molds are non-chlorophyll-containing entities, which require organic matter, living or dead, for survival. Mold is a critical participant in the "recycling" of dead organic material on the planet. Mold's relationship with humans range from the positive (e.g., food, antibiotics) to the negative (e.g., pathogens, antigens, toxins).

Molded brick See Brick

Molding

A decorative profile that is given to architectural members and subordinate parts of buildings, whether cavities or projections, such as cornices, bases, door and window jambs and heads.

backband molding

A piece of millwork used around a rectangular window or door casing to cover the gap between the casing and the wall, or used as a decorative feature.

band molding

A small, broad, flat molding, projecting slightly, of rectangular or slightly convex profile, used to decorate a surface, either as a continuous strip or formed into various shapes.

bar molding

A rabbeted molding applied to the edge of a bar or counter to serve as a nosing.

base molding

A molding used to trim the upper edge of an interior baseboard.

bead and reel molding

A classical molding consisting of alternate small, egg-shaped beads and semicircular disks set edgewise.

bead molding

A narrow wood drip molded on one edge against which a door or window closes, a stop bead; a strip of metal or wood used around the periphery of a pane of glass to secure it in a window frame.

beak molding

A molding ornamented with carved birds or fantastic animal-like heads or beaks.

bed molding

A molding or group of moldings that support the corona of a classical style entablature, often made up of a bottom ogee, a band, a quarter round, and a top band; a similar molding used as the bottom of a cornice.

beveled molding

Milled molding with an inclined plane surface.

billet molding

A common Norman or Romanesque molding formed by a series of circular cylinders, arranged alternately with notches in single or multiple rows.

blind stop molding

The molding used to stop an outside door or window shutter in the closed position.

bolection molding

A molding projecting beyond the surface of the work which it decorates, such as between a panel and the surrounding stiles and rails; often used to conceal a joint when the joining surfaces are at different levels.

cable molding

An ornamental molding formed like a cable showing twisted strands; the convex filling of the lower part of the flutes of classical columns.

calf's-tongue molding

A molding consisting of a series of pointed tongue-shaped elements all pointing in the same direction or toward a common center when around an arch.

cant molding

A square or rectangular molding having the outside face beveled.

cap molding

Trim at the top of a window or door; above the casing trim.

cavetto molding

A hollow member or round concave molding used in cornices and column bases, containing at least one quadrant of a circle in its profile.

chain molding

A molding carved with a representation of a chain.

corner bead

A vertical molding used to protect the external angle of two intersecting wall surfaces; a perforated metal strip used to strengthen and protect an external angle in plaster work or gypsum wallboard construction.

Molding

cove molding
A concave or canted interior corner molding, especially at the transition from the wall to a ceiling or floor.

cover molding
Any plain or molded wood strip covering a joint, as between sections of paneling, or covering a butt joint.

crenellated molding
A molding notched or indented to represent merlons and embrasures in fortifications.

crown molding
Any molding serving as a corona or otherwise forming the crowning or finishing member of a structure.

cyma molding
A molding that has a profile with a double curvature, or ogee.

cyma recta molding
A molding of double curvature that is concave at the outer edge and convex at the inner edge.

cyma reversa molding
A molding of double curvature that is convex at the outer edge and concave at the inner edge.

cymatium molding
The crowning molding of a Classical cornice, especially in the form of a cyma.

dovetail molding
A molding consisting of decorated fretwork in the form of dovetails.

drip molding
Any molding so formed and located as to act as a drip.

drop molding
A panel molding recessed below the surface of the surrounding stiles and rails.

egg-and-dart
An egg-shaped ornament alternating with a dart-like ornament used to enrich ovolo and echinus moldings and also on bands.

fillet molding
A molding consisting of a narrow flat band, often square in section; the term is loosely applied to almost any rectangular molding, usually used in conjunction with other moldings or ornaments.

flush bead molding
A molding whose surface is on the same plane as that of the wood member or assembly to which it is applied.

flush molding
An applied door or window molding that is flush with or below the surface of the rails and stiles.

guilloche
An ornament in the form of two or more bands that are twisted together in a continuous series, leaving circular openings that are filled with round ornaments.

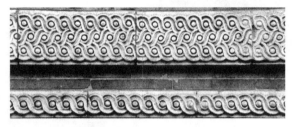

half-round molding
A convex strip or molding of semicircular profile.

hollow molding
A concave, often circular molding; a cavetto.

hollow square molding
A common molding consisting of a series of indented pyramidal shapes having a square base, found in Norman architecture.

hood mold
The projecting molding of the architrave over a door or window, whether inside or outside; also called a dripstone.

indented molding
A molding with the edge toothed or indented in triangular tooth-like shapes.

lattice molding
A wood molding, rectangular in section and broad in relation to its projection, resembling latticework.

medallion molding
A molding consisting of a series of medallions, found in the later examples of Norman architecture.

notched molding
An ornament produced by notching the edges of a band or fillet.

ovolo molding
A common convex molding consisting of a quarter circle in section.

pellet molding
Any small, round decorative projection; one of a series of small, flat disks or hemispherical projections.

quarter-round molding
A convex molding, with a projection that is exactly or nearly a quarter of a circle.

quirk bead molding
A molding containing a bead with a quirk on one side.

quirked molding
A molding characterized by a sudden and sharp return from its extreme projection or set off and made prominent by a quirk running parallel to it.

raised molding
A molding that extends above the adjoining surface, such as applied door moldings that overlaps and covers the joints between panels and the rails and stiles.

raking molding
Any molding adjusted at a slant, rake, or ramp.

reed molding
A small convex molding, usually one of several set close together to decorate a surface.

reticulated molding
A molding decorated with fillets interlaced to form a network or mesh-like appearance.

roll molding
Any convex rounded molding, which has a cylindrical or partially cylindrical form.

rover molding
Any member used as a molding that follows the line of a curve.

Scotia molding
A deep concave molding, especially one at the base of a column in Classical architecture.

scroll molding
An ornamental molding consisting of a spiral design or a terminal similar to the volutes of the Ionic capital or the S-curve on consoles.

square billet molding
A Norman molding consisting of a series of projecting cubes, with spaces between the cubes.

struck molding
A molding cut into rather than added to or planted onto another member.

sunk fillet molding
A molding slightly recessed behind the surface on which it is located; a fillet formed by a groove in a plane surface.

tresse molding
Flat or convex bandelets that are intertwined, especially such interlacing ornamentation used to adorn moldings.

Venetian molding
A molding with repetitive individual projecting elements similar to dentils.

wave molding
A molding decorated with a series of stylized representations of breaking waves.

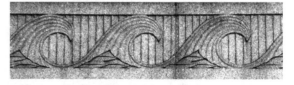

Moneo, José Rafael (1937–)
Spanish architect, who won the Pritzker Prize for architecture in 1996. Works include the Cathedral of Our Lady of the Angels in Los Angeles, CA; the Davis Art Museum at Wellesley College, Wellesley, MA; and the Audrey Jones Beck Building, an expansion of the Museum of Fine Arts, Houston, TX. Moneo also designed the Chace Center for the Rhode Island School of Design, Providence, RI.

Monitor
A superstructure that straddles the ridge of a roof or crowns a dome: often glazed to provide light or louvered to provide ventilation

Monochromatic
Consisting of only one color.

Monolith
An architectural member such as an obelisk or the shaft of a column consisting of a single stone.

Monolithic
Shapes usually formed of a single block of stone, or cast in one piece without construction joints; they are massive and uniform.

Monolithic column See Column

Monopteral
Describes a temple the roof of which is supported by columns, but without walls.

Monostyle
Having the same style of architecture throughout the structure; a single shaft applied to medieval pillars.

Montage
A composition made by overlapping parts or objects.

Monterey Style
An architectural style found in the southwest United States, especially California, that combines Spanish Colonial and Greek Revival style features; includes adobe or stucco stone walls, pitched roofs with shingles or clay tiles, Greek Revival wood trim at doors and windows, and cantilevered second-story porches.

Monument
A stone, pillar, megalith, structure or building erected in memory of the dead, an event, or an action.

Monumental scale See Scale

Moon gate See Gate

Moore, Arthur Cotton (1935–)
Architect in Washington, DC, known for a style called "industrial baroque" and probably best known for the Washington Harbour development on the Potomac River in Georgetown, the newer building of the Phillips Collection in Washington, and the renovation of the Jefferson Building of the Library of Congress. He also led the successful restoration of Washington's tallest residential building, the Cairo Hotel. He wrote *The Powers of Preservation: New Life for Urban Historic Places.*

Moore, Charles (1925–1993)
A leading figure of postmodernism, who was also notable as a university teacher and writer. Architectural history played a part in his designs, which are tempered with fancy, myth, and evocative motifs. These are typified in the Piazza d'Italia (illus.), New Orleans, LA (1975). Other work includes houses at Sea Ranch, CA (1970), and the Beverly Hills Civic Center (illus.), Los Angeles, CA (1985).

Moorish arch See Arch

Moorish architecture (500–900)

Prevalent in Spain and Morocco, the style was influenced by Mesopotamian brick and stucco techniques with frequent use of the horseshoe arch, along with Roman marble columns and limestone carved capitals. Vaults developed into highly complex ornate forms. Brick was used decoratively and structurally in combination with marble, with extensive use of stucco to build up the richly molded surfaces, painted with bright colors and sometimes gilded.

Moorish capital See Capital

Moorish revival

A revival style using horseshoe arches and multifoil window tracery.

Moretti, Luigi (1907–1973)

Italian architect whose early work was in the Neoclassical style of Rationalist architecture. His most notable later work is the Watergate complex (illus.), Washington, DC (1960).

Morgan, Julia (1872–1957)

American architect/engineer; the first woman to study at the École des Beaux-Arts, Paris; and California's first licensed woman architect. Her work includes several buildings for Mills College, Oakland, CA, including the reinforced-concrete campanile (1903), the library, and gymnasium (both in 1907). She designed the buildings at San Simeon, CA (1919), for William Randolph Hearst.

Moriyama, Raymond (1929–)

Japanese-Canadian architect who designed several buildings at Brock University, St. Catharines, Ontario, from the 1970s through the latest campus expansion. Raymond Moriyama is a founding partner of the Toronto firm, Moriyama & Teshima Architects.

Morris, Ailyn (1922–2009)

A Southern California modernist architect. His many residential projects are products of a unique vision. His first residential design was 1956's concrete, glass, and steel Brubeck House in Eagle Rock, CA, whose multileveled interlocking spaces created a dynamic effect.

Morris, William (1834–1896)

Although not trained as an architect, Morris had a profound influence on modern and contemporary art as a designer of wall coverings, stained glass, carpets, and tapestries. He was also a painter, poet, political publisher, typeface designer, and furniture maker. He is considered a pioneer of the Arts and Crafts movement.

Mortar joint

A tooled joint between the units in a brick or masonry wall.

beaded joint

Recessed mortar joint in the form of a quirked bead; a joint with a raised bead in the center that projects past the surface of the brick or stone.

bed joint

The horizontal joint between two masonry courses.

concave joint

A recessed masonry joint formed in mortar by the use of a curved steel jointing tool; because of its curved shape, it is very effective in resisting moisture.

Mortar joint

flush joint
A masonry joint finished flush with the surface.

ground Joint
A closely fitted joint in masonry, usually without mortar; also a machined metal joint that fits tightly without packing or employing a gasket.

head joint
A vertical joint between two masonry units that is perpendicular to the face of the wall.

raked joint
A joint made by removing the surface of mortar with a square-edged tool while it is still soft; produces marked shadows and tends to darken the overall appearance of a wall; not a weather-tight joint.

rustic joint
In stone masonry, a deeply sunk mortar joint that has been emphasized by having the edges of adjacent stones chamfered or recessed below the surface of the stone facing.

struck joint
A masonry joint from which excess mortar has been removed by a stroke of the trowel, leaving a flush joint; a weather-struck joint.

tooled joint
Any masonry joint that has been prepared with a tool before the mortar in the joint has set rigidly.

troweled joint
A mortar joint finished by striking off excess mortar with a trowel.

v-shaped joint
A horizontal V-shaped mortar joint made with a steel jointing tool; very effective in resisting the penetration of rain.

weather-struck joint
A horizontal masonry joint in which the mortar is sloped outward from the upper edge of the lower brick, so as to shed water readily; formed by pressing the mortar inward at the upper edge of the joint.

Mortise
A rectangular slot cut into one piece of timber, into which a tenon or tongue from another piece is fitted to form a joint.

Mortise and tenon joint See Joint

Mosaic
A process of inlaying small pieces of stone, tile, glass or enamel into a cement or plaster matrix, making a pattern, design, or representational picture.

Mosaic

Florentine mosaic
A kind of mosaic made with precious and semiprecious stones, inlaid in a field of white or black marble or similar material, generally displaying elaborate floor patterns.

Venetian mosaic
A type of terrazzo topping, containing primarily large chips, with smaller chips filling in between.

Mosque
A Muslim house of worship.

342

Moss, Eric Owen (1934 -)
Deconstructivist architect whose projects include the Petal House, Los Angeles (1984), Hayden Tower, Culver City (1992), and The Box, Culver City(1995) all in California.

Motif
A part or element repeated in an ornamental design.

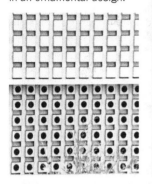

Motion sensors
Installing motion sensors for lights, particularly those installed outdoors, improves energy consumption because lights are only on when they are needed—and off when they are not.

Mouchette
Gothic tracery and derivatives, a typical small motif; pointed, elongated, and bounded by elliptical and ogee curves; a dagger motif with a curved axis.

Mouse-tooth gable See **Gable**

Mozarabic architecture (800–1400)
A northern Spanish style built by Christian refugees from Moorish domination, characterized by the horseshoe arch and retaining all other Moorish features.

Mudejar architecture (1200–1300)
A Spanish style created by the Moors while under Christian domination, characterized by a fusion of Romanesque and Gothic styles but retaining some Islamic elements, such as the horseshoe arch.

Mulch
A layer of material such as wood chips, straw, and/or leaves, placed around plants to hold moisture, prevent weed growth, and enrich or sterilize the soil.

Mulgardt, Louis Christian (1866–1942)
American architect of German descent; influenced by the Arts and Crafts movement. In the 1920s, he produced a series of fantastic proposals for San Francisco, including habitable piers and bridges connected by 24-lane tiered motorways. None were ever realized.

Mullet, Alfred B. (1834–1890)
Supervised architects in the federal program and designed many new public buildings. He was the architect of the State, War and Navy Building, now the Old Executive Office

Mullet, Alfred B.

Building, adjacent to the White House, and the St. Louis Post Office and Customs House, now since adapted. He also designed offices for the Baltimore Sun (1856), Baltimore, MD, in the Beaux-Arts Renaissance manner.

Mullion
A dividing piece between the lights of windows, usually taking on the characteristics of the style of the building.

Multicentered arch See **Arch**

Multicurved gable See **Gable**

Multifoil See **Foil**

Multifunctional
Something that serves more than one purpose. In product and furniture design, multifunctional pieces reduce the need for multiple products, thus using fewer raw resources and reducing clutter in modern homes and waste in landfills.

Multipane window
A window with more than one pane of glass. Dual-pane windows are fairly common, and triple-pane windows can sometimes be found in cold climates.

Multiple dwelling
A building for residential use that houses several separate family units.

Multiple-use development
The use of a piece of land for different purposes simultaneously, such as the use of riverside land for water filtration by plants, recreation, flood buffers, and wildlife habitat.

Multistory frame
A building framework of more than one story, in which loads are carried to the ground by a system of beams and columns.

Munsell book of color
One of the standard methods for specifying colors by means of an atlas of color chips, arranged according to the hue, value and chrome; developed (1905) by A. H. Munsell and now produced by the Munsell Color Company, Baltimore, MD.

Muntin
A secondary framing member to hold panes in a window, window wall, or glazed door; an intermediate vertical member that divides panels of a door. Each pane of glass or each insulated glazing unit separated by a muntin is a light.

Muqarnas
An original Islamic design involving various combinations of three-dimensional shapes featuring elaborate corbeling.

Mural
A wall painting; fresco is a type of mural technique.

Murcutt, Glenn Marcus (1936–)
British-born Australian architect and winner of the 2002 Pritzker Prize and 2009 AIA Gold Medal. Murcutt's early work experience was with various architects, such as Neville Gruzman, Ken Woolley, and Bryce Mortlock, which exposed him to their style of organic architecture, which focused on relationships to nature. Murcutt's motto is, "Touch the earth lightly." Murcutt considers environmental elements such as wind direction, water movement, temperature, and the light surrounding sites before he designs the building itself.

Murphy, Charles F. (1890-1985)
Worked with Daniel Burnham and the Chicago school. He produced several Chicago office buildings, and with partner Helmut Jahn designed O-Hare International Airport (1965) and the Exhibition building at McCormick Place (1971).

Museum
A public or private nonprofit institution organized on a permanent basis for essentially educational or aesthetic purposes that owns or uses tangible objects and works of art, cares for them, and exhibits them to the public on a regular basis through its own or other facilities.

Museum village
A site in which several or many structures have been restored, rebuilt, or moved, and whose purpose is to interpret a historical or cultural setting, often within the context of daily trades and activities of a past time. Often called an outdoor museum.

Mushroom column See **Column**

Muslim architecture (600–1500)
In this style a new domed mosque was developed from the Christian basilica. There were many variations of the basic elements such as arches, domes, cross ribs, and crenellations. Surfaces are covered with an abundant geometric, floral and calligraphic decoration, executed in stone, brick, stucco, wood and glazed tile.

Mutule
A sloping flat block on the soffit of the Doric cornice; usually decorated on the underside with rows of six guttae each, which occurs over each triglyph and metope of the frieze.

Mycenaean architecture (1600–1200 B.C.)
The earliest phase was exemplified by masonry sidewalls and a timber roof. Monumental beehive-like tombs were constructed of superimposed layers of corbeled stones to create a parabolic vault. Stone-faced, inclined access passages led to the entrance, which had sloping jambs; overhead, a stone lintel which was supported by a characteristic triangular sculptured panel.

N n

NAHB Green Building Program
The program includes the NAHB Model Green Home Building Guidelines, the National Green Building Conference,
the NAHB Green website, the Green Building Program
Hotline, the National Green Building Program Awards, the
Certified Green Professional Designation, and the green
building training and education that support the designation and guidelines. NAHB also offers a Certified Green
Professional (CGP) designation, designed to teach building
industry professionals strategies for incorporating green
building principles into homes, including an understanding
that demonstrates enhanced environmental impact and
increased performance and health benefits.

NAHB Model Green Home Building Guidelines
These guidelines were finalized in 2004–2005 and are a
toolkit for home builder associations to create new programs and to ensure that those programs expand and
flourish. Builders can sign up on the NAHB Green website
to use their web-based scoring tool to assess their
project(s) according to the Guidelines.

Nail
A slender piece of metal pointed at one end for driving into
wood and flat at the other end for striking with a hammer;
used as a board fastener. The size of the nail is indicated
by the term "penny" and the letter "d," which refers to the
length of the nail ranging 2d (1 inch) to 60d (6 inches).

Nailer
A block of wood installed within construction to provide a
means of attaching other pieces of wood or materials.

Naos
Inner sanctuary of a Greek temple.

Nara (710–794)
A period in Japanese history characterized by the adoption
of Chinese culture and form of government, named after
the first permanent capital and the chief Buddhist center
in ancient Japan.

Narthex
An arcaded porch or entrance hall to an early Christian
basilican church.

Nash, John (1753–1835)

Planned Regent Park and
Regent Street, London, as
a picturesque scheme. He
also designed the Brighton
Pavilion (1815) for the
Prince of Wales in a mix-
ture of Indian, Chinese, and
Gothic styles.

National ambient air quality standards
The standards established by the EPA (Environmental Protection Agency) that apply to outdoor air throughout the
country.

National Archives and Records Administration
Administers the National Archives and its collections, and
provides grants for the publication of historical papers and
preservation projects.

National Association of Home Builders (NAHB)
A third-party verified program for single-family construction, remodeling, and multifamily construction. Divided
into two parts, the Guidelines cover seven areas, or guiding principles: Lot Design; Resource Efficiency; Energy
Efficiency; Water Efficiency, Indoor Environmental Quality;
Homeowner Education; and Global Impact. There are three
levels of green building available to builders to rate their
projects—Bronze, Silver, and Gold.

National Center for Healthy Housing
Sponsors research on methods to reduce residential
environmental hazards and to scientifically assess risks.

National Center for Preservation Technology and Training (NCPTT)
A National Park Service center whose mission is to promote technology and training for historic preservation professionals and conservators, and training for information
management in architecture, archeology, and landscape
architecture.

National Electric Code
A nationally accepted guide to the safe installation of wiring and equipment, unrelated to design specifications, but
for safeguarding persons and buildings and contents from
the hazards arising from the use of electricity for heat,
light, and power.

National Environmental Policy Act of 1969
The act set up the Council on Environmental Quality to
assure consideration for environmental factors, including
historic preservation, in federal project planning. It included preservation among the considerations requiring
environmental impact statements for all major federal
actions affecting the quality of human environment, such
as urban quality, historic and cultural resources, environmental sensitivity, and the responsible design of the built
environment.

National Fenestration Rating Council (NFRC)
National organization that sets standards for windows and
doors.

National Foundation on the Arts and Humanities
Provides grants for design, historical and public programs
and museum operations.

National Green Building Standard (NGBS)
The NGBS is based on the NAHB Model Green Home
Building Guidelines and passed through ANSI. This standard can be applied to new homes, remodeling projects,
and additions. An overview of the points structure: Bronze:
222 points minimum, Silver: 406 points minimum, Gold:
558 points minimum, Emerald: 697 points minimum, and
total points available: 1300–1500. Number of mandatory
requirements (varying based on project specifics and level
of rating): 40.

NGBS - remodel
The NGBS is designed to work with remodeling projects in the following ways: 1. If the remodeling project involves 75 percent or more of the conditioned floor area, than this is considered a gut rehab and follows the program structure for new homes. 2. If the remodeling project is less than 75 percent of the home's conditioned floor area and the home was built prior to 1980, then the remodeling project can follow the "Remodel Path," meaning a measured percentage of improvements in energy and water efficiency and compliance with a list of 28 mandatory practices. 3. If the remodeling project is less than 75 percent of the home's conditioned floor area and the home was built during or after 1980, then the project must follow the "Green Building Path," which uses the new home point system with notes for line items that differ for remodeling projects.

National Historic Landmark
A designated district, site, building, structure or object of exceptional significance to the country as a whole, rather than just to a particular state or locality; each is listed in the National Register of Historic Places.

National Historic Landmarks Program
Established by the National Sites Act of 1935, surveys sites of national significance based on a series of theme studies, from prehistoric archeology, architecture, politics, religion and science. The effort is to promote a set of general concepts on what is worth saving.

National Historic Preservation Act of 1966
Expanded the National Register of Historic Places to make it a nationwide inventory of districts, sites, structures, and objects of state and local as well as national importance, maintained by the National Park Service, U.S. Department of the Interior. The act also created a review agency, the Advisory Council on Historic Preservation, made up of federal officials and private citizens to advise the president and Congress on historic preservation matters for highway or utility construction, if they are likely to have any effect on historic structures.

National Historic Sites Act of 1935
An act which authorized the Secretary of the Interior to acquire national historic sites and designate National Historic Landmarks.

National Institute for the Conservation of Cultural Properties (NIC)
An organization that promotes the conservation and preservation of U.S. heritage, including works of art, anthropological artifacts, documents, historic objects, architecture and natural science specimens.

National Institute of Occupational Safety and Health (NIOSH)
An agency of the Centers for Disease Control and Prevention (CDC) of the U.S. Department of Health and Human Services. NIOSH is the research arm of OSHA (Occupational Safety and Health Administration).

National Institute of Structural Technologies (NIST)
A U.S. federal agency that performs scientific research on building materials and systems.

National Park
A federally owned area administered by the National Park Service, often extensive in size, individually designated by Congress for purposes of recreation, culture, or scenic or historic preservation, such as Yosemite National Park, Yellowstone, and the Grand Canyon.

National Park Service
Conducts research and salvage programs in archeological areas as well as historic site surveys for the National Historic Landmarks Program.

National Preservation Week
In 1973 the National Trust succeeded in making preservation official for at least one week held annually in early May. The activities include tours and seminars, exhibits, film showings and children's programs, and ceremonial start and finishes to preservation projects.

National Register of Historic Places
A list, maintained by the National Park Service, of U.S. places of significance in history, archeology, architecture, engineering and culture, on a national, state or local level.

National Trust for Historic Preservation
A nonprofit organization that is the leader of the national preservation movement, committed to saving American diverse historic environments, and to preserving and revitalizing communities nationwide. It has seven regional offices, owns 18 historic sites and works with thousands of community groups in all 50 states.

Native plants
Plants that grow locally in the wild are preferred for use in green landscaping. These plants are adapted to the local conditions and will have less need for soil amendments, watering, and fertilizing and are the best to adapt to local conditions.

Native vegetation
A plant whose presence and survival in a specific region is not due to human intervention.

Natural cooling
Use of environmental phenomena to cool buildings, such as natural ventilation and evaporative cooling.

Natural environment
Includes all natural land forms, rivers and lakes, trees and plants, but does not include the built environment.

Natural forms
Those forms that include artificial foliage as well as derivations of the acanthus leaf, flowers and fruit festoons; also animal forms, such as the lion and eagle, and human forms, such as heads and figures.

Natural forms

Natural latex
Natural latex is produced from the milky white sap of the rubber tree. No harmful chemicals are used in its production; it is biodegradable and has a life expectancy of 20 years or more. It has a superior feel to polyurethane foam and provides unsurpassed resiliency and comfort.

Natural materials
A product that is made from materials and ingredients found in nature, with little or no human intervention. Natural materials include stone, glass, lime or mud plasters, adobe or rammed earth, bricks, tiles, untreated wood, cork, paper, reeds, bamboo, canes and grasses as well as all natural fibers. From both an aesthetic and health point of view, building with natural materials also helps *sustainable* development.

Natural siting criteria
An approach where the locations of roads, buildings, and other structures are selected to be where the geological and biological factors are most favorable; essentially working with nature when selecting locations for human-made structures.

Natural stone See **Stone**

Natural ventilation
A method for reducing energy use and cost and for providing acceptable indoor environmental quality rather than using mechanical ventilation. Natural ventilation systems rely on pressure differences to move fresh air through buildings. Openings between rooms such as transom windows, louvers, grills, or open plans are techniques to complete the airflow circuit through a building. Strategies include placement and operability of windows and doors, thermal chimneys, operable skylights, and exterior landscaped berms to direct airflow on a site.

Naturescaping
A landscaping method that uses native plants to conserve and create natural habitats that provide nurturing environments for wildlife.

Nave
The principal or central part of a church; by extension, both middle and side aisles of a church, from the entrance to the crossing of the chancel; that part of the church intended for the general public.

Neat cement
A cement used without sand, as opposed to cement mortar.

Neck
In the Classical orders, the space between the bottom of the capital and the top of the shaft, usually marked by a sinkage, or a ring of moldings.

Needle beam
A steel beam used to support an existing structure while it is being repaired, or to provide support when moving a structure, or when removing a portion of the wall below the beam; usually installed by removing a section of the wall or area, and inserting it into position.

Negotiated contract
An agreement between a design professional and owner, or between a contractor and owner, reaching a mutually agreed upon price based on the scope of professional services, or extent of construction work.

Neighborhood
A relatively small area in a town with sufficient binding character, such as architectural or social unity or clear boundaries, so that it is recognized as an entity by residents and outsiders alike. Usually an area in which all parts are within easy walking distance from one another, and having a diverse range of building types, thoroughfares, and public open spaces accommodating a variety of human activity.

Neoclassical
Refers to a rebirth of classicism in the architecture of Europe and America during the late 18th and 19th centuries. Characterized by the widespread use of Greek and Roman architectural orders and decorative motifs, strong geometric compositions, and shallow relief in ornamental detail.

Neo-Classicism (1900–1920)
A revival stylebased primarily on Greek and to a lesser extent on Roman orders, producing symmetrically arranged buildings of monumental proportions. Colossal pedimented porticos were flanked by a series of pilasters. The arch was not used, and enriched moldings are rare. The preference was for simple geometric forms and smooth surfaces. The design was based on the assembly of separate volumes, each dedicated to a single function.

Neo-Expressionism (1964–1975)

Structures which express continuity of form by sweeping curves characterize this style These structures were primarily the result of using reinforced concrete to create smooth shapes and seamless soaring forms.

Neo-Formalism (1964–1970)

A style which combines the Classical symmetrical forms and smooth wall surfaces with arches of precast concrete and decorative metal grilles, very often delicate in appearance, typical of the work of Minuro Yamasaki.

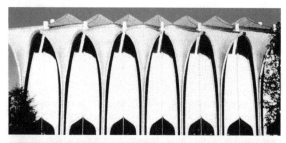

Neolithic era (9000–8000 B.C.)

The last phase of the Stone Age characterized by the cultivation of crops and the use of technically advanced stone implements.

Neon lamp

Consists of a glass tube from which the air is extracted and neon gas added. The tube glows when current flows through it.

Nervi, Pier Luigi (1891–1979)

Italian civil engineer; he was best known for reinforced-concrete structures. He created a structure for the Exhibition Hall, Turin, Italy (1947), and the Palazzo Dello Sports (1958), in Rome; where an immense concrete dome floats over the space.

Net metering

A method of crediting customers for electricity that they generate onsite in excess of their own electricity consumption. If customers generate more than they use in a billing period, their electric meter turns backwards to indicate their net excess generation. Depending on individual state or utility rules, the net excess generation may be credited to their account, or carried over to a future billing period.

Net positive impact

Refers to a project outcome where whole-of-life considerations of the project result in an increase in natural and social capital, in addition to favorable economic returns to all concerned in the inception, delivery, operation, and disposal of the facility.

Net vault See Vault

Netsch, Walter A. (1920–2008)

Architect Netsch spent nearly his entire career at Skidmore, Owings & Merrill, where he led the design team for the Air Force Academy Chapel (illus.), in Colorado Springs, CO, completed in 1963. He was also responsible, with SOM's Bruce Graham, for the Inland Steel Building in Chicago, recognized as one of the mid-20th-century's best modernist buildings. He designed 15 libraries and other academic buildings for Northwestern University and buildings for Grinnell College, Miami University of Ohio, the Illinois Institute of Technology, and the University of Chicago, among others. He also designed the east wing of the Art Institute of Chicago.

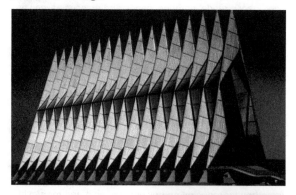

Network

Any set of interconnected elements that form an overall organization; also a diagram representing a series of interconnected events, as in the representation of the critical tasks in a building project.

348

Net-zero energy
Producing as much energy on an annual basis as one consumes onsite, usually with renewable energy sources, such as photovoltaics or small-scale wind turbines.

Neumann, Balthasar (1687–1753)
German Rococo architect whose work exhibits swirling curves, fluid spaces, and rich but delicate decoration. Work includes the Bishop's Palace (Residenz), Wurzburg, Germany.

Neutra, Richard (1892–1970)
Vienna-born architect working with Adolf Loos and Eric Mendelsohn. He met Louis Sullivan and Frank Lloyd Wright, and in 1925 formed an association in Los Angeles with Rudolph Schindler. He designed many homes for Hollywood notables.

New Brutalism (1953–1965)
This style was representative of buildings which expressed materials, structure and utilities honestly, in the tradition of Le Corbusier's *béton brut*; it featured rough, honest brickwork and exposed concrete imprinted with the grain of the wooden forms.

New Classicism (1982–)
A final phase of Postmodernism that led to a new form of Classicism, a freestyle version of the traditional language. It combines two purist styles—Classicism and Modernism—and adds new forms based on new technologies and social usage. Previous rules of composition are not disregarded but rather extended and distorted. Among those architects identified with this style are James Stirling, Robert Venturi, Michael Graves, Charles Moore and Arata Isozaki.

New Classicism

New England Colonial style (1600–1700)
A local style characterized by a use of natural materials in a straightforward manner. The box-like appearance is relieved by a prominent chimney and sparse distribution of small, casement-type windows. The characteristic shape, formed by extending the rear roof to a lower level than the front roof, was called a saltbox. In larger structures the upper floor projected beyond the lower floor, creating an overhang called a jetty.

New Urbanism

An approach to designing cities, towns, and neighborhoods that tries to reduce traffic and eliminate sprawl. Although the term *New Urbanism* emerged during the late 1980s and early 1990s, its principles are much older.

Newel

The central post or column which provides support for the inner edges of the steps in a circular staircase and around which the steps wind.

Newel cap

The terminal feature of a newel post, often molded or turned in a decorative manner.

Newel post

A tall post at the head or foot of a stair supporting the handrail, often ornamental.

Niche

A recess in a wall; usually semicircular at the back, terminating in a half-dome, or with small pediments supported on consoles, often used as a place for a statue.

angle niche

A niche formed at a corner of a building; common in medieval architecture.

Niemeyer, Oscar (1907–)

Brazilian architect who designed the Ministry of Health and Education, Rio de Janeiro (1945), a Modern-style tower block modified by louvers, devised by Le Corbusier, the consultant architect. Niemeyer was interested in buildings as sculpture. Brasilia, Brazil (1960), was planned in the shape of a bird, with the Parliamentary Building at its head. The influence of Le Corbusier is evident.

Night flushing

The process of removing hot air from a building during the cool evening hours, to cool elements with thermal mass within the building and flush stale air.

Nighttime ventilation

A strategy of flushing building structures with cool nighttime air to minimize the next day's cooling load.

Nishizawa, Ryue (1966–)

Japanese architect who works collaboratively with Kazuyo Sejima and is praised for designing powerful, minimalist buildings using common everyday materials. Nishizawa and Sejima shared the 2010 Pritzker Architecture Prize.

Nogging

Brick or miscellaneous masonry material used to fill the spaces between the wooden supports in a half-timber frame.

Noise pollution

Noise caused by traffic, car alarms, boom box radios, aircraft, industry or other human activity.

Noise reduction

Reducing the level of unwanted sound by any of several means of acoustical treatment.

Nominal size

The measurement used in naming a component, not necessarily its actual size when finished or milled. The size of timber is usually given in nominal size, and the actual size is slightly smaller, for example, a 2x4 actually measures 1 5/8 x 3 5/8 inches.

Nomland Jr., Kemper (1919-2009)
Southern California architect. Designed Case Study House No. 10 with his father as part of a program sponsored by *Arts & Architecture* magazine that promoted the design of modernist postwar housing. The house was constructed in 1947 in Pasadena, CA.

Noncombustible
In building construction, a material that will not ignite, burn, support combustion, or release flammable vapors when subjected to fire or heat.

Nonconforming
Said of any building or structure that does not comply with the requirements set forth in applicable code, rules, or regulations.

Nonconforming building
A building that does not meet the current requirements of a regulation in the building code or restrictions in the zoning code.

Nonconforming use
A building or use that is inconsistent with an area's zoning regulations. They may be "grandfathered" in subsequent zoning changes, but upon conversion to a new or adaptive use, will be required to adhere to the applicable regulations, unless a variance is granted.

Noncontributing resource
A building, site, structure, or object that does not add to the historic significance of a property or district.

Nondestructive investigation
Examination of the existing conditions of a structure without damaging or destroying it in the process. By using careful means of testing under controlled and specific loading conditions, stresses can be determined before the loads have any effect on the integrity of the structure.

Nondestructive testing
Testing that does not destroy the object being tested. Techniques include the use of strain gauges, x-rays, and ultrasound.

Nonhabitable area
The area of a building that cannot be utilized; includes the structure, partitions and ducts.

Non-load-bearing wall See **Wall**

Nonnative
A plant that was directly or indirectly introduced to a given region by humans. Nonnative plants were not present in the region beforehand and would not have spread into the area without human interference.

Nonpotable water
Not fit for consumption without treatment that meets or exceeds EPA (U.S. Environmental Protection Agency) drinking water standards. Graywater and rainwater are nonpotable waters that can be used in toilets or for washing cars, and the use of either reduces the demand for potable water, conserving this vital resource.

Nonrenewable resources
Those in finite supply that cannot be regenerated or renewed by synthesizing the energy of the sun. Such resources include fossil fuels, metals, and plastics.

Non-water-using urinal
A urinal that replaces the water flush with a specially designed trap that contains a layer of buoyant liquid that floats above the urine layer, blocking sewer gas and urine odors from the room.

Norman architecture (1066-1180)
A Romanesque form of architecture that predominated in England from the Norman Conquest to the rise of the Gothic style. It was plain and massive, with moldings confined to small features; archways were plain and capitals devoid of ornament. As the style advanced, greater enrichment was introduced, and later examples exhibit a profusion of ornament. Windows resemble small doors without mullions. Pillars were slender and channeled.

Norman cottage
A large asymmetrical house in the style of the farmhouses of Normandy, built in the early 1930s; typical elements include a round tower with a conical roof, steeply pitched roof with dormers, mixed brick, stone, and stucco walls, multipaned casement windows; often employs half-timbering.

Northeast Sustainable Energy Association (NESEA)
A regional membership organization promoting sustainable energy solutions. NESEA is committed to advancing three core elements: sustainable solutions, proven results, and cutting-edge development in the field. States included in this region extend from Maine to Maryland.

Nosing
The rounded edge of a horizontal surface that projects beyond the vertical surface below, such as the projection of a tread beyond the riser.

Notched molding See **Molding**

Notching
In carpentry, a method of joining timbers by cutting notches at the ends of a piece, then overlapping the notched pieces to form a joint.

Notman, John (1810-1865)
American architect in Philadelphia; helped introduce the Italianate style; designed the Athenaeum (1847), Philadelphia, PA and Prospect Villa (1852), Princeton, NJ.

Notter, George M. (1933-2007)
A pioneering preservationist, his many projects included Old Boston City Hall, Boston, MA and Mechanics Hall in Worcester, MA. In Washington, he worked on the renovation and expansion study for the Russell Senate Office Building and the master plan for the International Cultural and Trade Center. An early advocate of sustainability, Notter chaired the Designing for Energy conference in Dallas, TX helping place the organization at the forefront of the energy conservation movement.

Nouvel, Jean (1945–)

French architect whose most well known work is the Institut du Monde Arabe, Paris, France (1987).

Nowicki, Matthew (1910–1951)

Polish architect who designed the Norton Arena, North Carolina State Fair, Raleigh, NC (1948), with two intersecting hyperbolic parabolas. He is regarded as a pioneer of such structural design. He also worked with Saarinen on the master plan for Chandigarh, India.

Noxious

A term for species or groups of species that have been legally designated as pests, for example, by a county, state, or federal agency.

Noyes, Eliot (1910–1977)

American architect and industrial designer; he worked with Gropius and Breuer, as well as IBM and Westinghouse. Work includes the Bubble houses (1953), created by spraying concrete on a large balloon; United States Pavilion, Expo 67, Montreal, Canada; and the IBM Management Development Center, Armonk, NY (1980).

Nursing home

A building used for the lodging, boarding, and nursing care for patients of mental or physical incapacity who require care and related medical services less intense than those given in a hospital.

Nylon See Plastic

Nympheum

A Classical building or room with a fountain, statues, and plants; a place for relaxation.

O o

Oak See Wood

Obata, Gyo (1923–)
Partner in the firm of Hellmuth Obata and Kassabaum, noted for the Chapel Priory of St. Mary, St. Louis, MO (1962); the National Air and Space Museum, Washington, DC (1976); the renovation of Union Station, St. Louis, MO (1985), and the Dallas/Ft. Worth Airport, TX (1973).

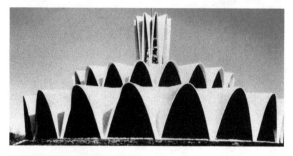

Obelisk
A four-sided stone shaft, either monolithic or jointed, tapering to a pyramidal top.

Oblique projection See Projection drawing

Oblique section See Projection drawing

Oblong
A right-angle plane figure with unequal pairs of sides; it can approach the dimensions of a square, on the one hand, or stretch out to express a band, on the other.

Obscure glass See Glass

Observatory
A structure in which astronomical observations are carried out; a place such as an upper room that affords a wide view, a lookout.

Obsidian See Stone

Obsolescence
A loss in value due to a decrease in the usefulness of property caused by decay, changes in technology, people's behavioral patterns and tastes, or environmental changes. Items or buildings that become out of date or practice and fall into disuse; also impairment of a building resulting from a change in the design or from external influences which tend to make the property less desirable for continued use.

Obsolete building
A building that for one reason or another has reached the end of its current useful life.

Obtuse angle arch See Arch

Occupancy
Acquiring title to a property by taking possession of it; the act of taking over anything which has no owner.

Occupancy sensing
Motion sensors that control lights according to space occupancy. Passive infrared sensors are the most common. Ultrasonic sensors work better in places with partitions. Sensors have sensitivity and time-out adjustments to fine-tune their use. Wall box sensors with internal switches or dimmers are used for small rooms. Multiple sensors are needed for larger spaces. Energy credit for their use are substantial, and the payback is short for retrofit applications.

Occupational Safety and Health Administration (OSHA)
A branch of the U.S. Department of Labor responsible for establishing and enforcing safety and health standards in the workplace.

Occupied zone
The volume of a conditioned space containing the occupants of the space, typically considered extending from floor level up to a height of 6 feet.

Octagon roof
A roof which fits over an octagonal-shaped structure.

Octagon style (1850–1870)
An architectural style characterized by an eight-sided shape, two or three stories in height, an encircling veranda, a cupola or belvedere, and minimum exterior detailing.

Octagonal
Refers to plane geometric figures containing eight equal sides and eight equal angles.

Octagonal

Octagonal house
An eight-sided house, usually
two to four stories high.

Octahedral
Forms that exhibit the characteristics of a regular polygon
having eight sides.

Octastyle
A portico having eight
columns in front.

Oculus
A roundel or bull's eye window opening, or an opening at
the crown of a dome.

Oculus

Off center
Applies to a structural member which is not properly
centered.

Off the grid
A system that runs on renewable energy sources independent of a conventional public utility grid.

Off-gas
The release of vapors (volatile organic compounds [VOCs]
or other chemicals) from a material through evaporation
or chemical decomposition. Many building materials and
furnishings off-gas, such as some structural panels,
paints, cabinetry, veneers, carpets, upholstery, and wall
coverings. Controlling indoor moisture, and specifying
prefinished materials, can reduce off-gas potential.

Office building
A building used for the transaction of public or private
business, especially clerical work that is associated with
business, government, legal or other professions.

Offset
Surface or piece forming the top of a horizontal projection
on a wall.

Offsite
Materials or equipment stored away from the main
construction site.

Off-white
White, with the addition of a small amount of another color, but insufficient to identify any color other than white; may be either a cool or warm color in appearance.

Ogee
A double curve resembling an S in shape, formed by the union of a convex and concave line.

Ogee arch See **Arch**

Ogee molding See **Molding**

Ogee pediment See **Pediment**

Ogee roof See **Roof**

Olhbrich, Joseph (1867–1908)
Cofounder of the Vienna Sezession movement. He designed the Ernst Ludwig Haus, Darmstadt, in the Art Nouveau style, and the Wedding Chapel and Exhibition Hall, Darmstadt, Germany.

Olive leaf cluster See **Ornament**

Olmec architecture (1200–500 B.C)
This architecture flourished in the tropical lowlands of the Mexican Gulf Coast; characterized by temple pyramids and vast ceremonial centers.

Olmsted, Frederick Law (1822–1903)
One of the most important landscape architects of the time, and an innovator in the design of public parks, much influenced by John Paxton's work in England. He designed Prospect Park in Brooklyn and Central Park in New York City. His last large scheme was the World's Columbian Exposition, Chicago (1893), where he created a sylvan setting for the Neoclassical buildings of McKim, Mead and White, Daniel Burnham, and others.

Omundson, Theodore (1921–2009)
A pioneering San Francisco Bay–area landscape architect, he was instrumental in establishing the Council of Landscape Architectural Registration Boards. His body of work included the Kaiser Center Roof Garden in Oakland, CA, with David Arbegast, and many residential projects, parks, playgrounds, college campuses, recreation areas, historic projects, and rooftop designs. He wrote *Roof Gardens, History, Design and Construction* (1999).

On grade
A building component placed and supported directly from the ground.

On-center
As used in house construction, the distance from the center of one framing member to the center of another. In wood-frame construction, studs are typically 16 or 24 inches on-center.

On-demand hot-water circulation
System that quickly delivers hot water to a bathroom or kitchen when needed, without wasting the water that has been sitting in the hot-water pipes, which circulates back to the water heater.

On-demand water heater
A device that heats water rapidly as it is dispensed from the faucet. Eliminates the need for a conventional tank water heater.

One-point perspective See **Perspective projection**

One-way slab
A concrete slab designed to span in one direction only.

Onigawara
Ornamental tiles at the ends of the main roof ridge of a traditional Japanese structure, at the lower ends of the roof slopes, and at the corner ends. The most common is an ogre mask, from which it gets its name.

Onion dome See **Dome**

Onsite renewable energy generation
Electricity generated by renewable resources using a system or device located at the site where the power is used. Onsite generation is a form of distributed energy generation.

Onsite stormwater management
Building and landscape strategies to control and limit stormwater pollution and runoff. Usually an integrated package of strategies, elements can include vegetated roofs, compost-amended soils, pervious paving, tree planting, drainage swales, and more.

Onsite wastewater system
Treatment and disposal of wastewater sewage from a house that is not connected to a municipal sewer system; most onsite systems include a septic tank and leach field.

Onsite wastewater treatment
Uses localized treatment systems to transport, store, treat, and dispose of wastewater volumes generated on the project site.

Opacity
Quality of being impenetrable by light; not reflecting light, or transmitting light, neither transparent nor translucent.

Opal finish
A coating of silica on the exterior surface of a bulb, which gives it a milky appearance.

Opal glass See **Glass**

Opalescent glass See **Glass**

Open cornice See **Cornice**

Open eaves
Overhanging eaves where the rafters and underside of the roof are visible from below.

Open pediment See **Pediment**

Open plan
A floor plan in which there are no internal walls or a minimum number of internal walls that subdivide the space, usually at a reduced height.

Open space
The area within a community that is not occupied by buildings or transportation networks; may be contained in a plaza, park, farmland, or part of the natural environment.

Open space area
If local zoning requirements do not clearly define open space, it is defined for the purposes of LEED (Leadership in Energy and Environmental Design) calculations as the property area minus the development footprint; and it must be vegetated and pervious, with exceptions only as noted in the credit requirements section. For projects located in urban areas, open space can also include nonvehicular, pedestrian-oriented hardscape spaces.

Open space preservation
The protection of natural areas both within and around communities that provide important community space, habitat for plants and animals, recreational opportunities, farm and ranch land, places of natural beauty, and critical environmental areas such as wetlands.

Open stair See **Stair**

Open web beam
A truss with parallel top and bottom chords formed by a pair of angles, employing a web of diagonal struts and used as a beam; the struts connecting the top and bottom chords are also composed of steel bars.

Open well stair See **Stair**

Open-grid pavement
Defined for LEED (Leadership in Energy and Environmental Design) purposes as pavement that is less than 50 percent impervious and contains vegetation in the open cells.

Open-grid paving systems
Form of pervious paving that allow space for vegetation; the vegetation's evapotranspiration reduces the heat island effect caused by pavement.

Opening light
The portion of a sash or casement window that may be opened for ventilation rather than a dead light which is fixed.

Open-loop recycling
A recycling system in which a product made from one type of material is recycled into a different type of product such as used newspapers into toilet paper. The product receiving recycled material itself may or may not be recycled.

Open-space easement
An easement requiring that a certain section of property remain undeveloped.

Open-string stair See **Stair**

Open-timbered
Heavy timber work which is exposed and not concealed by sheathing, plaster, or other covering.

Open-web joists
A lattice joist welded from light steel sections and mass-produced to certain standard lengths, used to support floor or roof loads.

Open-web wood joists
Wood joists built as flat trusses, using small-dimension lumber for web pieces. These are also available with stamped steel webs.

Openwork
Any work characterized by perforations, especially of an ornamental nature.

Opera house
A theater intended primarily for the performance of opera.

Operable transom
A pane of glass above a door, which may be opened for ventilation.

Operable window See Window

Operating cost
The cost of operating a device or building; including energy, maintenance, and repairs.

Operational energy
The energy used in a building, landscape or site during its operational phase, such as heating, cooling, ventilation, hot water, lighting, and other electrical appliances.

Operations and maintenance (O&M)
The activities related to the performance of routine, preventive, predictive, scheduled, and unscheduled actions aimed at preventing equipment failure or decline with the goal of increasing efficiency, reliability, and safety.

Operations and maintenance manual (O&M manual)
A manual developed to assist building occupants in maintaining and operating a green building and its features. The effectiveness of many of a building's green features can enhanced by the occupants and maintenance crews following the recommendations in the O&M manual. An operations manual usually includes product and system information and warranties, contact information, and other information required for effective O&M facility.

Operative temperature
In determining thermal comfort, operative temperature is roughly the average of the air and mean radiant temperature (MRT) a person is experiencing.

Opposition
The state or position of being placed opposite another or of lying in corresponding positions from an intervening space or object.

Optimum value engineering
Sometimes referred to as "advanced framing," a framing technique that uses less lumber and therefore improves a structure's level of insulation. Techniques include 24-inch on-center stud layout, single top plates, engineered header sizes, and special corner and wall configurations.

Opus incertum
Ancient Roman masonry consisting of small stones set irregularly in mortar.

Opus quadratum
The ancient Roman term for ashlar, or squared stones, frequently laid without mortar.

Opus reticulatum
Permanent formwork for ancient Roman concrete, consisting of stones or bricks set diagonally.

Opus spicatum
Ancient Roman brickwork set in a herringbone pattern.

Opus testaeceum
Ancient Roman facing of broken tiles set horizontally in mortar.

Oral history
Primary source material obtained by recording spoken words, generally by means of a planned tape recorded interview.

Orange peel
In painting, a term applied to a pebble effect in sprayed coats of paint or lacquer similar to the peel of an orange; may be caused by too much pressure by holding the spraying device too close to the surface, or using a thinner which dries too quickly and prevents the flow of solids.

Orb
A plain circular boss, used as a decorative accent, where two or more ribs of a vault cross each other.

Orchestra

A circular area in a Greek theater, where the chorus sang and danced.

Orchestration

To organize and combine harmoniously so as to achieve a desired or effective combination of form, color and texture of the materials used.

Order

A logical and regular arrangement among the separate components or elements of a group; a unity of idea, feeling and form.

Orders

In classical architecture, a style of columns and capitals with standard details appearing on the entablatures. Greek orders are the Doric, Ionic, and Corinthian; the Romans added the Tuscan and the Composite.

Organic

Forms that have a structure that perfectly fulfills their own functional requirements; intellectually integrated by a systematic connection and coordination of the parts to the whole.

Organic architecture (1985–)

The principles of organic architecture rely on the integration of form and function, in which the structure and appearance of a building are based on a unity of forms that stresses the integration of individual parts to the whole concept, relating it to the natural environment in a deliberate way with all forms expressing the natural use of materials.

Organic architecture

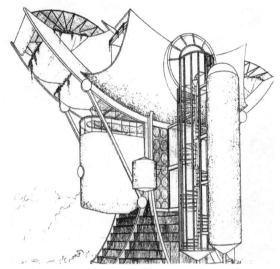

Design/Illustration: Bart Prince, Architect

Organization

An arrangement of elements or interdependent parts with varied functions into a coherent and functioning entity.

centralized organization

Spaces gathered around or coming together at a large or dominant central area.

clustered organization

Spaces that are grouped, collected, or gathered closely together and related by proximity to each other.

embedded organization

A space incorporated as an integrated and essential part of a larger space.

grid-based organization

Spaces that are organized with reference to a rectangular system of lines and coordinates.

interlocking organization
Two spaces interwoven or fit into each other so as to form an area of common space.

linear organization
Spaces that are extended, arranged, or linked along a line, path, or gallery.

linked organization
Two spaces that are joined or connected by a third intervening space.

radial organization
Spaces arranged like radii or rays from a central space or core.

Oriel window See Window

Oriental style
Adaptations of Middle Eastern or Asian architecture built in the United States, typically employing some variation of the Italianate style, using hipped roofs with multifoil arches; oriental features may include a Turkish dome, structural polychromy, and Moorish ornamentation.

Orientation
The placement of a structure on a site with regard to local conditions of sunlight, wind, drainage, and an outlook to specific vistas.

Orientation

Orientation (solar)
Orientation of a structure toward the sun for controlled solar gain. An essential ingredient in the success of passive and active solar design elements. Sun charts and software assist in orienting a building for maximum solar benefit, which can substantially reduce both heating and cooling loads.

Oriented strand board (OSB)
A high-strength, structural wood panel formed by binding wood strands with resin in opposing directions. OSB is beneficial in that it uses small-dimension and waste wood for its fiber; however, resin type should be considered for human health impact, and the production process monitored for air-pollutant emissions.

Original construction
The portion of a building or structure that was present when it was first built; may remain intact or have been subsequently altered or obscured by addition.

Ornament
Anything that embellishes, decorates, or adorns a structure, whether used intentionally and integrated into the structure or applied separately to enhance the building's form and appearance.

Ornament: Animal forms
The use of animals in a natural or idealized form for ornamental details, such as sculptured or relief figures on friezes, capitals or columns, and bas-relief panels.

aegricranes
Sculptured representations of the heads and skulls of goats and rams, once used as decoration on altars and friezes.

aegricranes

bestiary

A collection of medieval allegorical fables about animals, each with an interpretation of its significance to good or evil; in medieval churches, a group of highly imaginative and symbolic carved creatures.

birds

Any member of the class Aves, which includes warm-blooded, feathered vertebrates with forelimbs modified to form wings.

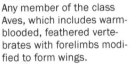

birds

bovine

Any of the Bovinae mammal species, such as the ox and cow.

bucranium

A sculptural ornament representing the head or skull of an ox, often garlanded, and most frequently used on Roman Ionic friezes.

canine

Any member of the dog family, including wolves and foxes.

centaur

In classical mythology, a monster, half man and half horse; a human torso on the body of a horse.

chimera

A fantastic assemblage of animal forms so combined as to produce a single but unnatural design; a creation of the imagination.

chimera

eagle

Any of various large birds of prey, characterized by a powerful hooked bill, and long broad wings; used as emblems, insignias, seals, and ornamental sculpture.

feline

Belonging to the cat family; includes lions, tigers, and jaguars.

griffin
A mythological beast with a lion's body and an eagle's head and wings, used decoratively.

horse
A large hoofed mammal, having a short-haired coat, a long mane, and a long tail, and domesticated since ancient times for riding and to pull vehicles or carry loads.

owl
Any of various nocturnal birds of prey, with hooked and feathered talons, large heads with short hooked beaks, and eyes set in a frontal facial plane.

sphinx
An Egyptian figure having the body of a lion and a male human head; the Greek version featured a female monster repre-sented with the body of a lion, winged, and the head and breasts of a woman.

wyvern
A two-legged dragon having wings and a barbed and knotted tail, used often in heraldry.

Ornament: Elements
Ornamental features that are integral with the structure and materials adorning buildings.

acorn
A small ornament in the shape of a nut of the oak tree; used in American Colonial architecture as a pendant, finial, carved on a panel, or as an element in the center of a broken pediment.

banderole
A decorative representation of a ribbon or long scroll, often bearing an emblem or inscription.

bouquet
The decorative ornament at the top of a finial or other projection in a floral or foliated form; similar to the anthemion.

bow knot
A decorative element in the stylized shape of a ribbon tied in a bow; often in the form of repetitive open loops which contain rosettes.

corner drop
A hand-carved or turned wood ornament that is attached to the bottom of an overhanging second-story post; often found in early American Colonial houses.

cornucopia
A goat's horn overflowing with fruits, flowers and corn, signifying prosperity; a horn of plenty; any cone-shaped receptacle or ornament.

crocket
In Gothic architecture and derivatives, an upward-oriented ornament, often vegetal in form, regularly spaced along sloping or vertical edges of emphasized features such as spires, pinnacles, and gables.

dog tooth
One of a series of projecting pyramidal ornaments resembling a row of teeth; used in Gothic Revival and Early English architecture.

fleur-de-lis
A stylized three-petal flower representing the French royal lily, tied by an encircling band and used as an ornamental device in late Gothic architecture and in later derivatives.

fret
An ornament usually in bands, but also covering broad surfaces, consisting of interlocking geometric motifs.

grapevine
A running ornament or carved panel which consisted of grapevines with bunches of grapes and grape leaves; popular in communities along the Rhine and elsewhere in Germany.

hip knob ornament
A finial or other similar ornament placed on the top of the hip of a roof or at the apex of a gable.

hollyhock
A tall plant, widely cultivated for its showy spike of large variously colored flowers; used as an ornamental motif by Frank Lloyd Wright on the Barnsdall residence in Los Angeles, CA.

honeysuckle ornament
A common name for the anthemion, common in Greek decorative sculpture.

knob
A protuberance, whether useful or ornamental, that forms the termination of an isolated member; also a handle that is more or less spherical, used for operating the mechanism for opening a door.

knot
In medieval architecture, a bunch of leaves, flowers, or a similar ornament, such as bosses at the intersection of ribs, and bunches of foliage in capitals; an ornamental design resembling cords that are interlaced.

knotwork
A carved ornamental arrangement of cord-like figures jointed together to form a type of fringe; used to decorate voussoirs and moldings.

linen scroll
A form of ornament for filling panels.

olive leaf cluster
Bunches of olive leaves sculpted to form the ornamentation of the Composite order.

palmette
A decorative motif based on the fan-shaped leaf of a palm tree.

pine cone
Oval drop that occurs in the open corner of the dentil course in the Composite order.

pineapple
A decorative carved ornament representing a pineapple, used as a terminal or finial for a hipped roof or as the central element of an ogee pediment.

rose
A stylized carving of a wild rose; used in Gothic style ornamentation and on Corinthian capitals.

scroll
Ornamentation that consists of a spirally wound band or a band resembling a partially rolled scroll of paper; S scrolls are found in ornamental brackets, window and door surrounds, and in other ornamental bands.

strapwork
Decoration formed by interlaced strips, either applied or carved in wood, stone, or plaster; used in screens, ceilings and cornices.

tooth
One of a series of carved ornaments, typically a pyramidal shape or a four-petal flower, usually set in a concave molding band; used in the Romanesque and Gothic Revival styles.

trefoil
An architectural ornament resembling a three-leaf clover.

wreath
A decorative element in the form of a garland or band of foliage; often intertwined with flowers, fruits, and ribbons.

Ornament in relievo
Decorative elements carved so that they are above the surface of a molding, such as scrolls or flowers and leaves.

Ornamental
Refers to forms that adorn or embellish a surface or any other part of a structure.

Ornamental cast iron
Decorative railings, brackets, spears and architectural elements molded from cast iron; popular in the lower Mississippi Valley.

Ornamental iron
Cast iron used for grilles, gates, finials, hardware, and innumerable architectural accessories.

Ornamental metals

Bronze, brass, copper, aluminum and stainless steel, not used for major construction, but as infill materials, including copper panels, sheet aluminum, stainless steel, and baked enamel metal alloy panels.

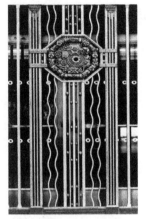

Ornamental plaster

Decorative moldings and ornamentation applied to plain plaster surfaces or used as integral designs.

Ornamental stone

Any type of stone used for ornamentation or interior finish work, as opposed to building stone.

Ornamentation

Any adjunct or detail used to adorn, decorate, or embellish the appearance or general effect of an object.

Orthographic projection See Projection drawing

OSHA See Occupational Safety and Health Administration

Ostberg, Ragner (1866-1945)

Swedish architect and designer known internationally for one work, the Stockholm Town Hall (1923), Sweden.

Otto, Frei (1925-)

German architect and pioneer of the suspended tent roof. He used the idea for the West German Pavilion, Expo '67, Montreal, Canada (1967), and for the Olympic stadium, Munich, Germany (1972). He published the book *Tensile Structures* in 1991.

Ottoman style (1350-1550)

The phase of Turkish Islamic architecture much influenced by Byzantine forms, under the rule of the Ottoman sultans in the Balkans, Anatolia, and the Middle East.

Ottonian architecture (960-1000)

The pre-Romanesque round-arched style in Germany during the rule of the Ottonian emperors, characterized by the development of forms derived from Carolingian and Byzantine styles.

Outbuilding

Any building that is detached from the main house or structure; which typically includes carriage houses, garages, sheds, stables or wood storage sheds.

Outdoor air supply

Air brought into a building from outside.

Outdoor museum

May be of any size from a small village to one covering hundreds of acres, and may or may not include buildings, whether located on the site or moved from their original location and reconstructed.

Outer string

The string at the outer and exposed edge of a stair, away from the wall.

Outgas

The emitting of fumes into the air. There are numerous building materials that have chemicals in them that outgas over time, particularly when exposed to high temperatures, moisture, or ozone levels.

Outlet

The termination of an electrical circuit for connection of a receptacle, switch, lighting fixture, appliance, or machine.

Outlet ventilator

A louvered opening in the gable end of a building that provides ventilation.

Outlooker

A member that projects beyond the face of a gable and supports the overhanging portion of a roof.

Out-of-plumb

A structural member which is not properly aligned in a true vertical fashion.

Out-of-true

In woodworking, irregularity caused by a twist or other deviation in the alignment of a form; also, a varying from exactness in a structural member.

Outrigger
A beam that extends from the ridge of a roof beyond the wall of the building; often serves as a support for hoisting tackle or for ornamentation.

Oval
Resembling an egg in shape, ellipsoidal or elliptical; it is duocentric with a long and short axis.

Oval window See Window

Overdoor
A wall area directly above a doorway containing a panel ornamented with carvings or figures.

Overhang
The horizontal distance that the upper story or roof projects beyond the story immediately below.

Overhang

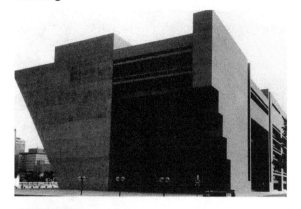

Overhanging
Projecting or extending beyond the wall surface below.

Overhanging eaves
The eaves of a roof that project past the line of a building rather than being flush.

Overhangs
Architectural elements on roofs and above windows that function to protect the structure from the elements or to assist in daylighting and control of unwanted solar gain. Sizing of over-hangs should consider their purpose, especially related to solar control.

Overhangs

Overhead door See **Door**

Overlapping

Forms extending over and covering part of an area or surface that has a common alignment; it may be slight or significant, as long as there is a common surface between the elements.

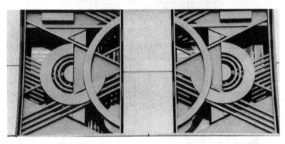

Overlapping counterpoint See **Counterpoint**

Overload

In electricity, more than a normal amount of electric current flowing through a device or machine, or a load greater than the device is designed to carry.

Oversailing

A process by which an arch, or course of bricks or stone, is made to project over a similar arch or course below; a type of repeated corbeling.

Overthrow

An ironwork hoop supporting a lantern, forming an arch in front of the door.

Ovolo

A common convex molding usually consisting of a quarter circle in section.

Ovolo molding See **Molding**

Ovum

An egg-shaped ornamental motif, used in ornamental bands in found in Classical architecture and Classical Revival styles.

Owings, Nathaniel (1903–1984)

Partner in the firm of Skidmore, Owings and Merrill (SOM). Noteworthy projects of the firm are Lever House, New York City (1952); Inland Steel Building, Chicago, IL (1958); One Chase Plaza, NYC (1962); Beinecke Rare Book Library, Yale University, New Haven, CT (1963); Air Force Academy Chapel, Colorado Springs, CO (1963); Circle Campus, University of Illinois, Chicago, IL (1965); John Hancock Center, Chicago, IL (1970); Weyerhauser Headquarters, Tacoma, WA (1971), and Sears Tower, Chicago, IL (1974).

Owl See **Ornament: animal forms**

Owner's project requirements

An explanation of the ideas, concepts, and criteria that are determined by the owner to be important to the success of the project, previously called the design intent. Also, the functional requirements of a project and the expectations of use and operations.

Ox-eye window See **Window**

Ozone

A form of oxygen found naturally that provides a protective layer shielding the Earth from ultraviolet radiation's harmful effects on humans and the environment. Ground-level ozone is the primary component of smog, produced near the Earth's surface through complex chemical reactions of nitrogen oxides, volatile organic compounds, and sunlight.

Ozone depletion potential

Amount of damage to the ozone layer a given chemical can cause compared to trichlorofluoromethane (CFC-11), which is given a value of 1.0 on this relative scale.

Ozone layer

Defined by the EPA (U.S. Environmental Protection Agency) as the protective layer of atmosphere 12–15 miles above the ground that absorbs some of the sun's ultraviolet rays, reducing the amount of potentially harmful radiation reaching the earth's surface. Ozone depletion is caused by the breakdown of certain chlorine-and/or bromine-containing compounds such as CFCs (chlorofluorocarbons) or halons.

P p

Package deal
A design-build or turnkey project.

Packaged chimney
A complete prefabricated chimney unit, usually made of metal, which comes in a range of specifications and types; some are coupled with a prefabricated fireplace unit.

Pad
An isolated mass of concrete forming a foundation.

Pagoda
A multistory shrine-like tower: originally a Buddhist monument crowned by a stupa. The stories may be open pavilions of wood with balconies and pent roofs of diminishing size with corbeled cornices.

Pai-lou See Gate

Paint
A protective finish for architectural elements, most often composed of a coloring agent ground in linseed oil or other synthetic base.

Paint remover
A liquid that softens paint and varnish so that it can be scraped or brushed off.

Painted glass See Glass

Painter
A craftsperson skilled in the preparation and application of paint, lacquer, and varnishes to wood, plaster, and other surfaces.

Paired brackets
Two brackets spaced close together to form a pair; also called coupled brackets. See also brackets.

Palace
Official residence of an important dignitary; or royalty; often an elaborate structure with many rooms.

Palazzo
In Italy, a palace; any impressive public building or private residence.

Paleolithic era
Prehistoric era beginning with the first chipped stone tools, about 75,000 years ago, and continuing until the beginning of the Mesolithic era, about 15,000 years ago.

Palladian (1508–1586)
A style named after Andrea Palladio, an Italian Renaissance architect, whose *Four Books of Architecture* set out the classic orders in detail, establishing the proportions between the various components in each one. Palladio studied the Roman architect Vitruvius and the laws of harmonic proportions. His villas were an inspiration for many of the later country houses, especially in England.

Palladian motif
A door or window opening in three parts, divided by posts, featuring a round-headed archway flanked by narrow openings with a flat lintel over each side; the arched area rests on the flat entablatures.

Palladianism
A mode of building following strict Roman forms, particularly popular in England, as set forth in the publications of Italian Renaissance architect Andrea Palladio (1508–1580).

Palladio, Andrea (1508–1580)

Italian architect. St. Giorgio Maggore and Il Rendentore are his two notable buildings in Venice (1566). His Villa Capra Vicenza (1569) is the most symmetrical of his villas, with porticoes on each side. His *Four Books on Architecture* was published in 1570.

Palm capital See **Capital**

Palmette ornament See **Ornament**

Pan forms

Pan-like metal or fiberglass structures used as forms for the bottom side of concrete floors. Reinforcing bars are placed in the recesses between the pans, which when filled with concrete form a waffle-like slab.

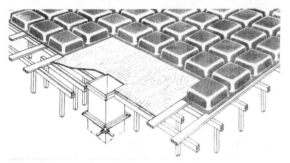

Pancharam

One of a number of miniature shrines located on the roof, cornices, or lintels of a Hindu temple, used as a decorative feature.

Pane

A relatively small piece of window glass set in an opening; also known as a light.

Panel

A portion of a flat surface recessed below the surrounding area set off by moldings or some other distinctive feature.

Panel box

A metal box in which electrical switches and fuses for branch circuits are located.

Panel divider

A molding that separates two wooden panels along their common edge.

Panel tracery See **Tracery**

Panel truss See **Truss**

Panel vault See **Vault**

Panelboard

A metal-cabinet-like housing for electrical equipment usually mounted into a wall and used to divide a feeder into separate distribution circuits to lights, receptacle outlets, and machinery. Each circuit is controlled by an overload protection device or circuit breaker.

Paneled ceiling

A ceiling divided into compartments by raised moldings.

Paneled door See **Door**

Paneling
A finished surface composed of multiple thin wood panels held by rails, stiles, or moldings.

Panic hardware
Door hardware that can be released quickly by pushing a horizontal bar; required for certain exit doors by building codes.

Pantheon
Temple dedicated to all the gods of a people; specifically the temple built in 25 B.C. in Rome by Emperor Hadrian, with a coffered concrete dome illuminated by an oculus at the top, set on a very thick circular drum, and having an octastyle portico attached to the drum outside.

Pantile
A roofing tile in the shape of an "S" laid on its side and overlapped in courses running up the slope of the roof.

Pantograph
A drafting instrument for copying drawings, or plans, either at the same scale or an enlarged or reduced scale.

Paperhanger
A painter or decorator who specializes in hanging wall coverings.

Papier-mâché
A material used for model making composed principally of paper; prepared by pulping a mass of paper and adding glue, to produce a dough-like consistency, and molding it into a desired form.

Papyriform capital See **Capital**

PAR lamp
A parabolic aluminum reflector lamp, which offers excellent beam control from a very narrow spot to a wide flood. PAR lamps can be used outdoors unprotected since they are made of hard glass that can withstand adverse weather.

Parabolic
Forms which resemble a parabola in outline.

Parabolic arch See **Arch**

Parabolic troffer
An inverted trough with parabolic shaped aluminum or plastic lenses that shield a lamp from direct view to improve visual comfort.

Paraboloid roof See **Roof**

Paraline drawing See **Projection**

Parallel
Always the same distance apart; thus two parallel lines never meet.

Parallel counterpoint See **Counterpoint**

Parallelogram
A quadrilateral having both pairs of the opposite sides parallel to each other.

Parapet
A low protective wall or railing along the edge of a raised platform, terrace, bridge, roof, balcony, and above cornices.

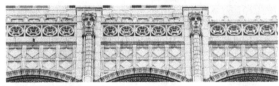

Parapet gutter
A gutter that is located behind a parapet wall.

Parapeted gable See **Gable**

Parasol
An umbrella form on top of a Chinese pagoda.

Parcel
A piece of land with its own metes and bounds.

Pargetting See **Plaster**

Parging
The application of mortar to the back of the facing material, or the face of the backing material. Also called backplastering.

Park
A tract of land set aside for public use; a landscaped city square; also an expanse of enclosed grounds for recreational use within or adjoining a town.

Parking area
An area, usually paved, set aside for the parking or storage of vehicles.

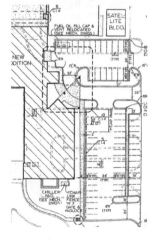

Parking garage
A garage for passenger vehicles only, exclusively for the purpose of parking or storing of automobiles and not for automobile repairs or service work.

Parking lot
An area on the ground surface used for parking vehicles; may be paved or unpaved.

Parlor
A multipurpose room for sitting and formal entertainment, situated on the main floor of most dwellings.

Parquet
A flat inlay pattern of closely fitted pieces, usually geometrical, for ornamental flooring or wainscoting; often employing two or more colors or materials such as stone or wood.

Parquetry
Small pieces of wood fitted together to form a geometrical design.

Parris, Alexander (1780–1852)
American architect based in Boston; Neoclassical builder-architect, associated with Bullfinch and others. Designed the Governor's Residence (1812), Richmond, VA; Charlestown Navy Yard and Harbor, (1829); and Quincy Market (1826), both in Boston, MA.

Parthian architecture (400 B.C.–200 A.D.)
The architecture developed while under Parthian rule in Iran and western Mesopotamia, combining Classical with indigenous features.

Parti See **Design drawing**

Particleboard
A panel product used in cabinets and furniture, generally made from wood fiber glued together with a binder. Similar to medium-density fiberboard, yet with a coarser texture. Most particleboard is made with formaldehyde-emitting urea-formaldehyde binder, although some wood particleboard and all straw particleboard use a nonformaldehyde methyl diisocyanate or low-emitting phenol-formaldehyde binder.

Particleboard See **Wood products**

Particulate pollution
The pollution made up of small liquid or solid particles suspended in the atmosphere or water supply.

Partition wall See **Wall**

Partner-in-charge
A member of a partnership who is charged with major responsibility for the firm's services on a project.

Partnership
A legal relationship existing between two or more persons contractually associated as joint principals in a business as co-owners.

Party wall See **Wall**

Passage
Any interior corridor connecting room in a building; also called a hallway.

Passageway
A space that connects one area of a building with another.

Passive building design
Building configurations that take advantage of a natural, renewable resource like sunlight and cooling breezes. Passive design strategies typically do not involve any moving parts or mechanical processes.

Passive cooling
Using passive building strategies to relieve the cooling load of a building by capitalizing on such things as predictable summer breezes or by shading windows from direct summer sunlight.

Passive design
Building design and placement in home construction that permits the use of natural thermal processes—convection, absorption, radiation, and conduction—to support comfort levels and reduce or eliminate the need for mechanical systems for these purposes.

Passive diffuser
An air supply outlet, without a fan, that relies on pressurized plenum or duct air to deliver air into the conditioned space of the building.

Passive heating
A system whereby a building's structure or an element of it is designed to allow natural thermal energy flows, such as radiation, conduction, and natural convection generated by the sun, to provide heat. The home relies solely or primarily on nonmechanical means of heating.

Passive solar
Technologies that convert sunlight into usable heat, cause air movement for ventilation or cooling, or store heat for future use, without the assistance of other energy sources. A type of design which takes maximum advantage of the sun's energy to help warm the home in winter and helps to redirect or block that energy to reduce cooling needs in the summer.

Passive solar design
Involves using nontechnical design methods, site conditions, local climate, sun angle, building massing, orientation, and daylight to save and retain energy within buildings. Unlike technologically advanced active solar counterparts, passive solar buildings do not rely on electrical or mechanical systems, control techniques, or other devices to operate. The two basic passive solar design methods are direct gain and indirect gain.

direct solar gain
A method that relies on the orientation of the building, the location of its openings, the building's materials and their attributes, the structure's heat storage capabilities, and its insulation systems. In this method, sunlight is allowed to enter the building through south-facing windows. Light is absorbed directly by the thermal mass, which stores and releases the heat as the building cools.

indirect solar gain
A method that requires a buffer thermal mass between the sun and the living space to be heated, where the thermal mass buffer can be a structure, a wall system, an absorp-

indirect solar gain
tion device, and/or another space. In an indirect gain system, the thermal mass acts as a collector, absorber, and distributor of the solar energy. Thermal distribution is accomplished by conduction. There are three main types of indirect gain systems: thermal storage wall, roof water, and sunrooms.

roofwater system
A system that absorbs and transfers heat from outside to inside during the winter, and from inside to outside during the summer. In order to heat and cool a building effectively during both seasons, the water stored in tanks or pipes, along with additional treatment materials such as antifreeze, must be insulated in reverse order from winter to summer. Used primarily in commercial buildings in low-humidity climates.

sunroom
A hybrid method of passive solar design, which includes the functions and benefits of both the direct and indirect gain methods. Also called "solar greenhouses" or "solariums," sunrooms have significant advantages over other indirect gain solutions, especially because of their ability to control the level of heat within a building.

thermal storage wall
Consists of a 10- to 16-inch thick masonry wall placed on the south side of a building where it will receive the most sunlight. Dark-colored, single- or double-glazed windows cover the exterior of the wall to absorb the solar energy, which is stored and then radiated to the living area after the space cools. For other thermal wall systems, the wall thickness varies depending on the material: 10–14 inches for brick, 12–16 inches for concrete, and 8–12 inches for adobe.

Passive solar energy system
A system by which energy is collected from the sun and stored; it is then distributed throughout the structure by natural means of conduction convection or radiation.

Passive solar heating
Using the sun's energy in the form of heat to diminish a building's heating load, usually through the use of large window areas that permit light penetration upon some massive material to use its thermal storage capacity.

Passive solar water heater
A water heating system that does not require mechanical pumps or controls to create hot water for domestic use.

Passive ventilation
Relies typically on using convective air flows that result from the tendency of warm air to rise and cool air to sink and taking advantage of prevailing winds. Many passive ventilation systems rely on the building users to control windows and vents as indicated by site conditions and conditions within the building.

Pastiche
Inappropriate architectural ornament added after the original work is completed.

Patera

A representation of a flat round or oval disk found in friezes.

Patina

A greenish-brown crust produced by oxidation that forms on the surface of copper and bronze, often multicolored and considered decorative; any thin oxide film which forms on a metal or other material.

Patio

An outdoor area, often paved and shaded, adjoining or enclosed by the walls or arcades of a house.

Pattern

The juxtaposition of repetitive elements in a design, organized so as to produce an arrangement of parts that are viewed as an unit; may occur at various scales and sizes.

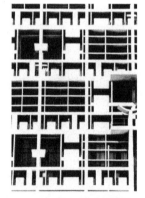

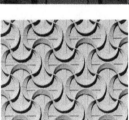

Pattern book

A book on architectural practice that serves as a builder's manual, buyer's guide, or handbook, containing plans and details for common building element such as columns, cornices, and doors and windows.

Pattern staining

Discoloration of plaster ceilings of composite construction, caused by the different thermal conductance of the backing; the air circulates more freely over the warmer parts, and deposits more dust on them.

Patterned brickwork

Bricks with more than one color or texture that are laid in different directions so as to form decorative designs.

374

Patterned concrete block

Concrete block with a customized, recessed decorative pattern on the front side.

Pavement

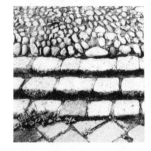

Patterned glass See **Glass**

Pavement

The durable surface of a sidewalk or other outdoor area, such as a walkway or open plaza.

Pavement light

Thick, translucent glass disks or prisms, set into a section of pavement to transmit light to a space below.

Paver

A paving stone, brick or tile.

Pavilion

An open structure or small ornamental building, shelter or kiosk used as a summer house or as an adjunct of a larger building. It is usually a detached structure for specialized activities, and is often located as a terminal structure with a hipped roof on all sides so as to have a pyramidal form.

Pavilion porch

Gazebo-type structure projecting out of a veranda or porch.

Pavilion roof See Roof

Paving brick

A hard vitrified clay brick with resistance to abrasion.

Paving stone

A block or slab of natural or prepared stone used as a paver.

Paving tile See Tile

Paxton, Sir Joseph (1801–1865)

English architect who designed the Crystal Palace, London, England (1851). It was the first prefabricated building constructed in iron, glass, and laminated wood.

Payback period

A popular nondiscounting project selection technique used when organizations require the capital investment of a project to be recovered within a specified period; the period it takes for the stream of net cash flows to equal the initial investment. Also, a term used in the evaluation of sustainable and renewable energy options, wherein greenhouse or greenhouse intensive energy savings that the technology may enable over its useful life are assessed in relation to the embodied energy required for its manufacture. For renewable energy systems, it can also refer to the period of time over which energy cost savings derived from accessing renewable energy offset the upfront capital costs of the system.

Peabody, Robert Swain (1845–1917)

American architect partnered with John Goddard Stearns, Jr. (1843–1917) in Boston. A prolific firm that trained numerous young architects. Designed Park Square Railroad Station (1874), Providence, RI; Mutual Life Insurance Company of New York (1875), NYC; Custom House Tower (1911), Boston, MA; Kragsyde (1884) Manchester-by-the-Sea, MA; and Groton School (1901), Groton, MA.

Peak watt

Highest possible unit of rated power output from a photovoltaic (PV) module in full sunlight, as distinct from its output at any given moment, which may be lower.

Peak-head window See Window

Pebble pavement

The pavement with a surface composed of water-rounded pebbles closely set in clay or concrete; different colors are often used in decorative designs.

Pebble pavement

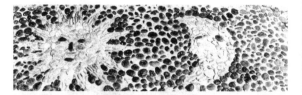

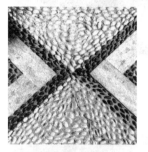
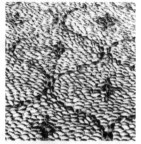

Pebble wall masonry See Masonry

Pebbledash
Small round stones applied to a fresh coat of plaster on an exterior wall to create a textured appearance.

Pedestal
A support for a column, urn, or statue, consisting of a base and a cap or cornice.

Pedestal

Pediment
A low-pitched triangular gable above a facade, or a smaller version over porticos above the doorway or above a window; a triangular gable end of the roof above the horizontal cornice, often decorated with sculpture.

broken pediment
A pediment with its raking cornice split apart at the center, and the gap often filled with a cartouche, urn, or other ornament.

broken pediment

ogee pediment
A pediment in the shape of an ogee.

open pediment
A form of broken pediment

round pediment
A round or curved pedi-
ment, used ornamentally
over a door or window.

round pediment

segmental pediment
A pediment above a door or
window that takes the form
of an arc of a circle.

swan's neck pediment
A broken pediment having a sloping double S-shaped
element on each side of the pediment; used often in
the Georgian style.

Peg
A tapered cylindrical wooden pin that is driven through
a hole to hold two or more members together.

Peg-braced frame
A timber frame featuring
pegged mortise-and-tenon
joints and diagonal corner
braces.

Pei, Ieoh Ming (1917–)

A Chinese-born American architect who studied with Walter Gropius at Harvard and worked with William Zeckendorf's contracting firm in New York City. His notable later works include Mile High Center, Denver, CO (1955); National Center for Atmospheric Research, Boulder, CO (1967); Christian Science Service Center, Boston, MA (1971); John Hancock Tower, Boston, MA (1975); Dallas Municipal Center (1977), Dallas, TX; National Gallery of Art, East Wing (illus.), Washington, DC (1978); Kennedy Library, Boston, MA (1979); Jacob Javits Convention Center, NYC (1986); Holocaust Museum, Washington, DC (1986); Bank of China, Hong Kong, China (1990); the entrance pyramid at the Louvre Museum (illus.), Paris, France (1993), and the Rock 'n' Roll Hall of Fame, Cleveland, OH (1995).

Pelli, Cesar (1926–)

Argentine-born American architect who worked for Eero Saarinen before becoming director of design for Daniel, Mann, Johnson & Mendenhall in 1964, and later the design partner, Gruen Associates. He set up his own practice in New Haven, CT (1977). The Pacific Design Center, Los Angeles, CA (1971), brought his name to notice. He designed the Winter Garden and World Financial Center at the World Trade Center, NYC (1981), the Canary Wharf Tower, London, England (1986), and the huge Petronas Twin Towers, Kuala Lampur, Malaysia (1997).

Pellet stove

Wood stove designed to burn pellets made from compressed sawdust or wood shavings that are fed into the firebox at a metered rate by a screw auger. An electric fan provides combustion air.

Pendant

A hanging ornament or suspended feature on ceilings or vaults.

Pendant fixture

A fixture that hangs from the ceiling, alone or in a series. It can provide both primary and background lighting. Some have a globe that provides diffused light, while others shine directly on a work surface, using incandescent or compact fluorescent lamps. The fixture itself may throw light upward, downward, or dispersed through a globe.

Pendentive

The curved triangular surface that results when the top corner of a square space is vaulted so as to provide a circular base for a dome.

Pendentive

Pendentive bracketing
Corbeling in the general form of a pendentive; common in Moorish and Muslim architecture.

Pendil
The projecting exposed lower end of a post of the overhanging upper story or jetty, often carved.

Penitentiary
Place for the imprisonment of inmates and for their reformation through discipline and work.

Pennsylvania barn
A bank barn found throughout Pennsylvania, employing an overshoot on the side opposite the bank; the livestock are housed in the masonry lower level, and the timber frame haymow occupies the upper levels.

Pent roof See Roof

Pentagonal
A plane figure with five equal sides and five equal angles, commonly found in nature.

Pentastyle
A portico having five columns in front.

Penthouse
A structure on a flat-roofed building, occupying usually less than half the roof area.

Pereira, William (1910–1990)
Founder of the Los Angeles-based firm of Pereira and Luckman, with Charles Luckman. Designed the control tower and restaurant at Los Angeles International Airport, Los Angeles, CA.

Perfluorocarbons
Potent greenhouse gases that accumulate in the atmosphere and remain there for thousands of years. Aluminum production and semiconductor manufacture are the largest known manmade sources of perfluorocarbons.

Perforated
Forms that exhibit holes or a series of holes in a pattern; formed by combining elements to produce voids, or through carving or casting materials containing pierced openings.

Performance approach
The practice of thinking and working in terms of "ends" rather than "means." It is about describing what the building is expected to do, and not prescribing how it is to be realized.

Performance assessment
The process of assessing or evaluating the performance of the whole building or its component parts, according to a set of performance targets, criteria, or requirements.

Performance criteria
Criteria against which the performance of a party to a contract is measured to determine if that party's obligations are fulfilled.

Performance indicators
A set of measures that reflect the environmental or sustainable credentials or performance of a building. It should be noted that, in research literature, a distinction is made between environmental or "green" assessment and "sustainable" assessment; the latter includes the indicators covered in the former and extends its scope to include social, economic, and other indicators.

Performance rating method
Requires the development of an energy model for the proposed building design.

Performance specifications
Specifications that delineate the results to be achieved rather than the specific methods or materials to be used.

Performance-based maintenance
Reliable delivery of defined services such as heating and cooling, as opposed to scheduled maintenance on specific plant items. The contractor is rewarded based on a reliable supply of those services.

Pergola
A garden structure consisting of an open wooden-framed roof, often latticed and supported by regularly spaced posts or columns and often covered by climbing plants to shade a walk or passageway.

Period architecture
Architectural styles relating to particular periods.

Period exhibit
A display that portrays a historical period of time, either in the furniture and wall coverings, or in a reconstructed setting depicting a historic time.

Period room
A collection of original furniture and furnishings, usually recreated in a historical era or presenting an aesthetically pleasing combination that may never existed in that form in a particular building.

Peripteral
Surrounded on the outside perimeter by a single row of columns.

Peripteral temple
A temple surrounded by a single row of columns.

Peristyle
A row of columns around the outside of a building or around the inside of a courtyard.

Perkins, Dwight (1867–1941)
Cousin to Marion Mahoney (Griffin); worked for Daniel Burnham in Chicago, and designed Steinway Hall (1896), the building that became headquarters for a number of Prairie School architects.

Perkins, Lawrence (1907–1982)
Founder of the Chicago-based firm of Perkins and Will.

Perlite
A lightweight, expanded mineral bead; highly flame resistant and with good insulating value.

Permaculture
An approach to designing human settlements, in particular, the development of perennial agricultural systems that mimic the structure and interrelationship found in natural ecologies.

Permanent formwork
The formwork that is not struck, but is left permanently in position after the concrete has been cast.

Permeability
A measure of the ability of a material, such as rock or soil, to transmit fluids or air through it.

Permeable pavement
A paving material that allows the penetration of water, thus significantly reducing runoff from stormwater.

Perpendicular
Line or plane that meets another at right angles.

Perpendicular style (1350–1550)
The last and longest phase of Gothic architecture in England, was characterized by a vertical emphasis and elaborate fan vaults displaying perpendicular tracery.

Perpendicular tracery See Tracery

Perrault, Claude (1613–1688)
French architect, one of the designers of the Louvre, Paris, France (1665), notably the east facade

Perret, Auguste (1874–1955)
French architect. His 25 Rue Franklin apartment house (illus.), in Paris, France (1903) was an early example of re-inforced-concrete frame construction. He designed the Garage Ponthiew (1906) and Notre-Dame du Raincy (1922), a hall-church design built with reinforced-concrete vaults and walls glazed with stained glass, both in Paris, France.

Perret, Auguste

Perron
A formal terrace or platform, especially one centered on a gate or doorway; an outdoor flight of steps, usually symmetrical, leading to a terrace.

Persian architecture (550–330 B.C.)
Architecture developed under the kings who ruled ancient Persia during the Achaemenid dynasty. It was characterized by a synthesis of architectural elements from surrounding countries, such as Assyria, Egypt, and Greece.

PERSIST construction
A method of construction including rainscreen cladding and foam insulation installed on the exterior of the building's frame. Developed by the National Research Council of Canada in the 1960s, PERSIST is an acronym for Pressure-Equalized Rain-Screen Insulated Structure Technique. Most PERSIST buildings have no insulation in the stud bays or rafter bays. Instead, 4 to 8 inches of rigid foam insulation is installed on the exterior of the wall and roof sheathing. The PERSIST system requires the installation of a rubberized asphalt membrane between the exterior foam and the wall and roof sheathing; this membrane acts as a water-resistant barrier, air barrier, and vapor barrier.

Persona
A mask of terra-cotta, stone or marble, designed to imitate the human face or the head of an animal, usually in the form of a grotesque; employed as an antic or as a gargoyle for discharging water.

Perspective drawing
Any one of a variety of techniques to represent three-dimensional objects and spatial relationships on a two-dimensional surface in the same manner as they would appear to the eye. With the given tools of object, station point, central visual ray and picture plane, all mechanical systems, photography and computer plotting of perspectives follow the exact same principals, developed during the Renaissance in Italy.

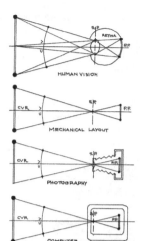

Perspective projection
A method of projection in which a three-dimensional object can be represented by projecting points upon a picture plane using straight lines converging at a fixed point, representing the eye of the viewer.

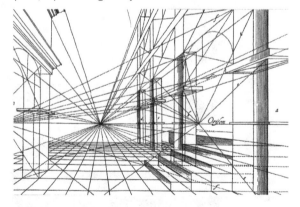

center of vision
A point representing the intersection of the central axis of vision and the picture plane in linear perspective drawing.

central visual axis
The sightline, which is perpendicular to the picture plane, indicating the direction in which the viewer is looking.

cone of vision
The field of vision radiating outwardly from the eye of the viewer in a more or less conical shape along the central visual axis.

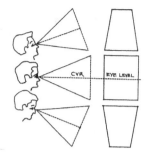

ground line
A horizontal line representing the intersection of the ground plane and the picture plane.

ground plane
A horizontal plane of reference in linear perspective from which vertical measurements can be taken; usually it is the plane supporting the object depicted or the one on which the viewer stands.

horizon line
A horizontal line in linear perspective representing the intersection of the picture plane and a horizontal plane through the eye of the viewer.

measuring line
Any line coincident with or parallel to the picture plane, as the ground line, on which accurate measurements can be taken.

one-point perspective
A rendition of an object with a principal face parallel to the picture plane; all horizontal lines parallel to the picture plane remain as is, and all other horizontal lines converge to a preselected vanishing point.

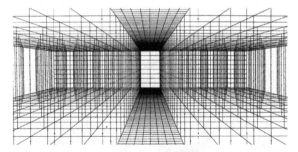

picture plane
An imaginary transparent plane, coexistent with the drawing surface, on which the image of a three-dimensional object is projected and on which all lines can be measured and drawn to exact scale.

station point
A fixed point in space representing a single eye of the viewer.

three-point perspective
A perspective of an object with all faces oblique to the picture plane; the three sets of parallel lines converge to three different vanishing points: one left, one right, and one above or below the horizon line.

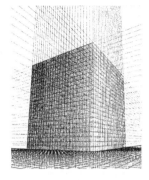

two-point perspective
A perspective of an object having two faces oblique to the picture plane. The vertical lines parallel to the picture plane remain vertical, and two horizontal sets of parallel lines oblique to the picture plane appear to converge to two vanishing points: one to the left and one to the right.

vanishing point
A point toward which receding parallel lines appear to converge, located at the point where a sight line parallel to the set of lines intersects picture plane.

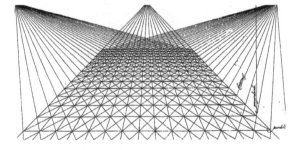

Peruzzi, Baldassare (1481–1536)
High Renaissance and Mannerist Italian architect; designed the Villa Farnesina, Rome (1505). Inside are brilliant frescoes by Raphael, Peruzzi, and others.

Pervious paving
A paving material that allows water to penetrate to the soil below; this reduces the amount of water that needs to be treated by the water infrastructure system and increases the water in the aquifer. It is a function of storm water management

PEX

Cross-linked polyethylene. Specialized type of polyethylene plastic that is strengthened by chemical bonds formed in addition to the usual bonds in the polymerization process. PEX is used primarily as tubing for hot- and cold-water distribution and radiant-floor heating.

Phenol-formaldehyde binder

Formaldehyde-based binder used for wood products, especially those made for exterior applications. It generally has lower formaldehyde emissions than urea-formaldehyde binder.

Phenolic laminate

A high-pressure laminated sheet made from paper and phenol-formaldehyde resin, commonly used for furniture and kitchen cabinet surfaces.

Phosphor

The internal coating of fluorescent bulbs, which glows when bombarded with electromagnetic radiation.

Photocells

A light-sensing device used to measures the amount of incident light present in a space and activate controllers at dawn or dusk.

Photodegradable

Substances that can be chemically decomposed by light. Photodegradable plastic, for example, becomes brittle and breaks into smaller pieces when exposed to sunlight.

Photogrammetry

The process of producing accurate maps by tracing outlines of contours that are produced by stereo aerial photographs with specialized stereoscopic plotters.

Photographic collections

Collections which document American architecture and its environment and held in libraries, and public and private archives, such as the National Archives, the Bettman Archive, and Avery Library at Columbia University, NYC.

Photometer

An instrument for measuring photometric quantities, such as luminance, luminous intensity, luminous flux, and illuminance .

Photometric report

A report that describes the manner in which light is emitted from a luminaire, presented in an industry standard format.

Photometry

The science of measuring visible light in a luminaire, with units that are weighted according to the sensitivity of the human eye.

Photovoltaic (PV)

Generation of electricity directly from sunlight. A photovoltaic cell has no moving parts; electrons are energized by sunlight and result in current flow.

Photovoltaic cell

An electronic device consisting of layers of semiconductor materials fabricated to form a junction (adjacent layers of materials with different electronic characteristics) and electrical contacts and capable of converting incident light into direct current electricity. PV cells are used in solar panels. When sunlight hits the cells, a chemical reaction occurs, resulting in the release of electricity.

Photovoltaic module

An integrated assembly of interconnected photovoltaic cells designed to deliver a selected level of working voltage and current at its output terminals, packaged for protection against environment degradation, and suited for incorporation in photovoltaic power systems.

Photovoltaic panels

PV devices that use semiconductor material to directly convert sunlight into electricity. Power is produced when sunlight strikes the semiconductor material and creates an electric current.

Phthalate plasticizer

Chemical added to polyvinyl chloride (PVC) and certain other plastics to make them more flexible.

Physical depreciation

The reduction in the property value due to deterioration of the physical fabric because of wear and tear, inadequate maintenance, and weathering and decay.

Phytoremediation

Low-cost option for site cleanup when the site has low levels of contamination that are widely dispersed. Phytoremediation (a subset of bioremediation) uses plants and trees to break down or take up contaminants.

Piano nobile

The principal floor of a great or noble house, usually containing state rooms on the first floor above the ground or basement.

Piano, Renzo (1937-)

Italian architect; worked with Richard Rogers designing the high-tech Centre Pompidou (illus.), Paris, France (1971). His later work included the Kansai International airport, Osaka, Japan (1994), one of the longest structures in the world, and the California Academy of Sciences (illus.), in San Francisco, CA (2008).

Piano, Renzo

Piazza
A public open space or square surrounded by buildings.

Pictograph
A stylized illustration used for communication; also includes symbols that were scratched or painted on rock surfaces by prehistoric people

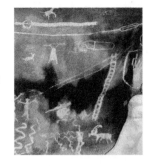

Picture plane See **Perspective projection**

Picture rail
A molding fixed to an interior wall; pictures may be suspended from it by means of small metal hooks, which fit over the top of the molding.

Picture window See **Window**

Picturesque movement
A reaction to the Classical Revival Style architecture that included irregularly planned landscapes, follies, grottos, and asymmetrical buildings, mostly in the Italianate style.

Pier
A freestanding support for an arch, usually composite in section and thicker than a column, but performing the same function; also, a thickened part of a wall to provide lateral support or bear concentrated loads.

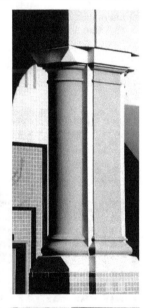

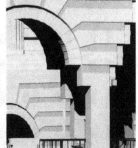

Pier arch See **Arch**

Pier buttress
A pier that receives the thrust of a flying buttress.

Pier foundation
Building foundation consisting of piers instead of continuous walls. Piers are resource efficient because they avoid the need for continuous foundation walls.

Pierced work
Decoration which consists mainly of perforations, such as a nonbearing masonry wall in which an ornamental pierced effect is achieved by alternating rectangular or any other shaped blocks with open spaces.

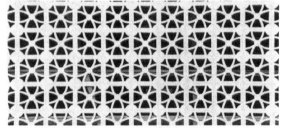

Pilaster
A partial pier or column, often with a base, shaft, and capital, that is embedded in a flat wall and projects slightly; may be constructed as a projection of the wall itself.

grouped pilaster
Two or more closely spaced pilasters forming a group, often on one pedestal.

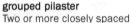

Pilaster face
The front surface of a pilaster that is parallel to the wall.

Pilaster mass
An engaged pier built up with the wall, usually without a capital or base.

Pilaster side
The form of the side surface of a pilaster perpendicular to the wall.

Pilaster strip
A slender pier of minimal projection.

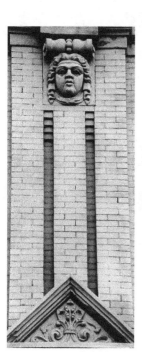

Pile
One of a series of large timbers or steel sections driven into soft ground down to bedrock to provide a solid foundation for the superstructure of a building.

Pillar
A column or post supporting an arch or other superimposed load. Clustered or compound pillars consist of a central shaft with smaller shafts that are grouped around it.

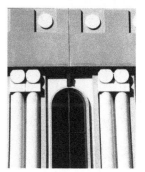

Pilotis
The freestanding columns, posts, or piles that support a building, raising it above an open ground level.

Pin
A peg or bolt of wood, metal, or any other material used to fasten or hold something in place, or serve as a point of support.

Pin connection
A structural connection of a truss or bridge made by a pin in an eyebar, rather than a rivet or turnbuckle.

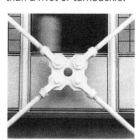

Pin joint
A flexible joint connected by a pin through a hole in each of the members to be joined; used primarily for large spans where complete freedom to rotate is required, such as roof forms secured with pin-joint buttresses as the main support.

Pine cone ornament See Ornament

Pineapple ornament See Ornament

Pinholing
In painting, a defect in a spray-painted surface caused by holes due to bubbles which persist until the film has dried. The bubbles may be caused by sealed air pockets, moisture or oil in the air lines or porous undercoating, or use of thinner that dries too quickly.

Pinnacle
An apex or small turret that usually tapers toward the top as a termination to a buttress.

Pintles
Square metal devices used to transfer loads of columns on upper floors by passing through intervening beams and girders to metal column caps on the column below.

Pinwheel
A shape that is fixed at the center with identical radiating arms, either angular or curvilinear, and repeated any number of times within the circumference of the circle from which they are generated.

Pipe
A long, tubular vessel used to carry a fluid or gas from a supply source to fixtures, or from plumbing fixtures back to sewer lines.

Pipe column See Column

Pipe coupling
In plumbing work, a short collar consisting of a threaded sleeve used to connect two pipes.

Pipe fitting
In plumbing, refers to ells, elbows, and various branch connectors used in assembling pipework.

Pipe hangers
In plumbing, applies to various types of supports, such as clamps and brackets, for soil pipes.

Pipework
An assembly of pipes and fittings used for the conveyance of fluids.

Piranesi, Giovanni Battista (1720–1778)
Italian architect and artist who produced etchings and engravings of imaginary scenes of antiquity. He published *Della Magnificenza ed Architecttura dei Romani,* and *Invenzioni Capric di Carceri*, the "Prisons" (c. 1745).

Pisé
A building whose walls are made of compressed earth, usually stiff clay formed and rammed into a movable framework; the building material itself; that is, stiff earth or clay rammed until firm to form walls and floors.

Pit dwelling
An excavated residence that is either partially or wholly below grade; found in the southwest United States and northern Mexico.

Pitch
Angle of a roof, or the proportion between the height and span of the roof.

Pitched roof See **Roof**

Pitch-faced masonry See **Masonry**

Pivoted door See **Door**

Pivoting light
A window that opens and closes by rotating on two pivots, either horizontally or vertically; generally located in the center of the window, or at the bottom in the case of a transom.

Pivoting window See **Window**

Plaid
A pattern created by regularly spaced bands at right angles to one another; the resultant checkered effects vary widely, depending on the relationship and intervals between lines and bands.

Plain
Unadorned; without any pattern or ornamentation.

Plan
A two-dimensional graphic representation of the design and dimensions of the project, or parts thereof, viewed in a horizontal plane from above.

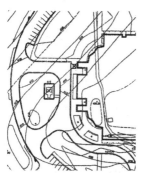

Plane
The simplest kind of two-dimensional surface, generated by the path of a straight line and defined by its length and width. The fundamental property of a plane is its shape and surface characteristics.

Plank
A long, flat piece of wood measuring at least 2 inches in thickness and more than 8 inches in width; used for flooring and sheathing.

Plank house
A type of timber construction consisting of sawn planks laid horizontally and notched at the corners.

Plank-frame
A type of house in colonial America constructed of heavy wood planks erected vertically into grooves in a sill plate, and drilled and pegged together.

Planned maintenance
A planned regular preventive maintenance program to keep a building and its services in good working order.

Planned Unit Development (PUD)
A zoning process that allows a developer to offer to build public amenities, such as roads or schools, or provide open space easements in exchange for looser restrictions on the number of dwelling units, building heights or other variances from existing restrictions.

Planner
The person and/or professional in architecture or interior design that deals with the layout, design, and furnishings of spaces within a proposed or existing structure, according to the requirements of the client.

Planning
The process of studying the layout of spaces within a building or other facility, or installations in open spaces, in order to develop the general scheme of a building or group of buildings. Also, the establishment of the project activities and events, their logical relations and interrelations to each other, and the sequence in which they are to be accomplished

Planning grid
An arrangement of one or more sets of regularly spaced parallel lines, with the sets at right angles or any other selected angles to each other, and used like graph paper to assist with modular planning.

Plans
The official approved plans, profiles, typical cross sections, working drawings, and supplemental drawings, or exact reproductions thereof, which show the location, character, dimensions, and details of the work to be performed.

Plantation
A large estate or farm on which crops such as cotton, tobacco, or sugar are grown and harvested, often by resident workers. These were mostly built with slave labor and one family's economic fortune. Usually contains a variety of outbuildings, such as smokehouses, livery stables, blacksmith shop and slave quarters. Also, a newly established colony or settlement.

Plantation house
The principal house of a southern plantation on which cash crops were cultivated; typically two stories with very tall windows, with a veranda across the entire front facade, supported by two-story columns.

Planter
A permanent, ornamental container to receive planted pots or boxes, often nonmovable and integral with the finish of a building.

Plaque
A tablet, often inscribed, added to or set into a surface on the exterior or interior wall.

Plaque

Plaques and markers
Used by many state historic programs to identify historic sites and National Register properties, usually affixed to the exteriors of buildings.

Plaquing programs
Programs developed by national, state, or local city and town governments, patriotic associations, or local preservation organizations and other agencies, to call attention to a place, by putting up a plaque. Individuals proud of their own buildings have erected their own plaques as well.

Plaster
A mixture of lime or gypsum, sand, Portland cement and water to produce a paste-like material which can be applied to the surfaces of walls and ceilings and which later sets to form a hard surface.

Plaster

daubing
A rough coating of plaster given to a wall by throwing plaster against it.

intonaco
The fine finish coat of plaster made with white marble dust to receive a fresco painting.

pargetting
A decorative feature in which flat wet plaster is ornamented by patterns either scratched or molded into it; sometimes decorated with figures either in low relief or indented.

rendering
A coat of plaster applied directly to an interior wall or stucco on an exterior wall; a perspective or elevation drawing of a project or portion thereof with artistic delineation of materials, shades, and shadows.

scagliola
Plaster work that imitates stone, in which mixtures of marble dust, sizing, and various pigments are laid in decorative figures routed into the surface.

shikkui
A plaster, mortar, stucco, or whitewash, made from a mixture of lime and clay and having the consistency of glue, used in traditional Japanese construction.

Plaster lath
Thin narrow strips of wood nailed to ceiling joists, studding, or rafters as a groundwork for receiving plastering.

Plaster of paris
A gypsum substance especially suitable for fine ornamental plasterwork because it fills a mold completely and dries quickly.

Plasterboard
A building board made of a core of gypsum or plaster, faced with two sheets of heavy paper.

Plastic
Any of the various synthetic complex organic compounds produced by polymerization; can be molded, extruded, or cast into various shapes or drawn into fibers.

acrylic fiber
A synthetic polymer fiber.

fiberglass
The generic term for a material consisting of extremely fine filaments of glass that are mixed with a resin to give the desired form in a mold. Layers of this combination are laid or sprayed into the mold.

nylon
A class of thermoplastics characterized by extreme toughness, strength, and elasticity and capable of being extruded into filaments, fibers and sheets.

Plastic

Plexiglas®
Used for windows and lighting fixtures.

polyethylene
A tough, light, flexible thermoplastic used in the form
of sheeting and film for
packaging, damp-proofing,
and as a vapor barrier.

polystyrene
A hard, tough, stable thermoplastic that is easily colored,
molded, expanded, or rolled into sheeting.

vinyl
Any of various tough, flexible plastics made from polyvinyl
resin.

Plastic laminate
A laminate consisting of paper, cloth or wood covered with
a phenolic resin; used for countertops and cabinets that
require a washable finish.

Plastic mortar
A mortar of a consistency that allows it to be readily used
without disintegrating; plasticity can be improved by additives called plasticizers.

Plastic veneers
Flexible plastic films with adhesive backs used to cover
various surfaces on which a fine finish is desired.

Plastic wood See Wood

Plasticizer
Admixture to mortar or concrete to increase its workability.

Plate
In wood-frame construction, a horizontal board connecting
and terminating posts, joists, or rafters; a timber laid horizontally on the ground to receive other timbers or joists.

Plate girder See Beam

Plate glass See Glass

Plate tracery See Tracery

Plateresque architecture (1474–1504)
The richly decorative style of the Spanish Renaissance in
the 16th century; also referred to as Isabelline architecture, after Queen Isabella.

Platform
A raised floor or terrace, open or roofed; a stair landing.

Platform framing
A system for framing wooden structures in which the studs
are only one-story high; the floor joists rest on the top
plates of the story below, and the bearing walls and partitions rest on the subfloor of each story.

Platt, Charles Adam (1861–1933)
American architect and landscape designer whose
Italianate gardens were famous; architect of the Freer
Gallery (1918), Washington, DC; designed the campus for
the University of Illinois (1930) , Champaign-Urbana, IL;
Coolidge Auditorium, Library of Congress (1925), Washing-
ton, DC; and the Deerfield Academy (1932), Deerfield, CT.
He published *Italian Gardens* in 1894.

Playhouse
A building in which plays and musical performances are
given; a theater; a small building serving children as a
make-believe home.

Plaza
An open square or market
place having one or more
levels, approached in
various ways by avenues,
streets, or stairs or a combination.

Plecnik, Joze (1872–1957)

Slovene architect who practiced in Vienna, Belgrade, Prague, and Ljubljana. He worked in Viennese architect Otto Wagner's office until 1900, and was affiliated with the Viennese Secession. From 1900 through 1910 Plecnik practiced architecture in Vienna, completing projects such as the Langer House (1900) and the Zacherlhaus (1905). These early projects featured organic motifs typical of the Secession. Plecnik's design for the Church of the Holy Spirit in Vienna (1913) used poured-in-place concrete as both structure and exterior surface. Most radical is the church's crypt, with its slender concrete columns and angular, cubist capitals and bases. His teachings influenced a generation of architects who would help define the avant-garde Czech Cubist movement of the 1920s. In 1920 he began work on Prague Castle, a medieval structure that dominates the historic capital. Renewed interest in Plecnik's work developed in the 1980s and 1990s, as postmodernism led to a reconsideration of classical forms and motifs in architecture.

Plenum

The air space in an integrated ceiling, which may be above the atmospheric pressure if used for the air supply, or below the atmospheric pressure if used for the air exhaust.

Plenum chamber

An air compartment maintained at a pressure slightly above atmospheric, to deliver and distribute the air to one or more ducts or outlets.

Plenum system

A method of heating or air conditioning whereby the air forced into the building is at a pressure slightly above atmospheric.

Plexiglas® See Plastic

Plinth

A square or rectangular base for column, pilaster, or door framing to support a statue or memorial; a solid monumental base, often ornamented with moldings, bas-reliefs, or inscriptions.

Plinth block

A flat, plain member at the base of a pillar, column, pedestal, or statue.

Plinth block

Plot

A parcel of land consisting of one or more lots or portions thereof, which is described by reference to a recorded plot by survey.

Plot ratio

The gross floor area of a building divided by the area of its site. The basic ratio permitted is frequently modified by providing a bonus for arcades, setbacks, plazas, and the incorporation of existing buildings of architectural significance.

Plug strip

An assembly of wire receptacle outlets. They are uniformly spaced and mounted in a metal strip enclosure for fastening to a wall. One such device is called wiremold.

Plug-in architecture (1964–1970)

A new type of architecture proposed by the English group Archigram. It consisted of a basic structure to contain transportation and communication services and a series of separate units—domestic environments, shops, and leisure activities—that are plugged into a central module.

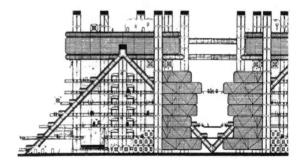

Plumb

Any method of lining up the building elements in a true vertical direction.

Plumber

A craftsperson skilled in the fitting of pipes, both for the supply of water and gas as well as for waste, soil, fire protection, and drainage.

Plumbing Code
A building code regulating the design of plumbing systems.

Plumbing fixture
Any of various devices for supplying and holding bathing water and removing human waste; types include basin, bathtub, lavatory, laundry sink, urinal and water closet.

Plumbing system
The combination of supply and distribution pipes for hot water, cold water, and gas and for removing liquid wastes in a building, including the water-supply distribution pipes; fixtures and traps; the soils, waste, and vent pipes; the building drain and building sewer; storm-drainage pipes; and all connections within or adjacent to the building.

Ply
A thickness of material used for building up several layers, as in plywood and built-up roofing.

Plywood See **Wood products**

Pneumatic architecture (1850–1880)
A term referring to a style of structures that are air-inflated, air-supported, and air-controlled. Structures generally consist of curved forms, domes, or half cylinders. Their rounded forms are organic and responsive to the technology which utilizes fabric and cables supported from within by air pressure.

Pneumatic structure
A structure held up by a slight excess of internal air pressure above the pressure outside; it must be sufficient to balance the weight of the roof membrane and must be maintained by air compressors or fans.

Pocket
A recess in a wall to allow passage of a sliding door hanging on a track.

Pocket door See **Door**

Podium
Any elevated platform; the high platform on which Roman temples were generally placed; a low step-like projection from a wall or building intended to form a raised platform for placing objects.

Podium

Poelzig, Hans (1869–1936)
Berlin-born architect who, as city architect of Dresden, Germany, designed the Grosse Schauspielhaus, Berlin (1919), and the Chemical Plant, Luban (1912). As a professor, he produced several fantastic expressionist designs, all unrealized.

Point
The smallest unit in a composition, depending on the scale of the work; it may be composed of straight lines and arcs, forms (flowing and curvilinear), or a combination.

Point method
A method of estimating the illuminance at various locations in a building, using photometric data.

Pointed arch See **Arch**

Pointed surface
The texture of stone blocks showing nicks made into the surface with a pointed tool, such as a pick or geologist's hammer.

Pointed work
The rough finish that is produced by a pointed tool on the face of a stone.

Pointing
The finishing operation on a mortar joint, without the addition of surface mortar; pressing surface mortar into an existing raked joint. Also, the process of filling a mortar joint after raking out the old mortar and working it to the desired joint profile; carried out for restoration purposes and not part of the original construction.

Points
Compliance with each LEED (Leadership in Energy and Environmental Design) credit earns one or more points toward certification. Compliance with prerequisites is required and does not earn points.

Pokorny, Jan Hird (1915–2008)
Pokorny led the restoration of the Schermerhorn Row Block in New York City's South Street Seaport, the Brooklyn Historical Society building, the National Lighthouse Museum on Staten Island, NY, and the Battery Maritime building in lower Manhattan. He designed four buildings and prepared the master plan for Lehman College in the Bronx, NY and was also in charge of the central campus plan for the State University of New York at Stony Brook, NY.

Pole
A slender log used as a structural member, with or without the bark removed.

Pole structure
A building or structure with a roof supported by round wood columns.

Police power
The inherent right of a government to restrict individual conduct or use of property to protect the public health, safety, and welfare. Police power is the basis for such regulations as zoning, building codes, and preservation ordinances.

Polk, Willis (1867–1924)
An American architect working in San Francisco; influenced by McKim Mead and White. He assisted Daniel Burnham in the city plan for San Francisco, CA (1904). The Hallidie Building, San Francisco, CA (1917), is his most distinctive work, having a fully glazed curtain wall hung from the main framed structure, the first of its kind.

Pollution
Any direct or indirect alteration to the environment which is hazardous, or potentially hazardous, to health, safety, and welfare of any living species.

Pollution prevention
Reducing the amount of energy, materials, packaging, or water in the design, manufacturing, or purchasing of products or materials in an effort to increase efficient use of resources, reduce toxicity, and eliminate waste.

Polshek, James Stewart (1930–)
Most noted for the New York State Bar Center, Albany, NY (1971), and most recently designer of the new Rose Planetarium at the Museum of Natural History (illus.), NYC (2000).

Polychromatic
Having or exhibiting a variety of colors.

Polychromy
The practice of decorating architectural elements or sculpture in a variety of colors.

Polyethylene See **Plastic**

Polygonal
Forms characteristic of a closed plane figure having three or more straight sides.

Polygonal masonry See **Masonry**

Polygonal vault See **Vault**

Polyhedron
A solid geometric figure bounded by plane faces.

Polyisocyanurate
Type of rigid foam insulation used in above-grade walls and roofs, typically with a foil facing on both sides. This kind of insulation was made with ozone-depleting HCFC-141b blowing agent, but manufacturers have switched to ozone-safe hydrocarbons.

Polymer concrete
A concrete in which a plastic is used as a binder, instead of Portland cement.

Polypropylene
A common flexible plastic usually spun into fiber for rope and woven goods.

Polystyle colonnade
Colonnade with many columns, such as one situated around a building.

Polystyrene See **Plastic**

Polysulfide
A thermosetting resin, used as a building sealant.

Polyurethane
A group of plastics used mainly as a light insulating material in the form of flexible or rigid foam and as a sealant.

Polyurethane foam
Insulation material made from polyol and isocyanate and a blowing agent that causes it to expand, typically sprayed into wall cavities or sprayed on roofs.

Polyvinyl chloride (PVC)
Most common plastic in building construction, widely used in such applications as drainage piping, flooring, exterior siding, window construction, and electrical wire.

Pompeii (1592)
An ancient city southeast of Naples that was destroyed by the eruption of Mount Vesuvius in 79 A.D. The ruins of Pompeii were first discovered in 1592. Excavations of the ruins did not begin until 1709.

Ponti, Gio (1891–1979)
Italian architect and designer, influenced by the Sezession movement and Otto Wagner. He was the founder-director of *Domus* magazine (1928). He worked with Nervi in the design of the Pirelli Tower in Milan, Italy (1956), and designed the Museum of Modern Art (illus.) in Denver, CO (1972).

Pop architecture (1962–1974)
A style which refers to structures that symbolically represent objects, to fantastic designs for vast sculptures on an architectural scale, or to any architecture produced more as metaphor than building.

Pop architecture

Pope, John Russell (1874–1937)
Disciple of McKim Mead and White; trained at the École des Beaux-Arts, Paris. Designed the Jefferson Memorial (1937) and the National Gallery of Art (1937), both in Washington, DC, and the Sculpture Hall, Tate Gallery, London (1937).

Poppelmann, Daniel (1662–1736)
German architect, who designed the Zwinger Palace, Dresden (1722), with Marcus Dietze. It is a Baroque structure with rich sculptural decoration.

Porcelain enamel
A glassy metal oxide coating bonded to a metal panel at an extremely high temperature and baked onto steel panels for large architectural applications. It is a very durable material that is scratch resistant.

Porch
A roofed entrance, either incorporated in a building or as an applied feature on the exterior.

Porch

Porch cornice
A continuous band at the cornice of a porch, filled with a spindlework design.

Porous paving
A paving material that allows rainfall to percolate through and infiltrate the ground, rather than contributing to stormwater runoff; can be asphalt, concrete, or grid paver.

Porta, Giacomo della (1533–1604)
Italian Mannerist architect who completed the dome of St. Peter's in Rome and also completed other designs by Michelangelo.

Portable lamp

A decorative source of light that can be moved around and plugged into any power source. Types include floor lamps and table lamps.

Portable luminaire

A luminaire equipped with a cord and plug and designed to be moved from space to space.

Portal

An entrance, gate, or door to a building or courtyard, often decorated; it marks the transition from the public exterior to the private interior space.

Porthole

A small window, usually circular, in a ship's side, or on an exterior part of a structure.

Portico

A range of columns or arches in front of a building, often merged into the facade, including a covered walkway of which one or more sides are open. It includes every kind of covered ambulatory.

Portico

Portman, John, Jr. (1924–)

American architect/developer known for his large urban buildings, including Peachtree Center (1961) and the Hyatt Regency Hotel (1967), both in Atlanta, GA. He also designed the Hyatt Regency Hotel (illus.), and Rockefeller West, both in San Francisco, CA (1975); and an office tower, One Peachtree Center, Atlanta, GA (1992).

Portland cement

Building material made from limestone, gypsum, and shale or clay that, when mixed with water, binds sand and gravel into concrete. Portland cement was invented in 1824 by Joseph Aspdin, a British stone mason, who named it after a natural stone quarried on the Isle of Portland off the British coast.

Post
Any stiff, vertical upright, made of wood, stone, or metal, used to support a superstructure or provide a firm point of lateral attachment.

angle post
The corner post in half-timbered construction.

crown post
Vertical member in a roof truss, especially a king post.

king post
A vertical member extending from the apex of the inclined rafters to the tie beam between the rafters at the lower ends of a truss, as well as in a roof.

queen post
One of the two vertical supports in a queen-post truss.

Post office
An office or building where letters and parcels are received and sorted, and from where they are distributed and then dispatched to various destinations.

Post, George B. (1837–1913)
American architect and engineer, designed early skyscrapers of 10 stories. His best-known work is the New York Stock Exchange, NYC.

Post-and-beam
A type of framing in which horizontal members are supported by a vertical post rather than by a bearing wall, or a system of arches and vaults.

Post-and-lintel
A type of construction characterized by the use of vertical columns, posts and a horizontal beam, or lintel to carry a load over an opening, in contrast to structural systems employing arches or vaults.

Postern
A minor, often inconspicuous, entry; a small door or gate near a larger one.

Postmodernism (1980–2000)
A reaction against the International style and Modernism was evidenced in this style. It reintroduced ornament and decorative motifs to building design, often in garish colors and illogical juxtaposition. It is an eclectic borrowing of historical details from several periods, but unlike previous revivals is not concerned with scholarly reproduction. Instead, it is a light-hearted compilation of esthetic symbols and details, often using arbitrary geometry, and with an intentional inconsistency of scale. The most prevalent aspect is the irony, ambiguity, and contradiction in the use of architectural forms. Those connected with the beginning of this movement include Aldo Rossi, Stanley Tigerman, Charles Moore, Michael Graves, Robert Krier, and Terry Farrell.

Postmodernism

Postoccupancy evaluation
A systematic way of comparing actual building performance with stated performance criteria usually undertaken by organizational or facility managers; usually done after the building has been occupied for at least one year. It seeks to measure and evaluate user satisfaction, fitness for purpose based on client requirements, technical performance, and value for the cost.

Potable water
The water suitable for drinking and supplied from wells or municipal water systems.

Potter, Edward T. (1831–1904)
American architect in New York who specialized in High Victorian Gothic churches and colleges. His polygonal Nott Memorial (1878), Schenectady, NY, featured an exposed iron interior structure, domed clerestory, and Moorish polychrome exterior. He also designed the Mark Twain House (1881) with Alfred H. Thorp, Hartford, CT, emphasizing the Stick style and French domestic architecture.

Potter, William A. (1842–1909)
American architect in New York; a supervising architect of the Treasury Department who also designed suburban estates and collegiate buildings. Projects include Chancellor Green Library (1873), Princeton University, Princeton, NJ; South Congregational Church (1875), Springfield, MA; and Customhouse and Post Office (1879), Evansville, IN.

Pounce
A fine colored powder, used to transfer mural designs to a wall surface; the design has been stenciled into heavy paper with a pounce wheel.

Pouncing
Using a pounce bag filled with a colored powder to transfer the dotted outlines of a mural design to a wall.

Prairie school
Refers to the work of Frank Lloyd Wright's contemporaries and followers, who adopted the style suited to the midwest, rather than to Wright's work itself.

Prairie style (1900–1940)
A style that is typical of the low horizontal house associated mostly with the work of Frank Lloyd Wright and his followers. Horizontal elements are emphasized in these one- or two-story houses, built with brick or timber and covered with stucco. The central portion that rises above the flanking wings are separated by clerestory windows. The eaves of the low-pitched roof extend well beyond the wall. A large chimney is located at the axis of intersecting roof planes. Casement windows are grouped into horizontal bands continuing around the corners. Exterior walls are highlighted by dark wood strips against a lighter stucco finish or by a coping of smooth stucco at the top of brick walls.

Prakash, Aditya (1923–2008)
Indian architect who worked with Le Corbusier on the Chandigarh Capital Project, commissioned by Indian prime minister Jawaharlal Nehru to build a new Punjabi capital. Prakash worked closely with Le Corbusier on the design of the School of Art in Chandigarh and collaborated with British architect Jane Drew on the General Hospital in Chandigarh, India. He later adapted the School of Art into the Chandigarh College of Architecture. In addition to his many Chandigarh projects, he designed the Panjab Agricultural University in Ludhiana, India.

Prandtauer, Jacob (1660–1726)
Austrian architect. Rebuilt Melk Abbey (1702), which is dramatically sited above the Danube River. Its facade is undulating Baroque in a mixture of alternating white and cream-colored bands.

Pratt truss See Truss

Precast
A building material that is molded before it is incorporated into a building; used for both structural and ornamental elements.

Precast concrete
Material that reduces the need for onsite formwork with a process known as tilt-up construction, in which precast panels are lifted into a vertical position and then attached to the structural frame.

Precast pile
A concrete pile that is cast and subsequently driven, as opposed to one cast in place.

Precipitation
All liquid or solid phase aqueous particles that originate in the atmosphere and fall to the Earth's surface.

Pre-Columbian architecture
The design and construction of Native American structures prior the voyage of Columbus and contact with any European culture.

Predevelopment hydrology
The combination of runoff, infiltration, and evapotranspiration rates and volumes that typically existed on a site before human-induced land disturbance occurred, such as construction of infrastructure on undeveloped meadows or forests.

Predock, Antoine (1936)
New Mexico-based architect using stark, abstract forms and natural desert materials, whose most well-known works are the Nelson Fine Arts Center, Arizona State University, Tempe, AZ (1989), and the Las Vegas Library and Discovery Museum (illus.), Las Vegas, NV (1990).

Prefabricated
Standardized building sections that are created in a factory to be shipped and assembled in another location.

Prefabricated construction
A building so designed as to involve a minimum of assembly at the site; usually comprising a series of large units or panels manufactured in a factory.

Prefabricated house
A house assembled from components cut to size at a factory, or assembled from entire building modules shipped to the construction site.

Prefabricated modular units
Units of construction which are prefabricated on a measurement base of 4 inches, or multiples of 4, and can be fitted together on the job with a minimum amount of adjustments.

Prefabrication
The manufacture of standardized units or components, usually at a mill or plant away from the site, for quick assembly and erection on the job site.

Prefinished
Materials, such as doors, moldings, cabinets, and paneling, which have been painted or stained and varnished in the shop.

Prefunctional performance test
Involves a series of tests for specified equipment or systems, which determine that the systems are installed correctly, start up, and are prepared for the functional performance tests. Often these tests are in a checklist format and may be completed as part of the normal contractor startup test.

Prehistoric archeology
The scientific study of human life and cultures that existed before recorded history.

Prehistoric architecture
Design of buildings before written records were kept.

Prehistoric era
The era before written or recorded human events, knowledge of which is gathered through archaeological investigations, discoveries and research.

Preliminary estimate
Project costs prepared by the architect during the development phase for the guidance of the owner.

Premises
The property, including the buildings, structures and grounds, that is included in a title to ownership, or a deed of conveyance.

Prequalification of prospective bidders
Investigating the qualifications of prospective bidders on the basis of their competence, integrity, and responsibility relative to the contemplated project.

Prerequisites
LEED Green Building Rating System component. Compliance is mandatory for achieving certification but does not count toward the accumulation of points.

Presentation drawing See **Design drawing**

Preservation
Protection of a material from physical deterioration due to natural elements or human activity; by various technical, scientific, and craft techniques

Preservation architect
Same as historical architect.

Preservation by deed
A method used by preservationists to use deed restrictions to curtail development or alteration of a structure or area. A preservation organization may acquire then sell a property subject to a covenant and reverter clause in the deed that the purchaser must comply with certain restoration guidelines. If not, the property reverts back to the original organization.

Preservation commission
A generic term for a municipal agency that designates and regulates historic districts and landmarks; it may be called a historic district review board or commission, landmarks commission, or architectural or design review board.

Preservation ordinances
Regulations regarding preservation. There are two kinds of statues: landmark commission ordinance and historic district ordinance. These ordinances define the architectural and historical standards of value in the area based on an architectural survey. Each sets up an agency with authority to review proposals to alter or demolish designated structures and to restrict such actions. Each provides a procedure by which decisions can be appealed by applicants, either to a higher administrative authority or zoning board, city council, or directly to a court.

Preservation plan components
Consist of a historical overview, treatment of architectural styles, construction , description of the community inventory of significant cultural resources, indication of problem areas; establishment of preservation planning goals and objectives; suggestion of cultural management program; and identification of potential funding sources.

Preservation restrictions
May involve easements, restrictive covenants, deed restrictions, leasing restrictions clauses and similar devices.

Preservation survey
An inventory of property that can be limited to a carefully defined geographic area, such as a neighborhood or enclave of buildings, or it can encompass an entire state or the nation. It can be limited to places of obvious cultural significance or individual building types, such as commercial, or it can cover all built and natural resources. The amount of information can include construction dates and architectural style, as well as an in-depth analysis of an individual structure, such as measured drawings, photographs, chain-of-title information, and technical architectural descriptions. Far-reaching surveys can include the area's social and economic factors, development patterns, visual relationships and design elements. A survey pro-duces a wealth of material such as historical data, development statistics, photographs, measured drawings and maps.

Preservationist
A person or professional who aspires to, agrees with, and takes action to preserve the natural and built environment.

Preservative
A substance that inhibits decay, infection, or attack by fungi and insects in timber.

Preset dimming
Multichannel save-dimming systems preset a dimming level in advance. The user needs only to push a button to bring all lighting in the space to the desired level for a specific function. Used for restaurants, conferencing facilities, auditoriums, hotel lobbies, and custom-built homes.

Presidential sites
U.S. presidents are commemorated by scores of sites, ranging from the humble to the palatial: birthplaces, residences, libraries, monuments and tombs.

Pressed brick See **Brick**

Pressed-metal ceiling
A thin sheet metal that is embossed with a decorative pattern and coated with a layer of tin or lead or prepared with an undercoat of paint to prevent oxidation; used as ceiling panels by nailing to the underside of beams or furring strips.

Pressure meter
A device for gauging the difference in air pressure between two spaces such as a garage and a crawl space.

Pressure-assist toilet
A toilet that uses air pressure, generated as the toilet tank refills, to produce a more forceful flush; some of the highest-performance high-efficiency toilets rely on pressure-assist technology.

Pressure-glued
A method of gluing which places the wood members under high pressure until the glue sets.

Pressure-treated wood
Wood that has been chemically treated to extend its life, especially when outdoors or in contact with the ground. Chromated copper arsenate was the most common pressure-treated wood until a few years ago, but has now been phased out for most applications because of health and environmental concerns. Other pressure-treating chemicals include ACQ (alkaline copper quaternary), copper azole, and sodium silicate.

Prestressed concrete See Concrete

Pretensioning
Prestressed concrete in which the tendons are tensioned before the concrete has hardened, and generally before it is cast; the concrete is cast around the wires which are flame-cut after the concrete has hardened.

Preventive maintenance
The maintenance contractor performs regular inspections, maintenance, and calibration according to an agreed schedule, and reports any obvious deterioration of the plant. The owner is responsible for the cost of all major repairs.

Price, Bruce (1845-1903)
American architect in New York City who designed Shingle style houses that influenced Frank Lloyd Wright among others. Projects include Tuxedo Park (1890), NY, and Chateau Frontenac Hotel (illus.), Quebec, Canada (1893).

Primary colors
Hues red, yellow, and blue. From these three colors, with the addition of white, it is possible to create the full color spectrum.

Prime contractor
The contractor who takes responsibility for the entire project; may build all of the project, or have subcontractors build the entire project, or may build part of it and subcontract out the rest.

Primer coat
Ground coat of paint applied to timber and other materials as a preservative and as a filler for the pores, which serves as a base for future coats of paint.

Priming coat
The first or ground coat of paint applied to wood to fill the pores of the surface. The priming paint is mixed with turpentine to make it thinner than normal consistency.

Prince, Bart (1947-)
American architect who studied under Bruce Goff, proponent of organic architecture, designed the Prince House, Albuquerque (1984), and the Price House, (illus.) Corona Del Mar, CA (1989).

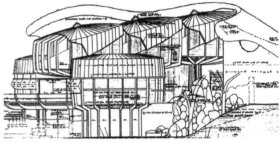

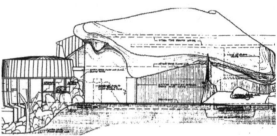

Principal purlin
A massive wood purlin that runs parallel to the ridge of the roof about midway between the ridge and top plate: it is framed into the principal rafters and supports a number of common rafters.

Principal rafter
One of several such wooden rafters that extend from the ridge uninterrupted down to the wall plate; more massive than common rafters, and framed into a tie beam that makes the whole assembly more stable.

Prism
A geometric solid with regular polygons at its ends and parallelograms on its sides connecting the ends.

Prismatic
Characteristic of a solid figure in which the two ends are similar and parallel, with parallelograms for sides: used extensively in space frames covering open areas or large atrium areas.

Prismatic glass
Rolled glass that has parallel prisms on one face. These refract the transmitted light and thus change its direction.

Prison
A place where persons convicted or accused of crimes are confined.

Pro forma
A statement of the economic analysis of the costs and value of a proposed real estate development; usually includes land costs, hard and soft costs, equity, financing and sales price.

Probst, Edward (1870–1942)
American architect and partner in the successor firm to Daniel H. Burnham in Chicago: Graham, Anderson, Probst and White.

Professional liability insurance
Insurance designed to insure an architect, engineer, or any person or firm undertaking design responsibility, such as design-build contractor, against claims for damages resulting from alleged professional negligence.

Profile
An outline of a form or structure seen or represented from the side, or one formed by a vertical plane passed through an object at right angles to one of its main horizontal dimensions.

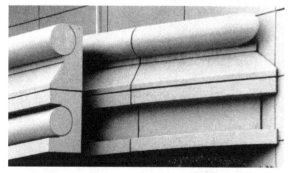

Program
A written document that defines the intended functions and uses of a building, or site, and is used to initiate and control an architectural design or preservation project.

Programmable thermostat
A thermostat that allows homeowners to set the temperature at different levels at different times of day. For example, in winter, it could be set to be colder while occupants sleep and warmer as occupants awaken.

Progress schedule
A diagram, graph, or other pictorial or written schedule showing proposed and actual times of starting and completing of the various elements of the work.

Progression
A gradual increase in the size or shape of a form or design, keeping the same basic theme or idea.

Project
The total activity that is to be undertaken, whether a construction or renovation project, that is the total construction work to be performed under the contract documents.

Project control
The ability to determine the project status as it relates to the timeline and schedule.

Project management
The practice of developing, planning, procuring, and controlling all the services and activities required for the satisfactory completion of a project according to an agreed time and cost.

Projecting window See Window

Projection
Any component, member, or part that juts out from a building; in masonry construction, stones or bricks that are set forward off the general wall surface to provide a rugged or rustic appearance.

Projection drawing

The process or technique of representing a three-dimensional object by projecting all its points by straight lines, either parallel or converging to a picture plane.

cutaway

A drawing or model having an outer section removed to display the interior space.

diametric projection

An axonometric projection of a three-dimensional object to the picture plane in such a way that two of its principal axes are equally foreshortened, and the third appears longer or shorter than the other two.

exploded view

A drawing that shows the individual parts of a structure or construction separately, but indicates their proper relationship to each other and to the whole.

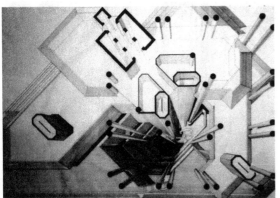

axonometric projection

The orthographic projection of a three-dimensional object inclined to the picture plane in such a way that its three principal axes can be drawn to scale but diagonal and curved lines appear distorted.

isometric projection

An axonometric projection of a three-dimensional object is created by having its principal faces equally inclined to the picture plane; so that its three principal axes are all equally foreshortened.

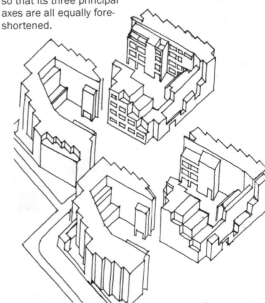

longitudinal section

An orthographic projection of a section made by cutting through the longest axis of an object.

oblique projection

A method of projection in which a three-dimensional object, having one principal face parallel to the picture plane, is represented by projecting parallel lines at some angle other than 90 degrees.

oblique section

An orthographic projection of a section made by cutting with a plane that is neither parallel nor perpendicular to the long axis of an object.

orthographic projection
A method of projection in which a three-dimensional object is represented by projecting lines perpendicular to a picture plane.

paraline drawing
Any of various single-view drawings characterized by parallel lines remaining parallel to each other rather than converging as in linear perspective.

section
An orthographic projection of an object or structure as it will appear if cut through by an intersecting plane to expose its internal configuration.

transverse section
An orthographic projection of a section made by cutting through an object along the shortest axis.

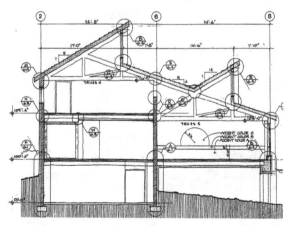

trimetric projection
An axonometric projection of a three-dimensional object inclined to the picture plane in such a way that all three principal axes are foreshortened at a different rate.

Promenade
A suitable place for walking for pleasure, as in a mall or pedestrian way.

Pronaos
Vestibule in a Greek temple that stands in front of the doorway to the naos.

Property
A plot or parcel of land, including buildings or other improvements. Also called real property.

Property insurance
Insurance on the work at the site against loss or damage caused by fire, lightning, wind, hail, vandalism, and additional perils that may occur.

Property line
A boundary line between plots of land, such as farms or lots; the location of a property line is recorded in the legal description of the piece of land.

Property title
A document stating right to ownership, or deed, especially of real estate, which is a court-recorded item.

Proportion
The ratio of one part to another, or its relationship to the whole; a comparison of parts as to size, length, width, and depth.

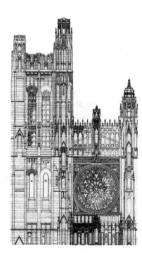

Proposal
A written offer from a contractor to perform work and/or supply materials; typically describes the extent of the work and its cost.

Propylaeum
The monumental gateway to a sacred enclosure; specifically, the elaborate gateway to the Acropolis in Athens.

Propylon
A monumental gateway, usually between two towers, in ancient Egyptian architecture. One, or a series, stood in front of the actual entrance or pylon of most temples and other important buildings.

Proscenium
The portion of a theater stage between the drop curtain and the orchestra.

Proscenium arch See Arch

Prostyle
Having a portico featuring columns at the exterior front of a building only.

Protection
The Secretary of the Interior's Standards indicates that *protection* means "to effect the physical conditions of a property by defending or guarding it from deterioration, loss or attack, or to cover or shield the property from damage or injury." In the case of buildings, it is generally temporary, anticipating future historic preservation treatment.

Pseudoperipteral
A Classical temple or other building which has columns all the way around; those on the flanks and rear are engaged rather than freestanding.

Public archeology
Archeological programs and projects funded by governments.

Public building
Buildings of, concerning, or affecting the community of people on the town, city, state, and federal level.

Public garden
An urban ground laid out with walks, plantings, and buildings, which offers a variety of entertainment and recreation to the public.

Public housing
Low-cost housing owned, sponsored, or administered by a municipal or other government agency.

Public transportation
Mass transit, including bus and light rail systems. Siting a building near public transit is considered a sustainable building strategy, as it facilitates commuting without the use of single occupancy vehicles.

Pueblo architecture (1905-1940)
A communal dwelling and defensive structure of the Pueblo Indians of the southwestern United States, built of adobe or stone, typically multistoried and terraced, with entry through the flat roofs by ladder.

Purcell, William (1880–1965)
Scottish-born American architect who, with partner George Grant Elmslie (1871–1952), helped establish the Prairie School style as applied to the many homes they designed. He had previously worked with Adler and Sullivan on the ornamentation for the Cage Building (1898) and the Carson Pirie Scott Store (1904), both in Chicago.

Putti
Chubby, usually naked infants represented in decorative sculpture or painting; used in Renaissance architecture and classical derivatives.

Pyramid
A polyhedron with a polygonal base and triangular faces meeting at a single common apex.

Q q

Quadra
Square border or frame enclosing a wall panel or painting; also refers to a square base or plinth at the bottom of a pedestal or podium.

Quadrafron
Having four fronts or faces looking in four directions.

Quadrafron capital See **Capital**

Quadrangle
A rectangular courtyard or grassy area enclosed by buildings, most often used in conjunction with academic or civic building groupings.

Quadrant
Arc of a circle, forming one-quarter of its circumference.

Quadratura
In Baroque interiors and derivatives, painted architecture, often continuing into the three-dimensional trim, executed by specialists in calculated perspective.

Quadrilateral
A plane figure bounded by four straight lines. If two of the lines are parallel, it is a trapezium. If two pairs of lines are parallel, it is a parallelogram. An equilateral parallelogram is a rhombus. A right-angled rhombus is a square.

Quadro-riportato
The simulation of a wall painting for a ceiling design in which painted scenes are arranged in panels that resemble frames on the surface of a shallow, curved vault.

Quaint
A term used to refer to antique or old-fashioned styles, such as English style cottages and Queen Anne houses.

Quality assurance
The system for assuring that the contract's technical requirements are met by the contractor. Also the management of quality at all stages in the manufacture of a product and delivery of services, as standardized internationally by ISO 9000.

Quality control
Tests and sampling techniques to see that the required quality of construction is provided.

Quantity survey
An inventory compiled for the purpose of estimating the amount of material and labor required to complete a construction operation.

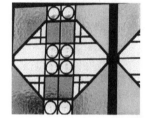

Quarry
A small, square or diamond-shaped pane of glass used in leaded windows.

Quarry tile See **Tile**

Quarry-faced masonry See **Masonry**

Quartered
Lumber that is sawn from a log at a perpendicular angle to the growth rings; the lumber sawn in this manner is easier to dry without warping and shrinking; used for flooring and weatherboards.

Quartered veneer
A veneer in which a log is sliced or sawed to bring out a certain pattern of the grain.

Quarter-round molding See **Molding**

Quartzite See **Stone**

Quatrefoil See **Foil**

Quattrocento
In the style of the early Italian Renaissance during the 15th century; from the Italian meaning 400, referring to dates starting with the number 14, that is, 1400 to 1499.

Queen Anne arch
An arch over the triple opening of the Venetian or Palladian window, flat over the narrow sidelights, round over the larger central opening.

Queen Anne style (1875–1890)
Similar to high Victorian, an exuberant and eclectic style in texture, form, and massing. Based on Elizabethan and

Queen Anne style

Jacobean precedents, brick and stone were combined in bold contrast with tall thin chimneys, multiple gables, complex roof shapes and turrets, towers, and bay windows, all with small-scale detailing.

rustic quoin
A stone quoin, projecting out from the main surface of the wall with rough, split faces and chamfered edges, to give the appearance of rugged strength.

Queen post See **Post**

Quirk
An indentation separating one element from another, such as between moldings.

Quirked bead molding See **Molding**

Quirked molding See **Molding**

Quoin header
A quoin which serves as a header in the face of a wall and a stretcher in the face of the return wall.

Quoining
Any material that forms a quoin.

Quoin
One of a series of stones or bricks used to mark or visually reinforce the exterior corners of a building; often through a contrast of size, shape, color or material, which may be imitated in non-load-bearing materials.

Quonset hut
A prefabricated military structure in the form of a barrel vault with flat ends, and composed of corrugated sheet steel that is reinforced with steel ribs.

R r

Rabbet
A long groove or channel that is cut into the edge or face of a board to receive another board that is fitted into the groove at a right angle to it.

Rabbeted
Two members joined together by interlocking grooves cut into each one.

Raceway
A continuous rectangular metal duct for encasing electrical work.

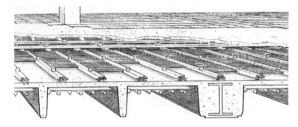

Radial
Forms radiating from or converging to a common center or developing symmetrically about a central point.

Radial dome See **Dome**

Radial organization See **Organization**

Radial symmetry See **Symmetry**

Radiant barrier
A material that reflects radiant heat, typically a foil-faced or foil-like material used in roof systems. Used properly in some climates, it can reduce cooling requirements but has no positive effect on heating requirements.

Radiant energy
Energy traveling in the form of electromagnetic waves, measured in units of energy such as joules or kilowatt hours. When this radiation is absorbed by an object in its path, then it may get absorbed partly or completely and transformed into some other form.

Radiant flux
Radiant flux is radiant energy per unit of time. The standard unit for measuring radiant flux is the watt (W), which is defined as Joules per second (J/sec).

Radiant heat
The heat transmitted to a body by electromagnetic waves, as distinct from heat transferred by thermal conduction or convection. In green building, radiant heat is generally defined as heat radiating from a floor or wall and being transferred from a hot source within the wall or floor, usually circulated water. Produces a uniform heat within the room. Zoning the heat is a method of concentrating heat only on areas that need it.

Radiant heating
Radiant heating systems involve supplying heat directly to the floor or to panels in the wall or ceiling of a house. When radiant heating is located in the floor, it is often called radiant floor heating or simply floor heating. Radiant heating is more efficient than baseboard heating and usually more efficient than forced-air heating because no energy is lost through ducts. Hydronic or liquid-based systems use little electricity and can also be heated with a wide variety of energy sources, including standard gas- or oil-fired boilers, wood-fired boilers, solar water heaters, or some combination of these heat sources.

Radiant temperature
The average temperature of the objects and surfaces that surround us, which radiate heat to, and absorb radiant heat from, our bodies.

Radiant-heated floors
Heating spaces using radiant energy that is emitted from a heat source. There are three types of radiant floor heat: air; electrical; and hot water (hydronic).

air
A system in which air is the heat-carrying medium.

electrical
A system that consists of electric cables built into the floor, usually made of concrete.

hydronic
A popular and cost-effective choice that pumps heated water from a boiler through tubing underneath the floor. In some systems, the temperature in each room is controlled by regulating the flow of hot water through each tubing loop.

Radiating chapels
The chapels added to an apse and fanning out in a radial pattern.

Radiator
A heating unit which transfers heat by radiation; usually fed by hot water or steam.

Radon
A naturally occurring gas, colorless and odorless, that has been shown to cause adverse health effects. Radon gas often enters a structure by seeping through cellar walls and floors.

Rafter
One of a series of inclined members that support the sheathing to which a roof covering is fixed.

hip rafter
A rafter located at the junction of the sloping sides of a hip roof.

jack rafter
Any rafter shorter than the full length of the sloping roof, such as one beginning or ending at a hip or valley.

valley rafter
In a roof framing system, the rafter in the line of the valley; connects the ridge to the wall plate along the meeting line of two inclined sides of a roof that are perpendicular to each other.

Rafter plate
A plate that supports the lower end of rafters and to which they are fixed.

Rafter tail
Portion of a rafter that projects beyond the exterior wall to support the eaves

Rail
A bar of wood or other material passing from one post or support to another support; a horizontal piece in the frame or paneling, as in a door rail or in the framework of a window sash.

Railing
Any open construction or rail used as a barrier, composed of one or a series of horizontal rails supported by spaced upright balusters.

Railing

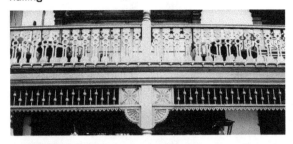

Railroad station
The structure where a train stops to load and unload passengers or freight; types range from a simple platform at grade to a large building with access to multiple raised covered platforms, which includes additional functions in the terminal itself.

Railroad stations
Buildings that were part of a transportation system of rail lines. Of the approximately 40,000 stations, which served the nation during the height of the railroad era, an estimated 20,000 still remain, of which most await rehabilitation. The railway station was usually built as the gateway into the city to impress arriving passengers with its economic and social character. With the demise of the passenger train as a prime means of transportation, railroad stations have become redundant structures. Today many are underused or abandoned. Because of their location in or near the centers of towns and cities they once served, they are excellent opportunities for adaptive use or even reused as passenger stations.

Rain barrels
Rain barrels promote water conservation by collecting rainwater from roofs for use in irrigation. They are often used in association with native plants and Xeriscape™ landscaping.

Rain chain
A water feature that is used as an alternative to a downspout. Rain chains guide runoff from a roof to either the ground, a cistern, or a rain barrel.

Rain garden
A shallow, constructed depression that is planted with deep-rooted native plants and grasses. It is located in the landscaping to receive runoff from hard surfaces, such as a roof, sidewalk, or driveway. Rain gardens slow the rush of water from these hard surfaces and hold the water for a short period of time, allowing the water to naturally filter into the ground. Rain gardens help control stormwater runoff.

Rain leader
A pipe that carries stormwater from the gutter to a downspout and eventually to the ground.

Rain/freeze sensors
Prevent automatic sprinkler systems from watering during rain or cold weather.

Rainforest Alliance
An organization that works to conserve biodiversity and ensure sustainable livelihoods by transforming land-use practices, business practices, and consumer behavior. Companies, cooperatives, and landowners that participate in its programs must meet rigorous standards that conserve biodiversity and provide sustainable livelihoods.

Rainproof
Constructed, protected, or treated to prevent rain from entering a building.

Rainscreen
Construction detail appropriate for all but the driest climates to prevent moisture entry and to extend the life of siding and sheathing materials; most commonly produced by installing thin strapping to hold the siding away from the sheathing by a quarter-inch to three-quarters of an inch.

Rainwater catchment
Onsite rainwater harvest and storage systems used to offset potable water needs for a building and/or landscape. Systems can take a variety of forms, but usually consist of a surface for collecting precipitation from a roof or other impervious surface and a storage system. Depending on the end use, a variety of filtration and purification systems may also be employed.

Rainwater catchment systems
Catches and stores rainwater by various means for uses such as irrigation, plumbing, or cooling towers, thereby reducing stormwater runoff. This is a fuinction of a stormwater management program.

Rainwater cistern
An underground basin of water, but it can also be an above-ground barrel or tank. Much like an artificial well, cisterns are used to make sure that water is collected and used for irrigation, thus conserving water resources.

Rainwater harvest
Systems implemented that capture and collect rainwater from roof drainage for use onsite. Some systems filter and purify the water, while others provide a means to distribute it, as in irrigation.

Raised floor See **Floor**

Raised molding See **Molding**

Raised panel
A panel whose center portion is thicker than the edges or projects above the surrounding frame or wall surface.

Raised skylight
A skylight with the glazing frame supported on a curb above the roof surface.

Raising
The process of physically lifting a framework into a vertical position by a team of workers, then connecting it to the other timbers. The framework was previously assembled on the ground, or on the floor decking; a technique frequently used for building the timber frames of barns.

Rake
The slope, or angle of inclination; the context usually indicates whether it is measured from the horizontal or the vertical axis.

Raked mortar joint See **Joint**

Raking bond See **Bond**

Raking coping See **Coping**

Raking cornice See **Cornice**

Raking molding See **Molding**

Rammed earth construction
A building technique involving dense compression of clay and dirt materials to create thick, flat surfaces, such as walls or floors.

Ramp
An inclined plane, used often as an alternative to stairs in a round or circular structure. It is also useful in complying with modifications to an existing structure for the ADA (Americans with Disabilities Act) requirements. See also Accessibility (handicapped).

Ramp

Rampant arch See Arch

Rampant vault See Vault

Rampart

An earthen wall located on the inner side of a ditch surrounding a bastion for purposes of defense.

Ranch style

A one-story house of rambling type, originally in the warmer climates of the U.S. West. The roof is low-pitched. The plan is geared to outdoor living with rooms frequently arranged around a court.

Random ashlar masonry See Masonry

Random coursed ashlar See Masonry

Random rubble See Masonry

Random shingles

Wood shingles of different widths banded together; often these vary from 3 to 12 inches in width.

Range

In masonry, a course of stonework laid in courses.

Rapidly renewable

Materials that are not depleted when used. These materials are typically harvested from fast-growing sources, do not require unnecessary chemical support, and take 10 or fewer years to grow or raise and to harvest in an ongoing and sustainable fashion. Examples include bamboo, flax, wheat, wool, and certain types of wood.

Rapson, Ralph (1914–2010)

Minneapolis-based architect, whose career began working with Eero Saarinen, Charles and Ray Eames, Harry Weese, Lazlo Moholy-Nagy, and at MIT with Alvar Aalto. Notable works include Cedar Square West (1974) and the Tyrone Guthrie Theater (illus.), both in Minneapolis, MN (1963). Rapson designed modernist embassies for the United States in Copenhagen and Stockholm. Other works include the Rang Center for the Performing Arts and the Fine Arts buildings for the University of Minnesota, the Cedar-Riverside complex, the State Capitol Credit Union, and the Performing Arts Center at the University of California, Santa Cruz.

Rastrelli, Bartolomeo F. (1700–1771)

Italian architect who settled in St. Petersburg, Russia, when his father, an architect and sculptor, was called there in 1715. His work in Russia includes the Andreas Church (1768), Kiev; renovation and extension to the Summer Palace, (1757), Peterhof, and the Strogonov Palace (1754), St. Petersburg; the Smolny Cathedral and Convent (1759), St, Petersburg; and the fourth Winter Palace (1762), now the Hermitage Museum, St. Petersburg, his last and finest work. Much of his interior work had been altered, but it is now being restored.

Ratio
A relationship in magnitude, quantity, or degree between two or more similar things.

Ravage
To lay waste or destroy by any means.

Raymond, Antonin (1888–1976)
Bohemian-born American architect who assisted Cass Gilbert in the design of the Woolworth Building (1912), NYC, then joined Frank Lloyd Wright in the building of the Imperial Hotel in Tokyo, Japan. He designed other structures in Japan, including the Reader's Digest Building in Tokyo (1950) and the campus of the Nanzan University, Nagoya (1966). He also designed the Pan Pacific Forum, University of Hawaii (1969), Honolulu, HI.

Raze
To tear down or demolish; to level to the ground.

Razing
The process of demolishing a structure without possible reconstruction.

Ready-mixed concrete
A concrete that is mixed at a central plant and transported to the building site.

Real estate
Land and any improvements such as buildings and other site features.

Real property
Land with or without buildings, fences and other features. Also called real estate.

Real property inventory
An itemized list and classification of the real property of a person or estate, showing the amount, character, and condition of such property at a given time; also, an objective listing of the supply, character, and condition of all real property in a given area at a certain time.

Rear arch See Arch

Rebuild
To build again. To make extensive structural repairs after a structure has fallen into disuse through neglect, vandalism, or natural disaster, such as fire or flood..

Receptacle
An electrical device to receive the prongs of a plug and which is connected to an electrical circuit. Available in several types to receive two prongs and a ground wire.

Receptacle load
Refers to all equipment that is plugged into the electrical system, from office equipment to refrigerators.

Recess
Receding part or space, such as a cavity in a wall for a door, an alcove, or a niche.

Recessed
Forms created by indentations or small hollows in an otherwise plain surface or straight line; can be angular, rectilinear or curvilinear.

Recessed arch See Arch

Recessed dormer See Dormer

Recessed lighting fixture
A fixture that is concealed above the finished ceiling and shines through an opening in the ceiling plane. Types include baffled downlight, adjustable eyeball, pinhole downlight, and baffled wall washer. Lamps for this type of fixture include flood, spot, and pinpoints.

Recessed luminaire
A luminaire recessed into a ceiling, so that its lower edge is flush with the ceiling, as compared with surface mounting.

Reciprocating saw
A handheld power woodworking tool with a reciprocating thin blade clamped to the chuck of the saw; used to cut fine curved work; similar to a saber saw.

Recirculated air
Return air that is diverted from the exhaust route, mixed with incoming outside air, conditioned, and delivered to the conditioned space. Recycling the air circulating through an HVAC system reduces energy requirements.

Recirculating hot water heater
Systems that use a thermostat or timer to automatically turn on the pump whenever water temperature drops below a set-point, or when the timer reaches a specific setting. Hot-water recirculation systems can be activated by the push of a button or by a thermostat, timer, or motion sensor. Such systems ensure that hot water is always available without any waiting time. Hot-water recirculation systems generally consist of a pump, an integrated electronic controller, and a zone valve.

Reclaim
To use a product again after its initial use. Reclaiming commonly refers to materials such as tile, brick, and stone.

Reclaimed material
The material that has been previously used in a building or project and is then reused in another project. The material might be altered, resized, refinished, or adapted, but is not reprocessed in any way and remains in its original form. This can include reclaimed lumber, flooring, tiles, beams, and other materials that are salvaged from a building under demolition or deconstruction and used in the construction of a new building.

Reclamation
Restoration of materials found in the waste stream to a beneficial use that may be other than the original use.

Recognition program
Programs that recognize buildings, places and areas of importance is a long-standing preservation technique. The largest is the National Register of Historic Places. There are also many private-sector programs that identify landmarks and signify their value.

Recommissioning
Refers to the periodic retesting of building systems, using the original functional performance tests to ensure that the equipment continues to operate as designed. The purpose of recommissioning is to ensure that the facility continues to perform as expected over its useful life.

Reconstituted
Taking smaller pieces of materials and binding them together to form a larger material item. Examples include wood chips that are adhered together to form trim, doors or panels.

Reconstruct
New construction following the exact forms and details as they once were.

Reconstruction
The process of duplicating the original materials, form, and appearance of a historic building that no longer exists, based on historical research; most often it is located at the original site.

Record drawings
Construction drawings revised to show significant changes made during the construction process, usually based on marked-up prints, drawings, and other data furnished by the contractor to the architect.

Recover
To regain a former state or condition; restore to a normal state.

Recovered materials
Waste materials and by-products that have been recovered or diverted from solid waste but not including materials and by-products generated from and commonly reused within an original manufacturing process.

Recreate
To refresh physically.

Rectangle
A four-sided parallelogram with four right angles; may be nearly square or stretched out to be nearly a band.

Rectified photographs
Prints of large-format field photography that were made using perspective control devices on a camera. Notes may be added to the photograph; dimensions can be added if control dimensions were taken in the field.

Rectilinear
Forming, formed by, or characterized by straight lines.

Rectilinear style
Similar to the Perpendicular style; it is characterized by perpendicular tracery and intricate stonework.

Recyclable
The ability of a product or material to be recovered from, or otherwise diverted from, the solid waste stream for the purposes of recycling. U.S. Federal Trade Commission (FTC) guidelines indicate that a product may not be advertised as "recyclable" unless a viable, active reclamation system exists that is available to a majority of end users and collects and processes the product for recycling.

Recyclable content
Materials that can be recovered or diverted from the waste stream for recycling or reuse.

Recycle
To extract and reuse substances found in waste.

Recycled components
Materials used in a home that were salvaged or are made from recycled materials. One example is wood flooring made from salvaged timber from old barns.

Recycled construction
A strategy in which homeowners who are remodeling a house recycle old materials and demolition waste rather than sending them to a landfill.

Recycled material
The material that has been reprocessed and reused in the building industry.

Recycled plastic lumber
Recycled plastic that is made into a material that looks and functions like lumber and is insect- and water-resistant.

Recycled water
Treated sewage that undergoes additional, advanced treatment to make it safe for certain uses such as landscape irrigation.

Recycling
The collection and reprocessing of materials to produce new products; can also be applied to the refurbishment or reuse of old buildings.

Recycling areas
Space dedicated to recycling activities is essential to a successful recycling program, both on the construction site and in the building after occupation.

Recycling bins
Containers to temporarily hold recyclable materials until transferred to a larger holding facility or picked up by a recycling service. Conveniently located bins increase recycling rates by allowing occupants to recycle more easily. Designing space for recycling bins is a physical reminder of a commitment to recycling.

Red lead
Commonly used as a rust-inhibiting primer on steel, but has been largely superseded by other primers due to the toxicity of lead.

Redesign
To alter the design toward a new scheme.

Redevelop
To restore to a better condition; to reclaim.

Redevelopment
The act or process of redeveloping or renovating a blighted area, as in urban renewal programs.

Reduced energy roofing
The "greenest" roof is a living roof, planted with native drought-resistant plants. It is both attractive and energy efficient. There are also many more conventional roofing materials that can help reduce energy usage, including reflective roofs, structured insulated panels (SIPs), and both recycled and recyclable metal roofs, among others.

Redwood See **Wood**

Reed molding See **Molding**

Reed, Charles A. (1857–1911)
American architect of St. Paul and NYC, who with Allen H. Stem (1856–1931) designed more than 100 railroad stations, including Grand Central Terminal (1913), while at Warren and Wetmore, NYC.

Re-entry
Air exhausted from a building that is immediately brought back into the system through the air intake and other openings.

Refectory
A room for dining in a monastery, college, or other institution.

Refinish
To apply a new finish to an existing surface.

Reflectance
The ratio of reflected light flux to incident light flux.

Reflected ceiling plan
The plan of a ceiling projected onto a flat plane directly below, which shows the size and location of light fixtures and other objects within the ceiling.

Reflected glare
The glare resulting from specular reflection of high luminances in polished or glossy surfaces in the field of view.

Reflected light
Light reflected from rough or matte surfaces neutralize the directional nature of the incident beam. Light is reflected from each part in all directions, eliminating bright spots. With semispecular reflection, light reflected from irregular surfaces disperses the reflected beam. With specular reflection, light reflects off a smooth, highly polished surface, which changes the direction of the beam of light without changing its form; the angle of incidence is the same as the angle of reflection.

Reflection
The process by which incident light flux leaves a surface, or medium, from the incident side, without a change in frequency.

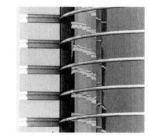

Reflection

Reflective glass See **Glass**

Reflective insulation
Metal that reflects infrared radiation and therefore reduces the amount of heat entering a building, particularly through the roof; aluminum foil is the most commonly used type.

Reflective roofing
A roofing material that reflects most of the sunlight striking it to help reduce cooling loads. The Energy Star Cool Roof program certifies roofing materials that meet specified standards for reflectivity.

Reflectivity
The ability of a material to turn away or deflect a fraction of the radiant heat generated by the sun. Also known as solar reflectivity.

Reflector
A device installed in luminaires used to direct light from a source via specular or diffuse reflection.

Refrigerant
Compound used in refrigerators, air conditioners, and heat pumps to transfer heat from one place to another, thus cooling or heating a space. Most refrigerants today are hydrochlorofluorocarbons (HCFCs), which deplete the ozone layer.

Refurbish
To make clean, bright or fresh again.

Refurbished
Products or materials that have been upgraded to be returned to active use in their original form. Refurbishing is considered a form of reuse, and it is preferable to recycling because it requires less processing and fewer inputs to return a product to useful service.

Refurbishment
To bring an existing building up to standard; or to make it suitable for a new use by renovation or by installing new equipment, fixtures, furnishings, and finishes.

Regence style (1715–1723)
The decorative and elegant Rococo style flourishing under the regency of Philip of Orleans during the reign of Louis XV.

Regency style (1811–1820)
A colorful Neoclassical style, often combined with oriental motifs prevalent in England during the reign of George IV, characterized by close imitation of ancient Greek and Roman, Gothic, and Egyptian forms.

Regenerate
To form, construct, or create anew.

Regeneration
Renewal of sites or habitats that have become unfit for human, animal, or plant habitation, bringing them back into productive use. The term most commonly refers to urban and industrial land.

Regenerative design
A system with absolute efficiency used for sustainable development. The concept behind this system is one in which all waste products of one system can be used in the same or in different systems with zero loss of input and output. The system thus sustains itself.

Regional manufacture
Goods produced within a certain radius of the project site. Using regionally produced goods is considered a sustainable building strategy because transportation impacts associated with the product are reduced, a better understanding of the production process is engendered, the likelihood that the product was manufactured in accordance with environmental laws increases, and regional economies are supported.

Regional planning
A profession which began in the United States in the 1970s whose central focus was the physical environment of cities and regions. Includes forces that shape the environment, plans and policies to ease and eliminate urban and regional problems, and availability of government and private resources. The purpose is to integrate the physical, social, economic, and political aspects of community development in finding solutions to urban and regional problems.

Register
The open end of a duct for warm or cool air; usually covered with an ornamental grille of metal or with vanes to control the distribution of air.

Regreen Program
A web-based resource consisting of best practice guidelines and targeted educational resources for sustainable residential remodeling projects.

Regula
In the Doric entablature, one of a series of short fillets beneath the taenia; each corresponding to a triglyph above.

Regular rhythm See Rhythm

Rehabilitate
To restore to a useful state by repair, alteration, or modification.

Rehabilitation
Slum areas and substandard buildings brought up to an acceptable living standard. The U.S. Department of the Interior's standards state: "returning a property to a state of utility through repair or alteration which makes possible an efficient contemporary use while preserving those portions or features of the property which are significant to its historical architectural, and cultural values."

Rehabilitation
To repair an existing building to good condition with minimal changes to the building fabric; may also include adaptive reuse or restoration; also called rehab.

Rehabilitation guidelines
Standards or recommendations to assist property owners in improving structures while preserving their special historical character.

Reinforce
To strengthen or make more effective by adding extra support.

Reinforced concrete
A concrete masonry construction, in which steel reinforcement is embedded in such a manner that the two materials act together in resisting forces.

Reinforcing
Steel or iron bars or wire mesh used to increase the tensile strength of reinforced concrete.

Reinforcing

Reinforcing bar
A steel bar used in concrete construction that provides additional strength; the bars are deformed with patterns made during the rolling process.

Reinhard, L. Andrew (1891–1964)
American architect, partner in the firm of Reinhard and Hoffmeister with Henry Hoffmeister (1891–1962).

Reinstate
To restore to a previous condition or position.

Reinvestment
The channeling of public and private resources into declining neighborhoods in a coordinated manner to combat disinvestment.

Rejuvenate
To restore to a youthful appearance.

Relative humidity
The ratio of the amount of water vapor in air to the maximum amount of water that the same volume of air can hold at the same temperature, expressed as a percentage.

Releasable adhesives
Dry, tacky adhesives that hold a carpet or other item in place but can be easily removed. After removal, no residue is left, and the item can be reattached.

Release agent
A substance that is applied to formwork in order to prevent adhesion between a concrete surface and the face of the form.

Relief
Carved or embossed decoration of a figure or form, raised above the background plane from which it is formed.

Relief

alto-relievo
Sculptural relief work in which the figures project more than half their thickness from the base surface.

anal glyph
An embellishment carved in low relief.

bas-relief
Sculptural decoration in low relief, in which none of the figures or motifs are separated from their background, projecting less than half their true proportions from the wall or surface.

cavo-relievo
Relief which does not project above the general surface upon which it is carved.

diaglyph
A relief engraved in reverse; an intaglio; a sunken relief.

glyph
A sculptured pictograph; a grooved channel, usually vertical, intended as an ornament.

high relief
Sculptural relief work in which the figures project more than half their thickness from the base.

high relief

in cavetto
The reverse of relief; differing from intaglio in that the designs are pressed into plaster or clay.

mezzo-relievo
Casting, carving, or embossing in moderate relief, intermediate between bas-relief and high relief.

stiacciato
In very low relief, as if a bas-relief had been pressed even flatter.

sunk relief
A relief in which the highest point of the forms does not project above the general surface from which it is modeled; also called cavo-relievo.

Relieve
To assist any overloaded member by any construction device, such as a discharging arch placed above an opening.

Relieving arch See **Arch**

Relite
Windows or translucent panels above doors or high in a partition wall intended to allow natural light to penetrate deeper into a building.

Relocate
To establish in a new place.

Relocating
Settling households or businesses in new locations, often necessitated by urban renewal or other governmental actions.

Relocation
The process of moving a building or structure to a new location, usually placing it on a totally new foundation.

Remake
To make new; to reconstruct.

Remanufacturing
A recycling concept by which an existing product can have its useful life extended through a secondary manufacturing or refurbishing process, such as remanufactured systems furniture.

Remediation
Efforts to counteract some or all of the effects of pollution after it has been released into an environment. Also, to clean a site that has been damaged by industry that is physical, chemical, or biological in origin.

Remodel
Revisions within an existing structure without increasing building areas.

Remodeling
The process of modifying an existing building or space for current use; usually involves replacing some of the existing building fabric or adding new components.

Renaissance
A rebirth, revival or renewal.

Renaissance architecture (1420–1550)
An architecture that developed during the rebirth of Classical art and learning in Europe and evolved through several periods. It was initially characterized by the use of the Classical orders, round arches, and symmetrical proportions. It represented a return to the models of Greco-Roman antiquity and was based on regular order, symmetry, and a central axis with grandiose plans and impressive facades. Silhouettes were clean and simple, with flat roofs replacing Gothic spires. Walls of large dressed masonry blocks gave buildings an imposing sense of dignity and strength. Gothic verticality was replaced with an emphasis on horizontality. Semicircular arches appeared over doors and windows and in freestanding arcades. Columns were used decoratively on facades and structurally in porticos, and ornamentation was based on pagan or Classical mythological subjects.

Renaissance architecture

Renaissance Revival style (1840–1890)
A revival style characterized by a studied formalism, its symmetrical compositions are reminiscent of early 16th-century Italian elements. Ashlar masonry is accented with rusticated quoins, architrave framed windows, and doors supporting entablatures or pediments. A belt or string course often divides the ground or first floor from the upper story, and small square windows are used on the top story.

Rendering See **Plaster**

Renew
To make new, or to restore to a former condition.

Renewability
The ability for resources, such as wood or water, to be replenished after being harvested for products. This ability depends both on the resource's natural rate of replenishment and the rate at which the resource is withdrawn for human use.

Renewable
A renewable product can be grown or naturally replenished or cleansed at a rate that exceeds human depletion of the resource.

Renewable content
The amount of a material that is made up of renewable resources.

Renewable energy
Renewable energy is obtained from sources that can be sustained indefinitely, such as photovoltaic solar collection, solar thermal turbine generation, and wind power.

Renewable Energy Certificates (RECs)
Also known as green tags, green energy certificates, or tradable renewable certificates, RECs, issued by the Green Power Partnership of the Environmental Protection Agency, represent the technology and environmental attributes of electricity generated from renewable sources. RECs are usually sold in 1-megawatt-hour (MWh) units.

Renewable energy sources

Energy sources that replenish themselves naturally within a short period of time. Sources of renewable energy include solar energy, hydroelectric power, geothermal energy, wind power, ocean thermal energy, wave power, wind power, and fuel wood.

Renewable energy technologies

Active, passive, and photovoltaic strategies integrated into building design.

Renewable materials

Materials that are not depleted when used. These materials are typically harvested from fast-growing sources and do not require unnecessary chemical support. Examples include bamboo, flax, wheat, wool, and certain types of wood.

Renewable resources

Resources that can be replenished at a rate equal to or greater than their rate of depletion; such as solar, wind, geothermal and biomass resources.

Renovate

To restore to an earlier condition by repairing or remodeling.

Renovation

The process of repairing and changing an existing building for modern use to make it functionally equivalent to that of a new building.

Renwick, James (1818–1895)

At 24, Renwick won the commission to design Grace Church in New York City (1843). Later he designed St. Patrick's Cathedral (1858), NYC, a vast Gothic church. He also designed the Smithsonian's "castle" (1847), in Washington, DC, and the original Corcoran Gallery (1859), now the Renwick Gallery in Washington, DC.

Repair

Any labor or material provided to restore, reconstruct, or renew any existing part of a building, its fixtures, or appurtenances; also, to bring to a sound condition after damage or injury in a fire.

Repetition

The recurrence of rhythmic patterns, forms, or accents that are separated by spaces of repeated formal elements or different forms.

Repetition

Replace

To put back in place; a substitution in place of the original.

Replacing

The renewing or restoring to a former place or condition, as the renewing of parts of a building which have been damaged or impaired by use or by the elements.

Replica

Exact copy of an original building or any building component.

Replicate

To duplicate, copy, or repeat; to reproduce in the likeness of the original.

Repose

Harmony in the arrangement of parts or colors that is restful to the eye.

Representative historic site

Intended to help the visitor understand a period of history or a way of life.

Repressed brick

Bricks that are pressed again while still in the plastic state, usually with the name of the manufacturer; produces a slightly denser uniform brick, but does not alter its original characteristics.

Reprocess
A process of changing the physical structure and properties of a waste material that would otherwise have been sent to landfill, in order to add financial value to the processed material.

Reproduce
To produce a counterpart, image, or copy of the original.

Repurpose
Using a structure for a purpose other than for what it was originally intended.

Requirements and submittals
LEED Green Building Rating System component that specifies the criteria to satisfy the prerequisites or credit, the total number of points available, and the documentation required for the LEED application.

Rescue
To save a structure from danger, such as demolition.

Researching structures
Examining past evidence of a building's history. Items to look for in researching buildings consist of architectural plans and drawings, assessment records, building inspector's records, deeds, film and videotapes, guide books, household inventories, insurance records, legal resources, magazines, manuscripts, maps, measured drawings, newspapers, oral history, photographs and slides, postcards, reports and feasibility studies, surveys, and trade catalogs.

Resemble
Be a likeness of another structure.

Resilient flooring
Various types of solid, flexible material used as finish flooring; types include asphalt tile, linoleum, vinyl asbestos tile, vinyl tile, and cork tile.

Resistance
The physical property of a material to resist or impede the conduction of electrical current, measured in ohms. High resistance means poor conductivity and vice versa.

Resort
A place frequented by people for relaxation and recreation; a customary or frequent gathering place.

Resource ecology
The calculation and control activities that focus on the building's natural resource use, such as water, energy, landscape, waste management, soil type, and groundwater.

Resource efficiency
A practice in which the primary consideration of material use begins with the concept of reduce–reuse–recycle–repair, stated in descending order of priority.

Respond
A pier or pilaster projecting from a wall as a support for an arch at the end of an arcade.

Restaurant
A commercial establishment where meals to order are served to the public, either in a separate building or within a hotel or other facility.

Restoration
The U.S. Department of the Interior's standards state: "accurately recovering the form and details of a property and its setting as it appeared at a particular period of time by means of the removal of later work or by the replacement of missing earlier work." Since authenticity is the primary goal, and this calls for extensive research, study, and money, restoration is frequently restricted to those structures intended for public use or those opened as house museums.

Restoration architect
An architect with special skills in conservation, restoration, and reconstruction methods and techniques.

Restore
To put back into a prior state, condition, or use.

Restrain
To limit or restrict; to deprive of freedom or liberty; to control, check, repress.

Restrictive covenant
A requirement to adhere to a specific restriction on the use or development of real property.

Retaining wall See **Wall**

Retarder
In concrete work, an additive that slows down the setting time, allowing for more time to place the concrete.

Retempering
In masonry work, restoring the workability of mortar that has stiffened due to evaporation by adding water and remixing.

Reticulated
Refers to surfaces that are marked with lines, resembling or forming a network of squares arranged on the diagonal.

Reticulated molding See **Molding**

Reticulated tracery See **Tracery**

Reticulated work
Masonry that is constructed with diamond-shaped or square stones placed diagonally or crossing in a network.

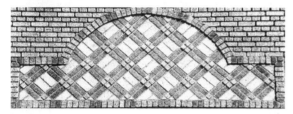

Retrieve
To get back again; regain, revive; to make good, put right, rectify.

Retro-commissioning
Commissioning performed on a facility that has been in service but not previously commissioned.

Retrofit
To make an improvement to an existing building, or to adapt it to a new use. Also, the replacement, upgrade, or improvement of a piece of equipment or structure in an existing building or facility.

Retrofitting
The process of installing new mechanical, fire protection, and electrical systems or equipment in an existing building; most often required to meet current building code requirements. Also, the process of rethinking a development plan after completion to include newer features, such as green or eco-friendly features.

Return
The continuation of a molding, projection, member, or cornice in a different direction, usually at right angles.

Return air
The air that has circulated through a building as supply air and has been returned to the HVAC system for additional conditioning or release from the building, in order to reduce the energy that would be consumed by using only fresh air as a source.

Return on investment (ROI)
An economic indicator that is used to evaluate the effectiveness of an investment. It is calculated as the ratio of the amount gained or lost relative to the amount invested. Simple ROI analyses do not take the time value of money into account; dynamic ROI analyses recognize that the value of money changes over time.

Returned nosing
A stair nosing that wraps around the side of an open string stair.

Reusability
A material's ability to be reused or salvaged in whole form without undergoing a reprocessing or remanufacturing process. Refers to how many times a material may be reused.

Reuse
Using a material, product or component of the waste stream in its original form more than once.

Reused components
Structural or finish materials removed from old buildings and reused in new or remodeled buildings.

Reveal
The visible side of an opening for a window or doorway between the framework and outer surface of the wall; where the opening is not filled with a door or window, the whole thickness of the wall.

Revell, Viljo (1910-1964)
A Finnish architect who studied with Alvar Aalto and made his name with the "Glass Palace" office building in Helsinki, Finland. He won the competition in 1958 for the Toronto City Hall, Canada (illus.), whose two curvilinear towers contrast with the neighboring office structures.

Revenue Act of 1978
Provided an investment tax credit for the rehabilitation of old buildings used for commercial and industrial purposes.

Reverberation
The buildup of sound within an architectural space, such as a room, as a result of repeated sound reflections at the surfaces of the room.

Reverberation time
The amount of time it takes for sound to decay 60 decibels in a given space. It is a function of room volume and amount of sound absorption provided by surface finishes in the room. Optimum levels are determined based on room volume and space usage.

Reversible disassembly
Reverse manufacturing, in which the removal of screws, clips, and other fasteners permits refurbishment and reuse of some or all of the components and modules of a product.

Revert
To return to a former condition.

Revise
To change or modify.

Revitalize
To impart new life or restore vitality in a residential or commercial areas through physical improvements and economic programs, such as a "Main Street" program.

Revival
A restoration to use after a period of obscurity.

Revival architecture
The use of older styles in new architectural movements, most often referring to the Gothic, Roman, Egyptian, Etruscan, Greek, Colonial, or revival styles of the 18th and 19th century.

Revival style
An architectural style which derives elements from a previous style and incorporates them, oftentimes with variations.

Revolving door See Door

Rezoning
To divide a municipality or piece of property into new or different zones to permit new uses.

RFP
Request for proposal, usually initiated by the owner.

Rheostat
An electrical device to vary the amount of resistance in a circuit and change the amount of current flowing through.

Rhythm
Any kind of movement characterized by the regular occurrence of elements, lines, shapes and forms; the flow of movement shown by light and heavy accents, similar to recurring musical beats.

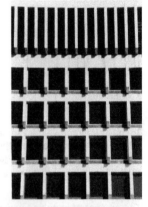

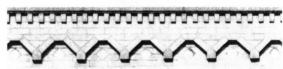

Rib
A curved structural member supporting any curved shape or panel; a molding that projects from the surface and separates the various roof or ceiling panels.

Rib

diagonal rib
A projecting rib that crosses a square or rectangular rib vault from corner to corner.

groined rib
A rib under a curve of a groin, used as a device to either mask or support it.

intermediate rib
A subordinate vault rib between primary ribs.

ridge rib
A continuous, projecting rib connecting the apexes of the intermediate ribs of a rib vault with the center of the vault.

tierceron
An intermediate rib between the main rib of a Gothic vault.

RIBA
Royal Institute of British Architects.

Ribbed arch See **Arch**

Ribbed fluting
Flutes alternating with fillets.

Ribbed slab
A panel composed of a thin slab reinforced by ribs in one or two directions, usually at right angles to one another; also called a waffle slab due to its appearance.

Ribbed vault See **Vault**

Ribbing
An assemblage or arrangement of ribs, as timberwork that sustains a vaulted ceiling.

Ribbon window See **Window**

Richardson, Henry Hobson (1838–1886)
American architect and pupil of Labrouste in Paris, France. Richardson was renowned for his massive Romanesque buildings. He designed Trinity Church, Boston (1877), inspired by French and Spanish Romanesque styles.

Richardsonian style (1870–1900)
Named for Henry Hobson Richardson, this style featured a straightforward treatment of stone, broad roof planes and a select grouping of door and window openings. It also featured a heavy, massive appearance with a simplicity of form and rough masonry. The effect is based on mass, volume, and scale rather than decorative detailing, except on the capitals of columns. The building entry includes a large arched opening without columns or piers for support.

Riddle, Theodate Pope (1868–1946)
American architect. When Riddle was refused admission to Princeton's architecture classes in the 1890s, she hired its faculty to tutor her privately.

Ridge
The horizontal lines at the junction of the upper edges of two sloping roof structures.

Ridge beam
A horizontal beam at the upper edge of the rafters, below the ridge of the roof.

Ridge board
A longitudinal member at the apex of a roof that supports the upper ends of the rafters.

Ridge cap
Any covering such as metal, wood, or shingles used to cover the ridge of a roof.

Ridge course
The last or top course of roofing tiles, roll roofing, or shingles.

Ridge crest
A linear ornamental device; usually composed of metal, attached to the crest of a roof, providing a transition to the sky.

Ridge roll
A wood strip, rounded on top, which is used to finish the ridge of a roof, often covered with lead sheathing; a metal or tile covering which caps the ridge of a roof.

Ridge tile See Tile

Ridge vent
An opening covered by a rainproof vent that follows the peak of the roof, typically required by code. Some insulating methods, however, negate the need for a ridge vent. (Local code requirements must be checked before eliminating the ridge vent.)

Ridge ventilator
A raised section on a roof ridge provided with vents to admit air currents.

Rietveld, Gerrit (1884–1964)
Dutch architect and furniture designer influenced by the De Stijl group. Rietveld's major work, the Schroder House (illus.), Utrecht (1924), was the first building to translate Cubist ideas of space and fractured planes into architectural forms.

Right of way
A strip of land over which a lawful right of passage exists for the benefit of persons who do not own the land.

Rightsizing
The process of correctly sizing equipment to the peak load.

Rigid arch
An arch without joints that is continuous and rigidly fixed at the abutments.

Rigid arch See Arch

Rigid conduit
Nonflexible steel tubing used to carry electrical conductors within the structure.

Rigid frame
A structural framework in which all columns and beams are roughly connected; there are no hinged joints, and the angular relationship between beam and column members are maintained under load.

Rigid frame

Rigid joint See **Joint**

Rinceau
An ornamental band of
undulant and curving plant
motifs, found in Classical
architecture.

Ring stone
One of the stones of an arch that shows on the face of the
wall or at the end of the arch: one of the voussoirs of the
face forming the archivolt.

Rio Bec style (550–900)
A style of Mayan architecture that was the transitional
style between that of sites at Tikal and at Uxmal; it was
characterized by lavishly decorated structures flanked by
soaring temple pyramids with steeply raked steps.

Rip saw
A handsaw used to cut lumber parallel to the fibers.

Riprap
Irregularly broken and random-sized large pieces of quarry
rock used for foundations; a foundation or parapet of
stones thrown together without any attempt at regular
structural arrangements.

Rise
Vertical height of an arch, roof truss, or rigid frame.

Rise and run
The ratio of vertical to horizontal distance of a slope, such
as a roof of 3 in 12; also, the total vertical and horizontal
distance of a flight of stairs, or an individual riser and
tread.

Riser
The vertical board under the tread of a step; a vertical
supply pipe for a sprinkler system; a pipe for water, drainage, gas, steam, or venting that extends vertically through
one or more stories and services other pipes.

Riser pipe
A vertical pipe which rises from one floor level to another
floor level, for the purpose of conducting steam, water, or
gas from one floor to another.

Risk
A measure of the probability of an adverse effect on a
population under a well-defined exposure scenario.

Risk assessment
An evaluation of potential consequences to humans, wildlife, or the environment caused by a process, product, or
activity; including both the likelihood and the effects of an
event. Also, qualitative and quantitative evaluation of the
risk posed to human health or the environment by the
actual or potential presence or use of specific pollutants.

Risk management
The management of an activity, accepting a level of risk
which is balanced against the benefit of the activity,
usually based on an economic assessment.

Rivet
A shank with a head that is inserted into holes in the two
pieces being joined and closed by forming a head on the
projecting shank. The rivets must be red hot to be formed
in such a manner and have generally been replaced by
welding or bolting.

Road diet
A technique of transportation planning in which the width
of a road or lane is narrowed in order to achieve improvements to the transportation system. A typical road diet
technique is to reduce the number of lanes on a roadway
cross section. The additional space that is freed up by
removing a vehicular lane can be converted into two bike
lanes on either side of the roadway.

Roadside architecture
Buildings designed along the side of a roadway, for the
servicing of the traveling public; may include diners, gas
stations, and the like.

Rocaille
A small ornament combining forms based on water-worn
rocks, plants and shells, characteristic of the 18th-century
Rococo period, especially during the reign of Louis XV.

Roche, Eamonn Kevin (1922–)

Irish-born American architect, formed the successor firm to Eero Saarinen in the 1950s with John Gerald Dinkeloo (1918–1981). Their first work was the Oakland Museum, CA (1961) a vast structure covering four city blocks. The best known is the Ford Foundation Headquarters (illus.), NYC (1963), with a 12-story indoor atrium. The later work includes the extension to the Metropolitan Museum of Art, NYC, including the Pavilion for the Ancient Egyptian Temple of Dendur.

Roche, Martin (1853–1927)

American architect and partner with William Holabird; designed the Tacoma Building, Chicago, IL (1889, later demolished). It included a structure of cast-iron columns and wrought-iron beams, clad in terra-cotta and glass, establishing the skeletal structure for the Chicago School style.

Rock rash

A patchwork appliqué of oddly shaped stone slabs used on edges as a veneer, often further embellished with small cobbles or geodes.

Rock wool

An insulating material that looks like wool, but is composed of substances such as granite or silica.

Rock-cut

A temple or tomb excavated in natural rock; usually represents an architectural front with dark interior chambers, of which sections are supported by masses of stone left in the form of solid pillars.

Rocklath

A flat sheet of gypsum used as a plaster base.

Rockwork

Quarry-faced masonry; any stonework in which the surface is left irregular and rough.

Rococo style (1720–1790)

A style of architecture and decoration, primarily French in origin, representing the final phase of the Baroque. It was characterized by a profuse, semiabstract ornamentation of shell work and foliage. It was associated with lightness, swirling forms, flowing lines, ornate stucco work, arabesque ornament, and the blending of separate members into a single molded volume.

Rococo style

Rodilla, Simon (1879–1965)

Italian-born mason who built the Watts Towers, an artistic fantasy on the outskirts of Los Angeles (1954). The Watts Towers consist of three steel-framed, freeform towers, constructed of concrete-covered reinforcing rods encrusted with inlaid tile, shells, and fragments of broken bottles. At several points in their lifetime, the Towers were threatened by building officials who tried to remove the towers, on the grounds of structural safety, but were unable to pull them down.

Rodilla, Simon

Roebling, John Augustus (1806–1869)
Pioneering suspension bridge engineer who, with Washington Augustus Roebling (1837–1926), designed the Brooklyn Bridge, NYC.

Rogers, Isaiah (1800–1869)
American architect who worked in Boston, New York, and Cincinnati; an innovative hotel designer and supervising architect of the Treasury Department. Designed the Tremont House (1829), Boston, MA; Merchant's Exchange (1842), NYC; and Burnett House (1850), Cincinnati, OH.

Rogers, James G. (1867–1947)
American architect who worked with LeBaron Jenny in Chicago and designed the Harkness Quadrangle at Yale, New Haven, CT (1917).

Rogers, John B. (1925–2010)
A cofounder of Denver-based architecture firm RNL. His many designs include the Colorado History Museum, the *Rocky Mountain News* building, and Colorado's Ocean Journey aquarium.

Rogers, Richard (1933–)
Italian-born English architect who was in partnership with Renzo Piano (1971–1981), and completed the high-tech Centre Pompidou (illus.), in Paris, and set up his own practice in London. Work includes Lloyds Building, London (1986), and Terminal 5 Heathrow Airport, London (1989).

Roll molding See **Molding**

Rolled form sections
Thin sheet-metal sections, formed from a continuous strip by passing through successive sets of rollers.

Rolled glass See **Glass**

Rolled steel section
Any hot-rolled steel section, including joists, angles, channels, and rails.

Rolling door See **Door**

Roll-up door See **Door**

Roman arch See **Arch**

Roman architecture (300 B.C.–365 A.D.)
An architecture influenced by the Etruscans, combining the use of the arch with Greek columns. The invention of concrete led to a system of vaulting and the development of the dome used to roof a circular area, demonstrated sophisticated engineering skills. The pilaster was used decoratively on walls instead of half-columns. Colonnades and arcades were both in use, and occurred one above the other at times. Doorway headers were both square and semicircular and became decorative features of importance in the exterior design of large public buildings. Window headers were generally semicircular. Orders were sometimes superimposed, and pedestals were developed to give the column additional height. The Romans relied on the abundant carving on their moldings rather than on the contours. Marble, granite and alabaster were the primary facing materials, as well as stucco and mosaics. The emphasis was on monumental public buildings, reflecting the grandeur of the empire. Many had very sophisticated building services, such as plumbing, heating and water supply. On an urban scale the Romans also produced an impressive array of planning elements: formal axial planning with whole communities and towns constructed on a grid plan were typical

Romanesque architecture (800-1180)

The style that emerged from Roman and Byzantine elements; characterized by massive articulated wall structures and semicircular arches and vaults. It showed an evolution of stone vaulting and of the rib method of construction. It was characterized by heavy masonry construction, sparse ornament, and smooth plain walls with decoration derived from the structure. It also featured thick molded piers, assembled from small stones individually carved to fit.

Romanesque Revival style (1840-1900)

A style characterized by monochromatic brick or stone buildings, highlighted by semicircular arches over window and door openings. The arch was also used decoratively to enrich corbel tables along the eaves and courses marking horizontal divisions. The arches and capitals of columns were carved with geometrical medieval moldings. Facades were flanked by polygonal towers and covered with various roof shapes.

Romanesque Revival style

Romano, Guilio (1492-1546)

Italian architect and painter who manipulated the rules of the Classical styles toward the Mannerist style. Works include the Palazzo Ducale, in Mantua, Italy (1538).

Rood

A large crucifix set above the chancel entrance.

Rood arch

The central arch in a rood screen.

Rood beam

A horizontal beam across the chancel to support the screen.

Rood loft

A gallery in which the rood is kept.

Rood screen

A carved wood or stone separating the nave and chancel.

Rood spire

A spire located over the crossing of the nave and transepts.

Rood stairs

Provide access to a rood loft.

Roof

The external covering on the top of a building, usually of wood shingles, slates, or tiles on pitched slopes, or a variety of built-up membranes for flat roofs. Roofs offer a variety of architectural expression,

Roof

barrel roof
A roof of semicylindrical section capable of spanning long distances, parallel to the axis of the cylinder.

bell roof
A roof whose cross section is shaped like a bell.

bell-cast roof
A form of mansard roof in which the lower roof slopes downward in a straight line, then curves outward at the eaves.

built-up roofing
Flat roof covered with multiple layers of roofing felt, secured with layers of hot tar, and topped with a layer of crushed stone.

butterfly roof
A roof shape that has two surfaces that rise from a valley at the roof's centerline to the eaves.

canopy roof
A roof that is in the shape of a suspended cloth canopy; often used over a balcony or porch.

compass roof
A convex-shaped roof formed either by curved rafters or by a combination of beams arranged in a vault.

conical roof
A roof in the shape of an inverted cone on top of a cylindrical tower; used in the Chateauesque and Queen Anne styles; also called a witch's hat.

curb roof

A pitched roof that slopes away from the ridge in two successive planes, as a gambrel or mansard roof.

double-gable roof

A roof composed of two parallel gables forming the shape of the letter M on the end wall.

double-hipped roof

A hipped roof having a double slope.

double-pitched roof

A roof having two flatter slopes on each side of a steep central ridge; similar to a gambrel roof.

Dutch gambrel roof

A type of gambrel roof that has two flat surfaces on each side of the ridge, each at a different pitch; the top slope is the flatter of the two, while the lower slopes often end in a flared eave.

Dutch roof

A gable roof divided into two sections of unequal slope; the flatter slope is from the top of the ridge, and the steeper slope connects to the top plate.

flat roof

A roof having no slope, or one with only a slight pitch so as to drain rainwater; a roof with only sufficient slope to effect drainage.

gable roof

A roof having a gable at one or both ends; a roof sloping downward in two opposite directions from a central ridge, so as to form a gable at each end.

gambrel roof

A roof with two pitches on each side.

hammer-beam roof

A roof without a tie beam at the top of the wall.

helm roof

Four faces rest diagonally between the gables and converge at the top.

hip roof

A roof which slopes upward from all four sides of a building, requiring a hip rafter at each corner.

hip-and-valley roof
A hip roof on a building with an irregular plan, with valleys at the inside corners.

hyperbolic roof
A shell roof in the form of a hyperbolic paraboloid, generated by two systems of straight lines.

jerkinhead roof
A combination of a gable roof and a hipped roof; the gable rises about halfway up the ridge, then the roof is tilted back at a steep incline.

lamella roof
A roof frame consisting of a series of skewed arches, made up of relatively short members, fastened together at an angle so that each is intersected by two similar adjacent members at its midpoint, forming a network of interlocking diamonds.

M roof
A roof formed by the junction of two parallel gable roofs with a valley between them.

mansard roof
A roof with a steep lower slope and a flatter upper slope on all sides, either of convex or concave shape.

ogee roof
A roof whose section is an ogee.

open-timbered roof
A roof construction in which there is no ceiling so that the rafters and roof sheathing are visible from below.

pavilion roof
A roof hipped equally on all sides, giving it a pyramidal form

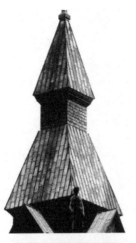

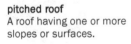

pent roof
A small sloping roof, the upper end of which butts against a wall of a house, usually above the first-floor window. If carried completely around the house, it is called a skirt roof.

pitched roof
A roof having one or more slopes or surfaces.

polygonal roof
A pavilion roof having more than four sides is a polygonal roof.

pyramidal hipped roof
Same as a pavilion roof.

rotunda roof
A circular roof with a low slope and overhanging eaves.

saddle roof
A roof having a concave-shaped ridge with gables at each end of the roof, suggesting a saddle.

saddleback roof
A ridged roof, with short gable ends and a straight ridge, typically found on the top of a tower.

sawtooth roof
A roof system having a number of small parallel roof surfaces with a profile similar to the teeth in a saw; usually the steeper side is splayed and faces north; usually asymmetrical with the shorter slope glazed.

shed roof
A roof shape having only one sloping plane.

single-pitched roof
A roof having only a single slope on each side of a central ridge, compared to a shed roof, which has a single slope, but without a central ridge.

sod roof
A roof composed of a thick layer of grassland containing roots; frequently it is pitched or barrel-shaped and supported by logs or other wall structure.

square roof
A roof where rafters on opposite sides of the ridge meet at a 90-degree angle; each side of the roof has a slope of 45 degrees.

stepped roof
A roof constructed of stones which are arranged in a stair-stepped fashion, diminishing toward the top in a peak.

suspended roof
A roof whose load is carried by a number of cables which are under tension from columns or posts that are in compression and that transmit the loads to the ground.

terrace roof
A roof that has been truncated so as to form a flat horizontal surface without a ridge.

thatched roof
A roof made of straw, reed, or similar materials fastened together to shed water and sometimes to provide thermal insulation.

435

translucent roof
A roof comprised of a structural system wherein the voids are filled with a material that lets light pass through, but is not transparent.

transparent roof
A structural system that is composed with an all glass infill, similar to those used in greenhouses.

truncated roof
A roof with sloped sides and a flat top for a terrace; may have a balustrade around the flat center section.

umbrella shell roof
A shell roof formed by four hypar shells.

undulating roof
A roof that exhibits a wavy.

undulating roof

visor roof
A relatively small section of roof that projects on brackets from a flat wall surface; sometimes it appears below a parapet, as in the Mission style.

Roof balustrade
A railing with supporting balusters on a roof; often near the eaves or surrounding a widow's walk.

Roof comb
A wall along the ridge of a roof that makes the roof appear higher.

Roof cornice
A cornice immediately below the eaves; also called an eaves cornice.

Roof curb
A pitched roof that slopes away from the ridge in two successive planes, as a gambrel or mansard roof.

Roof drain
A drain designed to receive water collecting on the surface of a roof and to discharge it into a leader or a downspout.

Roof garden
A rooftop terrace with planters and potted plants, especially when on a tall building.

Roof guard
Any of various devices installed near the bottom of a sloped roof to prevent snow from sliding off.

Roof gutter
A channel of metal or wood at the eaves or on the roof of a building for carrying off rainwater.

Roof pitch
The slope of a roof usually expressed as a ratio of vertical rise to horizontal run, or in inches of rise per foot of run.

Roof plate
A wall plate that receives the lower ends of roof rafters.

Roof rake
A slope or inclination; the incline from the horizontal of a roof slope.

Roof ridge
The horizontal line at the junction of the upper edge of two sloping roof surfaces.

Roof ridgebeam
A beam at the upper ends of the rafters, at the ridge of the roof.

Roof ridgecap
Any covering such as metal, wood, shingles, or tile used to cover the top course of materials at the ridge.

Roof ridgecrest
The ornamentation of the roof ridge.

Roof ridgeroll
A wood strip, rounded on top, which is used to finish the ridges of a roof; often covered with lead sheeting; a metal, tile, or asbestos-cement covering which caps the ridge of a roof.

Roof scupper
An opening in a wall or parapet that allows water to drain from a roof.

Roofing felt
Originally made with cotton rag content, this asphalt-saturated product is now made of paper. At times confused with building paper, roofing felt and building paper differ in two ways: (1) building felt is made of recycled-content paper, versus virgin paper for building paper; (2) felt is made of a heavier stock paper, versus a lighter stock for building paper.

Room
An enclosure or division of a house or other structure, separated from other divisions by partitions.

Room air conditioner
A self-contained machine inserted in a window or exterior wall of a room and used to circulate air cooled by refrigeration into the room while rejecting the heat outdoors. May also be fitted to supply heat if appropriately circuited.

Root, John Wellborn (1850–1891)

American architect who, in a partnership with Daniel Burnham, influenced the development of the glass curtain wall typical of the Chicago School and defined the characteristics of the skyscraper. Works include the Monadnock Building (1891) and the Rookery (1886), both in Chicago.

Rose ornament See Ornament

Rose window See Window

Rosette

A round pattern with a carved conventionalized floral motif; a circular decorative wood plaque used in joinery, such as one applied to a wall to receive the end of a stair rail.

Rossant, James (1928–2009)

Architect and planner Rossant's many projects include the planning of Reston, VA, and the 1966 master plan of lower Manhattan that led to the building of Battery Park City. His first significant project was the codesign of 1962's Butterfield House, a modern apartment building in a Greenwich Village neighborhood of historic townhomes. His partner for that project was William J. Conklin, with whom he practiced until 1995. Their design for Reston, a city of 75,000 outside Washington, DC, for developer Robert E. Simon, included planning commercial and residential districts, landscaping, recreation, and culture. Rossant also created a series of visionary architectural drawings. See Utopian Architecture.

Rossant, James

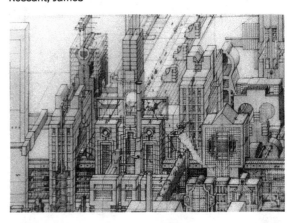

Rossi, Aldo (1931–1997)

Italian architect, and the most eminent protagonist of Rational Architecture. His work embraced aspects of International Modernism and a surrealism reminiscent of the paintings of Giorgio de Chirico, in projects such as the Carlo Felice Opera House, Genoa (1982), IBA Social Housing, Berlin (1989), and the Bonnefantin Museum, Mastricht (1994), and three office buildings (illus.) for the Disney Company in Celebration, FL (1998).

Rotary-cut veneer

Veneer in which the entire log is centered in a lathe and turned against a broad cutting knife, which is set into the log at a slight angle.

Rotated
Refers to forms created by revolving a shape on an axis and duplicating it in another location with the same relationship to the central point; as the forms are rotated they may be transformed in some manner.

Roth, Emery (1871–1948)
Designed the Helmsley Palace Hotel, in New York City (1980), and the Pan Am Building, now the Met Life Building, in New York City, with Walter Gropius and Pietro Belluschi (1963).

Rotunda
A building that is round both inside and outside, usually covered with a dome.

Rotunda roof See **Roof**

Rough arch See **Arch**

Rough carpentry
The wood framing and sheathing of a building, including blocking, joists, rafters, stringers, studs, and subflooring.

Rough floor
A subfloor serving as a basis for the laying of the finished floor, which may be of wood, linoleum, tile, or other suitable material.

Rough hardware
All the concealed hardware in a building such as bolts, nails, and spikes which were used in the construction, and may not be visible in the completed work.

Rough opening
An unfinished window or door opening; any unfinished opening in a building.

Rough-hewn
Timbers or lumber with a rough, uneven surface made by trimming with an axe or adze; sometimes used for decorative effect.

Rough-in
The early stage of construction in which the framing and rough carpentry is in place and electrical systems are roughed in, along with ductwork for the heating and air conditioning.

Rough-sawn
Wood that has been sawn to shape without planing or sanding; typically with saw marks on the surface; usually a preliminary step to being surfaced on all four sides.

Round arch See **Arch**

Round barn
A barn having a circular plan. See circular barn.

Round pediment See **Pediment**

Rounded
Refers to forms that may be spherical, globular, shaped like a ball, or circular in cross section.

Roundel
A small circular panel or window; in glazing, a bull's-eye or circular light like the bottom of a bottle.

Roundel window See **Window**

Routing
The cutting away of wood to shape a molding or other piece of millwork.

Rover molding See **Molding**

Row house
One of a series of identical buildings in a continuous row on one or both sides of a street, sharing one or more sidewalls in common with its neighbors, usually consisting of uniform plans, fenestration, and other architectural treatments.

Rowen, Daniel (1953–2009)
A New York architect with a client list that included hotelier Ian Schrager, art dealer Larry Gagosian, and Martha Stewart Living Omnimedia. Rowen was a protégé of modernist Charles Gwathmey.

Rowlock
A brick laid on its edge so that its end face is visible; one ring of a rowlock arch.

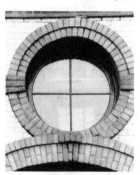

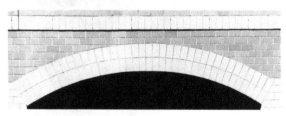

Rowlock arch See Arch

Royston, Robert (1918–2008)
San Francisco Bay–area landscape architect influenced by artists like Kandinsky and Miro, making art out of the transition between structure and landscape. Royston's numerous California projects include St. Mary's Square in San Francisco, Mitchell Park and Bowden Park, both in Palo

Royston, Robert

Alto, and Central Park in Santa Clara. This work led to the development of his landscape matrix theory of planned community design, where open spaces link together throughout the development.

Rubbed
A decorative finish that is obtained by rubbing bricks with a stone, brush, or abrasive tool, so as to produce a smooth surface of consistent color; used to highlight door, window, and arcade openings; arches; medallions; and bands and corners of facades.

Rubble
Rough stones of irregular shapes and sizes, used in rough, uncoursed work in the construction of walls, foundations, and paving.

Rubble wall
A rubble wall, either coursed or uncoursed.

Rubblework
Stone masonry built of rubble.

Rudolph, Paul (1918–1997)
Studied with Walter Gropius at Harvard. He was the chairman of Architecture at Yale University and designed the monumental Art and Architecture Building at New Haven, CT (1958–1965), which was typical of his Brutalist architecture. Other works include the Government Center, Boston, MA (1962), and other college structures. Most of his later work was done in Indonesia.

Runoff
When soil is infiltrated to full capacity with excess water from rain, snowmelt, or other sources. This is a major component of the water cycle. When runoff flows along the ground, it can pick up soil contaminants such as petroleum, pesticides, herbicides and insecticides, and fertilizers that pollute water sources.

Ruskin, John (1819–1900)
English writer, artist, and philosopher who championed the Gothic Revival style and paved the way for the Arts and Crafts movement in architecture. He wrote the *Seven Lamps of Architecture,* an influential book advocating functional planning and honesty in the use of materials in construction.

Russo-Byzantine architecture (1000–1500)
The first phase of Russian architecture, derived from the Byzantine architecture of Greece; it consisted mainly of stone churches characterized by cruciform plans and multiple bulbous domes.

Rust
A substance, usually in powder form, accumulating on the face of steel or iron as a result of oxidation; it will ultimately weaken or destroy the steel or iron on which it forms.

Ruin
The remains of something destroyed or demolished.

Run
Stonework having irregularly shaped units and no indication of systematic coursework; also the horizontal distance covered by a flight of stairs.

Run of rafter
In building, the horizontal distance from the face of a wall to the ridge of the roof; this distance is represented by the base of a right-angle triangle.

Running bond See Bond

Running measure
One-dimensional measurement of a piece of any material; also called linear measurement.

Running ornament
Any ornamental molding in which the design is continuous in intertwined or flowing lines, as in the representation of foliage and meanders.

Rustic
Descriptive of rough hand-dressed building stone, intentionally laid with high relief; used frequently in modest rural structures.

Rustic arch See **Arch**

Rustic brick See **Brick**

Rustic mortar joint See **Joint**

Rustic slate
One of a number of slate shingles of varying thicknesses, yielding an irregular surface when installed.

Rustic stone
Any rough broken stone suitable for rustic masonry, most commonly limestone or sandstone, usually set with the longest dimension exposed horizontally.

Rustic stone masonry See **Masonry**

Rustic style
A broad term applied to hunting lodges, ranger stations, or log cabins in mountainous areas, featuring log-cabin siding, peeled logs, and rough-cut lumber, a fieldstone chimney, a steep roof covered with hand-split shingles, and an overhanging roof with exposed rafters.

Rustic woodwork
Decorative or structural work constructed of unpeeled logs or poles.

Rustic work
In ashlar masonry, grooved or channeled joints in the horizontal direction, to render them more conspicuous.

Rusticate
To give a rustic appearance by beveling the edges of stone blocks to emphasize the joints between them; used mainly on the ground floor level.

Rusticated masonry See **Masonry**

Rusticated wood
A wood incised in block shapes to resemble rough stone.

Rustication
Masonry cut in large blocks with the surface left rough and unfinished, separated by deep recessed joints. The border of each block may be beveled or rabbeted on all four sides or top and bottom. It is used mainly on the lower part of a structure to give a bold , exaggerated took of strength.

R-value
A unit of thermal resistance. A material's R-value is a measure of the effectiveness of the material in stopping the flow of heat through it. The higher a material's R-value, the greater its insulating properties and the slower the heat flow through it. Examples of higher R-values for residential buildings: R-50 ceilings, R-21 walls, and R-30 floors.

R-value upgrades
Improved or added insulation in the attic or exterior walls to improve the R-value of the building envelope.

S s

Saarinen, Eero (1910–1961)
Finnish-born American architect and the son of Eliel
Saarinen. Projects include the General Motors Technical
Institute (illus.), Warren, MI (1951); Kresge Auditorium, MIT,
Cambridge, MA (1952); David S. Ingalls Ice Hockey Rink
(illus.), Yale University, New Haven, CT (1953); TWA Termi-
nal at Kennedy International Airport (illus.), NYC (1956);
Ezra Stiles and Morse College (illus.), Yale University, New
Haven, CT (1958); Dulles International Airport, Washington,
DC (1958), and the Gateway to the West Arch (illus.), St.
Louis, MO (1965).

Saarinen, Eliel (1873–1950)
Finnish-born American architect. Designed the Helsinki
railway station, built of rugged granite and inspired by the
designs of the Vienna Sezession and the Art Nouveau movement in Austria.

Saarinen, Eliel

Saddle
The ridge covering on the back of a chimney to carry water
back to the main roof surface. Also called a cricket.

Saddle notch
A round notch cut near one end in either the lower or upper
surfaces of a log that forms an interlocking joint when
matched with a similarly notched log set at right angles to it.

Saddle roof See Roof

Saddleback roof See Roof

Safde, Moshe (1938–)
Israeli-born Canadian architect who designed the Habitat
Housing (illus.), Montreal, Canada (1967), for Expo '67,
which consisted of prefabricated concrete housing units
fitted together in an experimental design. Later work includes Vancouver Library Square (illus.) Vancouver, Canada
(1991), the National Gallery of Canada, Ontario, and the
Ontario City Hall, Ontario, Canada.

Safde, Moshe

Safe carrying capacity
In construction, the capacity to carry the load a building is designed to support without failing in any way; in electrical work, the maximum current a conductor will carry without becoming overheated.

Safety glass See **Glass**

Sailor course bond See **Bond**

Sakha
In the architecture of India, a door jamb or door frame.

Salient
Any part or prominent member projecting beyond a surface.

Salmona, Rogelio (1929–2007)
One of Columbia's most significant architects and a former apprentice of Le Corbusier, Salmona was the first Latin American to win the Alvar Aalto Medal (2003) and is credited with making Bogota a model for a successful urban city. His work there includes the Virgilio Barco public library, the Archivo General, the vice president's compound, and a pedestrian mall in the city's historic center.

Salon
French word for a formal room in which exhibitions of art were held.

Saltbox
A wood-framed house, common to colonial New England, which has a short roof pitch in front and a long roof pitch sweeping close to the ground in the back.

Salvage
The controlled removal of construction or demolition debris, or other waste, from a permitted building or demolition site for the purpose of recycling, reuse, or storage for later recycling or reuse. Commonly salvaged materials include structural beams and posts, flooring, doors, cabinetry, brick, and decorative items.

Salvage archeology
The systematic rescue of archeological material and data threatened by damage or destruction; also called rescue archeology.

Sanctis, Francesco de (1693–1740)
Italian architect who, along with Alessandro Specchi, designed the Spanish Steps in Rome, which consist of an elegant ensemble of the piazza, the triple set of steps, and Bernini's Fountain.

Sanctuary
In a church, the immediate area around the principal altar.

Sand finish
Colored, textured plaster surface, similar in appearance to sandpaper.

Sandblasting
Abrading a surface, such as concrete, by a stream of sand ejected from a nozzle by compressed air; used for cleaning up construction joints, or carried deeper to expose the aggregate.

Sand-floated
A slightly textured stucco finish formed by sprinkling the partially hardened finished coat with sand and floating with a wooden trowel.

Sanding
Finishing surfaces, particularly wood, with sandpaper or some other abrasive.

Sandpaper
An abrasive paper, made by coating a heavy paper with a fine sand or other abrasive held in place with glue; used for polishing surfaces and finished work.

Sandstone See **Stone**

Sandwich beam
A built-up beam composed of two joists with a steel plate between them. The joists and plate are held together by bolts.

Sandwich construction
Composite construction consisting of a core of insulation and outer layers of higher density materials with greater strength.

Sandwich panel
A panel made of two sheets of plywood facing glued over a honeycomb core.

Sangallo, Antonio de (1485–1546)
Italian architect. Designed the Palazzo Farnese, the largest Renaissance palace in Rome.

Sanitary fixtures
In public, commercial or industrial buildings include toilets, urinals, basin taps, and showerheads.

Sanitary plumbing
The assembly of pipes, fittings, fixtures, and appliances that removes sanitary waste.

Sanitary sewers
Underground pipes that carry off only domestic or industrial waste, but not stormwater.

Sanitary survey
An onsite review of the water sources, facilities, equipment, operation, and maintenance of a public water system to evaluate the adequacy of those elements for producing and distributing safe drinking water.

Sanitary waste
The dirty water from kitchens, bathrooms, and laundries.

Sanitary water
The water discharged from sinks, showers, kitchens, or other nonindustrial operations, but not from commodes.

Sant'Elia, Antonio (1888–1916)
Italian architect who studied architecture in Milan and was influenced by Otto Wagner's Vienna School. Sant'Elia exhibited work with fellow student Mario Chiattone and published a manifesto on Futurism in the exhibition catalog.

Sanzio, Raphael (1483–1520)
High Renaissance architect and painter of great distinction. After Bramante's death, he was appointed master of works of St. Peter's, Rome, and proposed a basilican version of Bramante's plan.

Sapwood
The outer layers of the wood of a tree, in which food materials are conveyed and stored during the life of a tree; they are usually of lighter color than the heartwood.

Sarcophagus
An elaborate coffin for an important person, of terra-cotta, wood, stone, metal, or other material, decorated with painting and carving and large enough to contain only the body. If larger, it becomes a tomb.

Sash
Any framework of a window; may be movable or fixed, may slide in a vertical plane, or may be pivoted.

Sash balance
In double-hung windows, a device usually operated with a spring, designed to counterbalance the window sash without the use of weights, pulleys, and cords.

Sash bars
Strips of wood that separate the panes of glass in a window composed of several panes; also called muntins.

Sash window
A window formed with glazed frames that slide up and down in vertical grooves using counterbalanced weights.

Sashless window
A window composed of panes of glass that slide along parallel tracks in the window frame toward each other to leave openings at the sides.

Sassanian architecture (200–600)
Architecture that was prevalent in Persia, primarily in palace complexes. It featured extensive barrel vaults and parabolic domes set on squinches and stuccoed with plaster mortar. One notable example is the Palace at Ctesiphon.

Satinwood See **Wood**

Saturation
The purity of a hue; the higher the saturation, the purer the color.

Saucer dome See **Dome**

Sauer, Louis Edward (1928–)
American architect and design theorist whose work focused on over 90 residential and urban design commissions in differing contexts, from central city urban infill to suburban and rural areas, and new town developments at Reston, VA; Columbia, MA; and Montreal, Quebec. The conceptual innovation of most of these housing designs was a 12-foot-wide structural and functional module, which was part of a grid.

Sawtooth roof See **Roof**

Scab
Short piece of lumber nailed over a splice or joint to add strength or prevent slippage or movement.

Scaffold
A temporary platform to support workers and materials on the face of a structure and to provide access to work areas above the ground; any elaborated platform.

Scaffold crane
A small self-contained motorized crane attached to the scaffolding for lifting relatively small loads.

Scagliola See **Plaster**

Scale
The relationship of one part of an object to an outside measure, such as a human body or some standard reference; a system of representing or reproducing objects in a different size proportionately in every part.

human scale
The size or proportion of a building element or space, or article of furniture; relative to the structural or functional dimensions of the human body.

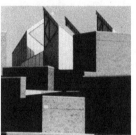

monumental scale
Impressively large, sturdy and enduring. Not related to the scale of the human body.

Scallop
One of a continuous series of curves resembling segments of a circle, used as a decorative element on the outer edge of a strip of wood used as a molding.

Scalloped capital See **Capital**

Scarf joint See **Joint**

Scarp
A steep slope constructed as a defensive measure in a fortification.

Scarpa, Carlo (1906-1978)
Italian architect practicing in Venice, who subscribed to the Modern movement. Later, his work focused on exhibitions, galleries, and museums.

Scharoun, Han (1893-1972)
German architect influenced by expressionism. His most celebrated work is the Hall for the Berlin Philharmonic Orchestra (1963), which illustrates his commitment to Organic architecture.

Schedule
A table or list included on working drawings giving the number, size, and placement of items such as doors and windows.

Schematic design phase
The first phase of the architect's basic services. The architect prepares schematic design studies, consisting of drawings illustrating the scale and relationship of the projected components for approval by the owner.

Scheme See **Design**

Schindler, Rudolph (1887-1953)
Austrian architect. Born in Vienna, he was influenced early by Otto Wagner. He later worked for Frank Lloyd Wright in Chicago, but his work remains reminiscent of the de Stijl movement, as shown in his Lovell Beach House, Newport Beach, CA (1926).

Schinkel, Karl Friedrich (1781-1841)
German architect of original Neoclassical buildings. His work was stylistically eclectic, but lyrical and logical. Schinkel's funeral in 1841 was a national event, and his grave is marked by a stele of his own design. King Friedrich Wilhelm IV (reigned 1840-1861) decreed that all of Schinkel's work be purchased by the state.

Schmidt, Richard E. (1865-1958)
American architect, in practice before Frank Lloyd Wright, but did occasional drawings for him. Schmidt was a partner in the firm of Schmidt, Garden, and Martin.

School
An institution for instruction in a skill or business. The abandonment or deterioration of an old school, which is a symbol of continuity and stability from one generation to the next, can have a negative effect on an entire community.

Schoolhouse
A building used for a school, especially an elementary school.

Scissors truss See **Truss**

Sconce
An electric lamp, designed and fabricated for mounting on a wall, resembling a candlestick or a group of candlesticks.

Scope
The extent or intention of the construction activity, also the portion of the plans, specifications, and addenda on which the contractor has based its bid.

Scope documents
Any document among the bidding or contracting documents setting forth the intent that the contract documents indicate only the general scope and performance criteria, and that the contractor shall be required to perform all work and furnish all materials necessary to accomplish construction according to the general scope standards.

Scoping
The involvement of local government agencies and the general public in the production of an environmental impact statement for a proposed project.

Scoring
Partial cutting of concrete flat work for the control of shrinkage cracking; also used to denote the roughening of a slab to develop mechanical bond.

Scotia molding See **Molding**

Scott Brown, Denise (1931-)
Zambian-born architect; married Robert Venturi in 1967 and is a partner in Venturi, Rauch & Scott Brown. She influenced Venturi's book, *Learning from Las Vegas* (1972).

Screed
A wooden or metal guide for leveling plaster or concrete, typically placed along the edge of the work at the desired level.

Screen
Any construction whose essential function is merely to separate, protect, seclude, or conceal, but not to support.

Screen facade
A nonstructural facing assembly used to disguise the form or overall size of a building.

Screen facade

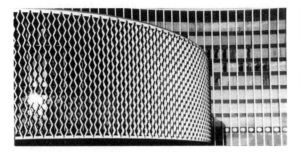

Screen wall See **Wall**

Scribbled ornament
A decorative effect produced by irregularly distributing lines and scrolls over a surface or on a panel.

Scroll
An ornamental molding consisting of a spiral design; or a terminal, such as the volutes of the Ionic capital or the S curves on consoles.

Scroll molding See **Molding**

Scroll ornament See **Ornament**

Scrollwork
Ornamental work of any kind in which scrolls, or lines of scroll-like character, are an element.

Sculpture
The art of shaping figures or designs in the round or in relief by carving wood, chiseling marble, modeling clay or casting in metal; any work of art that is created in this manner.

Sculpture

Sculpture in the round
Freestanding figures carved or molded in three dimensions.

Scutt, Der (1935–2010)
Is best known for his early 1980s addition to Fifth Avenue in NYC for Donald Trump—a bronzed glass modern tower. Scutt was a developer's architect, and his many New York projects included One Astor Plaza overlooking Times Square; 100 United Nations Plaza Tower; and the Corinthian apartment building, NYC.

Scuttle
A small opening in a ceiling or roof; usually installed on top of a built-up frame.

Sealed combustion furnace
Furnaces or boilers that draw air for combustion from outside the home directly onto the burner compartment and vent exhaust gases directly to the outside. The systems eliminate the possibility of backdrafting.

Sealed crawl space
A crawl space under a home that has been properly air-sealed to conserve energy.

Sealed ducting
A way to save energy and avoid moisture damage by repairing improperly installed ducts or by sealing the seams in ductwork.

Season
To dry wood through exposure to the air or the heat of a kiln, thus lowering its moisture content.

Seasonal energy efficiency ratio (SEER)
Energy performance rating of a central air conditioner or heat pump operating in the cooling mode, calculated as the ratio of the estimated seasonal cooling output divided by the seasonal power consumption in an average climate.

Seasonal performance tests
Include the full range of test procedures carried out to determine if all components, equipment, systems, and interfaces between systems function according to design intent during heating or cooling design days. When it is not practical to perform the test during an actual design day, these conditions may be simulated.

Seasoning
Removal of moisture from green wood in order to improve its workability and serviceability.

Seasoning check
Separation of wood extending a few inches in length longitudinally, formed during the drying process; commonly caused by the immediate effect of a dry wind or hot sun on freshly sawn timber.

Seat
A chair, stool, or bench on which to sit; may be built-in, such as a window seat.

Secondary colors
The colors green, orange, and purple that result from a mixture of pairs of primary colors.

Secondary element
The nonessential elements of a structure, such as trim around doors and windows, which are essentially finishing elements.

Section
The representation of a building or portion thereof, cut vertically at some imagined plane, so as to show the interior of the space or the profile of the member. See also Projection drawing.

Sediment
Transported and deposited particles or aggregates derived from rocks, soil, or biological material.

Sediment basin
A depression in the soil that is placed to retain sediment and debris onsite.

Sedimentation
The natural process of depositing sediment.

Segmental arch See **Arch**

Segmental pediment See **Pediment**

Seiberling, John (1919–2008)
Environmentalist and preservationist Seiberling was an eight-term congressman from Ohio who helped draft more than 60 park-related bills, including the one that established the Cuyahoga Valley National Park. President Bill Clinton honored him with the Presidential Citizens Medal in 2001, calling him an "environmental hero." Seiberling masterminded the Alaska Lands Act of 1980, doubling the size of U.S. national parks and wildlife refuges and tripling the amount of federally designated wilderness throughout the United States. He also helped create the Historic Preservation Fund, and he received the National Trust for Historic Preservation's top honor, the Louise du Pont Crowninshield Award, in 2002.

Seidler, Harry (1923–2006)
Austrian-born Australian architect who is considered to be one of the leading exponents of Modernism's methodology in Australia and the first architect to fully express the principles of the Bauhaus in Australia. Seidler attended the Harvard Graduate School of Design under Walter Gropius and Marcel Breuer in 1946, during which time he worked with Alvar Aalto in Boston drawing up plans for the Baker dormitory at MIT. He then attended Black Mountain College under the painter Josef Albers, and worked for Marcel Breuer in New York. Seidler also worked in the studio of the architect Oscar Niemeyer in Rio de Janeiro.

Seismic code
A building code that defines the minimum earthquake resistance of a structure; usually requires seismic reinforcing of existing buildings that are altered.

Seismic load
The design load for potential seismic forces acting on a building during an earthquake that is used to determine the extent of seismic reinforcing.

Seismic reinforcing
The structural strengthening of a building to resist seismic forces, using shear walls or partition trusses or increasing the size of structural members.

Sejima, Kazuyo (1956–)
Japanese architect who works collaboratively with Ryue Nishizawa. Sejima has been praised for designing powerful, minimalist buildings using common, everyday materials. Sejima and Nishizawa share the 2010 Pritzker Architecture Prize.

Self-closing door See **Door**

Seljuk architecture (1000–1200)
An early phase of Turkish Muslim architecture, influenced by Persian architecture, consisting mainly of mosques and minarets.

Semiarch See **Arch**

Semicircular
Describing a form that exhibits an arrangement of objects in the shape of a half-circle, as divided by its diameter.

Semicircular arch See **Arch**

Semicircular dome See **Dome**

Semidetached house
One of a pair of houses with separate entrances that share a party wall between them.

Semidome See **Dome**

Semigloss paint
A paint or enamel that, when dry, has some luster but is not very glossy.

Semiglyph
Half glyphs at the edge of a triglyph.

Seminary
A school, academy, college, or university, especially a school for the education of men for the priesthood.

Semipervious surface
A surface that can be partially penetrated by water and air.

Semirecessed fixture
A fixture that is partially recessed into the ceiling, but has part of its housing, reflectors, or lenses projecting below the surface.

Semirigid joint See **Joint**

Sense of place
The sum of attributes of any place that gives it a unique and distinctive character.

Septic tank
A concrete or steel tank where sewage is partially reduced by bacterial action. The liquids from the tank flow into the ground through a tile bed.

Seraph
A celestial being or angel of the highest degree, usually represented with six wings.

Serpentine
A form that resembles a serpent, showing a sinuous winding movement; a greenish brown or spotted mineral used as a decorative stone in architectural work.

Serpentine wall See **Wall**

Serrated
Consisting of notches on the edges, like a saw.

Sert, Josep Luis (1902–1983)
Catalonian architect who worked with Le Corbusier; he settled in the United States in 1939 and became dean of faculty at the Graduate School of Design, and chairman of Architecture at Harvard University (1953). He designed the U.S. Embassy, Baghdad, Iraq, and the Miro Foundation Building, Barcelona, Spain (1972).

Service area
Those areas of a building set aside for functions that support the operation of the building, such as loading, parking and waste collection.

Service cable
A cable supplying electricity to a building, either by an overhead wire or in an underground conduit.

Service conduit
A pipe that carries electrical service from the street to the service main in the building.

Service core
A vertical element in a multistory structure, containing the elevators, vertical runs of most of the mechanical and electrical services, and the fire stairs.

Service head
A device that connects an overhead electric service cable to a building; often at the top of a vertical conduit that connects directly to the electric meter.

Service stair See **Stair**

Service switch
Main disconnect switch at the location where the electric service cable enters a building.

Setback
The upper section of a building, successively recessed, that produces a ziggurat effect, admitting light and air to the streets below.

Setback line
The legal distance required between a property line and the face of a building; may be dictated by a deed restriction or local ordinance.

Set-point
The desired temperature, humidity, or pressure in a space, duct, or similar area.

Setting out
The process of establishing pegs, profiles and levels for excavation and positioning buildings or marking out the position of walls on a floor slab.

Sezession
A term adopted by several groups of artists in Germany who seceded from the traditional conservative academies to show their work; the most celebrated was the one founded in Vienna in 1897, and included the artist Gustav Klimt and the architects Joseph Olbrich and Otto Wagner.

Sgraffito
Decoration produced by covering a surface, such as plaster or enamel, of one color with a thin coat of a similar material of another color and scratching through the outer coat to show the color beneath.

Shading devices
Reduce building peak heat gain and cooling requirements and improve the natural lighting quality of building interiors. They also improve user visual comfort by controlling glare and reducing contrast ratios, which leads to increased user satisfaction and productivity.

Shaft
The main body of a column, pilaster or pier between the capital and the base, or a thin vertical member attached to a wall or pier, often supporting an arch or vaulting rib.

Shafting
In medieval architecture, an arrangement of shafts, combined in the mass of a pier or jamb, so that corresponding groupings of archivolt moldings above may start from their caps at the impost line.

Shake See **Wood products**

Shallow trench system
A type of drain field used in conjunction with a graywater system that allows for shallow placement of distribution pipes and use of the graywater for irrigation.

Shape
Implies a three-dimensional definition that indicates outline and bulk of the outlined area.

Shared street
A common space created to be shared by pedestrians, bicyclists, and low-speed motor vehicles. They are typically narrow streets without curbs and sidewalks, and vehicles are slowed by placing trees, planters, parking areas, and other obstacles in the street.

Shaw, Howard Van Doren (1869–1926)
American architect in Chicago who designed fashionable houses in the Midwest, many in English Revival styles. Designed Ragdale (1898) and Market Square (1916), Lake Forest, IL.

Shear braces
A bracing system, usually using metal brackets or straps, that eliminates most structural wall sheathing.

Shear wall See **Wall**

Sheathing
Material, usually plywood or oriented strand board (OSB), but sometimes wooden boards, installed on the exterior of wall studs, rafters, or roof trusses; siding or roofing installed on the sheathing, sometimes over strapping to create a rainscreen.

Shed
A rough structure for shelter, storage or a workshop; it may be a separate building or a lean-to against another structure, often with one or more open sides.

Shed dormer See **Dormer**

Shed dormer window See **Window**

Shed roof See **Roof**

Sheet glass See **Glass**

Sheet metal
A flat, rolled metal product, rectangular in cross section and formed with sheared, slit, or sawn edges.

Sheet pile
A pile in the form of a plank, driven in close contact with others to provide a tight wall to resist the lateral pressure of water or adjacent earth.

Sheetrock®
A proprietary name for gypsum wallboard.

Shelf angle
A steel angle that supports a brick veneer.

Shell
The basic structure and enclosure of a building, exclusive of any interior finishes and mechanical, plumbing, and electrical systems; often describes a deteriorated structure that has lost much of its original fabric.

Shell construction
A thin, curved, structural outer layer that distributes loads equally in all directions; most commonly used for roofs composed of concrete; also called thin-shell concrete.

Shell ornament
Any decoration where a shell form is a characteristic part; coquillage.

Shellac
A wood finisher and resin used in varnish, which produces a transparent shiny surface; often used to enhance and protect wood grain.

Shells
Hollow structures in the form of thin curved slabs, plates or membranes that are self-supporting. They are called form-resistant structures because they are shaped according to the loads they carry.

Shepley, George Foster (1860–1903)
American architect in Brookline, MA, who, with Charles H. Rutan (1851–1914) and Charles A. Coolidge (1858–1936), formed the successor firm to H. H. Richardson. Designed Stanford University (1892), Stanford, CA; Ames Building (1892), Boston, MA; and the Art Institute of Chicago, IL (1897).

Shikkui See **Plaster**

Shim
A thin piece, usually wedged-shaped, placed or driven into a joint to level or plumb a structural member.

Shiner course bond See **Bond**

Shingle nails
Used to install cedar shingles; have small heads and thin shanks to avoid splitting the shingle.

Shingle See **Wood products**

Shingle style (1880–1895)
A style that featured an eclectic American adaptation of New England forms to the structuralism of the Victorian era. Structures were de-emphasized by a uniform covering of entire surfaces of the roof and walls with monochromatic shingles; the eaves of the roofs are close to the walls so that they emphasize the homogeneous shingle covering. The houses in this style were rambling and horizontal and featured wide verandas and hipped roofs. In 1876 the centenary of the American Revolution encouraged a revival of Colonial Georgian with shingle cladding, gambrel roofs, dormers, oriels and other elements of the Queen Anne style, resulting in the shingle style; many had open planning inside, anticipating later works by Frank Lloyd Wright and Greene and Greene.

Shiplap siding See **Wood products**

Shipporeit-Heinrich

Originally from the office of Mies van der Rohe, G. D. Shipporeit and John C. Heinrich designed the first skyscraper enclosed entirely in glass, the Lake Point Towers (illus.) in Chicago, IL (1968).

Shopping mall

Shoddy

Materials or standards of workmanship that are of a low quality; slang term for any inferior quality work.

Shoe

A piece of timber, stone, or metal, shaped to receive the lower end of any member, also called a "soleplate"; a metal base plate for an arch or truss shaped to resist the lateral thrust.

Shoe mold

The small molding against the baseboard at the floor.

Shoji

A very lightweight sliding partition used in Japanese architecture, consisting of a wooden lattice covered on one side with translucent white rice paper. The lattice is most often composed of small horizontal rectangles.

Shop drawings

Drawings, diagrams, illustrations, schedules, performance charts, brochures, and other data prepared by the contractor, manufacturer, supplier, or distributor that illustrate how specific portions of the work shall be fabricated or installed.

Shopping center

A group of retail stores and service establishments in a suburban area, with parking facilities usually at grade level.

Shopping mall

A shopping center enclosed within a large structure and placed around a central atrium; may have numerous stores, entertainment facilities such as movie theaters, fast-food outlets, restaurants, and public areas.

Shore

A temporary support used in compression as a temporary support for excavations or formwork, or propping up unsafe structures.

Shoring

The use of timbers to prevent the sliding of earth adjoining an excavation; also, the use of timbers and adjustable steel or wooden devices placed in a slanted position as bracing against a wall, or used as temporary support during restoration.

Shoro

A small structure in a Japanese temple compound, from which a bell is hung.

Shotblasting

Cleaning a steel surface by projecting steel shot against it with compressed air.

Shotcrete

Cement mortar or concrete placed under pressure through the nozzle of a cement gun. Also called Gunite.

Shoulder
A projection or break changing the thickness or width of a piece of shaped wood, metal, or stone.

Shouldered arch See Arch

Shower facilities
In buildings where people work, shower facilities are considered a green building feature in that they encourage workers to travel by bicycle and other human-powered modes of transportation.

Shreve, Raymond (1877–1946)
American architect and partner in the firm of Shreve, Lamb and Harmon, designers of the Empire State Building, New York City (1932).

Shrine
A place, building, or structure made sacred by association with a historic event or holy personage; an altar, tomb, or chapel.

Shrinkage
A decrease in dimensions occurring when a material experiences temperature changes.

Shrinkage crack
A rift that appears in the surface of building materials such as concrete, masonry, and mortar joints, due to shrinkage.

Shrinkage reinforcement
Secondary reinforcement designed to resist shrinkage stresses in concrete.

Shut-off valve
A valve in a gas or water supply line that can completely seal off the flow to the building, a branch line, or a fixture; the main water shutoff typically has a gate valve.

Shutter
One of a pair of movable panels used at window openings to provide privacy and protection from the elements when closed.

Shutter bar
A hinged bar that can be fastened across the interior side of a pair of shutters, providing a measure of security.

Shutter box
A pocket or recess located along the interior side of a window to receive shutters when folded.

Shutter fastener
A pivoted device often made of decorative wrought iron that holds a shutter in the open position on the exterior side of a window.

Shutters
Protective covering for the outside of windows; louvered shutters are used for ventilation.

Siamese (Thai) architecture (1350–1500)
An architecture consisting of stupas and temples. The most characteristic forms are the eaves of overlapping roof planes, which are terminated with sculptural finials. The Temple of the Emerald Buddha in Bangkok is the most notable example.

Siamese (Thai) architecture

Siamese fixture
A plumbing fixture with a Y-shaped end, found on the exterior of a building for connecting a fire hose to a standpipe.

Sibyl
A woman in Greek and Roman mythology reputed to possess powers of prophecy and divination.

Sick building
A building which causes a higher than normal level of minor illness to its occupants. Typical symptoms are irritation to the eyes, nose, or throat; shortage of breath; dizziness; and general fatigue. Sick building syndrome (SBS) is mainly associated with air-conditioned buildings and those that have no user control of ventilation or heating or lighting levels.

Sick building syndrome
A human health condition in which infections linger, caused by exposure to contaminants within a building as a result of poor ventilation. A sick building is one whose occupants experience acute health and/or comfort effects that appear to be linked to time spent inside, but where no specific illness or cause can be identified. Complaints may be localized in a particular room or zone or may be spread throughout the building and may abate on leaving the building.

Side aisle
One of the corridors parallel to the nave of a church or basilica, separated from it by an arcade or colonnade.

Side gable See Gable

Side lap
The overlap required for two adjacent building components to prevent the penetration of rain.

Side lighting
The source of light coming primarily from one side only, primarily to emphasize the modeling or texture of an object.

Sidelight
A framed area of fixed glass, usually comprising a number of small panes; commonly one of pair of such lights, set vertically on each side of a door.

Sidewalk
Walkway along the side of a road, or leading to a building, usually constructed of flagstone, brick, or concrete.

Sidewalk bridge See Bridge

Siding
Long-lasting, low-maintenance exterior finish products reduce replacement frequency, which means cost savings, reduced landfill impact, and fewer resources and time devoted to maintenance and replacement. Fiber-cement siding is highly durable; it doesn't split, holds paint longer, and is more moisture-resistant than typical hardwood siding. Other green building siding options include recycled-content hardboard, natural or synthetic stucco, locally produced brick, and natural or faux stone.

Siding See **Wood**

Siegel, Robert
With Charles Gwathmey, modernist architect Siegel is a partner in the New York firm Gwathmey Siegel & Associates. The firm had extensive residential experience designing homes, apartment buildings, lofts, and penthouses for private clients. Projects include the addition to the Guggenheim Museum, New York City.

Signature stone
A stone, found on many 18th- and 19th-century dwellings, carved with the date of completion and the name or initials of the owner, usually embedded in the wall over an entry door or in a gable.

Sikhara
The spire or tower over the shrine of an Indian temple.

Sill
The horizontal exterior member at the bottom of a window or door opening which is usually sloped away from the bottom of the window for drainage of water, and overhanging the wall below.

Sill plate
A heavy timber plate at the bottom of the frame of a wood structure resting directly on the foundation.

Sillcourse
In stone masonry, a stringcourse set at the windowsill level; commonly differentiated from the wall by its greater projection, finish, or thickness.

Sillsbee, J. Lyman (1845–1880)
American architect who attended Harvard and MIT, and moved to Chicago. Hired initially by Frank Lloyd Wright's uncle, the Reverend Jenkins Lloyd Jones; Wright went to work for him briefly, prior to working for Adler and Sullivan.

Silo
A tall, enclosed structure used primarily to store grain or chopped plants, commonly constructed of wood, masonry, or concrete.

Siloe, Diego de (c. 1495–1563)
Spanish architect and sculptor who introduced Italian Renaissance forms to Spain and played a major role in the development of the distinctive Plateresque style. Work includes the Granada Cathedral (1549).

Simple beam See **Beam**

Simulated architecture
An architecture that consists of holographic images of forms and monuments projected into space by laser beams.

Simulated architecture

Simulation
The process of representing or modeling a situation.

Sinan, Koca (1489–1578)
One of the greatest Turkish architects and chief architect to the Ottoman court; brought to full development the classic Ottoman domed mosque, including the Suleymaniye Mosque, Istanbul (1551–1581).

Single-hung window See **Window**

Single-pass cooling
During this process, water is circulated once through a piece of equipment and then disposed down the drain. To remove the same heat load, single-pass systems use 40 times more water than a cooling tower operated at five cycles of concentration.

Single-phase power
A two- or three-wire distribution system commonly used for residences and small commercial buildings.

Single-pitched roof See **Roof**

Sinking groove
The incised groove that separates the shaft from the capital in the Doric order.

Sinks

Surfaces that tend to capture volatile compounds from air and release them later. Carpets, gypsum board, ceiling tiles, and upholstery are all sinks.

Siras

In Indian architecture, the capital of a column or pillar.

Sisal

A durable natural fiber used as a floor covering, derived from leaves of the sisal plant.

Site

A contiguous area of land, including a lot or lots or a portion thereof, upon which a project is developed or proposed for development; an area of property that is experiencing land development and management.

SITE (Structures in the Environment)

American group launched by James Wines, and best known for the designs for the Best Products chain of stores, where unique manipulation of architectural elements made the buildings notorious. One of the most unique, but never realized, was the "highrise of houses," wherein a neighborhood of complete single-story residences were stacked within a steel superstructure. The later works include exhibits for the Expo '92, Seville, Spain (1992).

Site and location evaluation

Part of a feasibility study which includes existing and planned uses of adjoining property; public services, transportation, and jobs; hidden assets of the area; entertainment and social offerings; and real estate sales and rental market.

Site assessment

The thorough environmental analysis conducted as a stage in planning to assess a variety of measures from soils, topography, hydrology, and environmental amenities such as wetlands, wind direction, solar orientation, animal and plant habitat, and connections to community. A geographic information system (GIS) can facilitate this task.

Site contamination

The degradation of land and buildings due to exposure to materials, processes or organisms detrimental to health.

Site development costs

All the costs needed to prepare land for building construction, such as the demolition of existing structures, site preparation, and offsite and onsite improvements.

Site marker

A plaque or sign located at or near a historic site or building.

Site plan

A plan of a construction site showing the dimensions and contours of the lot and the dimensions of the building or portion thereof to be erected.

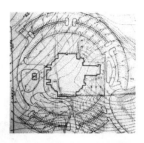

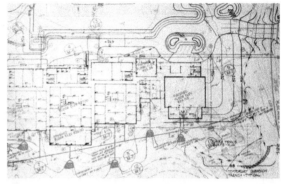

Site planning

Involves the organization of land-use zoning, access, circulation, privacy, security, shelter, land drainage, and other factors, by arranging the compositional elements of landform, planting, water, buildings, and paving. Generally begins by assessing a potential site for development through site analysis. Information about slope, soils, hydrology, vegetation, parcel ownership, and orientation are assessed and mapped.

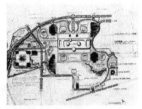

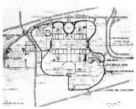

Site preservation

Minimizing the disruption of a building on its surrounding environment by reusing existing structures on a site, rather than building upon unused land; or avoiding building on top of environmentally fragile land that could interfere with natural ecosystems.

Site relationship

The plot of land where something was, is, or will be located; the situation or position of a place, town, or building, especially with reference to the surrounding locality.

Site relationship

Site selection and preparation
The complete sequence or series of activities and actions that begins with the natural environment and results in some specific geographic location defined in terms of boundaries, then altered and modified until it becomes a building site ready for construction to begin.

Site work
Construction outside of a building, including earthwork, landscaping utilities, and paving.

Siting
The process of selecting where and how a proposed structure will be situated on land. This process is as important as the components of the building itself and should, ideally, take into account such factors as landscape, soil, vegetation, water supply, and the position of the sun relative to the site at different points during the day. Good siting avoids unnecessary disturbance to the environment during and after building, and among other factors, takes advantage of natural light and existing vegetation for interior illumination and temperature moderation, requiring much less energy use over the life of the structure.

Sizing
Calculation of the heat loss and heat gain for a building at "design temperatures" (close to the maximum and minimum temperatures anticipated for a given location) in order to select heating and cooling equipment of sufficient capacity. Installing excess equipment capacity, or oversizing, is common but leads to inefficient operation and, for air conditioners, decreases the dehumidification. Also, A thin, pasty substance used as a sealer, binder, or filler; generally consisting of a diluted glue, oil, or resin; often required as an underlayment in applying gold leaf.

Skeletal frame
Refers to a structural framework of members, originally concealed within a building, or as a self-supporting grid of timber, steel, or concrete.

Skeleton construction
Any construction in which the loads are transmitted to the ground by a frame, as opposed to construction with load-bearing walls.

Sketch plan
A freehand sketch of a floor plan of an existing building done as part of site investigation for restoration or renovation, to be translated later into a measured drawing and incorporated into the contract documents.

Skew corbel
A stone built into the bottom of a gable to form an abutment for a wall cornice or eaves gutter.

Skewback
The sloping surface of a member that receives the component materials of an arch.

Skewed
Having an oblique position, or twisted to one side.

Skidmore, Louis (1897–1962)
Founded Skidmore, Owings and Merrill (SOM) in Chicago in 1936, which was organized on teamwork and used other ideas incorporated from business practice. SOM won fame for the Lever House in NYC (1951). Later work included the John Hancock Center, Chicago, IL (1968); projects in Saudi Arabia (1982) and the Canary Wharf developments in London, England (1990).

Skin
A non-load-bearing exterior wall.

Skintled bond See **Bond**

Skintled brickwork
In masonry, an irregular arrangement of bricks, with respect to the normal face of a wall. The bricks are set in and out so as to produce an uneven effect on the wall; also a rough effect caused by mortar being squeezed out of the joints.

Skirt roof See Roof

Sklarek, Norma (1928–)
First Black woman to become a U.S. architect. Her many projects include a new terminal for Los Angeles International Airport.

Sky lobby
An elevator lobby at an upper floor; the sky lobby is served by express lifts only, thus saving an appreciable amount of space.

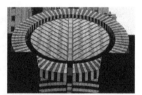

Skylight
An opening in a roof which is glazed with a transparent or translucent material used to admit natural or diffused light to the space below.

Skyscraper
A building of extreme height containing many stories, constructed of a steel or concrete frame that supports the exterior walls, as opposed to a load-bearing structure.

Skyscraper construction
A method of construction developed in Chicago, in which all building loads are transmitted to a metal skeleton, so that any exterior masonry is simply a protective cladding.

Skywalk See Bridge

Slab
The upper part of a reinforced concrete floor, which is carried on beams below; a concrete mat poured on subgrade, serving as a floor rather than as a structural member.

Slab-on-grade
Concrete floor that is supported directly on the earth or fill.

Slat
A thin, narrow strip of wood, often one of a series used within a framework to regulate the passage of light and air into an area.

Slate See Stone

Sleeper
A horizontal timber laid on a slab or on the ground and to which the subflooring is nailed; any long horizontal beam, at or near the ground, which distributes the load from the posts to the foundation.

Sleeve
A tube or tube-like part fitting over or around another part; a pipe used to provide openings for the installation of electric and plumbing services, used in solid concrete floors through which the services must pass. Also called a sleeve chase.

Sliced veneer

Wood veneer in which a log or sawn flitch is held securely in a slicing machine and the wood is sheared off in sheets.

Sliding door See Door

Sliding sash window See Window

Slip joint

In brickwork, a type of joint made where a new wall is joined to an old wall by cutting a channel or groove in the old wall to receive the brick of the new wall. This method of joining the two walls forms a kind of telescopic, nonleaking joint.

Slipforms

Concrete forms that are advanced for another pour after the concrete has set.

Sloan, Samuel (1815–1854)

American architect in Philadelphia who authored many plan books and designed public buildings, schools, hospitals for mentally ill people, and churches. Works include the Alabama Insane Hospital (1852), Tuskaloosa, AL, and Longwood (1862), Natchez, MS.

Slum

An overcrowded, dirty, neglected, unhygienic area of decaying buildings normally inhabited by economically disadvantaged people.

Slurry wall

A foundation wall in an excavation that is heavily reinforced with steel and filled with a liquid concrete mixture by pouring or pumping.

Small

All else being equal, smaller is usually preferable in sustainable building. Larger buildings and spaces require more materials and energy to construct and use more resources to heat, cool, and maintain.

Smart building

A focus on the use of a building automation system (BAS) that controls building heating, air conditioning, and security systems. These systems are often implemented indepen-

Smart building

dently; one cabling system used to support the building security system, another to support the HVAC systems, and so forth. In addition to cabling installed to support a BAS, telephone wiring is then installed to all work spaces. Finally, data networks are installed, usually on a tenant-by-tenant basis given the wide gaps between tenants in the use of networking technology.

Smart growth

A collection of urban development strategies meant to reduce sprawl that are fiscally, environmentally, and socially responsible. Smart growth is development that enhances quality of life, protects the environment, and uses tax revenues wisely.

Smart house

A home that has programmable electronic controls and sensors that regulate heating, cooling, ventilation, lighting, and appliance and equipment operation in a way that responds to interior climate conditions in order to conserve energy.

Smart materials

Engineered materials that sense and react to environmental conditions or have properties that can be altered in a controlled fashion by light, temperature, moisture, mechanical force, or electric or magnetic fields. All changes are reversible, given that the materials return to their original states once the external stimulus expires.

electrochromic

The ability of a material to transmit light due to a change in electrical current. The optical properties are reversible, and the material reverts to its original state once the electrical current is removed. As such, electrochromic materials are the primary choice for visual devices, such as smart windows, light shutters, information displays, reflectance mirrors, and thermal radiators. In green architecture, electrochromic materials are mainly used in "smart windows" for their energy efficiency and thermal comfort. The transparency/opacity level is adjusted by an applied voltage.

electrostrictive

Materials that change in size in response to an electric field and produce electricity when stretched are called electrostrictive. These materials are primarily used as precision control systems such as vibration control and acoustic regulation systems in engineering, vibration damping in floor systems, and dynamic loading in building construction.

light-responsive materials

Smart materials that transform due to a change in light are called light-responsive.

photochromic

Materials that change their ability to reflect color when exposed to light are called photochromic, and the color change is proportional to their level of UV light absorption. This results in reversible color reflections. When the light source is terminated, the material changes back to a clear state.

photoluminescent

Materials that absorb radiation from light and convert it into visible light are called photoluminescent. In green architecture, they are widely used for exit signs and other self-luminous emergency egress indicators, because they do not rely on external energy sources and they require only minimal maintenance.

photovoltaic

In a photovoltaic system, an electrical current is produced in a solid material. The physical process of this conversion is called the photovoltaic effect. Currently, there are two types of solar cell technologies: (1) crystalline materials and (2) thin-film materials. In architecture, photovoltaic materials are used in custom panels, shingles, solar tiles, and window film applications.

shape memory

Changing shape from a rigid form to an elastic state when thermal energy is applied is called shape memory. When the thermal stimulus is removed, the material reverts back to its original rigid state without degradation, which is called superelasticity.

thermochromic

Materials that change color in response to temperature differences are called thermochromic. In architecture, these materials are currently used mostly for interactive visual effects, although they could have additional applications for green architecture in the future. Thermochromic window films alter their color structures as well as reduce solar heat transmission by blocking UV radiation. Thermochromic paints change color and thus heat absorption based on temperature changes in the outdoor environment.

thermoelectric

A thermoelectric module is a small, light-weight, and silent solid-state device that can operate as a heat pump or as an electrical power generator with no moving parts. Thermoelectric modules are durable, reliable, silent, lightweight, and compact green materials; they do not include compressed gases, chemicals, or toxic agents.

thermoresponsive materials

Thermoresponsive materials are smart materials that transform in response to temperature changes.

thermotropic

Materials that undergo various property transformations in response to heat and temperature changes, including conductivity, transmissivity, volumetric expansion, and solubility. Visibility in thermotropic windows is directly controlled by climatic temperature changes. However, there are no visual changes to the window at low temperatures. Therefore, during the winter these windows allow solar light and heat to penetrate undiminished into the building.

Smith, Adrian D. (1944–)

American architect who designed skyscrapers including the Burj Khalifa, in Dubai; Jin Mao Tower, Shanghai; and the Trump International Hotel and Tower, Chicago. The Burj Khalifa is the world's tallest building. Smith was the design partner at SOM for the Burj Khalifa, which officially opened

Smith, Adrian D.

in 2010. With Gordon Gill, he designed the Masdar City Headquarters Building, in Abu Dhabi.

Smithmeyer, John L. (1832–1908)

American architect in Washington, DC, who with Paul J. Pelz (1841–1918) designed in the Classical styles. Pelz was also known for lighthouses and the Library of Congress (1892), Washington, DC.

Smithson, Peter (1923–)

Husband to partner Alison (1928–1993), member of CIAM, and leader of the Modern Movement in Britain in the 1950s and 1960s. The use of exposed raw concrete earned Smithson the nickname "Brutus."

Smithsonian Institution

U.S. public institution that maintains collections of historic and cultural significant objects in various history and art museums; preserves its own historic public buildings; and administers document collections useful in research.

Smoke chamber

The portion of a chimney flue located directly over a fireplace.

Smoke control system

A system to control the movement of smoke during a fire within a building.

Smoke control system, active

A system that uses fans to produce airflows and pressure differences across smoke barriers to limit and direct smoke movement.

Smoke control system, passive

A system that shuts down fans and closes dampers to limit the spread of fire and smoke.

Smoke venting

Provision for allowing smoke from a building fire to escape rapidly into the atmosphere.

Snow guard

A board or other device that prevents snow from sliding off the roof.

Snow load

The superimposed load assumed to result from severe snowfalls in any particular region.

Soane, Sir John (1753–1837)

British architect. Designed Number 3 Lincoln's Inn Fields, London.

Society for Commercial Archeology

A group that focuses on the preservation of roadside architecture.

Society of Architectural Historians

A scholarly organization of architectural historians, architects, educators, and others interested in architectural history. It sponsors a Historic Preservation Committee, promotes preservation of significant architectural landmarks, and conducts tours of historic buildings and districts.

Socle

A low, plain base course for a pedestal, column, or wall; a plain plinth.

Sod roof See **Roof**

Sodium silicate

A liquid used in asbestos encapsulation, concrete and mortar waterproofing, and high-temperature insulations; also called water glass. This substance is nontoxic when cured but caustic when wet.

Soffit

A ceiling or exposed underside surface of entablatures, archways, balconies, beams, lintels or columns.

Soffit

Soffit ventilation

Intake ventilation installed under the eaves or at the roof edge.

Soft costs

Development expenses other than those devoted to land and actual construction, such as interest on borrowed funds, architectural and other fees, marketing costs, and incidental expenses.

Softwood See **Wood**

Soil stack

The main vertical pipe which receives waste water from all fixtures in a building.

Solar access

The access to the sun's rays by, for instance, restricting the location of shade trees or siting a building to maximize the usefulness of solar energy.

Solar altitude

In solar analysis, the vertical angular distance of a point in the sky above the horizon. Altitude is measured positively from the horizon to the zenith, from 0 to 90 degrees.

Solar azimuth

In solar analysis, the horizontal angular distance between the vertical plane containing a point in the sky (usually the sun) and true south.

Solar collector

A device designed to absorb radiation from the sun and transfer this energy to air or fluid passing through a collector.

Solar energy

Passive energy from the sun. If designed ecologically, good passive solar energy provides just enough sunlight into the rooms to be absorbed by the surrounding thermal mass, which acts as a heat battery and gives the warmth back into the room when the sun goes down. Crushed volcanic rock and straw bales make for good thermal mass insulation and designs in a green building.

Solar gain

Increase in temperature contributed to a space by the sun's rays.

Solar heat

Active solar space-heating systems consist of collectors that collect and absorb solar radiation combined with electric fans or pumps to transfer and distribute that solar heat. Active systems also generally have an energy-storage system to provide heat when the sun is not shining. The two basic types of active solar space-heating systems use either liquid or air as the heat-transfer medium in their solar energy collectors.

Solar heat gain coefficient (SHGC)

An indicator of the amount of solar radiation admitted through and absorbed by a window and subsequently released as heat indoors. SHGC is expressed as a number between 0 and 1; the higher the number, the more solar heat the window transmits.

Solar heating system

An assembly of subsystems and components which converts solar energy into thermal energy and uses it for heating.

Solar heating: active

A solar heating system using mechanical means such as solar collectors, fans, or pumps, to collect, store, and distribute the energy from the sun.

Solar heating: passive

A solar heating system using a building's site orientation, design, window placement, and wall and flooring materials to collect, store, and distribute heat with a minimal use of fans or pumps, relying on the natural flow of heat.

Solar house

A building designed so that the sun's rays are used to maximum advantage in heating, supplementing, or replacing other heating methods, including a vast array of solar panels on the roof.

Solar orientation

The placement of a building in relation to the sun to maximize the amount of heat gained during the coldest months and minimize the amount of heat gained during the warmest months. Sun charts and software assist in orienting a building for maximum solar benefit.

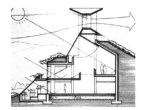

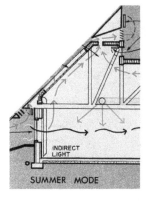

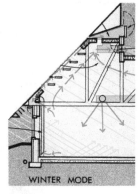

SUMMER MODE   WINTER MODE

Solar panels

General term for an assembly of photovoltaic modules. Use of solar panels is a sustainable building strategy in that it lessens a building's reliance on nonrenewable sources of power distributed through the grid system.

Solar pool equipment

A pool filtration system that pumps pool water through a solar collector so that the collected heat is transferred directly to the pool water. Solar pool-heating collectors operate just slightly warmer than the surrounding air temperature and typically use inexpensive, unglazed, low-temperature collectors made from specially formulated plastic materials. Glazed solar collectors are not typically used in pool-heating applications, except for indoor pools, hot tubs, or spas in colder climates.

Solar power

A source of renewable energy that uses radiation emitted by the sun. Also known as solar energy.

Solar radiation

The full spectrum of electromagnetic energy including visible light from the sun. When solar radiation strikes a solid surface or a transparent medium such as air or glass, some of the energy is absorbed and converted into heat energy, some is reflected, and some is transmitted. All three of these effects are important for effective passive solar design.

Solar reflectance index (SRI)

A value that incorporates both solar reflectance and emittance in a single value to represent a material's temperature in the sun. SRI quantifies how hot a surface will become relative to standard black and standard white

Solar reflectance index

surfaces. It is calculated using equations based on previously measured values of solar reflectance and emittance as laid out in the ASTM Standard E 1980. It is expressed as a decimal (0.0 to 1.0) or percentage (0 percent to 100 percent).

Solar rough-in

A method of installing plumbing and/or electrical systems that allows a later addition of a solar photovoltaic or a solar hot water system.

Solar water heater

Heat from the sun is absorbed and then transferred by pumps to a storage unit, which is then transported to the hot water supply of a home through a heat exchanger.

Solar-assisted hot water

A solar hot-water array on a roof that supplements (but does not exclusively supply) the hot-water supply to a home. Water warmed by solar energy is pumped into the hot-water heater, where less energy is required to heat it than if only cold water was in the tank.

Solarium

In ancient architecture, a terrace on top of a flat-roofed house or over a porch, surrounded by a parapet wall but open to the sky; a sunny room with more glass than usual, and often used for therapy.

Soldier course bond See **Bond**

Soleri, Paolo (1919–)

Italian-born American architect; worked for Frank Lloyd Wright in 1947 before returning to Italy to build the Ceramics Factory in Salerno (1953). He later established the Cosanti Foundation in Scottsdale, AZ (1955), and evolved the concept of Arcology, in which architecture and ecology are merged. Arcosanti, in Mesa, AZ (outside of Phoenix), was founded in 1970; it represents the development of a city guided by Soleri's visionary ideas.

Soleri, Paolo

Solid bridging

A form of lateral stiffening between timber joists to prevent them from twisting, and to provide a fire stop. Short lengths of wood similar in depth are fixed between each adjacent pair of joists.

Solid-core door See **Door**

Solid-web joist

A conventional joist with a solid web formed by a plate or rolled section, as opposed to an open-web joist.

Solomon, Willard (1783–1861)

American architect influenced by the work of Latrobe; brought Greek Revival to the Boston area, in the Suffolk County Courthouse (1835), Boston, and the Town Hall (1844), Quincy. He published *Plans and Sections of the Obelisk on Bunker's Hill* in 1883.

Solvent

A liquid that dissolves and cleans surface of other materials.

Sopraporta

Means "above the door" in Italian; an illustrative painting above a doorway, typically framed by a continuation of the door moldings, is an example.

Soriano, Rafael (1907–1988)

American architect who worked with Richard Neutra in California before establishing his own office in the San Francisco Bay area.

Sorin

The crowning spire on a Japanese pagoda, usually made of bronze.

Sottsass, Ettore (1918–2007)

An Italian architect, designer, and founder of the Memphis postmodern design group, best known for flamboyant product designs, like the 1969 red Olivetti Valentine typewriter, bright plastic tabletop pieces for Alessi, and light fixtures for Artemide. The controversial Memphis design movement of

Sottsass, Ettore

the early 1980s featured odd shapes and angles, discordant materials, and form before function, demanding that everyday objects get a new review. It was not until his mid-60s that Sottsass began to regularly produce works of architecture. From nearly 20 luxury homes to yachts and a golf resort for the People's Liberation Army in China, Sottsass brought his unique vision to projects worldwide. In his home city of Milan, he completed the interiors of the Malpensa International Airport in 2000.

Soufflot, Jacques (1713–1780)

One of the greatest French Neoclassical architects, Soufflot combined the monumentality of ancient Roman models with the structural lightness admired in Gothic architecture. Work includes the Panthéon, Paris.

Sound

The sensation stimulated in the auditory organs by a vibratory disturbance.

Sound barrier

Any solid obstacle that is relatively opaque to sound that blocks the line of sight between a sound source and the point of reception of the sound.

Sound insulation

The use of structures and materials designed to reduce the transmission of sound from one room or area of a building to another, or from the exterior to the interior of the building.

Sound lock

A vestibule or entranceway that has highly absorptive walls and ceilings and a carpeted floor; used to reduce transmission of noise into an auditorium, rehearsal room or studio, or from the area outside.

Sound transmission

Sound passing from one room to another, normally through an air return plenum, or through a material, construction or other medium.

Sound-insulating glass See Glass

Soundproofing

The application of sound-deadening material to walls, ceilings and floors to prevent sound from passing through into other areas.

Southern Colonial style

A brick or wood-frame house or building, usually one room deep and crowned with a high gabled roof; Classical details, such as arches over windows and doorways, are common.

Southern exposure

In northern latitudes, a home with a southern exposure can take advantage of collecting the sun's energy either for production of electricity through photovoltaics, heating water, or as part of a passive solar design. The approach collects the sun's energy in the winter and offers natural lighting during the summer.

Space
The unlimited continuous three-dimensional expanse in which all material objects exist; all the area in and around a structure, or volume between specified boundaries, and the interval between two objects.

Space allocation
The process of generating spatial designs by allocating units of functional spaces to physical space locations.

Space audit
A physical survey and record of the space that is occupied and its functional use.

Space frame
A three-dimensional structural framework made up of interconnected triangular elements that enclose a space; as opposed to a frame where all the elements lie in a single plane.

Space planning
The definition of space in terms of size, type, activity, and adjacency for any particular type of space.

Space planning
The process of converting the needs expressed by the client into the drawings and supporting documentation that outline the plan for the design and building team. For the interior designer, space planning includes the placement of all fixed and movable elements, from plumbed and wired appliances to furniture and the way people move around in the space. Plumbing and HVAC contractors are concerned primarily with hidden elements. With comprehensive space planning, the team works together to ensure that the design is optimized across disciplines.

Space utilization
The ratio of the number of people using a space to its potential use capacity, multiplied by the ratio of hours of actual usage to the total available hours, and then expressed as a percentage.

Space-age sites
The space age has created small-town and historic sites, such as the Apollo Launch Complex (1965) at Kennedy Space Center, Cape Canaveral, FL.

Spackle
A paste to fill holes, cracks and defects in the surfaces of various materials.

Spall
A small fragment or chip dislodged from the face of a stone or masonry unit by a blow or by the action caused by the elements, such as a freezing and thawing cycle.

Span
The interval between any two consecutive supports of a beam, girder, or truss or between the opening of an arch.

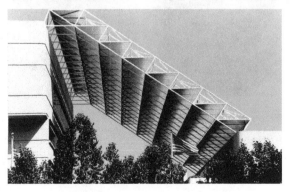

466

Span

Spandrel

The triangular space that is formed between the sides of adjacent arches and the line across their tops; in a skeletal frame building, the walls inside the columns and between the top of the windows and the sill above.

Spandrel bracket

A pair of curved brackets that form an arched shape.

Spandrel glass

An opaque glass used in curtain walls to conceal spandrel beams, columns, or other internal structural construction.

Spandrel panel

A panel covering the spandrel area between the head of a window on one level and the sill of the window immediately above.

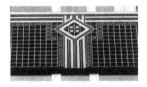

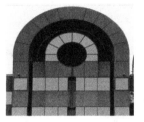

Spandrel wall See Wall

Spanish Colonial Revival style (1915–1940)

A unique feature of this revival style is the ornate low-relief carvings highlighting arches, columns, window surrounds, cornices, and parapets. Red-tiled hipped roofs and arcaded porches are typical. Exterior walls are left exposed or finished in plaster or stucco. Iron window grilles and balconies are prevalent. A molded or arcaded cornice highlights the eaves, and large buildings have ornamental parapets and a symbolic bell tower.

Spanish Colonial style (1650–1840)

Adobe-brick wall construction covered with a lime wash or plaster characterized this style. Rounded roof beams were extended over porches, which were covered with tile roofs. Missions of the U.S. Southwest were richly ornamented vernacular interpretations of this Baroque-like style.

Spanish tile See **Tile**

Spar varnish
A varnish that is resistant to salt, sun, and water.

Specifications
A written document that accompanies the drawings describing the materials and workmanship required to carry out the works for each particular trade.

Spectrally selective glazing
A glazing that has a high transmittance of visible light but low transmittance of solar heat gain.

Specular
Having the reflective properties of a mirror. This means that any incident light will be reflected in the mirror angle direction. Most surfaces are a mix between a specular and lambertian surface.

Speculative builder
One who develops and constructs building projects on the speculation that they will be sold or leased in the future.

Spherical
Refers to a three-dimensional surface, all parts of which are equidistant from a fixed point.

Sphinx See **Ornament: animal forms**

Spill light
Any light that spills over from the main pattern of the beam established for that light.

Spindle
A turned-wood architectural element, produced on a lathe, and used as a banister or ornamental spindlework on porches and other locations.

Spindlework
A Queen Anne decoration, featuring short, turned parts, similar to balusters.

Spindlework

Spiral
Refers to forms that are generated by a continuous curve, traced by a point moving around a fixed point in a fixed plane, while steadily increasing the distance from that point.

Spiral stair See **Stair**

Spire
A slender pointed element on top of a building, generally a narrow octagonal pyramid set above a square tower.

Spire

Spirit level
A leveling instrument used for testing the horizontal or vertical position of any structural member.

Splay
A sloped surface that makes an oblique angle with another at the sides of a door or window, with the opening larger on one side than the other; a large chamfer; a reveal at an oblique angle.

Splayed arch See Arch

Splayed jamb
Any jamb whose face is not at right angles to the wall in which it is set.

Splayed lintel
A horizontal lintel above a window or doorway which slants downward toward the centerline; often containing a keystone at its center.

Splayed mullion
A mullion that joins two glazed units which are at an angle to each other, such as the mullion of a bay window.

Splayed window See Window

Splice
To connect, unite, or join two similar members, wires, columns or pieces; usually in a straight line, by fastening the lapped ends with mechanical end-connectors, or by welding.

butt splice
A butt joint, which is further secured by nailing a piece of wood to each side of a butt joint.

lap splice
A splice made by placing one piece on top of another and fastening them together with pins, nails, screws, bolts, rivets, or similar devices.

Spline
A thin flat piece of wood used between two pieces of heavy subflooring, taking the place of a tongue-and-groove joint; also used as a means of stiffening a miter joint.

Spline joint See Joint

Split
A cleft in a piece of wood that goes all the way through the member.

Split ring connector
A method of connecting members of a wood truss or built-up beam, by means of installing a metal ring, often fitted with grooved teeth for driving into one member, installed into a bored circular slot in the other piece, and then connected with a bolt drilled through the center of the ring.

Split-level house
A house with a kitchen, dining, and living room area on the main floor, with stairs leading up to the bedrooms at a half-story higher; other stairs may lead downward from the main floor to a family room or utility room.

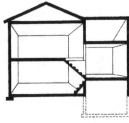

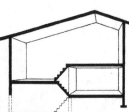

Spontaneous combustion
The instantaneous bursting into flames of a substance due to a chemical reaction of its own constituents, such as oily rags in an unventilated pile of rubbish.

Spotlight
A bulb with a strong directional beam, used for accent lighting; may be surface or track-mounted on a wall, ceiling, or floor.

Spout
A short channel or tube used to spill stormwater from gullies, balconies, or exterior galleries, so that the water will fall clear of the building; a gargoyle.

Spray foam insulation
An insulation that is sprayed into place and then expands to fill cavities. It acts as both an insulator and a sealant and is an alternative to the standard insulation bats. The two types of spray foam are open cell (isocyanurate) and closed cell (polyurethane). Closed-cell foams typically have a higher R-value than open-cell foams.

Sprayed asbestos
Spraying steel structural members with a fire-resistant coating of asbestos fibers mixed with cement and water. Used on structures prior to 1986, then banned and removed from any structure where it presented a health risk, or when renovating a structure containing the material.

Sprayed fireproof insulation
A mixture of mineral fiber with other ingredients, such as asbestos, applied by air pressure with a spray gun; used to provide fire protection or thermal insulation.

Spread footing See Footing

Spreckelsen, Johan Otto von (1929–1987)
Danish architect, who used pure forms in his work, such as the cylinder and sphere. He was (with Paul Andrew) winner of the competition for the design of the Grande Arche de la Défense (illus.), Paris, France (1983), a huge trapezoidal structure with a hollow center, which terminates the axis that runs from the Tuileries Gardens in the east through the Arc de Triomphe to La Défense.

Spreckelsen, Johan Otto von

Springer
The impost or place where the vertical support for an arch terminates and the curve of the arch begins; the lower voussoir, or bottom stone of an arch, which lies immediately on an impost.

Springing
The point at which an arch rises from its supports.

Springing line
The imaginary horizontal line at which an arch or vault begins to curve; the line in which the springers rest on the imposts.

Sprinkler system

A system, usually automatic, for protection against fire; when activated, it sprays water over a large area in a systematic pattern.

Spruce See **Wood**

Spur

A decorative appendage on the corners of the base of a round column resting on a square or polygonal plinth, in the form of a grotesque, a tongue, or leafwork.

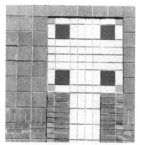

Square

A regular four-sided figure with equal sides and four equal right angles; may be subdivided along the diagonals or oblique lines connecting the corner angles and the lines that connect the center of each side. Also, an open area at the intersection of streets in an urban setting.

Square billet molding See **Molding**

Square roof See **Roof**

Square rubble masonry See **Masonry**

Squared stone See **Stone**

Square-headed

Cut off at right angles, as an opening with upright parallel sides and a straight horizontal lintel, as distinguished from an arched opening.

Square-headed window See **Window**

Squat

To use land or a building without having ownership or permission from the owner; often used as a means of free accommodations in slum areas which are slated for rehabilitation.

Squatter's rights

The privilege, without legal right or arrangement, of occupying a property by virtue of a long and continuous use of the land.

Squinch

Corbeling built at the upper corners of a structural bay to support a smaller dome or drum; a small arch across the corner of a square room which supports a superimposed octagonal structure above.

Squinch arch See **Arch**

Squint brick
A brick manufactured with an angular corner; commonly used in the construction of oblique corners.

Stabilization
The process of temporarily protecting a historic building until restoration or rehabilitation can begin; it typically includes making the building weathertight, structurally stable, and secure against intrusion. The U.S. Department of the Interior's standards state: "to reestablish a weather resistant enclosure and the structural stability of an unsafe or deteriorated property while maintaining the essential form as it exists at the present."

Stabilized soil
Earth that has been treated with a binder such as Portland cement, bitumen, resin, or a more soluble soil to reduce its movement.

Stack bond See **Bond**

Stack effect
Also referred to as the chimney effect, this is one of three primary forces that drive air leakage in buildings. When warm air is in a column, such as a building, its buoyancy pulls in the colder air that is low in buildings, as the buoyant, warmer air exerts pressure to escape out the top. The pressure of stack effect is proportional to the height of the column of air and the temperature difference between the air in the column and ambient air. Stack effect is much stronger in cold climates during the heating season than in hot climates during the cooling season.

Stack vent
Extension of a waste or soil stack above the highest horizontal drain which is connected to the stack.

Stadium
A sports arena, usually shaped like an oval, or in a horseshoe shape.

Staff
Ornamental plastering, made in molds and reinforced with fiber, usually nailed or wired into place.

Staggered
Two or more rows of objects that are offset from one another in such a manner as to form a zigzag pattern.

Staggered partition
Two rows of studs that are thinner than the top and bottom plates, and are alternately offset from the surface of the plates; each side of the wall is attached to the alternate stud; used to prohibit the passage of sound through the wall.

Staging
The sequencing and physical positioning of building materials on a construction site. Sustainable building pays particular attention to staging in order to minimize the impact to the construction site and protect materials from damage.

Stain
A coloring liquid or dye for application to any porous material, most often wood; thinner than paint and readily absorbed by the wood so that the texture and grain of the wood is enhanced, and not concealed.

Stained glass See **Glass**

Stained glass window See **Window**

Stainless steel See **Metal**

Stair
A series of steps or flights of steps for going between two or more successive levels with connecting landings at each level, either on the exterior or in the interior.

box stair
An interior staircase constructed with a closed string on both sides, often enclosed by walls or partitions with door openings at various floor levels.

circular stair
A stair having a cylindrical shape.

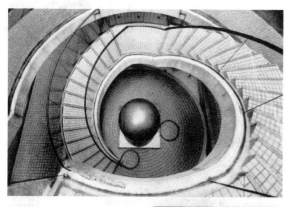

closed-string stair
A staircase whose profile of treads and risers is covered at the side by a string or sloping member which supports the balustrade.

dogleg stair
Two flights of stairs that are parallel to each other with a half-landing in between.

dogleg stair

elliptical stair
Winding stair in which the plan of its inside edges is in the shape of an ellipse.

geometrical stair
A winder stair in which the outside of the winders is supported by walls forming three sides of a rectangle, and the inner side is unsupported and without a newel post.

hanging stair
A stair supported by individual hanging steps projecting horizontally from a wall on one side, with corbels or brackets.

helical stair
A stair whose treads are wrapped around a helix: commonly called a spiral stair.

open stair
A stair or stairway, whose treads are visible on one or both sides and open to a room in which it is located.

open-string stair
A staircase whose profile of treads and risers is visible from the side; the treads support the balusters.

open-well stair
A stair built around a well, leaving an open space.

service stair
A stairway that is not used for general purposes; it provides access to specific areas such as the roof and equipment rooms.

spiral stair
A flight of stairs, circular in plan, whose treads wind around a central newel, or post.

straight stair
A stair with a single straight flight of steps between levels.

winding stair
Any stair constructed chiefly or entirely of winders.

Stair flight
A continuous series of steps with no intermediate landings.

Stair half-landing
A landing located halfway up a flight of stairs.

Stair landing
Horizontal platform at the end of a flight of stairs or between two flights of stairs.

Stair rail
A bar of wood or other material that connects the balusters on a stair.

Stair riser
Vertical portion of a stair step; may be open or closed.

Stair tower
A part of a structure containing a winding stair which fills it exactly; a stair enclosure which projects beyond the roof of a building.

Stair turret
A domed or conical roof over a stair that rises above the main roof surface

Staircase
A vertical element of access in a structure for ascending or descending from one level to another. The form of the staircase is often expressed on the exterior of the building, if it is located adjacent to an exterior wall.

Staircase

Stairwell
A vertical space in a building that encloses a staircase.

Stakeholder
Any organization, governmental entity, or individual that has a stake in or may be impacted by a given approach to environmental regulation, pollution prevention, or energy conservation.

Stalactite work

An Islamic decorative design consisting of multiple corbeling that resembles natural stalactites, in either plaster, marble or wood.

Stamped metal
Sheet metal that has been shaped by stamping or pressing it with dies to form a raised or recessed decorative design on the surface.

Stanchion
A column, particularly of structural steel; an upright bar placed between the mullions to strengthen a leaded light.

Standard 55
The ASHRAE (American Society of Heating, Refrigerating and Air-Conditioning Engineers) standard for thermal comfort, titled "Thermal Environmental Conditions for Human Occupancy." It takes into account all of the factors that affect human thermal comfort: air temperature, mean radiant temperature, relative humidity, air speed, local discomfort, and temperature variations over time, with the first two parameters being the most influential and, when combined, called operative temperature.

Standby lighting
A lighting system that will supply the adequate illumination if the normal system should fail; usually supplied by an emergency generator.

Standby power
The power that is available within one minute of a power failure to operate life safety equipment and continuously operating equipment; emergency power is the power available within 10 seconds.

Standing lamps
Tall fixtures on poles that put light above the user, usually directing the light upwards toward the ceiling.

Standing seam joint See Joint

Standing seam roof
A sheet-metal roof with vertical folded seams joining adjacent flat panels; the parallel seams run along the slope.

Standpipe
A vertical pipe riser with multiple interior connections for fire hoses, terminated at the building facade with a fitting for a fire engine hose; may also have a connection to the main water supply.

Standpipe system
An arrangement of piping, valves, hose connections and allied equipment installed in a building with the hose connections located so that water can be discharged in streams or spray patterns through attached hoses and nozzles; to extinguish a fire and protect a building, its contents, and occupants.

Stapler
A device for applying staples; may be mechanical or air operated which is more capable of driving staples into hard materials at a high rate of speed.

Statehouse
Capitol building of a state government.

Station point See Perspective projection

Statuary
Freestanding sculpture, as opposed to relief work.

Statue
A form of likeness sculpted, modeled, carved, or cast in material such as stone, clay, wood, or bronze.

Statue

Statuette
Diminutive statue, especially one that is less than life size

Statute of limitations
A statute specifying the period of time within which legal action must be brought for alleged damages or injury; in construction industry cases.

Stave
Wedge-shaped timber.

Stave church
A Scandinavian church of the 12th and 13th centuries constructed entirely of wood with few windows and a steep roof; highly original in structure with fantastic semipagan decorative features.

Steam boiler
A boiler in which water is raised to or above saturation temperature at a desired pressure, and the resultant steam is drawn off for use in the heating system.

Steamboat Gothic style (1850–1880)
A richly ornamental mode of Gothic Revival building in the Ohio and Mississippi River Valleys characterized by the gingerbread and ornamental construction found on riverboats of the Victorian period.

Steel casement
A casement type window made of steel sections, incorporated most often into masonry type structures.

Steel forms
Removable pieces of steel which hold wet concrete in the desired shapes for casting foundations, footings and window frames on the spot. Some formwork comes with interlocking modular hardware. They are long-lasting; produce a clean, accurate face; and are easier to set up, take down, and clean than wooden forms.

Steel frame
A skeleton of steel beams and columns providing all that is structurally necessary for a building to stand.

Steel See **Metal**

Steel shutters
Used in fortifications, or in ordinary buildings when security is an issue, or if strong winds are a regular occurrence.

Steel square
An instrument having at least one right angle and two or more straight edges; used by carpenters for testing the accuracy of right angles and for laying out work.

Steel stud
A bent steel sheet-metal stud with holes punched out in the widest face; used as interior framing for drywall construction.

Steel wool
A mass of fine steel threads matted together and used principally for polishing and cleaning surfaces of wood or metal.

Steeple
A tall ornamental structure terminating in a spire and surmounting the tower of a church or public building.

Stein, Clarence S. (1883–1975)
American architect in New York; a proponent of garden cities and planned housing; designed the planned community of Radburn, NJ (1929), with Henry Wright.

Steiner, Rudolf (1861–1925)
An Austria-Hungarian philosopher, artist, scientist, and architect. His Geotheanum (illus.), Dornach, Switzerland (1913), was the epitome of Expressionism, with a strong Symbolist and Jugendstil flavor. It was built of reinforced concrete.

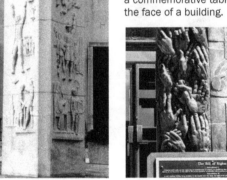

Stele
An upright stone slab or pillar with a carved or inscribed surface, used as a monument or marker, or as a commemorative tablet in the face of a building.

Stellar vault See **Vault**

Stenciling
Decorative painting on interior walls, consisting of repetitive patterns applied by brushing paint onto the surface through openings cut in the stencil.

Step
A stair unit that consists of one tread, the horizontal upper surface, and one riser, the vertical face.

bull-nosed step
A step, usually the lowest in a flight, having one or both ends rounded to a semicircle and projecting beyond the face of the stair string. The projection extends beyond and around the newel post.

cantilever step
A step built into the wall at one end, but supported at the other end only by the steps below.

curtail step
A step, usually the lowest in the flight, of which one or both ends are rounded into a spiral or scroll shape which projects beyond the newel.

riser
The vertical face of a stair step.

tread
The horizontal upper surface of a step; includes the rounded edge or nosing which extends over the riser.

Stepped
Refers to forms that are increased or decreased by a series of successive increments or modulated by incremental stages or steps.

Stepped arch See **Arch**

Stepped flashing
Individual pieces of flashing used where a sloped roof abuts a vertical masonry wall or chimney; each piece is bent and inserted into a mortar joint.

Stepped gable See **Gable**

Stepped pyramid
An early type of pyramid having a stepped and terraced appearance.

Stepped voussoir
A voussoir that is squared along its upper surfaces so that it fits horizontal courses of masonry units.

Stepped window See **Window**

Stepped-back chimney
Any exterior brick chimney that is wide enough at the base to enclose a large fireplace in the kitchen, then decreasing in area in several steps, possibly collecting other flues from the upper floor.

Stereobate
The substructure, foundation, or solid platform upon which a building is erected. In a columnar building, it includes the uppermost step or platform upon which the columns stand.

Stereogram
A drawing or photograph that can be viewed in three-dimensions.

Stereography
The science of perspective projection.

Stereoscope
An instrument for viewing a stereoscopic pair of photographs three-dimensionally, consisting of two lenses set at the correct distance apart to correspond with the separation of the stereoscopic camera lens.

Stereoscopic camera
A camera designed to produce two displaced images, called a stereoscopic pair, by means of two matched lenses and shutters, so that when viewed by both human eyes in a stereoscope gives a three-dimensional view of the object.

Stern, Robert A.M. (1939 –)
Worked with Richard Meier before setting up his own practice in 1977. He is seen as one of the influences in Postmodernism, advocating a study of history and the eclectic use of forms. Among his later works are the Walt Disney Casting Center (illus.), Lake Buena Vista, FL (1989), and Celebration Health (illus.), Celebration, FL.

Stern, Robert A.M.

Stewardship
Custodial management of historic resources in museums or other facilities.

Stiacciato See **Relief**

Stick style
An eclectic wooden-frame style of the late 1800s that was usually asymmetrical in plan and elevation. It had wood trim members applied as ornamentation on the exterior that expressed the structure of the building, as corner posts and diagonal bracing; also featured porches and towers and ornamented gable apexes.

Stile
One of the upright structural members of a frame, such as at the outer edge of a door or a window sash.

Stilted arch See **Arch**

Stippled
A texture imparted to plaster with the bristles of a stiff brush, applied by driving the bristles directly at the surface, rather than by using painting strokes.

Stippling brush
A brush with short stiff bristles used to give a stippled texture to plaster; can be made by cutting short the long bristles of a paint brush.

Stirling, James (1926–1992)

Scots architect; influenced by Le Corbusier. Fell into the category of Brutalism in the Engineering Block building, University of Leicester, England (1959). His later work became increasingly eclectic and expressive and contained illusions to historical themes. Works include the engineering building, Leister University (1959), Olivetti Building Haslemere, Surrey (1972), and Braun Headquarters, Melsungen, Germany (1991).

Stirrup

In concrete construction, reinforcement to resist shear; normally consists of a U-shaped bar, anchored to the longitudinal side and placed perpendicular to it.

Stoa

A covered colonnade in ancient Greek and Roman cities, flanking the agora (an open market and meeting place); either one- or two-storied, with an open front and shops or offices built into the rear wall.

Stock

Any material such as lumber or masonry that is normally readily available from a supplier's stock.

Stock size

Material or product available in a variety of sizes readily available for purchase, and not specially ordered.

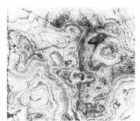

Stockade

A defensible space that is enclosed by a fence or palisade with loopholes.

Stone

Native rock that has been processed by shaping, cutting, or sizing for building or landscaping use. It is fire resistant and varies according to type, from fairly porous to impregnable. There are three basic types of stone: igneous, such as granite, is long-lasting and durable; sedimentary, such as limestone, is made up of organic remains; and metamorphic rock, such as marble, is either igneous or sedimentary rock that has been transformed by pressure and heat or both.

alabaster

A fine-grained, translucent variety of very pure gypsum, white or delicately shaded, and used for ornamental work.

basalt

A dense, dark gray volcanic rock, often full of small cavities, used as a building stone.

Belgian block

A hard paving stone, typically granite, roughly cut to the shape of a truncated pyramid, where the top is slightly smaller than the base.

bluestone

A dense fine-grained sandstone that splits easily along bedding planes to form thin slabs.

brownstone

A dark brown or reddish-brown sandstone, used extensively for building in the United States during the middle and late 19th century.

cobble

Stone that is smaller than a boulder but larger than gravel.

cobblestone

A naturally rounded stone used in paving, wall construction, and foundations.

dolomite

Limestone consisting principally of the mineral dolomite.

fieldstone

Loose stone found on the surface or in the soil, flat in the direction of bedding and suitable for use as drywall masonry.

flagstone

A naturally thin flat stone, normally used as a stepping stone or as outdoor paving; sometimes split from rock that cleaves easily.

gneiss

A coarse-grained, dark metamorphic rock; composed mainly of quartz, feldspar, mica, and other minerals corresponding in composition to granite, in which the minerals are arranged in layers.

granite
An igneous rock having crystals or grains of visible size; consists mainly of quartz and mica or other colored minerals.

limestone
Rock of sedimentary origin composed principally of calcite, dolomite, or both; used as a building stone or crushed-stone aggregate, or burnt to produce lime.

marble
Metamorphic rock made up largely of calcite or dolomite; capable of taking a high polish, and used especially in architecture and sculpture; numerous minerals account for its distinctive appearance.

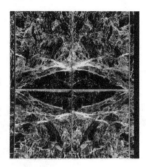

natural stone
A stone that has been quarried and cut, but not crushed into chips or reconstituted into cast stone.

obsidian
A natural volcanic glass, usually black with a bright luster, that is transparent in thin slabs.

quartzite
A variety of sandstone composed largely of granular quartz cemented by silica, forming a homogeneous mass of very high tensile and crushing strengths; used as a building stone and as an aggregate in concrete.

sandstone
Sedimentary rock that is composed of sand-sized grains naturally cemented by mineral materials.

serpentine
A group of minerals consisting of hydrous magnesium silicate or rock largely composed of these minerals; commonly occurs in greenish shades; used as decorative stone.

slate
A hard, brittle metamorphic rock characterized by good cleavage along parallel planes; used as cut stone in thin sheets for flooring, roofing, and panels, and in granular form as surfacing on composition roofing.

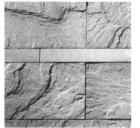

squared stone
Roughly dressed stone blocks with rectangular faces.

travertine
A variety of limestone deposited by springs, usually banded, commonly coarse and cellular, often containing fossils; used as building stones, especially for interior facing or flooring.

undressed stone
Not trimmed or rendered smooth.

verde antique
A dark green serpentine rock marked with white veins of calcite that takes a high polish; used for decorative purposes since ancient times; sometimes classified as a marble.

volcanic stone
A low-density, high-porosity rock composed of volcanic particles, ranging from ash size to small pebble size, which are compacted or cemented together; used as a building stone or as a thermal insulation material.

Stone Age
The earliest known period of human culture, characterized by the use of stone tools and weapons.

Stone building
Any type of stone suitable for use in exterior construction; including granite, limestone, sandstone, and marble.

Stone, Edward Durrell (1902–1978)
Absorbed the Modern movement working on Rockefeller Center, and designed the interior of Radio City Music Hall, NYC. His U.S. Embassy, New Delhi, India (1954), and the Kennedy Center for the Performing Arts, Washington, DC (1961), were axial and symmetrical, and paraphrased the Classical movement.

Stone house
A house constructed entirely of stone.

Stone Jr., Edward D. (1923–2009)
Landscape architect and son of architect Edward Durrell Stone. Stone's projects included the Atlantis in the Bahamas; Madinat Jumeriah in Dubai, UAE; and PepsiCo's corporate headquarters, in Rye, NY, which won the ASLA Landmark Award and was honored by the National Trust for Historic Preservation. Specializing in tourist destinations and recreation-based communities, his firm, EDSA, was also responsible for Disney World's West Side, Euro Disneyland, and the Hyatt Regency Aruba Resort and Casino.

Stonemason
An artisan skilled in dressing and laying stone for buildings and other purposes.

Stonorov, Oskar (1905–1978)
American architect who worked with Louis Kahn and became partner in the firm of Kahn and Stonorov.

Stool
The flat piece upon which a window closes, corresponding to the sill of a door.

Stoop
A platform or small porch at the entrance to a house, usually up several steps.

Stop
The molding or trim on the inside face of a door or window frame against which the door or window closes.

Stopped flute
A column flute that stops approximately one-third of the shaft height above its base.

Storefront
Ground-level shop employing large sheets of glass display windows with minimal-sized mullions; often having a recessed entrance.

Storeroom
A room set aside for the storage of goods and supplies.

Storm door See Door

Storm drain system
A system of pipes, channels, or other conveyance that collect and carry runoff to receiving waters or treatment systems.

Storm porch
Enclosed porch that protects the entrance to a house from severe weather.

Storm window See **Window**

Storm windows
Single-pane windows often installed on the interior of the main windows of home to improve insulation. When window replacement is cost prohibitive, adding storm windows can be an alternative for saving energy.

Stormwater
Runoff due to rainfall collected from roofs, impervious surfaces, and drainage systems.

Stormwater detention systems
Regulate and control runoff by slowing the rate of discharge to reduce impacts downstream.

Stormwater infiltration
The process through which stormwater runoff penetrates into soil from the ground surface.

Stormwater management
Building and landscape strategies to control and limit stormwater pollution and runoff. Usually an integrated package of strategies, including elements like vegetated roofs, compost-amended soils, pervious paving, tree planting, drainage swales, and more.

Stormwater retention systems
Store water indefinitely, until it is lost through percolation or evaporation, or is taken up by plants.

Story
The space in a building between floor levels; in some codes a basement is considered a story, generally a cellar is not; a major architectural division even where no floor exists, as a tier or a row of windows.

Straight joint See **Joint**

Straight line gable See **Gable**

Straight stair See **Stair**

Straight-flight stair See **Stair**

Straight-line depreciation
A depreciation deduction calculated by subtracting any anticipated salvage value from the initial cost or value of the improvements, and then dividing the estimated economic life of the improvements into that figure.

Strap
A thin piece of metal of any required dimension used to attach, secure, or fasten an object to another.

Strap hanger
Metal bar with a rectangular cross section that is bent into a channel shape and used to attach beams to purlins; also called a stirrup.

Strap hinge
A surface-mounted hinge having two leaves; one is fastened to the door and the other is secured to the frame or post.

Strap pipe hanger
In plumbing work, a metal strap or band nailed or screwed to the ceiling or rafter and slung around a suspended pipe.

Strapwork ornament See **Ornament**

Straw bale construction
An annually renewable agricultural waste product made of wheat, oats, barley, rye, or rice, used to build thick, super-insulated, stucco-covered walls. Straw bales are traditionally a waste product that farmers do not till under the soil, but do sell as animal bedding or landscape supply. Straw is the dry plant material or stalk left in the field after a plant has matured, been harvested for seed, and is no longer alive. Bales can be taken directly from a baling machine or recompressed for higher density. In contrast, hay bales are made from short species of livestock feed grass that are green/alive and not suitable for this application. Bales are placed over a "stem wall" to protect the straw from the ground soil, and the straw bales are stuccoed and plastered over for finishing.

Streamline Modern
A phase of Art Deco that emphasizes the horizontal aspects of design; elements include curved end walls, round corners, horizontal stainless-steel railings, flush windows, round windows, and use of glass block.

Street

A paved way on which vehicles travel and park.

Street furniture

Manufactured elements, such as benches, streetlights, fire hydrants, and light fixtures, found in public spaces.

Streetlight

One of a series of lamps at the top of posts along a street or similar public location.

Streetscape

A pictorial character of a scene made up of buildings grouped in combination with landscaping, and street furniture, paving, fences, benches, lampposts, mailboxes, utility poles, transit shelters, trash receptacles, street signs, and commercial signs.

Stress

Internal forces per unit area; when the forces are tangential to the plane, they are shear stresses; when they are perpendicular, they are called either tensile or compressive stresses, depending on whether they act toward or away from the plane of separation. The deformation caused by stress is called strain.

Stressed skin

A structural panel with the sheathing permanently bonded to the frame or core to increase its strength.

Stressed-skin construction

A form of construction in which the outer skin acts within a framework to contribute to the membrane and strength of the unit, instead of being just a cladding to protect the inside from the weather.

Stressed-skin panel See **Wood**

Stresses
Forces on a member caused by loads; consisting of torsion or twisting, compression or pushing, tension or pulling, and shear or cutting.

Stretcher
A masonry unit laid horizontally with its length in the direction of the face of the wall.

Striation
Fine, narrow ridges or grooves parallel to each other.

Strickland, William (1788–1854)
American architect. A pupil of Latrobe; designed mostly in the Greek Revival style, as well as the Egyptian Revival style.

String
In a stair, an inclined board that supports the end of the steps; also called a stringer.

face string
An outer string, usually of better material or finish than the rough string which it covers; may be part of the actual construction or applied to the face of the supporting member.

outer string
The string at the outer and exposed edge of a stair, away from the wall.

Stringcourse
A horizontal band of masonry, extending across the facade to mark a division in a wall, often encircling decorative features such as pillars or engaged columns; may be flush or projecting, molded or richly carved.

Stringer
A horizontal piece of timber or steel that connects the uprights in a framework and supports the floor; the inclined member that supports the treads and risers of a stair.

Strip flooring
Long narrow strips of wood, usually made with a tongue-and-groove along its sides, and sometimes along its ends as well.

Stripping
Removal of forms from poured concrete after it has hardened.

Struck molding See Molding

Struck mortar joint See Mortar joint

Structural analysis
Part of a feasibility study which includes examining the structural stability of a building, the mechanical systems, and the cost of code compliance.

Structural engineering
A branch of engineering concerned with the design and construction of structures to withstand physical forces or displacements without danger of collapse or without loss of serviceability or function. It includes inspection, analysis, design, planning, and research of structural components and structural systems. Structural engineers generally look at technical, economic, and environmental concerns.

Structural glass See Glass

Structural lumber See Wood

Structural shape
A hot-rolled steel beam of standardized cross section, temper, size, and alloy; includes angle iron, channels, tees, I-beams, and H sections; commonly used for structural purposes.

Structural stability
The relative structural soundness of an existing building or structure based on current and projected dead load, live load, wind, and seismic load factors.

Structure
The completed building envelope on the site, externally and internally complete, including all operating systems ready for their interior furnishings.

Structured insulated panels (SIPs)
Panels made from a thick layer of foam (polystyrene or polyurethane) sandwiched between two layers of oriented strand board (OSB), plywood, or fiber-cement. SIPs are often used in panelized construction. They are an alternative to foam core and are available with a core of agriculture fibers (such as wheat straw) that provides similar thermal and structural performance. The result is an engineered panel that provides structural framing, insulation, and exterior sheathing in a solid, one-piece component. Can be used for walls, roofs, or flooring, and result in a structure very resistant to air infiltration. SIPs can be erected very quickly with a crane to create an energy-efficient, sturdy home.

Strut
A bracing member, or any piece of a frame which resists thrusts in the direction of its own length, whether upright, horizontal, or diagonal.

Strut

Stuart style (1603–1688)
A style typifying the late English Renaissance.

Stubbins, Hugh (1912–)
In 1939, became assistant to Walter Gropius at Harvard, established his own practice in Cambridge (1940), and succeeded Gropius as chairman of the Department of Architecture. He designed the Congress Hall in Berlin, Germany (1957), and Citicorp Center, NYC (1978).

Stucco
An exterior fine plaster finish composed of Portland cement, lime, and sand mixed with water, used for decorative work or moldings, and usually textured.

Stud
One of a series of upright posts or vertical structural members that act as the supporting elements in a wall or partition.

Stud cavity
The space between the vertical members of a conventionally framed wood structure.

Stud partition
An interior partition with stud framing, normally covered with various materials, such as plaster or drywall.

Studio
Working space of an artist, photographer, or craftsperson; may have special lighting provisions, such as skylights.

Studio apartment
An apartment with a single living space that includes a galley kitchen, plus an enclosed bathroom.

Study See Design

Stupa
A Buddhist memorial site, consisting of an artificial mound on a platform, surrounded by an outer ambulatory with four gateways, and crowned by a multiple sunshade, erected to enshrine a relic.

Sturgis, John H. (1834–1888)
American architect educated in England, where he learned the Gothic Revival and Arts and Crafts styles. He won the competition for the Museum of Fine Arts (1870), Boston, MA. His most notable domestic designs were the Ames House (1882) and the interiors of the Gardner House (1882), both in Boston, MA.

Sturgis, Russell (1838–1909)
American architect who worked for Eidlitz before studying medieval style in Munich, Germany. He set up office in New York, and he designed Farnum Hall, Yale University (1870), New Haven, CT. He also designed the Farnum House (1884), New Haven, CT, in the Queen Anne style. He later compiled the important three volume *Sturgis' Illustrated Dictionary of Architecture and Building* (1902) and built up the collection in the Avery Library at Columbia University, NYC.

Style
The overall appearance of the design of a building including form, space, scale, materials, including ornamentation; it may be either a unique individual expression or part of a broad cultural pattern.

Stylobate
The single top course of the three steps forming a foundation of a Classical temple upon which the columns rests; any continuous base, plinth, or pedestal upon which a row of columns rests directly.

Subbasement
A story one or more levels below the basement level.

Subcontractor
A specialized building contractor; typically works for the general contractor performing work in a particular trade, such as electrical, masonry, or millwork.

Subdivision
A tract of land divided into residential lots.

Subfloor
Structural floor that supports the finish floor, such as a concrete floor or rough wood floor.

Suborder
A secondary architectural order, introduced chiefly for decoration, as distinguished from a main order of a structure.

Subsidies

Economic incentives to engage in an activity or purchase a product. Subsidies can work for or against environmental protection. Governments and utilities will sometimes offer subsidies for technologies that decrease energy or water use.

Substitution

A material or process offered in lieu of, and equivalent to, the specified material or process.

Substructure

The foundation and footings as opposed to the superstructure.

Suburb

An outlying neighborhood or town near a city center, used primarily as a residential community, with single-family homes.

Suburban sprawl

The spreading of a city's population out into the surrounding countryside, forming suburbs.

Sullivan, Louis H. (1856–1924)

Leader of the Chicago School of Architecture and a pioneer in skyscraper design. The Auditorium Building, Chicago, IL (1887), was his first major work. It was followed by the Getty Tomb, Chicago, IL (1890); Wainwright Building, St Louis, MO (1890); Schiller Theater Building, Chicago, IL (1892); and the Guaranty Building (illus.), Buffalo, NY (1895), notable for its rich terra-cotta ornamentation and an early use of Art Nouveau interior decoration. The Carson Pirie Scott Department Store (illus.), Chicago, IL (1900), was designed with a horizontal emphasis and demonstrated his interest in organically inspired facade ornament. His many bank buildings include the National Farmer's Bank (illus.), Owatonna, MN (1908). The Bayard-Condict Building (illus.), Sullivan's only NYC building (1897), also exhibits a high degree of ornamentation.

Sullivan, Louis H.

Sullivanesque style (1890–1920)

A style named after Louis Henry Sullivan, noted for his stylized ornamentation and simple multistory forms, designed as uninterrupted elements to express height, much the same as fluting in Classical columns. The windows were separated by ornamental terra-cotta panels, and the massive decorative cornice resembled the capital. An intricate weaving of linear forms with stylized foliage was highlighted in the cornice with low-relief ornamentation in terra-cotta.

Sumerian architecture (5000–2000 B.C.)

An architecture made of locally available materials: clay-tied bundles of reeds used as structural framing for huts and halls, with sun-dried bricks for the walls between these buttresses. Monumental temples and palaces were built around a series of courtyards; the ziggurat of Ur is the most famous. Large cities had well-developed drainage and sewer systems, and were protected by strong ramparts.

Summerbeam See Beam

Sump

Reservoir or pit in the basement of a house into which water can drain, especially during flooding. A sump pump is used to pump collected water out of this reservoir to the sewer pipes.

Sun deck

A roof area, balcony, or open porch that is exposed to the sun.

Sun disk

A disk representing the sun with wings, especially used in Egyptian antiquity as emblematic of the sun god.

Sun tempering
Practice of using a modest area of south-facing windows to provide limited passive solar heating to a house.

Sun-bearing angle
The solar azimuth angle relative to the horizontal direction a building surface is facing. Often referred to as the relative solar azimuth.

Sunk draft
A margin cut into a building stone that is sunk below the face of the stone to give it a raised appearance.

Sunk molding See Molding

Sunk panel
A panel recessed below the surface of Its surrounding framing or carved Into solid masonry or timber.

Sunk relief See Relief

Sunlight
The source of all natural daylight is the sun, which varies from time, location, season, and atmospheric conditions. Increasing the amount of sunlight into a building decreases the use of energy and provides solar heat in cold weather. However, it also increases the need for air conditioning in warmer weather. Another problem associated with sunlight is glare, caused by excessive contrast between the brightness of a window opening and the darker wall surfaces.

Sunscreen
Awning or window treatment that effectively blocks the sun's heat.

Sunshades
Devices for blocking unwanted solar gain.

Superimposed orders
Use of one order on top of another on the face of a building of more than one story. Usually in the following sequence: Doric (first story), Ionic, and Corinthian (top). The upper order is usually lighter in form than the lower ones.

Superimposition
Placing one classical order above another; generally placing the simpler orders at the bottom and the increasingly ornate orders above.

Superinsulation
The superinsulation of the building envelope minimizes heat gain during the summer and heat loss during the winter. Superinsulation involves substantially increased R-values combined with proper detailing for minimized thermal bridging and thorough air sealing for minimized infiltration. This strategy must be paired with controlled ventilation in order to maintain a healthy indoor environment.

Supermarket
A large, self-service, retail market that sells food, household goods, and household merchandise.

Superplasticizers
Chemical additives for concrete that increase the fluidity of the mix without excess water.

Superstructure
Any structure built upon something else, such as a building on its foundation; any structure above the main supporting level, as opposed to the substructure or basement.

Superwindow
A window with a very low U-value achieved through the use of multiple glazings, low-E (low-emissivity) coatings, and inert gas fills, usually argon or krypton, placed between sealed panes of glazing in order to provide resistance to heat flow.

Supply air
The total quantity of air supplied to a building or part of a building for thermal conditioning and ventilation. Typically, supply air consists of a mixture of return air and outdoor air that is appropriately filtered and conditioned.

Surbase
The molding or cornice at the top edge of a pedestal.

Surbased arch See Arch

Surface mount
General term for a fixture directly mounted on a ceiling. Many hold up to four bulbs and are used in hallways and bathrooms. Used primarily in rooms with low ceilings.

Surface runoff

The precipitation, snow melt, or irrigation water in excess of what can infiltrate the soil surface and be stored in small surface depressions; a major transporter of non-point-source pollutants in rivers, streams and lakes.

Surface-mounted fixture

A fixture attached to a recessed junction box, extending from the finished ceiling. Wall-mounted fixtures can be used to provide task lighting or add general light to an area.

Surplus property

A building or site no longer in demand for its current use. Unused property owned by the federal, state, or local government may be transferred with minimal or no charge to another owner for rehabilitation and new use.

Surround

An encircling border or decorative frame around a door, window or other opening.

Survey

A boundary or topographic mapping of a site; a compilation of the measurements of an existing building; an analysis of a building for use of the interior space.

Suspended ceiling See Ceiling

Suspended forms

Refers to forms that are hung so as to allow free movement and appear to be supported without any attachment to objects underneath.

Suspended forms

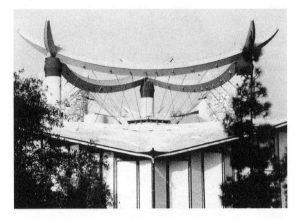

Suspended luminaire

A lighting fixture hung from the ceiling by rigid or flexible supports, leaving an air space above the luminaire.

Suspended luminaire

Suspended roof See **Roof**

Suspension bridge See **Bridge**

Sustainability

The concept of sustainability can be traced back to President Theodore Roosevelt, who stated in 1910, "I recognize the right and duty of this generation to develop and use the natural resources of our land; but I do not recognize the right to waste them, or to rob, by wasteful use, the generations that come after us." In 1987 the United Nations World Commission on Environment and Development (the Brundtland Commission) defined a sustainable development as one that "meets the needs of the present without compromising the ability of future generations to meet their own needs." Sustainability has three interdependent dimensions—the environment, economics, and society—often referred to as the triple bottom line.

Sustainability report

Differs from an environmental report or an environmental, health, and safety (EHS) report by presenting a holistic picture of company activities and providing a balanced view of benefits and trade-offs among social, economic, and environmental impacts.

Sustainable

Being able to meet the needs of present generations without compromising the needs of future generations. To be truly sustainable, a human community must not decrease biodiversity, must not consume resources faster than they are renewed, must recycle and reuse virtually all materials, and must rely primarily on resources in its own region.

Sustainable architecture

A term used to describe technologically, materially, ecologically, and environmentally stable building design. Sustainable architecture can be measured by its durability, maintenance level, and recyclability, as well as economic issues related to its construction, profitability, and building stock value. Resource sustainability can be measured by site conditions; cost-effectiveness of the operational and life cycle of the building; accessibility and favorable natural forces; and the creation of healthy, habitable and safe environments with social capacity.

Sustainable buildings

Those that rely on sustainable resources, that is, supply their own graywater and power, have proximity to water resources and existing waste management systems, enjoy favorable climate conditions, and provide access to public transportation and bicycle paths.

Sustainable Buildings Industry Council (SBIC)

A nonprofit organization whose mission is to advance the design, affordability, energy performance, and environmental soundness of residential, institutional, and commercial buildings nationwide.

Sustainable community

Preserves or improves quality of life while minimizing its impact on the environment, and achieving these goals using fiscally and environmentally responsible policies.

Sustainable construction

Construction in which the impacts of building activity on resources, biodiversity and ecosystem services are accounted for as part of the life cycle of the project. The desired outcome of sustainable construction is for no negative impact on financial, natural, and social capital over the lifetime of the project.

Sustainable design

A design that values the unique cultural and natural character of a given region, and protects and benefits ecosystems, watersheds, and wildlife habitat in the presence of human development. It conserves resources and maximizes comfort through design adaptations to site-specific and regional climate conditions. It creates comfortable interior environments that provide daylight, views, and fresh air. It conserves water and protects water quality, conserves energy and resources, and reduces the carbon footprint while improving building performance and comfort. Sustainable design anticipates future energy sources and needs, and includes the informed selection of materials and products to reduce product-cycle environmental impacts. It also optimizes occupant health and comfort and seeks to enhance and increase ecological, social, and economic values over time.

Sustainable design

The art of designing physical objects and the built environment to comply with the principles of economic, social, and
ecological sustainability in a way that reduces use of nonrenewable resources, minimizes environmental impact, and
relates people with the natural environment.

Sustainable environment

An environment that is protective, healthy, and habitable
and that promotes social and institutional networks. The
environment should make occupants feel safe and secure
against the elements and be free from physical and psycho-
logical effects, such as sick building syndrome.

Sustainable flooring

Flooring, such as bamboo and cork, that is made from reclaimed or rapidly renewable sources.

Sustainable forestry

The practice of managing forest resources to meet the long-
term forest product needs of humans while maintaining the
biodiversity of forested landscapes. The primary goal is to
restore, enhance, and sustain a full range of forest values—
economic, social, and ecological.

Sustainable materials

Materials that are generated from resources that are managed such that they are, for all practical purposes, sustainable over an extended period of time.

Sustainable resources

Resources that include energy, water, and waste management systems.

Sustainable sites

A LEED (Leadership in Energy and Environmental Design)
Rating System category. Prerequisites and credits in this
category focus on the selection of a LEED Certified Building
and/or buildings that have incorporated key sustainability
concepts and practices into design, construction, operations, and maintenance activities.

Sustainable technologies

Those technologies using durable, low-maintenance, recyclable, and economical materials and technologies, including using abundant, local elements involving little or no
transportation costs. Facilities that can be dismantled and
reused or recycled at the end, or can be salvaged, refurbished, renovated, repaired, or generally improved in appearance, performance, quality, functionality, or value.

Sustainably managed forest

The use of forests in a way and at a rate that maintains
their biodiversity, productivity, regeneration capacity, and
vitality without compromising their potential to fulfill, now
and in the future, relevant ecological functions, and that
does not cause damage to other ecosystems.

Swag

A festoon, hung between
rosettes or other terminals.

Swag

Swale

Low area of ground used for drainage and often the
infiltration of stormwater.

Swan's neck pediment

A broken pediment having a sloping double S-shaped
element on each side of the pediment; used often in
the Georgian style.

Sway brace See Brace

Sway bridge See Bridge

Sweat equity

The investment of property owner's or occupant's own
labor in rehabilitation work as a form of payment.

Sweat joint

In plumbing work, a type of joint made by the union of two
pieces of copper tubing which are coated with solder containing tin; the pipes are pressed together into sleeves or
other joints, and heat is applied until the solder melts.

Swinging door See Door

Swing-stage scaffold

A scaffold supported from
hooks over the parapet
above, using ropes and
pulleys or wire cables,
which permit it to be raised
or lowered as needed; used
on exterior work.

Swiss Chalet style

An architectural house style loosely based on the Swiss Chalet prototype; typical elements include two stories with a front gable, low-pitched roof, front balcony or porch with scrollwork railings, stick work or board-and-batten siding, or stucco with painted ornamentation or scenery.

Switch

In electrical work, a device for closing, opening, or changing the connections of the circuit in which it is placed.

Switchboard

Similar in function to an electric panelboard but considerably larger and standing on the floor, usually with access from the back. Sometimes referred to as switchgear.

Switches

Standby single-throw switches are the most common. Three-way and four-way switches permit control from multiple locations. Examples include toggle, electronic, and touch switches

Symbol

Something that stands for or represents something else by association, resemblance, or convention, deriving its meaning chiefly from the structure on which it appears.

Symbol

Symmetry

The exact correspondence of forms of similar size and arrangement of parts, equidistant and on opposite sides of a dividing line or plane about the center line or axis.

bilateral symmetry
A balanced arrangement of identical similar elements about a central axis.

radial symmetry
Balanced elements that radiate from a central point.

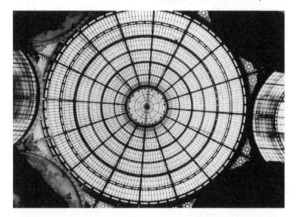

Synagogue
A place of assembly, or a building for Jewish worship and religious instruction.

Synagogue

Synergy
Action of two or more substances to achieve an effect of which neither is capable individually.

Syrian arch See Arch

System commissioning
Statically and dynamically testing the operation of equipment and building systems to ensure they operate as designed and can satisfactorily meet the needs of the building throughout the entire range of operating conditions.

System furniture
Modular office furniture with provision for integrated electrical and data connections, including desktops, shelving, and drawers; they can be interconnected in a variety of combinations to create individual or linked workstations.

Systems design
Three definitions characterize this approach to problem solving. The first is the design of a range of components to be prefabricated in factories and combined in different ways to yield different types of structures. The second is the application of analysis to the supply of materials and assembly processes. The third is a conceptual overview of design where each building is regarded as part of a greater whole and each project is seen in its social, cultural and economic context.

T t

Tabernacle
A freestanding ornamental canopy above an altar, tomb, or ornamental niche.

Table
Applied generally to all horizontal bands of moldings, base moldings, and cornices.

Table lamp
Most common form of residential lighting, containing a base, light socket, and lamp shade.

Tablero
A rectangular framed panel which is cantilevered over an outward sloping apron with which it is always used; characteristic of the temples at Teotihuacan In Mexico.

Tablet
A rectangular separate panel or flat slab, often bearing an inscription or carving of an image.

Tack weld
A weld designed not to carry a load, but to make a non-structural connection.

Tafel, Edgar A. (1912–2011)
American architect who began his career as an apprentice to Frank Lloyd Wright at Taliesin. and worked on several of Wright's most notable projects, including Fallingwater, Wingspread, and the Johnson Wax Headquarters. One of his best known works as a solo practitioner is the Mellin Macnab Building for the First Presbyterian Church in New York City. Tafel's design combined Prairie School influences with the Gothic style of the sanctuary. Other designs included the Protestant Chapel at Kennedy International Airport; St. John's in the Village Episcopal Church in Greenwich Village (1974); the 1964 master plan for the campus of the State University of New York at Geneseo, and the North Wing expansion to the Allentown Art Museum in Allentown, Pennsylvania. Tafel wrote *Apprentice to Genius: Years with Frank Lloyd Wright* (1979), *About Wright: An Album of Recollections by Those Who Knew Frank Lloyd Wright* (1993), and *The Frank Lloyd Wright Way*, a film that won first prize at the 1995 Houston International Film Festival.

Tail cut
The cutting of the lower end of a projecting rafter, which is often trimmed to an ornamental profile.

Takamatsu, Shin (1948–)
Japanese architect, exploring high-tech forms, such as the Kirin Plaza Building (illus.), Osaka, Japan (1987).

Talus
Sloped architectural feature, such as a battered wall or an inclined retaining wall against an embankment.

Tange, Kenzo (1913–2005)
He came from Kunio Mayekawa's office and was influenced by Le Corbusier, but drew on Japanese themes. The Hiroshima Peace Center and Museum (1955) was his first major project, followed by St. Mary's Cathedral, Tokyo (1961). Later work includes the gymnasium for the Tokyo Olympics (illus.) Tokyo, Japan (1964), covered by a gigantic tensile catenary roof structure, and Tokyo City Hall (1991).

Tange, Kenzo

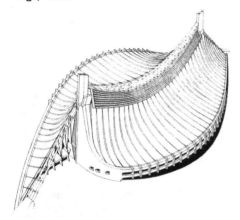

Tankless coil
Heat exchanger used for heating water that is integrated into a boiler. Effective in the winter months when the boiler is operating for space heating, tankless coils waste energy in warmer months, since they require the boiler to fire up every time hot water is drawn.

Tankless water heater
A system that delivers hot water at a preset temperature when needed, but without requiring the storage of water. The approach reduces or eliminates energy standby losses. Tankless water heaters can be used for supplementary heat, such as a booster to a solar hot water system, or to meet all hot water needs. Tankless water heaters have an electric, gas, or propane heating device that is activated by the flow of water.

Tapering
Forms exhibiting a gradual diminution in thickness, or reduction in cross section, as in a spire or column.

Tapestry
A large woven illustration hung as a wall decoration.

Tapestry brick See **Brick**

Task lighting
Lighting directed to a particular spot or work space, as opposed to general lighting for the area.

Tatami
A thick straw floor mat in a Japanese house covered with finely woven reeds and bound with plain or decorated bands of silk, cotton, or hemp; its size of 3 feet by 6 feet is used as a standard unit of measurement.

Tatlin, Vladimir (1885–1953)
Designed the Memorial to the Third International. It was a Constructivist architectural fantasy, a spiral leaning tower. Constructivism was a short-lived ideal in Russia.

Taut, Bruno (1880–1938)
German architect and advisor to the German Garden City movement. His glass pavilion, Werkbund Exposition, Cologne, Germany (1914), is his most celebrated work, a paradigm of Expressionism. He also published many books.

Tavern
An establishment licensed to sell liquor and beer to be consumed on the premises. A public house or inn for travelers; saloon, bar.

Tax Reform Act of 1976
Created tax incentives for the rehabilitation of income-producing historic structures; penalized demolition, and codified deductions for charitable transfers of preservation easements.

T-beam See **Beam**

Teahouse
A Japanese garden house used for the tea ceremony.

Teak See **Wood**

Tectiforms
Shapes resembling manmade structures found painted on the walls of Paleolithic caves.

Tee
A finial in the form of a conventionalized umbrella; found in Japanese architecture on stupas and pagodas.

Telescope house
A house composed of several sections, each of descending height, giving the appearance of fitting together like a collapsible telescope.

Tempera
A rapidly drying paint consisting of egg white, gum, pigment, and water, used in painting murals.

Tempera painting
A mural painting technique used widely in the Middle Ages and the Renaissance, which uses transparent colors on gesso.

Temperature bars
Steel rods placed horizontally in concrete slabs for prevention of cracks due to temperature changes or drying; placed parallel to the reinforcing rods. The steel rods are placed at right angles to the main reinforcing bars.

Tempered glass See **Glass**

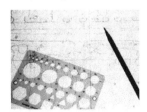

Template
A pattern of sheet material, used as a guide for setting out work and in repeating patterns of painted ornamentation. Also, an outline of fixtures and graphic shapes and symbols.

Temple
An edifice dedicated to the service of a deity or deities, and connected with a system of worship; an edifice erected as a place of public worship.

Tendril
A long, slender, coiling extension, such as a stem, serving as an ornamental device; used primarily by Art Nouveau architects.

Tenement
A building with multiple dwelling units accessed by a single stairway, with two or more apartments on each floor.

Tenon
The projecting end of a piece of wood or other material, reduced in cross section so that it may be inserted into a corresponding mortise in another piece to form a secure joint.

Tensile structures
Those that stretch or extend a member or other ductile material such as a fabric or membrane; some forms express this quality even if the material is not fabric, such as concrete shells.

Tensile structures

Tension
A pulling or stretching force in line with the axis of the body; the opposite of compression, which is a pushing, crushing stress.

Tension column See Column

Tepee
A tent of the American Indians, made usually from animal skins laid on a conical frame of long poles and having an opening at the top for ventilation and a flap door.

Tepidarium
Room of moderate heat in a Roman bath.

Terminal
A terminus occurring at the end of a series of incidents, as a resting point; a point of emphasis, as in an object situated at the end of an element.

Terminal figure
A carving in the form of a pedestal that tapers outward at the top and merges into an animal, human, or mythological figure; used for columns, statues, and consoles. See also Gaine.

Terminal pedestal
A tapered pedestal for a bust, with both objects forming a terminal figure.

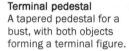

Termination

An ornamental element that finishes off an architectural feature.

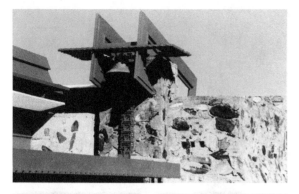

Terminus

A bust or figure of the upper part of the human body, terminated in a plain rectangular block, sometimes attached to a wall as a pillar or found springing out of a column.

Termite

A wood-devouring insect that eats the woodwork of a structure, which can ruin a building; resembling ants in appearance and in their habit of living in colonies.

Termite barrier

A barrier to prevent termites traveling from the ground through holes or cracks in a concrete floor slab into a building; usually consists of a termite shield.

Termite shield

A protective shield made of noncorroding metal, placed in or on a foundation mass of masonry, or around pipes entering a building to prevent passage of termites into the building.

Terne coating

A corrosion-protective coating for steel sheet used for roofing, consisting of a combination of lead and tin.

Terrace

A flat roof or raised space or platform adjoining a building, paved or planted, especially one used for leisure enjoyment.

Terrace roof See Roof

Terra-cotta

A hard-burnt glazed or unglazed clay unit, either plain or ornamental, machine extruded or hand-molded, usually larger in size than a brick or facing tile, used in building construction.

Terra-cotta

Terrazzo
Marble-aggregate concrete that is cast in place, or precast and ground smooth; used as a decorative surface for walls and floors.

Terry, Quinlan (1937–)
British architect, whose most well-known projects include the Howard Building, Downing College, Cambridge, England (1989) and Kansai International Airport, Osaka, Japan (1994).

Tessellated
Formed of small square pieces of marble, stone, or glass in the manner of an ornamental mosaic.

Tessera
A small square piece of colored marble, glass, or tile, used to make geometric or figurative mosaic patterns.

Test pit
A pit dug for subsurface exploration.

Tetrahedron
A polygon with four plane surfaces.

Tetrastyle
Having four columns; such as a tetrastyle portico, which has four columns in front.

Textile mill
A factory in which woven fabrics are manufactured; many early mills were of timber and masonry construction, and were sometimes operated by water power to run machinery.

Texture
The tactile and visual quality of a surface as distinct from its color or form; as showing a grainy, coarse, tactile or dimensional quality, as opposed to uniformly flat or smooth.

Texture

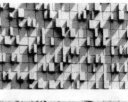

Texture paint
A paint that can be manipulated after application to give a textured finish.

Thatch
A roof covering made of straw, reed, or similar materials fastened together to shed water and sometimes to provide thermal insulation; in tropical countries palm leaves are widely used.

Thatched roof See Roof

The Architect's Collaborative
A firm started by Walter Gropius in Cambridge, MA. They designed the American Institute of Architect's National Headquarters Building in Washington, DC (1973).

The Green Blue Institute
Nonprofit organization dedicated to speeding the adoption of sustainable, cradle-to-cradle thinking, and informing how people design, manufacture, and utilize goods and services.

The grid
Refers to an interconnected network for delivering electricity from suppliers to consumers.

Theater

A building or outdoor structure providing a stage and associated equipment for the presentation of dramatic or musical performances and seating for spectators.

Theater Historical Society of America

An organization dedicated to documenting the rich heritage of historic theaters in the United States from small-town theaters to ornate movie palaces, drive-ins, and multiplexes.

Theme

An idea, point of view, or perception embodied in a work of art; an underlying and essential subject of artistic expression.

development

To disclose by degree or in detail; to evolve the possibilities by a process of growth; to elaborate with the gradual unfolding of an idea.

variation

Repetition of a theme with embellishments in rhythm, details, and materials while keeping the essential characteristics of the original.

Thermae

Ancient Roman buildings for public baths, which also incorporated places for sports, discussions, and reading.

Thermal barrier

An element of low heat conductivity placed on an assembly to reduce or prevent the flow of heat between highly conductive materials; used in metal window or curtain wall designs in cold climates.

Thermal break

Method of increasing the thermal performance of a material or assembly by reducing conductive heat loss. By inserting a less thermally conductive material in a material or assembly that bridges conditioned and unconditioned space, the conductive path is reduced or broken. An example is the thermal break featured in aluminum-framed windows.

Thermal bridging

Heat flow that occurs across more conductive components in an otherwise well-insulated material, resulting in disproportionately significant heat loss. For example, steel studs in an insulated wall dramatically reduce the overall energy performance of the wall, because of thermal bridging through the steel.

Thermal buffer

A space or other element that reduces the heating and cooling load on another space located between the space and the exterior.

Thermal bypass

An opening between a conditioned and unconditioned space that heated or cooled air can move through, therefore violating the airtightness of the building envelope.

Thermal comfort

The appropriate combination of temperatures, warm or cool, combined with air flow and humidity, that allows one to be comfortable within the confines of a building. This comfort is not usually achieved by the fixed setting of thermostats but through careful design and planning.

Thermal conduction

The process of heat transfer through a material medium, in which kinetic energy is transmitted by particles of the material without displacement of the particles.

Thermal energy storage

Technologies used for storing energy in a thermal reservoir for later use. For cooling, ice is produced at night to provide daytime cooling, thereby reducing peak daytime electricity demands. For heating, solar collectors are most commonly used to gather heat, which is then stored for later use.

Thermal envelope

The shell of a building that essentially creates a barrier from the elements. A highly insulated thermal envelope allows maximum control of interior temperatures without outdoor influence.

Thermal expansion

The temporary increase in volume or linear dimensions of materials when heated.

Thermal flywheel

A space or other element such as a solid masonry wall that collects heat during one period and releases it during another in a repetitive pattern.

Thermal gradient

The difference in temperature between the floor and the ceiling.

Thermal imaging camera

A camera that provides an image showing radiation in the infrared range of the electromagnetic spectrum. Since the amount of infrared radiation emitted from a surface varies with temperature, a thermal imaging camera is a useful tool for detecting hot or cold areas on walls, ceilings, roofs, and duct systems. When used to scan a building envelope, a thermal imaging camera can detect missing insulation or locations with high levels of infiltration. Thermal imaging cameras can provide useful information when the difference in temperature (delta T) between the indoors and the outdoors is as low as 18 degrees Fahrenheit. However, the higher the delta T, the easier it is to see building defects.

Thermal insulation
The reduction of the flow of heat; it is measured by thermal resistance.

Thermal mass
A lot of heat energy is required to change the temperature of high-density materials like concrete, bricks, and tiles. They are therefore said to have high thermal mass. Eco designers will strategically place these materials to create heat sinks that absorb heat in the winter and remain cool in the summer. Lightweight materials such as timber have low thermal mass.

Thermal resistance
The resistance (R-value) to the passage of heat provided by the roof, walls, or floors of a building.

Thermal shock
The strain produced in a material due to sudden changes in temperature.

Thermal stress
A stress produced by thermal movement that is resisted by the building; if the thermal stresses are higher than the capacity of the materials to resist them, expansion or contraction joints are required.

Thermosiphon solar water heater
A solar water heater that operates passively through natural convection, circulating water through a solar collector and into an insulated storage tank situated above the collector. Pumps and controls are not required.

Thermostat
An instrument, electrically operated, such as a bimetal strip, for automatically maintaining a constant temperature; commonly used in conjunction with heating and air-conditioning plants.

Thinner
Any volatile liquid that lowers the viscosity of a paint or varnish, and thus make it flow more easily; it must be compatible with the medium of the paint; the most common thinner is turpentine.

Thinset
A modified Portland cement and sand mortar used for tile setting, which may contain an acrylic additive for strength.

Third-party commissioning
Use of a commissioning agent that is independent of other parties to the delivery of the building project.

Tholos
Domed building with a circular plan, such as the Jefferson Memorial in Washington, DC.

Tholos

Thompson, Benjamin (1918–2002)
Noted American architect/developer; he designed the Design Research Building, Cambridge, MA (1969); the large retail complex Harbor Place, Baltimore, MD (1980); and Faneuil Hall Marketplace and Quincy Market Restoration, Boston, MA (1977), for the Rouse Company.

Thomson, Alexander (1817–1875)
Scottish Neoclassical architect; an original designer influenced by Schinkel. He worked mostly in Glasgow. His buildings drew on a variety of sources, including ancient Egyptian, Persian, and Indian architecture.

Thornton, William (1759–1828)
British West Indian–born American architect who designed a submittal for the U.S. Capitol. He designed the Octagon (1800), Washington, DC, home of the American Institute of Architects; a house at Tudor Place (1810), Washington, DC; and Pavilion VII, University of Virginia (1821), Charlottesville, VA.

Three-centered arch See Arch

Three-dimensional graphics
The process of drawing and displaying objects in three dimensions, using computer graphics programs.

Three-dimensional modeling
The process of producing three-dimensional objects using constructive geometry and rendering them as solids in computer graphics programs.

Three-hinged arch See Arch

Three-pinned arch See Arch

Three-point perspective See Perspective projection

Three-way switch
A switch used in house wiring when a fixture is to be turned on or off from two different locations in the house. A three-way switch must be used at each location.

Three-way switch system
A convenient outlet switching system whereby any two switches can turn an outlet on or off. Commonly used where it is necessary to switch lights from two different locations.

Threshold
A strip fastened to the floor beneath a door, to cover the joint where two types of floor materials meet or to provide weather protection.

Throat
A groove which is cut along the underside of a member, such as a stringcourse or coping on a wall, to prevent water from running back toward the wall.

Thrust
The force exerted by beams against a wall; or the outward force of an arch, dome, or vault, counterbalanced if necessary by buttresses.

Tie beam See **Beam**

Tie rod
A rod in tension, used to hold parts of a structure together.

Tier
One of a series of rows arranged one above the other.

Tierceron
An intermediate rib between the main ribs of a Gothic vault.

Tiffany, Louis Comfort (1864–1933)
American designer; best known for his designs in the Art Nouveau style and developing interiors for McKim Mead and White. He also designed many fine artifacts and light fixtures in stained glass.

Tigerman, Stanley (1930–)
American architect, who opened his office in 1964; where the early work was reminiscent of SOM and Mies Van der Rohe, his growing eclecticism produced many controversial structures.

Tight buildings
Buildings that are designed to let in minimal infiltration air in order to reduce heating and cooling energy costs. In actuality, buildings typically exhibit leakage that is on the same order as required ventilation; however, this leakage is not well distributed and cannot serve as a substitute for proper ventilation.

Tile
A ceramic surfacing unit, usually thin in relation to the facial area; made from clay or a mixture of clay and other ceramic materials; has either a glazed or an unglazed face.

acoustic tile
Rectangular sound-absorbing tile, normally used as a ceiling, whether glued to a backing or used in a grid as a suspended ceiling.

ceramic mosaic tile
An unglazed tile, usually mounted on sheets to facilitate setting; may be composed of porcelain or natural clay.

clay tile
A roofing tile of hard, burnt clay. In flooring it is called quarry tile.

503

crest tile
A tile which fits like a saddle on the ridge of a roof.

encaustic tile
A tile for pavement and wall decoration, in which the pattern is inlaid or incrusted in clay of one color in a ground of clay of another color.

floor tile
A ceramic tile that can be used as a floor finish, such as encaustic tile, quarry tile, and glazed tile.

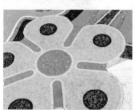

glazed tile
Ceramic tile having a fused impervious glazed surface finish, composed of ceramic materials fused into the body of the tile; the body may be nonvitreous, semi-vitreous, or impervious.

hollow tile
A structural clay tile unit with vertical hollow cells; used to build interior masonry partitions and as a backup block for brick veneer.

mission tile
A clay roofing tile, approximately semicylindrical in shape; laid in courses with the units having their convex side alternating up and down.

paving tile
Unglazed porcelain or natural clay tile, formed by the dust-pressed method; similar to ceramic mosaic tile in composition and physical properties, but thicker.

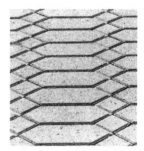

quarry tile
A dense, unglazed, ceramic tile, used most often for flooring.

ridge tile
A tile which is curved in section, often decorative, used to cover the ridge of a roof.

Spanish roof tile
An interlocking terra-cotta roof tile with a convex, curved top that adjoins a narrow flat valley.

unglazed tile
A hard, dense ceramic tile for floor or walls; of homogeneous composition, and deriving its color and texture from the materials and the method of manufacture.

vinyl-asbestos tile
A resilient, semiflexible floor tile; composed of asbestos fibers, limestone, plasticizers, pigments, and a polyvinyl chloride resin binder; has good wearing qualities, high grease resistance, and relatively good resilience.

wall tile
Thin tile used as a wall finish, it is glued to the wall with mastic, then the joints are grouted. Types include glazed ceramic, terra-cotta, glass, mosaic, or plastic; may be square, rectangular, or other geometric shape.

Tile hanging
Slate or tile shingles that are hung nearly vertically on a wall or roof for weatherproofing.

Tile setting adhesive
Specifically formulated glues or mastics used for setting tile. They are cleaner, more waterproof, less expensive, and faster than mortar bed.

Tilt-up construction
Any construction of concrete wall panels which are cast horizontally, adjacent to their final positions, and then tilted up into a vertical position when hardened.

Timber
Uncut trees that are suitable for construction or conversion to lumber.

Timber connector
A steel device, used in conjunction with a bolt, that makes a connection between timber members that overlap each other. There are several types of timber connectors.

Timber-framed building
A building having timbers as its structural elements, except for the foundations.

Time and material contract
Work agreed to between the owner and the contractor, with payment based on the contractor's cost for labor, equipment, materials, and an add-on factor to cover overhead and profit; or based on material costs plus a specified amount per hour of labor, which includes direct labor, overhead and profit.

Time scheduling
Lighting controls that use clocks to operate lighting systems on predictable schedules; the most commonly used form of automatic light control. Some energy codes require automatic controls as a minimum standard. Devices vary from individual time-clock switches to electric programmable timers to large-scale energy management systems.

Time switch
A clock that has built-in switches to control the time at which lights or other electrical devices go on or off.

Time-lapse photography
The study of work processes by use of a series of photographs taken at specifically timed intervals; useful for collecting video images of a construction site for posting on an intranet, to monitor construction progress.

Tin ceiling
Embossed metal panels nailed to furring strips or ceiling beams, and painted as a finished ceiling.

Tin See Metal

Tinted glass See Glass

Title search
Investigation conducted to see that the ownership of a property is free of liens and other encumbrances.

Tolerance
The permitted variation from a given dimension. The limits of size are the two extreme sizes between which the actual size must lie, and the difference between them is the tolerance.

Toltec architecture (750–1200)
An austere geometric architecture that formed the basis for the Aztec style and others. It was characterized by the use of colonnades, square carved roof supports, monumental serpent columns, and narrative relief panels set in plain wall surfaces. Tula was one of the major sites in this style, which featured colossal statues of warriors and stone panels carved with human-headed jaguars and carved symbols of Quetzalcoatl.

Tomb
In architecture, a memorial structure over or beside a grave.

Tombstone light
A small window with lights in the shape of an arched tombstone, usually in the transom above a doorway.

Tongue
A projecting rib cut along the edge of a piece of timber so it can be fitted into a groove in an adjoining piece.

Tongue-and-groove joint See **Joint**

Tooled mortar joint See **Mortar joint**

Tooth ornament See **Ornament**

Top hung window See **Window**

Top plate Continuous horizontal piece of lumber across the top of an exterior wall that supports the foot of the rafters, or other framework above it.

Topcoat
The final coat of paint or plaster applied to a surface over a primer or undercoat.

Topiarium opus
A wall painting representing trees, shrubs, and trellis work, as at Pompeii.

Topiary work
The clipping or trimming of plants, trees, and shrubs, usually evergreens, into ornamental and fantastic shapes.

Topographic map
A map that indicates the shape of the surface of the ground, and other salient physical features such as buildings and roads; typically with a series of contour lines that indicate the elevation above sea level.

Topography
Physical features of a particular location, including the shape of the surface of the ground.

Torana See **Gate**

Torii See **Gate**

Torre, Susana (1944–)
Feminist architect and planner. Through her teaching, writing, and architectural practice, she has worked to improve the status of women in architecture.

Torroja, Eduardo (1899–1961)
Spanish architect, and engineer of concrete shells, who studied civil engineering. Most of his designs employed folded, undulating, or warped shapes.

Torsion
The force tending to twist an architectural member.

Torus molding See **Molding**

Total run
The total of all the tread widths in a stair.

Totem
The emblem of an individual family or clan, usually represented by an animal.

Totem pole
A wooden post carved and ornamented with the emblem of a clan or family, and erected in front of the homes of the Indians of the northwestern United States.

Tou-kung

A cantilevered bracket in traditional Chinese construction; tiers or clusters of brackets are used to carry rafters that support purlins far beyond the outermost columns of a building.

Tourelle

A small projecting turret with a conical roof, especially one located at an outside corner of a building.

Tout ensemble

From the French phrase for "all together"; used to describe the cohesive ambience of an entire area. Derived from a New Orleans ruling (1941) regarding preserving the antiquity of the entire French Quarter. The elements of the *tout ensemble* include location, design setting, materials, workmanship, feeling, and associations to lives, events, or visual qualities.

Tower

A tall structure designed for observation, communication, or defense. A bell tower is synonymous with the term "campanile"; church towers were used for hanging bells, hence the use of the term "belfry."

Tower

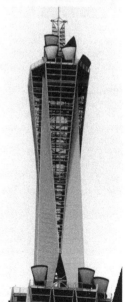

Tower crane

A crane in which the lifting mechanism is mounted on top of a tower that is structurally independent of the structure it is working on.

Town

A concentration of residential and related buildings surrounded by countryside; typically smaller than a city and larger than a village.

Town hall

A public hall or building, belonging to a town, where public offices are established, the town council meets, and the people assemble for town meetings.

Town house

An urban building, without sideyards; containing one residence on one or more floors.

Town plan

A comprehensive map of a town or city that delineates its streets, important buildings and other urban features.

Town, Ithiel (1784–1844)

American architect who worked in New Haven, CT, and New York; and associated with Alexander J. Davies. Early leader in Greek and Gothic Revival styles and bridge engineering. Designed Trinity Church (1816), NYC; the Connecticut State House (1831), New Haven, CT; City Hall and Market House (1829), Hartford, CT; Indiana State Capitol (1835), with Davies, Indianapolis, IN; and the North Carolina State Capitol, with Davies, at Raleigh, NC.

Townscape

A view of a town or city from a single vantage point; the planning and construction of buildings within a town or city with the objective of achieving overall aesthetically pleasing relationships.

Township

A political and geographic area within the boundaries of a municipal government; sometimes separate from a larger county.

Trabeated

Descriptive of construction using beams or lintels, following the principle of post-and-lintel construction, as distinguished from construction using arches and vaults.

Trabeation

Construction using upright posts and horizontal beams and lintels, rather than arches or vaults.

Tracery

The curvilinear ornamental branch-like shapes of stone or wood, creating an openwork pattern of mullions; so treated as to be ornamental; found within the upper part of a Gothic window or opening of similar character.

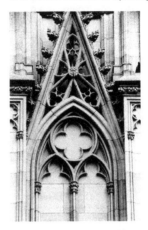

bar tracery

A pattern formed by interlocking branching mullions within the arch of Gothic window tracery.

blind tracery

Any tracery that is not pierced through.

branch tracery

A form of Gothic tracery in Germany in the late 15th and early 16th century made to imitate rustic work with boughs and knots.

fan tracery

A tracery on the soffit of a vault whose ribs radiate like the ribs of a fan.

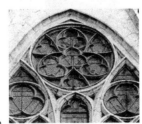

geometric tracery

Gothic tracery characterized by a pattern of geometric shapes, as circles and foils.

intersecting tracery
Any tracery formed by the upward curving, forking and continuation of the mullions, springing from alternate mullions or from every third mullion and intersecting each other.

panel tracery
Gothic style window tracery in sections within a large opening.

perpendicular tracery
Tracery of the Perpendicular style with repeated perpendicular mullions, crossed at intervals by horizontal transoms, producing repeated vertical rectangles which often rise to the full curve of the arch.

plate tracery
Tracery whose openings are pierced through thin slabs of stone.

reticulated tracery
Gothic tracery consisting mainly of a net-like arrangement of repeated geometrical figures.

Y-tracery
Tracery where the head of the mullions split into a Y-shape pattern.

Tracery vaulting See **Vault**

Track lighting
Consists of a series of separate lamps that fit into a metal track secured to the ceiling. The lamps can be moved along the track, but the track cannot be moved.

Track-mounted fixture
Consists of a fixture mounted into a continuous track that carries electrical current. The fixture can be moved anywhere along the track and adjusted to throw light in any direction. Many types of fixtures can be used within the track, including spots, downlights, and pendant fixtures.

Track-off mat
A mat at a entrance across which people scuff their shoes to remove moisture, dirt, and other particulates, keeping contaminants out and reducing cleaning requirements.

Trade union
A combination of trades organized for the purpose of promoting their common interests with regard to wages, hours of work safety measurements, unemployment compensation, and other benefits.

Traditional neighborhood development
A development that is based on human-scale design for comfortable walking and may include such elements as alleys, streets laid out in a grid system, buildings oriented to the street, front porches on houses, pedestrian orientation, compatible and mixed luses, and village squares.

Trajanic column

A large commemorative single freestanding column, with an internal winding stair, spiral bands of relief sculpture winding around the shaft, a large pedestal base, and a commemorative statue on top; named after the antique column of the Emperor Trajan in Rome (c. A.D. 112).

Transept

The space that crosses at a right angle to the nave of a building; may be the same size as the nave in a cruciform building, or larger.

Transfer column See Column

Transformation

The metamorphosis that occurs where primary shapes and forms are changed into additive or subtractive forms.

Transformer

A device that carries electrical current from one source to another, usually with an increase or decrease in voltage.

Transition piece

A sheet metal device shaped to form a transition from one shape of duct to different shape or size of duct.

Transitional style

An architectural style characterized by elements of an older style blended with elements of a more modern style in the same building, such as the evolution of Romanesque into Gothic.

Transit-oriented development (TOD)

Moderate and high-density housing concentrated in mixed-use developments located along transit routes. The location, design, and mix of uses in a TOD emphasize pedestrian-oriented environments and encourage the use of public transportation.

Translation

The movement of a point in space without rotation.

Translucency

The quality of a material that transmits light sufficiently diffused to eliminate any perception of distinct images beyond.

Transmission

A general term for the passage of energy from one side of a medium to the other.

Transmissivity

The fraction of light that passes through the interior of a glass pane at normal incidence in one traversal. This does not consider light lost to reflection by the front or back surface, or multiple internal reflections.

Transmittance

The ratio of the total radiant or luminous flux transmitted by a transparent object to the incident flux, usually given for normal incidence.

Transom

A horizontal bar of wood or stone across a door or window; the crossbar separating a door from the fanlight above it; a window divided by a transom bar.

Transom bar

A horizontal bar of wood or stone across a door or window; the crossbar separating a door from the fanlight above it; a window divided by a transom bar.

Transom light

A glazed light above the transom bar of a door.

Transparency

The quality of a material that is capable of transmitting light so that objects or images may easily be seen on the other side.

Transverse arch See **Arch**

Transverse loading

Loading that is perpendicular to a structural member, such as a vertical loading on a horizontal beam.

Transverse section See **Projection drawing**

Trap

A bend in a soil drain, arranged in such a manner that it is always full of water, which provides a water seal and prevents odors from entering back through the pipes.

Trapdoor

A door that is flush with the surface; located in a floor, roof, or ceiling, or in the stage of a theater.

Trapezoid

A four-sided figure with unequal sides; a parallel trapezoid has two unequal parallel sides and two equal nonparallel sides; a symmetrical trapezoid has two pairs of adjacent equal sides.

Travertine See **Stone**

Tread

The horizontal upper surface of a step; includes the rounded edge or nosing which extends over the riser.

Tread See **Step**

Trefoil arch See **Arch**

Trefoil ornament See **Ornament**

Trellis

A structural frame supporting an open latticework or grating constructed of either metal or wood, used to support plants or vines or left exposed.

Tresse

Flat or convex bandelets that are intertwined; especially such interlocking ornamentation used to adorn buildings.

Tresse molding See **Molding**

Triangle

A plane geometrical figure with three sides and three angles; the equilateral triangle has both equal sides and equal angles.

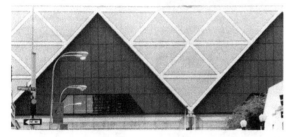

Triangular arch See **Arch**

Triangulated

Refers to any construction based on a continuous series of triangles for stability; particularly evident in designs for atrium skylights and space frames.

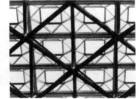

Triangulated

Triangulation

A method used in the design of space frames to ensure that they are static.

Tribunal

In an ancient Roman basilica, a raised platform for the chair of the magistrates; a place of honor.

Triforium

The space above the vaulting, below the roof, and under the clerestory windows, on the side aisle of a church; typically containing three arched openings in each bay forming an arcade into the nave space.

Triglyph

A characteristic ornament of the Doric frieze, consisting of raised blocks of three vertical bands separated by V-shaped grooves, alternating with plain or sculptured panels called "metopes."

Triglyph

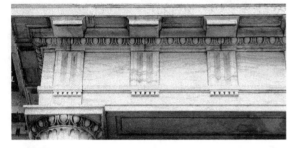

Trilith

A monument, or part of a monument, consisting of two upright stones supporting a horizontal stone, as at Stonehenge and other ancient sites.

Trim

The visible woodwork on moldings, such as baseboards, cornices, and casings around doors and windows; any visible element, which covers or protects joints, edges, or ends of another material.

Trimetric projection See **Projection drawing**

Trimmer

A piece of timber in a roof, floor, or wooden partition, to support a header which in turn supports the ends of the joists, rafters, or studs; a small horizontal beam, into which the ends of one or more joists are framed.

Trimming joist

A joist supporting one end of a header at the edge of an opening in a floor or roof frame, parallel to the other common joists.

Tripartite scheme

A type of design for a multi-story commercial building often associated with the work of Louis Sullivan. The building's facade is characterized by three major divisions: a base, consisting of the lower three stories; a cap, of one to three stories; and a shaft, consisting of the floors between the base and the cap.

Triple bottom line

Measuring the economic, social, and environmental performance of a project. This method of assessment aims for synergy among these three aspects rather than compromise or trade-offs among them.

Triple-hung window See **Window**

Triton

A sea monster, half man and half fish; often used in Classical and Renaissance ornamentation.

Triumphal arch

An arch commemorating the return of a victorious army, usually located along the line of march during a triumphal procession.

Troffers

Inverted troughs designed to be installed into acoustic tile ceilings, with the fixture's face matching the size of the tile. Different mounting types are made for "T" grids or metal pan ceilings. Some troffers are designed to interface with other building HVAC systems, such as heat-extraction troffers with vents to allow return air to be pulled into them, or air-handling fixtures for supply air.

Trombe wall

Consists of an 8- to 16-inch thick masonry wall on the south side of a house. A single or double layer of glass is mounted about 1 inch or less in front of the wall's surface. Solar heat is absorbed by the wall's dark-colored outside surface and stored in the wall's mass, where it radiates into the living space. Also referred to as a solar wall.

Trompe l'oeil

A phrase meaning "that which deceives the eye"; it was originally used to describe precisely rendered views of earlier architectural styles, wherein painters produced a convincing illusion of reality. This has been applied to exterior and interior mural design where architectural elements and entire facades have been painted on blank expanses of buildings, indicating a particular architectural style, period, or design.

Trompe l'oeil

Trophy
A sculptural composition of arms and armor as an emblem of, or a memorial to, victorious battles or triumphant military figures.

Troweled joint
A mortar joint finished by striking off excess mortar with a trowel.

Troweling
To smooth and compact the surface of a fresh plaster or concrete by strokes of a trowel.

Truck tire wash-down area
A strategy for removing dirt and other contaminants from construction vehicles in order to prevent stormwater contamination related to transport of contaminants offsite on vehicle tires. A specified area is created for wash-down, with structural controls in place to prevent wash-down water from entering the storm system or the larger environment.

Trumbauer, Horace (1868–1938)
American architect who designed luxurious houses in Philadelphia, New York, and Washington, DC. An accomplished Beaux-Arts Renaissance Revivalist architect; noted for two main campuses at Duke University (1938), Durham, NC, and the Philadelphia Free Library (1927), Philadelphia, PA.

Trumeau
A stone mullion supporting the middle portion of a tympanum.

Truncated
Forms that have been cut off at one end, usually the apex, often with a plane parallel to the base.

Truss
A composite structural system composed of straight members transmitting only axial tension or compression stresses along each member, joined to form a triangular arrangement.

arched truss
A truss with an arched upper chord and a straight bottom chord, with vertical hangers between the two chords.

bollman truss
A bridge truss with tension rods that radiate from the top of the two end posts to the bottom of the evenly spaced vertical chords; the roadbed is supported between the bottoms of two such trusses.

bowstring truss
A truss with one curved member in the shape of a bow and a straight or cambered member, which ties together the two ends of the bow.

curved truss
A truss, with either a flat howe or pratt configuration, but in a curved profile.

howe truss
A truss having upper and lower horizontal members, between which are vertical and diagonal members; the vertical web members take tension, and the diagonal web members are under compression.

king truss
A triangular truss with a single vertical king post that connects the apex of the triangle with the middle of the horizontal tie beam.

lattice truss
A truss consisting of upper and lower horizontal chords, connected by web members which cross each other, usually stiffened by joining at the intersection of the braces.

panel truss
A structural truss having rectangular divisions with diagonal braces between opposite corners.

pratt truss
A statically determinate truss, consisting of straight top and bottom chords, regularly spaced vertical compression members, and diagonal tension members; used for medium to long spans in buildings and for small bridges.

scissors truss

A type of truss used to support a pitched roof; the ties cross each other and are connected to the opposite rafters at an intermediate point along their length.

suspended arch truss

A truss with a straight upper chord and an arched lower chord, with vertical members between the two chords.

vierendeel truss

A steel open web truss composed of rectangular panels without diagonals, and with rigid joints between all the members.

warren truss

A truss having parallel upper and lower chords, with connecting members which are inclined, forming a series of approximately equilateral triangles.

Trussed

Facades with decorative forms derived from trusses that support the structure either horizontally or vertically; featuring triangular patterns or diagonal bracing, expressed in exterior materials, either subtly or boldly.

Trussed

Trussed arch See Arch

Trussed beam See Beam

Truth window

An exposed section of a wall that reveals the layered components within it that usually would not be seen once the wall is finished.

Trylon

A tall, narrow, three-sided pyramid; the term was used to describe the symbol of the 1939 New York World's Fair, the Trylon and Perisphere.

Bernard, Tschumi (1944–)

Swiss-born American architect, one of seven identified with Deconstructivism. His best-known work is the Parc de La Villette (illus.), Paris, France (1990), a series of red toy-like sculptural forms set into the intersection points of a large grid that covered the entire site.

Tube

A long, hollow cylinder, or other hollow shape; typically a flexible or thin-walled metal cylinder, as opposed to a pipe.

Tube structure

Structural system for a tall building which considers the columns and spandrels on the facade as forming a pierced tube that is cantilevered from the ground.

Tubular scaffold

A scaffold for both interior and exterior work, made of tube steel. Tubular scaffolds are lightweight, offer low wind resistance, and are easily assembled and dismantled. They are available in several lengths for varying heights and types of work.

Tubular scaffolding

Scaffolding that is built up from galvanized-steel or aluminum tubes, which are held together with clamps.

Tubular skylight
Round skylight that transmits sunlight down through a tube with internally reflective walls, even through an attic space; it delivers daylighting through a ceiling light diffuser. Most tubular skylights are 12 to 16 inches in diameter and deliver daytime lighting comparable to several 100-watt incandescent light bulbs.

Tuck pointing
The repairing of worn or damaged mortar joints, by raking out the old mortar and replacing with fresh mortar.

Tudor arch See **Arch**

Tudor Revival style (1910–1940)
A residential style loosely based on English Tudor style architecture; typical elements include asymmetrical massing; steeply pitched, cross-gabled roofs; half-timbered patterns on upper exterior walls; tall, multipaned casement windows; massive chimneys with chimney pots; and Tudor arches. A typically asymmetrical style, with the exterior clad in brick or stucco and employing a false half-timbering treatment. It typically employs steeply pitched gables with little overhang at the eaves; bargeboards on the gables; tall, massive, elaborate chimneys; and narrow casement windows usually set with a number of small, diamond-shaped panes, often within a Tudor arch.

Tudor style (1485–1547)
The final development of English Perpendicular Gothic architecture, during the reigns of Henry VI and Henry VIII, preceding Elizabethan architecture, and characterized by the use of four-centered arches.

Tulipwood See **Wood**

Tumbling course
A sloping course of bricks that are set perpendicular to a straightline gable in Dutch Colonial architecture, in imitation of a similar brick construction found in medieval houses in Flanders.

Tungsten bulb
A British term for incandescent bulb, named for the tungsten filament that the bulb contains.

Turkish dome See **Dome**

Turnbuckle
A coupling between the ends of two rods, with one having a left-hand and the other a right-hand thread. Rotation of the buckle adjusts the tension in the rods. A simpler type has only one right-hand thread, and a swivel at the other end.

Turned
Having a circular cross section that has been produced by turning on a lathe.

Turnkey system
A complete system supplied by a single supplier.

Turret
A diminutive tower, characteristically projecting out on corbels from a corner of the structure.

Tuscan order
One of the five classical orders; a simplified version of the Roman Doric order, with fewer and bolder moldings, unfluted columns, and a plain frieze, and without triglyphs.

Twist
A feature with a curve or turn, specifically a curved stair railing that makes a radial turn with the change of direction. Also, a distortion in a wood member caused by the turning of the edges of a board so that the four corners of any face are no longer in the same plane.

Twisted column See **Column**

Two-centered arch See **Arch**

Two-dimensional graphics
The process of drawing and displaying graphical objects in two dimensions, using elements such as lines and circles.

Two-hinged arch See **Arch**

Two-point perspective See **Perspective projection**

Two-tiered porch
A porch whose first and second stories are similar; each floor is supported by a separate row of columns.

Two-way switch
An electrical switch used to control a light or lights from two locations; two two-way switches are required, one in each location.

Tympanum
The triangular space between the horizontal and sloping cornices immediately above the opening of a doorway or a window, or the space between the lintel above a door and the arch above.

U u

U.S. Department of Agriculture
Provides aid for the rehabilitation of historic structures in rural areas; manages and interprets historic areas in national forests.

U.S. Department of Commerce
Administers economic development and small business programs that can be used for the renovation, reuse, and management of historic buildings and commercial revitalization programs.

U.S. Department of Defense
Component services administers historical records, maintains museums, and issues publications. The Army Corps of Engineers, a branch of the DoD, recovers archeological data from project sites.

U.S. Department of Energy
Responsible for research, development, and demonstration of energy, technology, and energy conservation, including solar technology.

U.S. Department of Housing and Urban Development
Administers housing, community development, and neighborhood programs, including aid to rehabilitate historic residential structures, home ownership and rental assistance, an urban homesteading program, community Development Block Grants, and Urban Development Action Grants for revitalization of distressed cities and urban counties.

U.S. Department of Labor
Administers ongoing and special public employment programs that may aid rehabilitation of historic properties.

U.S. Department of the Interior
The principal federal preservation and conservation agency. Administers National Parks, national historic sites and other public lands, as well as the major federal preservation program.

U.S. Department of Transportation
Establishes national transportation policy, including highway planning, and supervises programs for urban mass transit; railroads; aviation; and the safety of waterways, ports, highways, and pipelines. Administers the Transportation Act of 1966, which protects historic sites from being adversely affected by transportation projects.

U.S. Department of Treasury
Develops and administers regulations affecting the tax aspects of federal preservation tax incentives and other tax issues affecting cultural property.

U.S. Green Building Council (USGBC)
National, nonprofit organization founded in 1993 whose mission is to accelerate the adoption of green building practices, technologies, policies, and standards. USGBC established the LEED (Leadership in Energy and Environmental Design) Certification guidelines. USGBC is com-

U.S. Green Building Council
posed of more than 12,000 organizations from across the building industry that are working to advance structures that are environmentally sustainable, profitable, and healthy places to live and work. Members includes building owners, real estate developers, facility managers, architects, designers, engineers, general contractors, subcontractors, product and building system manufacturers, government agencies, and nonprofit organizations. Leaders from every sector of the building industry work to promote buildings that are environmentally responsible, profitable, and healthy places to live and work.

U-bolt
A steel bar bent into U-shape and fitted with screw threads and nuts on each end.

U-factor
Measure of the heat conducted through a given product or material—the number of British thermal units (Btu) of heat that move through a square foot of the material in one hour for every 1 degree Fahrenheit difference in temperature across the material. U-factor is the inverse of R-value.

Ultimate strength
The highest load that a piece can sustain before failing.

Umbra penumbra
The shadow volume behind an object lit by an area light source (in contrast to a point light source) that does not have sharp boundaries, given that each point in the boundary area is only partially shadowed. The area or volume in full shadow is the umbra, the boundary area the penumbra.

Umbrella shell roof See Roof

Unbraced frame
A structural framework in which the resistance to lateral load is provided by the bending resistance of its structural members and their connections.

Unbundled parking
A parking strategy in which parking spaces are rented or sold separately, rather than automatically included with the rent or purchase price of a residential or commercial unit. Tenants or owners can purchase only as much parking as they need and are given the opportunity to save money and space by using fewer parking stalls. Unbundled parking is more equitable and can reduce the total amount of parking required for the building.

Unconditioned space

Area within the outermost shell of a house that is not heated or cooled (that is, the area outside the thermal envelope). Such areas typically include crawlspaces, attics, and garages.

Underbuilt

Construction that is of insufficient strength or quality.

Undercoat

A coat of paint applied on new wood or over a primer or over a previous coat of paint; improves the seal and serves as a base for the top coat, for which it provides better adhesion.

Undercut

In stonework, to cut away a lower part, leaving a projection above that serves the function of a drip. To rout a groove back from the edge of an overhanging member.

Underfloor air distribution

A system using an underfloor plenum (that is, the open space between the structural concrete slab and the underside of a raised floor system) to deliver conditioned air directly into the occupied zone of the building. Air is delivered through supply outlets typically located at floor level or integrated as part of the office furniture and partitions. Return grilles are located above the occupied zone. This upward convection of warm air is used to efficiently remove heat loads and contaminants from the space. This system saves 10 percent more energy than a traditional ceiling-based system, reduces ductwork lengths and sizes, and allows for individual vent control.

Underfloor electrical duct

A raceway consisting of a sheet-metal enclosure within a floor; usually a slab-on-grade, with provisions for access through their finished floor.

Underfloor heating

Heating provided below the finished floor by electric cables or hot-water pipes; they are usually cast into a concrete slab.

Underlay

Material laid directly on the floor to act as a padding; generally for the installation of carpet.

Underlayment

A layer of roofing felt, composition board, or other thin sheet material that covers the subfloor as a base for a finish floor.

Underpin

To provide a new foundation for a wall or column in an existing building without removing the superstructure.

Underpinning

The rebuilding or deepening of the foundation of an existing building to provide additional or improved support, as the result of an excavation in adjoining property that is deeper than the existing foundation.

Underwriters Laboratories (UL)

Independent laboratories that test appliances to determine if they meet certain established safety requirements. The UL stamp indicating approval should be affixed to any fixture that meets the requirements.

Undressed stone See **Stone**

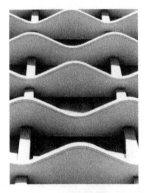

Undulating

Forms that have a wave-like character or depict sinuous motion with a wavy outline or appearance.

520

UNESCO
An organization whose function it is to protect the world's landmarks, through administration of the International Council on Monuments and Sites (ICOMOS) and the World Heritage Committee.

Ungers, Oswald Mathias (1926–2007)
German architect and theorist known for his public buildings in Germany. Ungers only completed one project in the United States—the German ambassador's residence in Washington, DC. His use of geometric forms was unadorned and rigorous, such as in the German Architecture Museum in Frankfurt, the state library in Karlsruhe, the Alfred Wegener Institute for Polar and Marine Research in Bremerhaven, and the Family Court building in Kreuzberg.

Unglazed tile See Tile

Uniform load
A load that is evenly distributed over an area.

Uniformity
The state of being identical, homogeneous, or regular.

Unit construction
A construction method which includes two or more preassembled walls, together with floor and ceiling construction, ready for shipment to the building site.

Unity
A oneness and absence of diversity; a combination or arrangement of parts and the ordering of all the elements in a work so that each one contributes to a total single aesthetic effect.

Universal design
A design that makes a building accessible to as many individuals as possible, including older people and those with physical handicaps.

Unvented gas heater
Gas-burning space heater that is not vented to the outdoors. Although unvented gas heaters burn very efficiently, indoor air quality experts strongly discourage their use because combustion gases, including high levels of water vapor, are released into the room or building.

Upjohn, Richard (1802–1878)
English-born American architect remembered primarily as a church architect and as a Gothic Revivalist. His most well-known building is Trinity Church, New York City (1841). He was the first president of the American Institute of Architects, which he helped to found.

Uplight
A fixture designed to project light up toward the ceiling. They can be tall or short; the tall ones are called torchieres.

Urbahn, Max (1912–1995)
Designed the Vehicle Assembly Building at the Kennedy Space Center, Cape Canaveral, FL (1966).

Urban design
Design of public urban environments.

Urban design
The aspects of architecture and city planning that deal with the design of urban structures and space.

Urban growth boundary
A boundary that identifies the urban lands needed during a specified planning period to support urban development densities and separate these lands from rural lands.

Urban heat-island effect
A microclimate created by the extra energy emissions and absorbed solar heat in a city. Buildings in urban areas retain far more heat into the night than the natural vegetation of rural areas, making it more difficult for urban areas to cool. This greatly increases amount of energy needed and consumed for air conditioning.

Urban infill
Redevelopment of sites in the core of metropolitan areas for commercial and residential purposes.

Urban planning
The design of urban structures and spaces.

Urban renewal
Revitalization of established urban areas to provide for a greater range of housing, employment, and social activities. The principal focus is to transform old and neglected buildings into vibrant communities with viable investment opportunities through effective planning, infrastructure improvements, and partnerships among government, the community, and local businesses.

Urban renewal
The improvement of a slum, or a deteriorated or underused area of a city; the rehabilitation of relatively sound structures, as well as conservation measures to arrest deterioration.

Urban sprawl
Uncontrolled spread of urban and suburban development farther and farther away from the urban core.

Urbanization
The process of change from a rural area to an urban area.

Urea-formaldehyde binder
Interior-grade, formaldehyde-based binder used for particleboard, medium-density fiberboard, and hardwood plywood. This type of binder generally has higher formaldehyde emissions than a phenol-formaldehyde binder.

Urea-formaldehyde foam insulation
A material once used to conserve energy by sealing crawl spaces and attics; no longer used because emissions were found to be a health hazard.

Urethanes
A family of plastics used for varnish coatings, foamed insulations, highly durable paints, and rubber goods.

Urn
A vase of varying size and shape, usually having a footed base or pedestal, and used as a decorative device; originally, to contain ashes from the dead.

Usable floor area
Space that can actually be occupied by a user.

Utopian architecture (1960-1993)
A style of architecture called "fantastic" or "visionary," produced without the constraints of clients, budgets, materials, or building and planning regulations. It is produced in the form of drawings or models that transcend limitations but are unlikely to be constructed, at least in the foreseeable future.

Utopian architecture

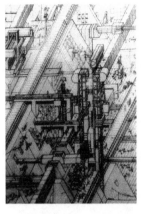

Utzon, Jorn (1918–2008)
Son of a naval architect. While many of his projects evoked the sea, none did so more than the nautical Sydney opera house with its five prow-like roof structures bowing toward the harbor. Utzon resigned from the project in 1966, and it was completed by government-appointed architects who made major alterations to his interior design. The building has become the symbol of modern Australia. Utzon also designed Kuwait City's tent-like National Assembly building, the Lutheran Church at Bagsvaerd near Copenhagen, as well as the Houses at Fredensborg, North Zealand, Denmark, and the Kingo Houses, at Elsinore, Denmark. He was the 2003 winner of the Pritzker Architecture Prize.

U-value
A measure of the amount of heat that flows in or out of a substance under constant conditions when there is a 1-degree difference between the air within and outside a building. U-values are used in determining the performance of a glazing system or window assembly.

V v

Vacancy rate
The ratio of the currently vacant area to the total available space.

Vacated street
A public thoroughfare which is abandoned through appropriate official action by a public authority.

Valance lighting
A method of indirect lighting, by concealing the fixture behind a suspended valance, whereby the light can be directed upward or downward.

Valley
The lower trough or gutter formed by the intersection of two inclined planes of a roof.

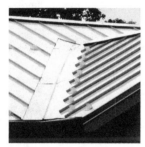

Valley flashing
Pieces of lead, tin, or sheet metal worked in with shingles or other roofing materials along the valley of a roof.

Valley rafter See Rafter

Valley shingles
Shingles which have been cut at the proper angle so that they will fit correctly along a valley.

Valley tile
Trough-shaped building tile made for use in valleys of roofs.

Value
The amount of light reflected by a hue. The greater the amount of light, the higher the value.

Value analysis
A technique used to examine the evolving design in order to achieve design objectives as economically as possible; it usually involves critiques of the design and subsequent analysis of alternative ideas.

Value engineering
An organized activity in which building systems, equipment, design features, and materials are analyzed in order to attain the lowest building life-cycle cost while maintaining the stated functional and performance goals including quality, reliability, and safety.

Value recovery
Redirecting materials typically targeted for landfill or incineration into practical end-use products; energy could be one of those products.

Van Alen, William (1883–1954)
American architect who designed the Chrysler Building, New York City (1930), in a Moderne style, which combined massive forms, inspired by ancient architecture, with Art Deco motifs and streamlined shapes. He used steel and aluminum to celebrate the machine age. The Art Deco upper part incorporates eagle-head and radiator-cap gargoyles, as well as a series of semicircular forms recalling the design of hubcaps.

Van de Velde, Henri (1863–1957)
Dutch Art Nouveau artist and architect.

Van der Rohe, Ludwig Mies (1886–1969)
Designed the German Pavilion, Barcelona Exhibition (1929), built on one level with carefully articulated space, all in high-quality materials. The Tugendhat House, Brno, Czech Republic (1930), had the living space divided only by screen walls. Mies moved to the United States in 1938, designing Crown Hall, the architecture school of the Illinois Institute of Technology (1956). Many tall glazed office buildings in the world bear his influence, such as the Lake Shore Drive Apartments, Chicago, IL (1948); and the Seagram Building, NYC (1958), designed with Philip Johnson. It was considered the culmination of Mies's streamlined style; it was a rectangular slab of bronze, marble, and gray-tinted glass.

Van der Rohe, Ludwig Mies

Vanbrugh, Sir John (1664–1726)
An architect with no formal training, who designed England's largest and most flamboyant Baroque country houses, with bold massing and dramatically varied skylines. Works include Blenheim Palace, Oxfordshire.

Vane
A metal banner that turns around a pivoted point, moving with the prevailing wind, to indicate the direction of the wind.

Vanishing point See **Perspective projection**

Vapor barrier
Airtight skin of polyethylene of aluminum that prevents moisture from warm damp air in a building from passing into a colder space, causing condensation.

Vapor diffusion
Movement of water vapor through a material; water vapor can diffuse through even a solid material if the material's permeability is high enough.

Vapor profile
A vapor profile is an assessment of the relative vapor permeabilities of each individual component in a building assembly and a determination of the assembly's overall drying potential and drying direction based on vapor permeabilities of all of the components. The vapor profile addresses not only how the building's enclosure assembly protects itself from getting wet, but also how it dries out when it gets wet.

Vapor retarder
Layer that inhibits vapor diffusion through a building envelope. Examples include polyethylene sheeting, foil facing, kraft paper facing on batt insulation, and low-permeability paints. Most building codes define a vapor retarder as 1 perm or less, with many common vapor retarders being significantly less than 1 perm.

Variable-air-volume (VAV)
A method of modulating the amount of heating or cooling effect that is delivered to a building by the HVAC system. The flow of air is modulated rather than the temperature. VAV systems typically consist of VAV boxes that throttle supply airflow to individual zones, some mechanism to control supply-fan flow to match box demand, and the interconnecting ductwork and components.

Variance
The waiving of a zoning restriction, such as the width of a side yard, for a particular site, typically approved by a Board of Zoning Adjustments, or similar body upon appeal by an owner for hardship reasons.

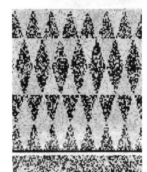

Variegated
Said to be of a material or surface that is irregularly marked with different colors.

Variety
The state or quality of having varied or diverse forms, types, or characteristics.

Varnish
A resin dissolved in oil or spirit, which dries to a brilliant, thin, protective film.

Vasari, Giorgio (1511–1574)
The Uffizi Palace, his only significant work, is Mannerist in style.

Vase

A decorative vessel, usually higher than it is wide; used as an architectural ornament, usually on a pedestal, balustrade, or an acroteria on roofs.

Vault

An arched roof or ceiling or a continuous semicircular ceiling that extends in a straight line over a hall, room, or other partially enclosed space.

barrel and groin vault

A vault formed by two identical tunnel-shaped vaults that intersect in the middle.

barrel vault

A masonry vault resting on two parallel walls having the form of a half cylinder, sometimes called a tunnel vault.

compound vault

Any vault formed by the intersection of two or more vaults; types include cloister vault, domical vault, groin vault, and segmental vault.

conical vault

A vault having a cross section in the form of a circular arc, which is larger at one end than the other.

corbel vault

A continuous corbel arch over a space, used by the ancient Mayas of Yucatan; also known as a Mayan arch.

cross vault

A vault formed by the intersection of two barrel vaults which meet at right angles.

domical vault
A dome-shaped vault, where the ribs or groins are semicircular, causing the center of the vaulted bay to rise higher than the side arches.

double vault
A vault, usually domical, consisting of an inner shell separated from a higher outer shell.

fan vault
A concave conical vault, whose ribs, of unequal length and curvature, radiate from the springing like the ribs of a fan.

groined vault
One covering a square bay where two barrel vaults of equal diameter and height intersect.

hemispherical vault
Masonry dome with a semicircular cross section.

net vault
Vault where the ribs form a network of lozenges.

panel vault
A rib vault having a central square panel connected with diagonal ribs to the corners of the larger square it covers.

polygonal vault
A vault with more than four intersecting vault surfaces; typically octagonal in plan.

rampant vault
A vault whose two abutments are located on an inclined plane, such as a vault supporting or forming the ceiling of a stairway; the impost on one side is higher than the impost on the other side.

ribbed vault
A vault in which the ribs support, or seem to support, the web of the vault.

stellar vault
A vault where the ribs are so arranged as to form a star-shape.

tracery vault
A type of solid vaulting with decorative ribs forming patterns on the surface.

Vault capital
The capital of a pier that supports a vault or a rib thereof.

Vault rib
An arch under the soffit of a vault, appearing to support it.

Vaulted
Constructed as a vault.

Vaulting course
A horizontal course made up of abutments or the springers of a vaulted roof.

Vaulting shaft
A colonnette in a membered pier that appears to support a rib in a vault.

Vegetated filter strip
A band of closely growing vegetation, usually grass, planted between pollution sources and downstream receiving water bodies that are designed to filter, slow flow velocities, and infiltrate runoff.

Vegetated roof
A roof that is partially or completely covered with vegetation and soil, or a growing medium, planted over a waterproofing membrane. These roofs reduce heating and cooling loads, reduce the urban heat-island effect, increase roof lifespan, reduce storm-water runoff, filter pollutants and CO_2 out of the air, filter pollutants and heavy metals out of rainwater, and increase wildlife habitat in built-up areas. Also referred to as a green roof, eco-roof, or living roof.

Vegetated swale
An open drainage channel lined with vegetation designed to detain or infiltrate stormwater runoff.

Velarium
A large tent-like arrangement drawn up over an amphitheater to protect spectators from the sun.

Veneer
The covering of one material with thin slices of another to give an effect of greater richness.

Veneered construction
A method of construction in which a facing material is applied to the external surface of steel, reinforced concrete, or frame walls.

Veneered walls
A wall with a masonry facing which is not bonded, but is attached to a wall so as to form an integral part of the wall for purposes of load bearing and stability.

Venetian arch See Arch

Venetian blind
A window shading device, composed of a series of thin slats, which can be turned at an angle to block out the view; they can be raised and lowered very easily.

Venetian door See **Door**

Venetian mosaic See **Mosaic**

Venetian window See **Window**

Vent pipe
A flue or pipe connecting any interior space in a building with the outer air for purposes of ventilation; any small pipe extending from any of the various plumbing fixtures in a structure to the vent stack.

Vent stack
A vertical pipe connected with all vent pipes carrying off gases from a building. It extends through the roof and provides an outlet for gases and contaminated air and also aids in maintaining a water seal in the trap.

Ventilated facade
A special type of curtain wall consisting of two glazed facades separated by a gap through which ambient air is allowed to flow. The flow of air removes a large amount of solar heat gain that would ordinarily enter the building, resulting in a reduction in space cooling needs and energy consumption. Also known as double facades, double-skin facades, and ventilated cavity curtain walls.

Ventilating duct
General ductwork involved with the process of supplying or removing air by natural or mechanical means, to or from any space.

Ventilating skylight
An opening in a roof that is glazed with transparent or translucent material that admits light or diffused light to the space below. The frame that holds the glazing is capable of being adjusted.

Ventilation
The supply of clean outdoor air to a space for the purpose of cooling; a process of changing the air in a room by either natural or artificial means; any provision for removing contaminated air or gases from a room and replacing it with fresh air.

Ventilation control by occupants
The ability of building occupants to control ventilation rates. A strategy for giving control of comfort back to occupants, this can be achieved through access to individual electronic controls or by operable windows in workspaces.

Ventilation effectiveness
The system's ability to remove pollutants generated by internal sources in a space, zone, or building. In comparison, air change effectiveness describes the ability of an air distribution system to ventilate a space, zone, or building.

Ventilation rate
The rate at which indoor air enters and leaves a building. Expressed as the number of changes of outdoor air per unit of time: air changes per hour, or the rate at which a volume of outdoor air enters in cubic feet per minute.

Ventilation systems
Ventilation circulates air within the building, as well as exchanges the inside air with the outside air. Ventilation systems control temperature and humidity levels, and remove airborne bacteria, odor, and dust. The two types of ventilation are natural and mechanical.

mechanical ventilation
A forced ventilation method that circulates the air, removes odors, and controls humidity within the building. It is often used in wet areas such as food preparation rooms and bathrooms to control odor. Ceiling and window fans or portable ventilation devices are used to circulate the air within the space. They cannot be used for air replacement unless a clear indoor/outdoor circulation pattern is established.

natural ventilation
A method that uses operable windows and direct outside air circulation, when the temperature, wind, precipitation, humidity, and pollution levels are acceptable. The amount of natural ventilation depends on the type, shape, placement, and size of the building and its openings. There are two primary natural ventilation methods: cross-ventilation and stack-ventilation. Cross-ventilation depends on wind-driven breezes, whereas stack-ventilation uses air density differences to create air movement across a building.

Ventilator
In a room or building, any device or contrivance used to provide fresh air or expel stale air.

Venturi, Robert (1925-)
An American Postmodernist who set up practice with John Rausch (1930-) and later with wife Denise Scott Brown (1930-), and later still with Steven Izenour (1930-). Early work included the Vanna Venturi House, Philadelphia, PA (1963); Franklin Court, Philadelphia, PA (1976); Gordon Wu Hall (illus.), Princeton University, NJ (1983); Seattle Art Museum (illus.), Seattle, WA (1991); and the Museum of Contemporary Art, San Diego, CA (1996).

Veranda
Similar to a balcony but located on the ground level; it can extend around one, two, or all sides of a building.

Veranda

Verde antique
A dark green serpentine rock marked with white veins of calcite that takes a high polish; used for decorative purposes since ancient times; sometimes classified as a marble.

Verdigris
The green copper carbonate formed on copper roofs and statues that are exposed to the atmosphere; the patina it produces can be carefully controlled.

Verge
Edge of a sloping roof that overhangs a gable.

Vergeboard
An ornamental board hanging from a projecting roof; a bargeboard.

Vermiculated
Ornamented by regular winding, wandering, and wavy lines, as if caused by the movement of worms.

Vermiculated masonry
A form of masonry surface, incised with discontinuous wandering grooves resembling worm tracks; a type of ornamental winding frets or knots on mosaic pavements, resembling the tracks of worms.

Vermiculite
A generic name for treated minerals which are used for insulation and fire protection, often as an aggregate in plaster or concrete.

Vernacular
In architecture, vernacular buildings reflect the traditional architecture of the region originally developed in response to the climate, land conditions, social and cultural preferences, scenery, and locally available resources and materials. The forms are native or peculiar to a particular country or locality. It represents a form of building that is based on regional forms and materials, primarily concerned with ordinary domestic and functional buildings, rather than commercial structures.

Vernacular architecture
Architecture that makes use of common regional forms and materials at a particular place and time; usually modest and unpretentious, and often a mixture of traditional and modern styles, or a hybrid of several styles.

Vertex
The highest point of a structure; the apex or summit.

Vertical
Pertaining to anything, such as a structural member, which is upright in position, perpendicular to a horizontal member, and exactly plumb.

Vertical clearance
The unobstructed vertical height between objects in any given space.

Vertical pivoting window See Window

Vertical shore
A pair of vertical struts used to provide temporary support to a wall.

Vertical siding
A type of siding consisting of matched boards 10 and 12 inches wide, or they may be of random widths, the joints being either a V-cut or covered with battens.

Vertical sliding window See Window

Vestibule
An intermediate chamber or passage located between the entrance and interior of a building that serves as a shelter or transitional element from exterior to interior space.

Vest-pocket park
A miniature park, usually in an urban area, that is built on a small plot of land.

Vestry
Attached building to a church, where the vestments and sacred vessels are kept; also called a sacristy.

Viaduct
An elevated structure supported on a series of arches, piers, or trestles that carries a roadway or railway over a valley or other depression.

Victorian (1837–1901)
Any style used during the reign of Queen Victoria typical styles of the period include Eastlake, Gothic Revival, Italianate, Queen Anne, and Richardsonian Romanesque.

Victorian architecture
A term that encompasses a number of ornate and highly decorative architectural styles, such as High Victorian Italianate, Shingle, Victorian Romanesque, Gingerbread, Queen Anne, and Gothic Revival.

Victorian Gothic style (1860–1890)
A colorful style, wherein materials of different colors and textures are juxtaposed, creating decorative bands and highlighting corners, arches, and arcades. Materials most often used are ornamental pressed bricks, terra-cotta tile, and incised carvings of foliated and geometric patterns. Openings have straight heads as well as pointed Gothic arched heads. In timber-framed buildings, the gable, porch, and eave trim are massive and strong.

Victorian Romanesque style (1870–1890)
A polychromatic exterior combined with the semicircular arch highlight this style. Different colored stone or brick for window trim, arches, quoins, and belt courses contrasts with the stone wall surface. Decorative bricks and terra-cotta tiles are also used. Round arches are supported by short polished stone columns. Foliated forms, including grotesques and arabesques, decorate the capitals and corbels. Windows vary in size and shape.

Victorian Romanesque style

Victory

A female deity of the ancient Romans or the corresponding deity the Greeks called Nike; representation of the deity, usually a woman in wind-blown draperies, holding a laurel wreath, palm branch, or other symbolic object.

Vierendeel truss See **Truss**

Vignette

A French design for an iron balconet, used as a protection at window openings.

Vignette

Vignola, Giacomo Barozzi de (1507–1573)

Born in Vignola, Italy, he became a leading architect in Rome following Michelangelo's death. He wrote the *Rules for the Five Orders of Architecture* in 1562.

Vihara

A Buddhist or Jain monastery in Indian architecture.

Villa

In Roman and Renaissance periods, a country seat with its dwellings, outbuildings, and gardens, often quite elaborate; in modern times, a detached suburban or country house of some pretension.

Village

A small group of houses and related facilities surrounded by countryside; typically smaller than a township.

Viñoly, Rafael (1944–)

Uruguayan-born American architect who founded the firm Rafael Viñoly Architects PC in 1983 in New York City. His first major project was the John Jay College of Criminal Justice, completed in 1988. In 1989, he won an international competition to design the Tokyo International Forum, completed in 1996. His firm's design was one of the finalists in the World Trade Center design competition. The firm expanded to include affiliate offices in London and Los Angeles as well as offices around the world, including the newest in Abu Dhabi and Bahrain.

Vinyl See **Plastic**

Vinyl-asbestos tile See **Tile**

Viollet-Le-Duc, E. Emmanuel (1814–1879)

Architect and medievalist; appointed to head the Ecole des Beaux-Arts in 1863; wrote a *Dictionary of Architecture* in 1854.

Virtual

Pertaining to conceptual rather than physical.

Virtual office

A workplace that is determined by where the employee happens to be engaged in carrying out work at a particular time; usually remote from the main workplace.

Virtual reality

The simulation of the real world in virtual space by computer programs, allowing for the virtual interaction of users, walking through a computer-generated environment.

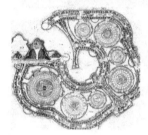

Visionary architecture

A term which applies to any imaginary scheme featuring fantastic or futuristic structures. Images from *Visionary Architecture: Unbuilt Works of the Imagination,* by Ernest Burden (McGraw-Hill 2000). Drawings by Paolo Soleri (below), Gilbert Gorsky (top right), and Lebbeus Woods (bottom right),

Visionary architecture

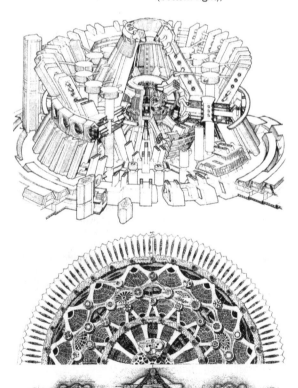

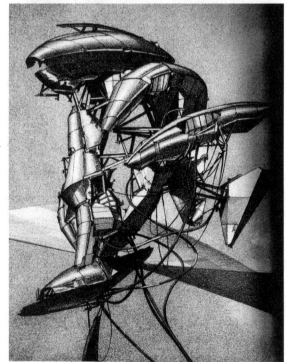

Visionproof glass See **Glass**

Visor roof See **Roof**

Visual pollution
Anything that is offensive to the sense of sight based on its concept, from garbage dumps to garish neon, or roadside signs; it is a subjective matter.

Visualization
The process of displaying realistic visual images of an object for evaluation.

Vitrified clay tile
Pipes and fittings made of clay baked hard and then glazed so that they are impervious to water; used especially for underground drainage.

Vitruvian
In the style of architecture recorded by Vitruvius, a first century B.C. Roman architect and engineer, who wrote about the components and rules of design in a series of ten books, *De Architectura*; translations of this treatise have been a reference for designers in the Classical styles.

Vitruvian opening
A wall opening, as described by Vitruvius as one with jambs that slope inward at the top.

Vitruvian scroll
A series of scrolls connected by a stylized wave-like continuous band; also called a wave scroll.

Vitruvius Pollio, Marcus (46–30 B.C.)
A Roman architect, engineer, and architectural theorist. His treatise, *De Architectura*, the only one from antiquity to survive, was published in Italy and became a basic Renaissance sourcebook.

Volatile organic compound (VOC)
Any organic compound that evaporates at room temperature and is hazardous to human health, causing poor indoor air quality. Many VOCs found in homes, such as paint strippers and wood preservatives, contribute to sick building syndrome because of their high vapor pressure. VOCs are often used in paint, carpet backing, plastics, and cosmetics. The U.S. Environmental Protection Agency (EPA) has found concentrations of VOCs in indoor air to be, on average, two to five times greater than that in outdoor air. During certain activities, indoor levels of VOCs may reach 1,000 times that of the outside air.

Volcanic stone See **Stone**

Voltage drop
The difference in voltage between the beginning and the end of a circuit, depending on the amount of resistance of the conductors.

Volute
A spiral, scroll-like ornament having a twisted or rolled shape, found most often on the capital of the Ionic column.

Vomitory
An entrance or opening, usually one of a series, piercing a bank of seats in a theater or stadium, permitting entry or egress by large numbers of people.

Voussoir
A wedge-shaped block whose converging sides radiate from a center forming an element of an arch or vaulted ceiling.

Voussoir

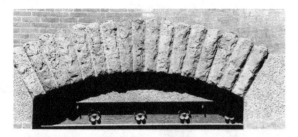

stepped voussoir
A voussoir that is squared
along its upper surfaces
so that it fits horizontal
courses of masonry units.

Vries, Hans Vredeman de (1526–1606)
Flemish architect and painter whose engravings, derived
from Serlio, and architectural pattern books were very influential on northern Europe.

V-shaped joint See **Mortar joints**

W w

Waffle slab
A two-way ribbed slab.

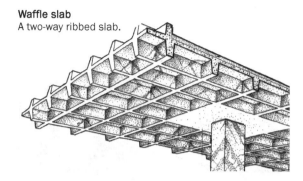

Wagner, Otto (1841–1918)
Austrian architect whose architecture predated the Art Deco style. He designed the Majolika Haus (illus.), Vienna, Austria (1898), an original Art Nouveau building in which a floral design covers the facade. He designed the Sezession Exhibition Building (illus.), Vienna (1899); and also the Postal Savings Bank, Vienna (1904), which had a high vaulted central hall with tapering metal supports.

Wainscot
A protective or decorative facing applied to the lower portion of an interior partition or wall, such as wood paneling or other facing material.

Walker, Ralph (1889–1973)
An American architect who was best known for his Art Deco skyscrapers, including the Barclay-Vesey Telephone Building (1923) and the Irving Trust Building (1929), both in NYC.

Walker, Thomas Ustick (1804–1887)
American architect of German descent; pupil of William Strickland and John Haviland; the second president of the American Institute of Architects. Work includes the U.S. Capitol, Washington, DC. The Grand College for Orphans was his most notable work. He also worked on John McArthur's design for the Philadelphia City Hall, PA.

Walking tour
Visits on foot by a group of people interested in viewing the architecture of a specific locale; may be organized by a cruise line tour company, local guides, preservationists, architects, or individuals. Specific sites of significance are pointed out, with names of the architects and dates of construction or other significant facts.

Walk-off mat
Design strategy for reducing the amount of contaminants introduced into an interior space by providing grating or other material to remove contaminants from shoes. A significant portion of contaminants in a building are brought in this way, impacting indoor environmental quality.

Walk-up apartment
Apartment without an elevator, usually five stories or less.

Walkway
A passage or lane designated for pedestrian traffic.

Wall
A structure that encloses or subdivides a space with a continuous surface; except where fenestration or other openings occur.

balloon frame wall
A system of framing a wooden building wherein the exterior bearing walls and partitions consists of single studs that extend the full height of the frame from the top of the soleplate to the roof plate.

bearing wall
Supports any vertical load in addition to its own weight.

cant wall
A wall canted in elevation from true vertical.

cantilever retaining wall
A wall retaining soil that acts as a cantilevered beam as opposed to one acting as a continuous beam spanning between supports.

cavity wall
An exterior wall, usually of masonry, consisting of an outer course and an inner course separated by a continuous air space connected by metal ties.

curtain wall
A method of construction in which all building loads are transmitted to a metal skeleton frame, so that the non-load-bearing exterior walls of metal and glass are simply a protective cladding.

exterior wall
A wall that is part of the envelope of a building thereby having one face exposed to the weather or to earth.

fire wall
Any fire-resistant wall that separates one building from another or that subdivides a large building into smaller spaces; it is usually continuous from the foundations extending above the roof.

foundation wall
A wall below, or partly below grade, to provide support for the exterior walls or other parts of the structure.

gable wall
A wall which continues to the roofline on the gable end of a structure.

gravity retaining wall
Retaining wall that relies on the weight of the masonry or concrete for its stability.

half-timbered wall
Descriptive of buildings of the 16th and 17th centuries, which were built with strong timber foundations, supports, knees, and studs, and whose walls were filled in with plaster or masonry materials such as brick.

interior wall
Any wall within a building; entirely surrounded by exterior walls.

load-bearing wall
A wall capable of supporting an imposed load in addition to its own weight.

masonry wall
A load-bearing or non-load-bearing wall consisting of hollow masonry units.

non-load-bearing wall
A wall subject only to its own weigh and wind pressure.

partition
An interior wall dividing a room or part of a building into separate areas; may be either non-load-bearing or load-bearing.

party wall
A wall used jointly by two parties under an easement agreement, erected upon a line dividing two parcels of land, each one a separate real estate entity; a common wall.

retaining wall
A wall, either freestanding or laterally braced, that bears against earth or other fill surface and resists lateral and other forces from the material in contact with the side of the wall.

screen wall
A movable or fixed device, especially a framed construction designed to divide, conceal, or protect, but not to support.

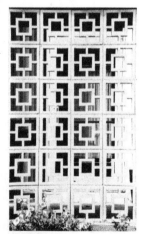

serpentine wall
A wall with a plan in the shape of a wavy line with alternating arcs of circles.

shear wall
A wall that resists shear forces in its own plane due to wind or earthquake forces.

spandrel wall
A wall built on the extrados of an arch, filling the spandrels.

Wall anchor
A type of anchor used to tie the walls to the floors and hold them firmly in place.

Wall box dimming
Popular in residential applications and commercial spaces where flexibility in illumination levels is desired. Incandescent dimming is inexpensive and simple to apply. Plate controls include linear slide, rotary, and touch.

Wall chase
A square or rectangular recess in a wall to accommodate pipes, heating ducts, and similar equipment.

Wall column See **Column**

Wall cornice
A cornice at the top of a masonry wall.

Wall dormer
A dormer window with its front wall flush with, or part of, the main building wall below.

Wall dormer See **Dormer**

Wall gable See **Gable**

Wall hanger
A support of steel or cast iron, partially built into a wall for carrying the end of a structural timber, when the timber itself is not to be built into the wall.

Wall mural
A large painting using a wall, typically blank, for a canvas.

Wall plate
A horizontal piece of timber, laid flat along the top of the wall at the level of the eaves, which carries the rafters.

Wall sconce
Fixtures mounted on a wall. They are both functional and decorative, and have a long history of centuries of use.

Wall shaft
Any engaged column or colonnette that is supported by a corbel or bracket and appears to support a vault rib or clustered ribs above it.

Wall tile
Thin tile glued to the wall with mastic and with grouted joints. Types include glazed ceramic, terra-cotta, glass, mosaic, or plastic; may be any geometric shape.

Wallboard
Large rigid sheets, made of gypsum, wood chips, or other filler material, that is fastened to the frame of a building to provide a surface finish.

Wall-mounted display fixture
Wired and mounted directly into a built-in cabinet or bookcase. The fixtures are usually visible.

Wall-mounted lamp
A fixture attached permanently to a wall with a stationary or movable swing-arm.

Wallpaper
A decorative wall and ceiling finish composed of printed sheets glued to the surface; wood pulp paper was the kind first used.

Wank Adams Slavin Associates (1890)
American historic preservation firm in New York City. Projects include Frank Lloyd Wright's Fallingwater, (1937), Bear Run, PA; Richard Morris Hunt's The Breakers, West Palm Beach, FL; the Municipal Building, designed by McKim Mead and White, restored (1983); Louis Sullivan's Bayard-Condict Building, New York City, restored (2001).

Ware, William R. (1832–1915)
Partner in the firm of Ware and Van Brundt. Both were pupils of Richard Morris Hunt in New York. Designed Memorial Hall, Harvard (1880), and the Stack addition to Gore Hall (1876), Harvard University, Cambridge, MA. He was the first professor in the first U.S. school of architecture at MIT (1865), and later he set up the architecture school at Columbia University, New York. The firm designed Union Station (1875), Worcester, MA, abandoned after disuse in 1950, and restored in 2001.

Warehouse
A place in which goods and merchandise are stored; a storehouse. Built usually of masonry, there was little need for windows, and as a result most restorations need to cut new windows or enlarge those in the existing structure.

Warm color
Red, orange, and yellow; optically they tend to advance.

Warm roof
A roof constructed with its insulating layer above the roof space rather than a cold roof which is at the ceiling level.

Warnecke, John Carl (1919–2010)
The son of architect Carl I. Warnecke, he rose to prominence as a leading urban contextualist, where a building relates to its neighbors and the general locale through its architectural details, historical overtones, and climatic considerations. Warnecke had developed a new master plan for the Naval Academy at Annapolis, MD, and was working with President John F. Kennedy on early designs for his presidential library when Kennedy was killed. He was then called upon to design the John F. Kennedy gravesite at Arlington. His firm's large-scale projects included the AT&T Long Lines Building in Manhattan; the Soviet Embassy and Hart Senate Office Building, both in

Warnecke, John Carl

Washington, DC; and the South Terminal at Logan Airport in Boston, MA. He also designed the Hennepin County Government Center (illus.), Minneapolis, MN (1976), consisting of twin 24-story towers with an enclosed 24-story atrium between them.

Warp
Distortion in the shape of a plane timber surface, due to the movement of moisture; may be caused by improper seasoning.

Warpage
The change in the flatness of a material caused by differences in the temperature or humidity on the opposite surfaces of the materials.

Warped
Any piece of timber that has been twisted out of shape and permanently distorted during the process of seasoning.

Warren truss See Truss

Warren, Whitney (1864–1943)
American architect trained in Paris at the École des Beaux-Arts. He partnered with Charles D. Wetmore (1867–1941), and they became known for their designs for railroad

Warren, Whitney

stations, which included Grand Central Terminal, with Reed and Stem, New York City. Also designed the New York Yacht Club (1898), NYC.

Waste management
The process of collecting, recycling, and disposing of waste materials produced by human activities. Waste management involves solid, liquid, and gaseous substances, some of which can be hazardous. As such, each requires a different method and procedure to process. Solid waste, such as wood, concrete, glass, drywall, and asphalt shingles, is the primary type of waste produced by buildings.

Waste management plan
A plan that addresses the collection and disposal of waste generated during construction or renovation, usually including the collection and storage of recyclable materials.

Waste management recycling
The process of extracting resources from waste, and/or adding additional use value to waste. The various methods to recycle waste materials include extraction, reprocessing, conversion, and reuse. Although most building material waste contains no harmful components, many states have strict regulations about its transportation, storage, and disposal. Common waste products include metal, lumber, masonry, glass, plastic, paper, appliances, asphalt, paints, and landscape-related materials. Most of these materials can be reused and/or recycled.

Waste pipe
The pipe that discharges liquid waste into the soil drain.

Waste reduction
Using source reduction, recycling, or composting to prevent or reduce waste generation.

Waste stack
A vertical pipe in a plumbing system which carries the discharge from any fixture.

Wastewater

Used water from toilets, showers, sinks, dishwashers, clothes washers, and other sources in the home, including all contaminants, which can either flow into a municipal sewer system or be treated with an onsite wastewater disposal system.

Watch tower

An elevated structure of any type used as a lookout; may be a separate structure or rise above the other portions of a building or wall.

Water and wastewater infrastructure

The network of pipes, systems, and facilities that provide fresh water supply and wastewater management for communities.

Water balance

An accounting of the inflow to, outflow from, and storage in a hydrologic unit.

Water conservation

Preservation and careful management of water resources.

Water efficiency

LEED (Leadership in Energy and Environmental Design) Rating System category. Prerequisites and credits in this category focus on the water efficiency of a building.

Water efficiency labeling scheme (WELS)

Provides mandatory water-efficiency labeling for appliances, including showerheads, washing machines, flow controllers, dishwashers, toilets, taps and urinals. WELS-labeled products are given a water efficiency star rating from 1 to 6.

Water harvesting

Collection of both runoff and rainwater for various purposes, such as irrigation or fountains.

Water management systems

For any building to be considered "green," water management issues should be managed using sustainable, ecological, and high-performance methods and approaches.

bioretention

Adds landscaping and vegetation to treat stormwater runoff; bioretention systems are often called rain gardens.

composting toilets

Using little or no water, composting toilets are designed to convert human waste into compost via aerobic decomposition and can be installed anywhere. Composting toilets require that the solids be removed periodically, depending on the size of the unit.

graywater systems

Graywater is collected wastewater from domestic processes, such as bathing, washing dishes, or laundry, and reused for flushing toilets, watering landscapes, and irrigation. Graywater makes up three-quarters of domestic wastewater, and is different from "blackwater," which is heavily polluted and has high levels of biological contaminants and toxic chemicals.

harvesting

Reducing the amount of runoff water before it can cause flooding.

infiltration

Collects and filters stormwater into an aquifer.

waterless urinals

Urinals that use cartridge inserts filled with a sealant liquid which collects and traps the urine without releasing the odor. The cartridge with sealant is replaced periodically.

Water reclamation

Reuse of effluent from wastewater treatment facilities through irrigation, land application, or other recycling methods.

Water side economizer

A water side economizer reduces energy consumption in cooling mode by allowing the chiller to be turned off when the cooling tower alone can produce water at the desired chilled water set point. The cooling tower system, rather than the chiller, provides the cold water for cooling.

Water stain
Discoloration on converted timber, caused by water.

Water table
A horizontal offset in a wall sloped on the top to throw off water running down the wall.

Water tower
An elevated structure located above the roof of a building to create sufficient pressure to supply the fixtures within the building. Also, can be freestanding and serving entire communities.

Water-based paint
Any paint that can be thinned with water. Includes oil-bound or emulsion paints, whose binder is insoluble in water, but which can be thinned with water.

Water-efficient landscaping
Using native plant species and landscape designs appropriate to the local climate reducing the amount of watering needed for maintenance. In dry climates this strategy can be particularly beneficial.

Waterfront
Improved or unimproved land abutting on a body of water, such as a lake, harbor, or the like. There is a high degree of waterfront revival today that would not have been possible before.

Waterproof adhesive
An adhesive that forms a bond that will withstand the full exposure to the weather and is unaffected by microorganisms.

Waterproofing
A coating or membrane applied to a surface, such as a foundation wall, to prevent the intrusion of water under pressure; materials may include asphalt, felt, tar, or various synthetic membranes.

Water-repellent preservative
A liquid designed to penetrate wood and repel water and a moderate preservative protection. It is used for millwork, such as sash and frames, and is usually applied by dipping.

Water-resistant adhesive
An adhesive that forms a bond that will retain practically all of its strength when subjected to wetting and drying.

Water-resistive barrier
Sometimes also called the weather-resistive barrier, this layer of any wall assembly is the material interior to the wall cladding that forms a secondary drainage plane for water that makes it past the cladding. This layer can be building paper, housewrap, or even a fluid-applied material.

WaterSense
Program developed and administered by the U.S. Environmental Protection Agency (EPA) to promote and label water-efficient plumbing fixtures. WaterSense-labeled products are analogous to Energy Star–labeled products and will perform well, help save money, and use less water.

Watershed
Area of land that, as a result of topography, drains to a single point or area.

Watertight
An enclosure that does not permit the passage of moisture.

Wattle and daub
A primitive construction consisting of a coarse basketwork of twigs woven between upright poles, and plastered over with mud.

Watts per square foot
A shorthand measure of the energy use of a building, often applied to indoor lighting. Energy codes often limit the watts per square foot based on building type and function.

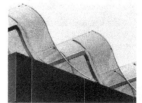

Wavy
Refers to forms that are arranged into curls or undulations, or any graphic representation of curved shapes, outlines or patterns that resembles such a wave.

Weatherboarding

Wood siding commonly used as an exterior covering on a frame building consisting of boards with a rabbeted upper edge that fits under an overlapping board.

Weathered

Descriptive of a material or surface that has been exposed to the elements for a long period of time; having an upper surface that is splayed so as to allow water to drain off.

Weathered siding See **Wood products**

Weathering

An inclination given to the surface of horizontal joints in masonry construction to prevent water from collecting in them.

Weatherization

Cost-effective energy-efficiency measures for existing residential and multifamily housing. When weatherization includes the house as a system, it is often called whole-house weatherization. It may involve caulking, weather stripping, adding insulation, and other similar improvements to the building shell.

Weatherproof

A general term indicating the ability to withstand natural elements.

Weather-struck joint See **Mortar joints**

Weathervane

A metal form, fixed on a rotating spindle that turns to indicate the direction of the wind, usually located on top of a spire, pinnacle or other elevated position on a building.

Webb, Philip (1831–1915)

Arts and Crafts architect and designer, who championed the Gothic Revival style and paved the way for the Arts and Crafts movement in architecture. Famous for comfortable country houses, furniture, tapestries, and stained glass.

Weep hole

A small opening in a wall or window member, through which accumulated condensation or water may drain to the building exterior, such as from the base of a cavity wall flashing or a skylight.

Weese, Harry (1919–1998)

Principal of the Chicago-based firm which designed office buildings for Chicago. He also designed the metro subway stations for Washington, DC. Weese is also well known for his firm advocacy of historic preservation and was remembered as the architect who shaped Chicago's skyline and the way the city thought about everything from the lakefront to its treasure-trove of historical buildings.

Weil, Martin (1941–2009)

Restoration architect and founding member of the Los Angeles Conservancy, responsible for the restoration of many Southern California historic properties including the El Capitan Theatre, Pasadena City Hall, Watts Towers, and the Frank Lloyd Wright–designed Storer House in Hollywood Hills. Weil helped found the L.A. Conservancy in 1978, which prevented the demolition of the Los Angeles Central Library.

Welded beam See **Beam**

Well opening

A floor opening for a stairway.

Wells, Malcolm (1926–2009)

A pioneer of the green roof movement, Wells refocused his priorities after realizing the pavilion he designed for RCA

Wells, Malcolm

at the 1964 World's Fair would soon be torn down and consigned to a landfill. He then looked for ways to build in harmony with the land and to leave as light a footprint on it as possible. He designed numerous homes and offices with green roofs, most with passive heat.

Welt

A seam in flat metal roof coverings where the edges are folded and dressed down; also called single and double welts, depending on the number of folds.

Western Stick style (1890–1920)

An adaptation of the Stick style characterized by a gently pitched gable roof that extends well beyond the wall, and by projecting balconies. A unique feature is the exposed stick-like rafters that project along the roof eaves. Window lintels, railings, and other beams extend beyond vertical posts. Pegs were used to join the members, and the ends were rounded off, as were corners of posts and beams. The exterior was finished in wood shingles.

Weston, Roy F. (1911–2007)

Founder and CEO of Weston Solutions, a large Pennsylvania-based environmental engineering firm and an early voice for sustainability. In the 1930s, Weston was a sanitary engineer, developing multidisciplinary strategies to address issues of public health and environmental reclamation. Weston and his wife Madeleen established the Roy F. Weston Center for Sustainability at his alma mater, the University of Wisconsin at Madison, to define, solve, and prevent environmental problems.

Wet laboratory

A laboratory where chemicals, drugs, or other materials or biological matter is tested and analyzed, and which requires water, direct ventilation, and specialized piped utilities.

Wet rot

A fungus that feeds on wood and destroys wet timber; most often found in cellars, neglected external joinery, and rafter ends. Causes timber to soften, darken and develop cracks along the grain and lose strength.

Wet weather green infrastructure

The infrastructure associated with stormwater management and low-impact development that encompasses approaches and technologies to infiltrate, evapotranspire, capture, and reuse stormwater to maintain or restore natural hydrologies.

Wetland

An area that is inundated or saturated by surface or groundwater at a frequency and duration sufficient to support a prevalence of vegetation typically adapted for life in saturated soil conditions. Wetlands generally include swamps, marshes, bogs, and similar areas.

Wexler, Donald (1926–)

A minimalist modernist, Wexler pioneered prefab construction when he designed steel houses for the Alexander Construction Company. From his influence, Palm Springs, California, became the center of midcentury Modernism.

Wheel window See Window

White cement

A pure white Portland cement; since the gray color of Portland cement comes from impurities, the white cement requires raw materials of low iron content. It is used for decorative surface finishes and is more expensive than ordinary cement.

White lead

An opaque white pigment, used extensively as an undercoat for exterior paint. Because it is toxic, it is now rarely used,

White oak See Wood

White pine See Wood

White, George M. (1921–2011)

The architect who oversaw myriad federal projects on Capitol Hill, including the construction of the Hart Senate Office Building and the restoration of the old Supreme Court and Senate chambers in the United States Capitol itself.

White, Howard (1870–1936)

American architect and partner in the successor firm to Daniel H. Burnham in Chicago: Graham, Anderson, Probst, and White.

White, Stanford (1853–1906)

Partner in the firm of McKim Mead and White. The buildings produced by the firm were the most appreciated of their time. He designed the Washington Square Arch in New York City, as well as houses for Louis Comfort Tiffany, Charles Dana Gibson, and Joseph Pulitzer.

Whole-house air conditioner
Air-conditioning system that serves an entire house; cooled air is delivered through a central system of ducts.

Whole-wall R-value
Average R-value of a wall, taking into account the thermal bridging through wall studs.

Wide-flange section
A structural section whose cross section resembles the letter H rather than the letter I; it is used for columns due to its capacity to avoid rotation or buckling.

Widow's walk
A rooftop platform or narrow walkway, used originally as a lookout for incoming ships in colonial coastal houses.

Wigwam
Eastern native American dwelling, round or oval in plan, with a rounded roof consisting of a bent pole framework covered by pressed bark or skins.

Wildscaping
Retention of native soil, vegetation, and other natural features when building on land, rather than the removal of soil, vegetation, and natural features followed by artificial landscaping once the building is completed.

Williams, Frank (1937–2010)
An architect of many New York City skyscrapers, including Trump Place and 515 Park Avenue. He designed high-rise luxury apartment towers for Zeckendorf, as well as Donald Trump, Madison Equities, and the Monian Group. Williams also designed the W Hotel at Times Square, the residential portion of World Wide Plaza on Ninth Avenue, and the Four Seasons Hotel on East 57th Street in collaboration with I. M. Pei. Williams was a consultant to the Regional Plan Association in New York City and, with Rai Okamoto, authored 1969's *Urban Design Manhattan,* an influential study on managing growth in Midtown Manhattan.

Williams, Paul Revere (1894–1980)
A Black American architect who became renowned for designing major buildings such the Los Angeles International Airport and homes in Southern California, including mansions in Hollywood.

Williams, Wayne R. (1919–2007)
Southern California Modernist architect who designed homes, commercial buildings, and schools throughout the booming Southern California basin with post-and-beam style with large expanses of glass and natural light. Commercial projects included the Blaisdell Medical Building and Friend Paper Co., in Pasadena, and the 1956 Mobil gas station, Anaheim.

Williams-Ellis, Sir Bertram Clough (1883–1978)
Williams-Ellis devoted his life to the cause of environmental preservation. His work on the Italian Style resort village of Portmeirion, a popular tourist village in Gwynedd, North Wales, represented his efforts to prove that it was possible to build beautiful and colorful housing without defiling the natural landscape.

Wimberly, George J. (1916–1996)
An architect known for his work in Honolulu and for his firm's designs of resorts. He was principal in the architectural firm of Wimberly, Allison, Tong & Goo. Wimberly came to Hawaii in 1940 as a journeyman architect doing naval work at Pearl Harbor. After the war he worked with Howard Cook in the architectural firm of Wimberly and Cook. The rehabilitation of the Royal Hawaiian Hotel in Honolulu was one of his first jobs. His work is typified by the use of local materials such as coral stone, lava rock, wood beams, thatch, bamboo, and glass; local forms such as flowing indoor/outdoor open spaces sheltered by big dramatic roofs with big eaves; and liberal use of figurations, patterns, and motifs derived from the cultures of the Pacific. Wimberly was instrumental in founding the Pacific Area Travel Association in 1952 with Bill Mullahey, the regional chief of Pan American Airlines in the 1950s. He found new hotel opportunities in Australia, New Zealand, Tahiti, Fiji, Jakarta, Singapore, and Bali.

Wind brace
Any brace, such as a strut, which strengthens a structure or framework against the wind; a diagonal brace that ties rafters of a roof together to prevent racking.

Wind energy
Electricity produced when wind is captured by turbines and converted into electricity. This is the cheapest, and fastest growing, renewable energy technology.

Wind farm
A vast tract of land covered with wind-powered turbines that are used to drive generators that produce electricity.

Wind load
The positive or negative force of the wind acting on a structure; wind applies a positive pressure on the windward side of buildings and a negative suction to the leeward side.

Wind power
A form of renewable energy generated by wind-spinning turbines.

Wind turbine
A device that converts the kinetic energy of the wind into mechanical energy that can be used to drive equipment such as pumps. The addition of a generator allows the wind's kinetic energy to be converted into electricity. There are two types of wind turbines: horizontal axis turbines, with blades that rotate about a horizontal axis, and vertical axis turbines, with blades rotate about a vertical axis.

Windbreak
One or more rows of trees or shrubs planted in such a manner as to provide shelter from the wind and to protect soil from erosion. Windbreaks around a home can reduce the cost of heating and cooling and save energy. Windbreaks are also planted to help keep snow from drifting onto roadways and even yards. Other benefits include providing habitat for wildlife, and in some regions the trees are harvested for wood products.

Winders
Treads of steps used in a winding staircase or when stairs are carried around curves or angles.

Winding stair See Stair

Windlass
A device for hoisting materials consisting usually of a horizontal cylinder turned by a lever or crank. A cable attached to the material winds around the cylinder as the crank is turned, raising the load to whatever position is desired.

Windmill
A tower structure with wind-powered vanes connected by a rotating shaft to a pump or generator for pumping water and generating electricity.

Window
An opening in an exterior wall of a building to admit light and air, usually glazed; an entire assembly consisting of a window frame, its glazing, and any operating hardware. The window has seen a significant increase in performance using new technologies, including double and triple panes, low-E coatings and gas-filled windows, which improve the insulation value. While high-performance windows may cost slightly more, the energy savings often result in a rapid payback.

angled bay window
A bay window that protrudes out over a wall and is triangular in plan.

awning window
A window consisting of a number of top-hinged horizontal sashes one above the other, the bottom edges of which swing outward; operated by one control device.

bay window
A window forming a recess in a room and projecting outwards from the wall either in a rectangular, polygonal or semicircular form. Some are supported on corbels or on projecting moldings.

bay window

bent window

A window that is curved in plan, typically with a bent sash; the jambs are typically parallel or radial.

blank window

A recess in an exterior wall, having the external appearance of a window; a window that has been sealed off but is still visible.

bow window

A rounded bay window projecting from the face of a wall; in plan it is a segment of a circle.

bow window

box-head window

A window constructed so that the sashes can slide vertically up into the head to provide maximum opening for ventilation.

bungalow window

A double-hung window with a single light in the bottom sash and rectangular divided lights in the upper sash.

cabinet window

A type of projecting window or bay window for the display of goods in shops.

camber window

A window arched at the top.

cant window
A bay window erected on a plan of canted outlines; the sides are not at right angles to the wall.

casement window
A window ventilating sash, fixed at the sides of the opening into which it is fitted, which swings open on hinges along its entire length.

Chicago window
A horizontal window consisting of a large square fixed central pane with narrow vertical sliding sashes on either side, typically the full width of the bay, as in the Carson Pirie & Scott store by Louis Sullivan.

circular window
A window in the shape of a full circle, often with decorative elements, and arranged in a radial manner.

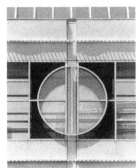

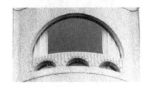

compass window
A rounded bay window that projects from the face of a wall; also a window having a rounded semicircular member at its head.

coupled window
Two closely spaced windows which form a pair.

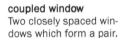

crippled window
A dormer window.

double lancet window
A window having mullions shaped to form two lancet windows that are side by side; found in Carpenter Gothic, Collegiate Gothic, and Tudor Revival styles.

dormer window
A vertical window that projects from a sloping roof, placed in a small gable.

double window
Two windows, side by side, which form a single architectural unit.

double-hung window
A window having two vertically sliding sashes, each closing a different part of the window; the weight of each sash is counterbalanced for ease of opening and closing.

eyebrow window
A bottom-hinged, inward-opening sash located in the window of an eyebrow dormer.

false window
The representation of a window that is inserted in a facade to complete a series of windows or to give the appearance of symmetry.

folding casement
One of a pair of casements, with rabbeted meeting stiles, which is hung in a single frame without a mullion and hinged together so that it can open and fold in a confined space.

French window
A type of casement window, similar to a door, where the sash swings from the jamb of the opening.

frieze-band window
One of a series of small windows that form a horizontal band directly below the cornice; usually continuing across the main facade.

gable window
A window in a gable; a window shaped like a multi-curved gable.

high-light window
A window or row of windows set high up in the wall; also called a clerestory window.

hopper window
A window sash which opens inward and is hinged at the bottom; when open, air passes over the top of the sash.

jalousie window
A window consisting of a series of overlapping horizontal glass louvers, which pivot simultaneously in a common frame and are actuated by one or more operating devices.

lancet window

A narrow window with a sharp pointed arch that is typical of English Gothic architecture; one pane shaped in the form of a lancet window.

oriel window

A bay window corbeled out from a wall of an upper story; a projecting bay that forms the extension of a room, used extensively in medieval English residential architecture.

landscape window

A double-hung window whose upper sash is highly decorated with small panes of colored glass; the lower sash contains a larger pane of clear glass.

lattice window

A window casement, fixed or hinged, with glazing bars set diagonally.

loop window

A long, narrow, vertical opening, usually widening inward, cut in a medieval wall, parapet, or fortification, for use by archers.

lozenge window

A window composed of lozenge-shaped panes set on the diagonal.

lucarne window

A small dormer window in a roof or spire.

oval window

A window in the shape of an ellipse, or in a shape between an ellipse and a circle.

operable window

A window which may be opened for ventilation, as opposed to a fixed light.

ox-eye window
A round or oval aperture, open, louvered, or glazed; an oculus or oeil-de-boeuf.

peak-head window
A window with a triangular head, most often found in Gothic Revival church architecture; also called a lancet window.

picture window
A large fixed pane of glass, often between two narrower operable windows, usually located to present the most attractive view to the exterior.

pivoting window
A window having a sash which rotates about fixed vertical or horizontal pivots, located at or toward the center, in contrast to one hung on hinges along an edge.

ribbon window
One of a horizontal series of windows, separated only by mullions, which form a horizontal band across the facade of a building.

rose window
A large, circular medieval window, containing tracery arranged in a radial manner.

roundel
A small circular panel or window; an oculus, a bull's eye, or a circular light like the bottom of a bottle.

sash window
A window formed with glazed frames that slide up and down in vertical grooves by means of counterbalanced weights.

shed dormer window
A dormer window whose eave line is parallel to the eave line of the main roof instead of gabled to provide more attic space.

single-hung window
A window with two sashes, only one of which opens.

sliding sash window
A window that moves horizontally in grooves or between runners at the top and bottom of the window frame.

splayed window
A window whose frame is set at an angle with respect to the face of the wall.

square-headed window
A window that has a straight horizontal lintel above it.

stained-glass window
A window with colored glass.

stained-glass window

stepped windows
A series of windows, usually in a wall adjacent to a staircase, arranged in a stepped pattern that generally follows the step's profile.

storm window
An auxiliary window, usually placed with the existing window in the same frame, to provide additional protection against severe weather.

top-hung window
A casement window hinged horizontally.

triple-hung window
A window with three vertical sliding sashes that allow the window to open to two-thirds of its height.

Venetian window
A large window, characteristic of Neoclassical styles, divided by columns or piers resembling pilasters into three lights and sometimes arched at the head.

vertically pivoted window
A window having a sash which pivots about a vertical axis at or near its center; when opened, the outside glass surface is conveniently accessible for cleaning.

wheel window
A large circular window in which the tracery radiates from the center; a variety of the rose window.

Window casing
The finished frame surrounding a window; the visible frame; usually consisting of wood, metal, or stone.

Window casing

Window crown
The upper termination of a window; such as a hood or pediment; also called a window cap.

Window frame
The fixed, nonoperable frame of a window, consisting of two jambs, a head, and a sill, designed to receive and hold the sash or casement and all necessary hardware.

Window head
The upper horizontal cross member or decorative element of a window frame.

Window head

Window light
A pane of glass which has been installed in a window frame.

Window lock
A hardware device that locks a window in a closed position; most often a sash lock

Window mullion
A vertical member between the lights of a window.

Window muntin
A rabbeted member for holding the edges of windowpanes within a sash.

Window opening
Any open space in a wall where a window is to be placed.

Window schedule
A table, usually located on the elevation drawings, which gives the symbol for each type of window. This code symbol is placed on the drawing by each particular window; the quantity, type rough opening, sash size, and manufacturer's number are also placed in the schedule. A remarks column gives information on the specific type of glass that goes in the sash.

Window seat
A seat built into the inside bottom of a window.

Window shading
Any device for reducing unwanted heat gain from a window.

Window unit
A complete window, with sashes or casements, ready for shipment or installation in a building.

Window stool
In a window trim, the nosing directly above the apron; the horizontal member of the window finish which forms a stool for the side casings and conceals the window sill.

Window valance
A spring that counterbalances the weight of a vertically sliding window.

Window wall
A type of curtain wall, usually composed of vertical and horizontal metal framing members containing fixed lights, operable windows, or opaque panels, or a combination thereof.

Windowpane
One of the divisions of a window or door, consisting of a single unit of glass set in a frame.

Windowsill
The horizontal member at the base of a window opening.

Window-to-floor ratio
The ratio of total, unobstructed window glass area to total floor area served by the windows, expressed as a percentage. This value can also be subdivided by solar orientation, such as the ratio of south-facing window area to floor area.

Windward
On the side exposed to the wind; the opposite of leeward.

Winery
A building on an estate that houses the production and processing of wine. Many early wineries were modest frame structures that perished long ago, but by the 1880s, stone wineries were being built.

Wines, James (1932–)
American artist/architect associated with environmental design, Wines is also a product designer and educator. He began his career as a sculptor and graphic designer, with a gallery in Rome and the Marlborough Gallery in New York City. He founded SITE (Structures in the Environment) Environmental Design in 1970, design firm of architecture, environmental art, interior design, and public-space and landscape-architecture projects, including ones sponsored by large corporations. His municipal clients have included the cities of Hiroshima, Yokohama, Toyama, Seville, Vienna, Vancouver, Le Puy en Velay, Chattanooga, and New York City. See also SITE.

Wing
Projection on the side of a building that is smaller than the main mass; often one of a symmetrical pair.

Winged bull
A winged human-headed bull of colossal size, usually in pairs, guarding the portals of ancient Assyrian palaces as a symbol of force and domination.

Winter garden
A large greenhouse, such as a courtyard with a glazed roof, for the permanent display of plants; often includes paved paths, fountains and sculptures.

Wire connector
In electrical wiring, devices used to join two wires; also called a wire nut.

Wire glass See Glass

Wire-cut brick See Brick

Wireframe model
In computer graphics, a three-dimensional display consisting of lines or polygons without any surface rendering.

Wiremold®
A trademarked two-piece metal molding system for electric wiring, installed on the exterior surface of walls.

Wireway
A metal wiring enclosure, usually rectangular in cross section, with a removable front or top cover to protect the conductors inside it. Also called a raceway.

Wiring
The connecting of electrical conductors to form circuits to fixtures, outlets, and equipment.

Witch's cap See Roof

Withers, Frederick C. (1828–1901)
American architect in Newburgh, NY; associated with Andrew Jackson Downing, Calvert Vaux, and Frederick Law Olmstead. Leading designer of High Victorian Gothic architecture and churches; Dutch Reformed Church (1859), Beacon, NY; Gallaudet College (1885), Washington, DC; Jefferson Market Courthouse, NYC (1878); and Chapel of the Good Shepherd (1888), NYC.

Wood
The hard, fibrous substance that composes the trunk and branches of a tree, lying between the pitch and the bark.

artificial wood
Any of the various mixtures that are molded to simulate wood; often using sawdust, paper, or other wood fiber as a major ingredient mixed with glue.

bald cypress
A deciduous softwood tree resistant to decay and often used in contact with the soil and for exposed elements such as wood shingles; also used for flooring and trim.

balsam fir
A softwood tree with coarse-grained wood, used for interior trim.

balsam poplar
A large hardwood tree, with soft straight-grained wood used for painted millwork.

bevel siding
Tapered boards used as siding, installed with the thinner part at the top.

birch
A moderately strong, high-density wood, yellowish to brown in color; its uniform texture is well suited for veneer, flooring, and turned wood products.

burl
A decorative pattern in wood caused by adjacent knots.

cedar
A highly aromatic, moderately high-density, fine-textured wood of a distinctive red color with white streaks; widely used for fence posts, shingles, and mothproof closet linings.

cherry
An even-textured, moderately high-density wood, rich red-brown in color; takes a high luster, and is used for cabinetwork and paneling.

chestnut
A light, coarse-grained, medium-hard wood, used for ornamental work and trim.

clapboard
One of a series of boards used for siding, with a tapered cross section, most commonly called beveled siding.

clapboard siding
A wood siding commonly used as an exterior covering on a building of frame construction, applied horizontally and overlapped, with the grain running lengthwise, thicker along the lower edge than the upper.

conifer
A tree belonging to the botanical group which bears cones; it includes all the softwoods used in building, particularly the pines and firs.

cypress
A moderately strong, hard, and heavy softwood; its heartwood is naturally decay-resistant, and is used for exterior and interior construction where durability is required.

dimensional timber
Rough-sawn wood with a rectangular or square cross section that exceeds the nominal dimensions of 4 by 5 inches.

Douglas fir
A strong, medium-density, medium-textured softwood; widely used for plywood and lumber in construction.

ebony
Wood of a number of tropical species, usually distinguished by its dark color, durability, and hardness; used for carving and ornamental cabinetwork.

elm
A tough, strong, moderately high-density hardwood of brown color; often has a twisted interlocked grain; used for decorative veneer, piles, and planks.

fir
A softwood of the temperate climates including Douglas fir, white fir, silver fir, and balsam fir; used for framing and interior trim.

folded plate
A thin skin of plywood reinforced by purlins to form tructures of great strength.

glue-laminated arch
An arch made from layers of wood that are joined with adhesives. The glued joints transmit the shear stresses, so the structure acts as one piece capable for use in structural arches and long-span beams.

gum
A moderately high-density hardwood, whitish to gray-green in color and of uniform texture; used for low-grade veneer, plywood, and rough cabinet work.

hardboard
A dense smooth-surfaced composition board composed of highly compressed fibers; one such type is called Masonite®.

hardwood
Timber from all trees except the conifers, which are called softwood.

heartwood
The center portion of a tree trunk that is no longer growing or carrying the sap; often harder and denser.

hemlock
Wood of a coniferous tree; moisture-resistant, soft, coarse, and uneven-textured; it splinters easily and is inferior for construction use.

hickory
A tough, hard, strong wood; has high shock resistance and high bending strength.

laminated timber
Timber beam or arch manufactured from four or more layers of wood, usually about 1 inch thick, bonded together with waterproof adhesive.

larch
A fine-textured, strong, hard, straight-grained wood of a coniferous tree; heavier than most softwoods.

lath
Narrow strips of wood that serve as a base for plaster, usually nailed to studs in walls or rafters in ceilings.

limba
A straight-grained, fine-textured wood used for interior paneling.

locust
Wood of the locust tree; coarse-grained, strong, hard, decay-resistant, and durable.

mahogany
A straight-grained wood of intermediate density, pinkish to red-brown in color; used primarily for interior cabinetwork and decorative paneling.

maple
A hard, tough, moderately high-density wood, light to dark brown in color, with a uniform texture; used for flooring and wood trim.

Masonite®
Trade name of a brand of tempered pressed board.

oak
A tough, hard, high-density wood; coarse-textured, ranging in color from light tan to pink or brown; used for both decorative and structural and applications, such as framing timbers, flooring, and plywood.

particleboard
A large class of building boards made from wood particles compressed in a binder; often faced with a veneer.

pine
A wood of a number of species of coniferous evergreens. The two classes, soft pine and hard pitch pine are an important source of construction lumber and plywood.

plastic wood
A paste of wood flour, synthetic resin, and a volatile solvent; used for filling holes and cracks in wood, it dries soon after application.

plywood
An engineered panel composed of an odd number of thin sheets permanently bonded together, sometimes faced with a veneer.

redwood
A durable, straight-grained, high-strength, low-density softwood; especially resistant to decay and insect attack; light red to deep reddish-brown; used primarily for construction, plywood, and millwork.

rusticated wood
Wood incised in block shapes to resemble rough stone.

satinwood
A hard, fine-grained, pale to golden yellow wood of the acacia gum tree; used in cabinetwork and decorative paneling.

shiplap siding
Wood sheathing whose edges are rabbeted to make an overlapping joint.

softwood
Wood from trees with needles and that produce cones, typically evergreen; includes cedar, cypress, Douglas fir, hemlock, pine, spruce, and tamarack.

spruce
A white to light brown or red-brown, straight and even-grained wood; moderately low density and strength; relatively inexpensive; used for general utility lumber.

stressed skin panel
A panel constructed of plywood and seasoned lumber; the simple framing and plywood skin act as a total unit to resist loads.

teak
A dark golden yellow or brown wood with a greenish or black cast, moderately hard, coarse-grained, very durable; immune to the attack of insects; used for construction, plywood, and decorative paneling.

tulipwood
A soft, close-textured durable wood that is yellowish in color; used for millwork and veneer.

veneer
A thin sheet of wood that has been sliced, rotary-cut, or sawn from a log; used as one of several plies in plywood for added strength or as facing material on less attractive wood.

white oak
A hard, heavy, durable wood, gray to reddish-brown in color; used for flooring, paneling, and trim.

white pine
A soft, light wood that works easily; does not split when nailed; does not swell or warp appreciably; is widely used in building construction.

yellow pine
A hard resinous wood of the longleaf pine tree, having dark bands of summerwood alternating with lighter-colored springwood; used as flooring and in general construction.

Wood brick
A wooden block, the size and shape of a brick, built into brickwork to provide a hold for nailing finish material; a nailing block.

Wood door See **Door**

Wood frame construction
Construction in which the exterior walls, bearing walls and partitions, floor and roof constructions, and their supports are of wood or other combustible material (compare to heavy-timber construction).

Wood, John (1704–1754)
English architect. Planned Bath, England, in the form of a Palladian-style Roman city, using crescent-shaped terraces.

Wood, John (the younger) (1767–1775)
English architect. Planned the Royal Crescent, Bath, England, on a circular plan.

Wood joint
A joint formed by two boards, timbers, or sheets of wood that are held together by nails, fasteners, pegs, or glue.

Wood shake
Any thick hand-split, edge-grained shingle or clap-board; formed by splitting a short log into tapered sections.

Wood shingle
A roofing unit of wood that is cut to stock dimensions and thicknesses and used as an overlapping covering over sloping roofs and side walls.

Wood shiplap siding See Wood

Wood veneer See Wood

Woods, Lebbeus (1940–)
American architect, artist, and educator. Woods is a revolutionary, experimental, visionary, and theoretical architect. He is the founder of the Research Institute for Experimental Architecture (RIEA), and a Professor of Visionary Architecture at the European Graduate School (EGS) in Saas-Fee, Switzerland. Woods is the author of numerous books including *Lebbeus Woods: System Wein* (2006), *The Storm and the Fall* (2004), and *Lebbeus Woods: Experimental Architecture* (2004).

Worked lumber
A piece of lumber that has been matched, shiplapped, or patterned for use as siding.

Working drawings
The drawings for use by a contractor, subcontractor, or fabricator that form part of the contract documents for a building; they contain the necessary information to manufacture or erect an object or structure.

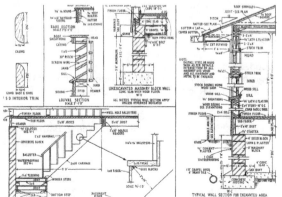

Working load
The normal dead, live, wind, and earthquake load that a structure is required to support in service. Also called the service load.

Workmanship
The quality of work that is executed by a craftsperson or a contractor.

Workplane
The plane at which work is usually done and on which the illuminance is specified and measured. Unless otherwise indicated, this is assumed to be a horizontal plane, 30 inches above the floor.

World Heritage Committee
The branch of UNESCO that administers the World Heritage Fund and selects sites for the World Heritage List.

World Monument Fund
A New York–based nonprofit dedicated to preserving and protecting endangered works of historic art and architecture around the world.

World Monument Watch
A global program launched in 1995 that calls attention to imperiled cultural heritage sites around the world and directs timely financial support to their preservation. A panel of leading international experts selects the List of 100 Most Endangered Sites from nominations submitted to the World Monument Fund every two years by governments and organizations active in the field of cultural preservation.

Wreath ornament See Ornament

Wreathed column See Column

Wrecking ball
A heavy ball of concrete suspended from the boom of a crane and swung against or dropped on a structure to demolish it. They vary in weight, averaging several tons.

Wrecking bar
A steel bar used for prying and pulling nails; one end is slightly bent with a chisel-shaped tip, and the other end U-shaped with a claw tip for pulling nails. Also called a pinch bar.

Wrecking strip
In concrete formwork, small pieces of panel fitted into the assembly in such a way that it can be easily removed ahead of the main panels or forms, making it easier to strip those major form components.

Wren, Sir Christopher (1632–1723)
One of England's greatest scientists and architects. He was active in rebuilding London after the fire of 1666. He rebuilt St. Paul's Cathedral, London, England (1673).

Wren, Sir Christopher

Wright, Frank Lloyd (1867–1959)

Originator of the Organic style, as demonstrated in many innovative works. Unity Temple, Oak Park (illus.), IL (1906), was a concrete church with a complex interior space on several levels. The Millard House, Pasadena, CA (1923), was built of decorative precast concrete blocks. His masterpiece residence, Fallingwater, Bear Run, PA (1937), was cantilevered out over a waterfall in horizontal sections. It was not unlike the European International style in elevation but was three-dimensional in actuality. The Johnson Wax Building, Racine, WI (1937), features the innovative use of materials, such as glass tubes for skylights. He built a second home, office, and school in the desert at Taliesin West (illus.), Scottsdale, AZ (1938). The Guggenheim Museum, (illus.), New York City (1943–1960), is one of his best-known works, with the exterior expressing the interior arrangement of the spiral ramp display area. A third period provided architectural forms based on geometric shapes, such as hexagons, octagons, circles, and arcs. He designed more than 400 buildings and an equal number of unrealized projects.

Wright, Frank Lloyd

Wright, Frank Lloyd

Wright, Lloyd (1890–1978)
American architect, and elder son of Frank Lloyd Wright. Trained in his father's studio, he helped prepare drawings for the *Wasmuth Portfolio* (1918). He worked on several houses including the Barnsdall Residence in Los Angeles. Later work included the Swedenborg Memorial Wayfarer's Chapel, Palos Verdes, CA (1946).

Wrightian style (1900–1959)
The architecture of Frank Lloyd Wright, for which this style is named, was characterized first by the Prairie-style house. The long, low buildings with broad, overhanging, low-pitched roofs and rows of casement windows emphasized the house's horizontal relationship with the site. It culminated in the building of Taliesin, his home and school in Scottsdale, AZ.

Writer's residence
The home of a famous author in history. These homes are important landmarks and are usually noted with plaques, or the residence may be used as a living museum.

Wrought iron
An easily forged iron containing carbon. It can be hammered into shapes, either when hot or cold, and is used as decorative grilles for window openings, entryways, or balcony railings.

Wurster, William (1895–1973)
American architect. As principal of Wurster, Bernardi and Emmons, he produced many works in what was called the Bay Region style. He designed many university buildings in California, and Ghirardelli Square in San Francisco (1962).

Wyatt, James (1746–1813)
British architect; designed Fonthill Abbey (1796), the most exotic of all Gothic buildings in England.

Wyvern See Ornament: animal forms

X x

X-brace See **Brace**

Xeriscape™
Derived from the Greek word "xeros," meaning dry and combined with "landscape," Xeriscape means gardening with less than average water. It is a trademarked term for water-efficient choices in planting and irrigation design that refers to seven basic principles for conserving water and protecting the environment: (1) planning and design, (2) use of well-adapted plants, (3) soil analysis, (4) practical turf areas, (5) use of mulches, (6) appropriate maintenance, and (7) efficient irrigation.

X-mark
A mark made by a construction worker to indicate where certain structural members are to be placed; also used to indicate the face side of a structural member.

XPS extruded polystyrene
Highly insulating, water-resistant rigid foam insulation that is widely used above and below grade, such as on exterior walls and underneath concrete floor slabs. In North America, XPS is made with ozone-depleting HCFC-142b. XPS has higher density and R-value and lower vapor permeability than EPS rigid insulation.

Y y

Y fitting
In pipe work, a short specialized section, one end of which branches or divides, usually at an angle of 45 degrees, resembling the letter Y and forming two separate openings.

Yamasaki, Minoru (1912–1986)
An American architect of Japanese descent; he and his partner George Hellmuth made their mark with the TWA Terminal at Lambert Airport, St. Louis. The Pruitt-Igoe public housing project, also in St. Louis (1954), won several awards, but was detested by its inhabitants and later demolished. He used aluminum grille screens and other intricate detailing in high-rise structures, such as the Michigan Gas Company Building in Detroit (1963). His twin towers for the World Trade Center in NYC (1972) were his landmark structures.

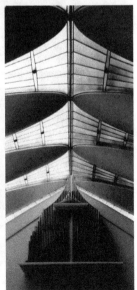

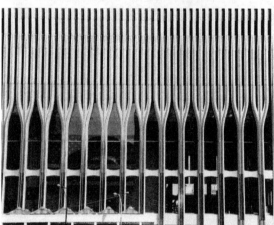

An area of uncultivated ground adjacent to a dwelling. In urban sites, yards are often paved with brick, stone, or tile.

Yard lumber
Lumber that is less than 5 inches thick and is intended for general building purposes.

Yardage
A term applied to cubic yards of earth excavated or installed.

Yeang, Ken (1948–)
Malaysian architect and writer best known for developing environmental design solutions for high-rise buildings in the tropics. As a principal in T. R. Hamzah and Yeang in Kuala Lumpur, Malaysia, since 1975, he has been instrumental in developing passive-mode low energy skyscrapers, called bioclimatic skyscrapers. Yeang is regarded as the father or inventor of the bioclimatic skyscraper, largely the result of his book, *The Skyscraper: Bioclimatically Considered*. Yeang's architecture is based on his theoretical ecological studies, published in *The Green Skyscraper: The Basis for Designing Ecological Sustainable Buildings*. Early experiments include the Roof-Roof House (illus.), in Kuala Lumpur, Malaysia (1985), providing a louvered umbrella structure over the building. As a consequence of his strong beliefs in ecomimicry, Yeang's design projects focus on achieving benign and seamless biointegration that includes reduced or zero dependency on nonrenewable sources of energy, enhanced ecological nexus through devices such as eco-land bridges, ecoundercrofts, vertical landscaping, ecocells, green living walls, ecological corridors and fingers that reach into the landscape and toward the sky at the same time. Some of the devices he uses in his builtforms include light shelves, light pipes, skycourts, vertical linked enclosed green atriums, and windscoops.

Yeang, Ken

Yoke

An arrangement of members used in formwork which encircles beam or column forms to secure them together and prevent movement.

York, Edward P. (1865–1927)

American architect. Employed with McKim Mead and White before setting up a partnership with Phillips Sawyer (1868–1949).

Young, Ammi B. (1798–1874)

American designer of classical and Italianate public buildings. Projects include the Vermont State House (1836), Montpelier, VT; the Custom House (1847), Boston; Reed Hall (1839), Dartmouth College, Hanover, NH; and the Custom House and Post Office (1858), Windsor, VT.

Y-tracery See Tracery

Yurt

A circular tent-like dwelling, usually movable, used by nations of northern and central Asia; constructed of skins stretched over a wooden framework.

Z z

Zapotec architecture (700–900)

This eclectic architecture is found in Oaxaca, Mexico. The Zapotecs assimilated influences from the Olmecs (700–300 B.C.) and especially from Teotihuacan (A.D. 30–900). It culminated in a recognizable regional style, characterized by pyramids having several stepped terraces, accented with balustrades whose tops were decorated. One of the most notable sites is Monte Alban, a carefully planned ceremonial complex.

Zero energy building

A building with a net energy consumption of zero over a typical year, because the energy provided by onsite renewable energy sources is equal to the energy used. Buildings approaching this goal may be called near zero energy build-ings or ultra-low-energy buildings.

Zero-energy design

Describes buildings and products that have no net energy consumption in a given year.

Zero-net-energy building

A building that provides all of its own energy on an annual basis from onsite renewable resources or offsite renewable energy purchases. In this way it is still connected to the grid, providing power when it has a surplus and drawing from the grid when it needs power, such as at night. This approach involves using solar energy for electricity, water heating, and space heating, and employing design measures like passive solar design, natural ventilation, and operable windows for cooling with electric power assist.

Zeidler, Eberhard (1926–)

German-born architect; trained at the Bauhaus and later settled in Canada. Best known for the enormous Eaton Center (illus.), Toronto, Canada (1969); Ontario Place (illus.), Toronto (1968); and Queens Quay Terminal Warehouse (illus.), Toronto (1981); and the Vancouver Convention Center (illus.), Vancouver, Canada (1997).

Zevi, Bruno (1918–1999)

Italian architectural theorist; studied at Harvard University before returning to Italy. He published many books, including *Towards an Organic Architecture* (1945) and the *Modern Language of Architecture* (1973). He was opposed to International Modernism, Postmodernism, Classicism, and Neoclassicism, advocating an organic approach.

Ziggurat

A Mesopotamian temple having the form of a terraced pyramid rising in three to seven successively receding stages in height; built of mud brick, featuring an outside staircase and a shrine at the top.

Zigzag

A line formed by angles that alternately project and retreat; occurring in bands, on columns, and in larger patterns on cornices.

Zimmerman, Bernard (1930–2009)

Architect and educator who cofounded the Los Angeles Institute of Architecture and Design in 2001 and helped launch the Architecture + Design Museum in Los Angeles, CA. He was a passionate modernist and preservationist, working to save numerous Los Angeles landmarks including the Hollywood sign, the Schindler House, and the Watts Towers.

Zinc See **Metal**

Zodiac

The imaginary band of the celestial sphere on either side of the yearly path of the sun, moon, and stars, divided since Babylonian times into 12 segments named after the 12 constellations, each with its own symbol.

Zone

A number of adjacent floors that are served by the same elevators; also applies to spaces that have different requirements for heating or cooling. Also, a space or group of spaces in a building having similar heating and cooling requirements throughout its occupied area, so that comfort conditions may be controlled by a single temperature sensor with corresponding controller.

Zone controls
A control system which uses two or more thermostats.

Zoned air conditioning
Systems with separate thermostat controls in different parts of a structure that allow for independent temperature control of each area.

Zoned heating
A heating system with separate thermostat controls in different parts of a structure to allow for independent temperature control in each area. Also maintaining different temperatures for different areas or zones.

Zoning
Political jurisdictions divided into geographic zones with different mixtures of allowable use, size, siting, and form of real property; typically applied in conjunction with a zoning code or review of permit applications for developments and variances. The allocation of land use by a statutory authority for planning purposes and the legal restriction that deems that part of cities be reserved for particular uses, such as residential, commercial, industrial, and recreational.

Zoning permit
A permit issued by the appropriate governmental authority permitting land to be used for a specific purpose.

Zonolite® concrete
A form of concrete that acts as an insulator.

Zoo
A public park or institution in which living animals are kept and exhibited to the public. In cultural history, the buildings provided for keeping wild animals in captivity have often been of elaborate design. Today, the housing of wild animals places the emphasis on the natural habitat of the species, and provides elements to protect the visitors and spectators.

Zumthor, Peter (1943–)
Swiss architect and winner of the 2009 Pritzker Prize.

Zoomorph
Boulders carved by the Maya with animals and other forms, as distinct from megaliths cut and carved into the form of a stele.

Zoophoric column
A column bearing a figure or figures of one or more men or animals.

Zoomorphism
Representation of gods in the form of animals; also, use of animal forms in art or symbolism.

563

Zoophorus

A horizontal band bearing carved figures, animals or persons, especially a sculptured Ionic frieze.

Zumthor, Peter (1943–)

Zumthor is the son of a cabinetmaker and apprenticed to a carpenter in 1958. In 1968, he became a conservationist architect for the Department for the Preservation of Monuments of the canton of Graubünden. This work on historic restoration projects gave him a further understanding of construction and the qualities of different rustic building materials. Zumthor was able to incorporate this knowledge of materials into Modernist construction and detailing. Projects are the Kunsthaus Bregenz (1997), a shimmering glass and concrete cube that overlooks Lake Constance in Austria; the cave-like thermal baths in Vals, Switzerland (1999); the Swiss Pavilion for Expo 2000 in Hannover,

Zumthor, Peter

Germany an all-timber structure intended to be recycled after the event; the Kolumba (2007), in Cologne, Germany; and the Bruder Klaus Field Chapel, on a farm near Wachendorf, Germany.

Zwinger

The protective fortress of a city; the modern name of several German palaces or parts of palaces.

CPSIA information can be obtained
at www.ICGtesting.com
Printed in the USA
BVHW012347100622
639514BV00007B/63

9 780071 772938